Statistics for Biology and Health

Series Editors

Mitchell Gail, Division of Cancer Epidemiology and Genetics, National Cancer Institute, Rockville, MD, USA
Jonathan M. Samet, Department of Epidemiology, School of Public Health, Johns Hopkins University, Baltimore, MD, USA

For further volumes:
www.springer.com/series/2848

Leonhard Held · Daniel Sabanés Bové

Likelihood and Bayesian Inference

With Applications in Biology and Medicine

Second Edition

Springer

Leonhard Held
Epidemiology, Biostatistics and Prevention
Institute
University of Zurich
Zürich, Switzerland

Daniel Sabanés Bové
Google
Zürich, Switzerland

ISSN 1431-8776
ISSN 2197-5671 (electronic)
Statistics for Biology and Health
ISBN 978-3-662-60794-7
ISBN 978-3-662-60792-3 (eBook)
DOI 10.1007/978-3-662-60792-3

Printed on acid-free paper

This Springer imprint is published by the registered company Springer-Verlag GmbH, DE part of Springer Nature
The registered company address is: Heidelberger Platz 3, 14197 Berlin, Germany

To Our Families:
Ulrike, Valentina, Richard and Lorenz,
Carrie and Ben

Preface

Statistical inference is the science of analysing and interpreting data. It provides essential tools for processing information, summarising the amount of knowledge gained and quantifying the remaining uncertainty. This book provides an introduction to the principles and concepts of the two most commonly used methods in scientific investigations: Likelihood and Bayesian inference. The two approaches are usually seen as competing paradigms, but we also emphasise their connections. In particular, both approaches are linked to the notion of a statistical model and the corresponding likelihood function, as described in the first two chapters. We discuss frequentist inference based on the likelihood in detail, followed by the essentials of Bayesian inference. Important practical topics, including model selection, prediction and numerical computation, are discussed from both perspectives.

This second edition also includes a new chapter on Markov models for time series analysis. We discuss the central Markov property, and describe observation- as well as parameter-driven models. These important time series models are used heavily in various biomedical applications.

The intended audience includes graduate students of Statistics, Biostatistics, Applied Mathematics, Biomathematics, Bioinformatics, as well as students from biology and medicine programmes which aim to understand the basis of statistical methods from the ground up. The reader should ideally be familiar with elementary concepts of probability, calculus, matrix algebra and numerical analysis, and these are summarised in detailed Appendices A–C. Several applications, taken from the area of biomedical research, are described in the Introduction and serve as examples throughout the book. We hope the data and `R` code provided at https://github.com/lheld/HSB will make it easy to apply the discussed methods to one's own statistical problem. Each chapter finishes with exercises, which can be used to deepen the knowledge obtained. The solutions are also available at https://github.com/lheld/HSB.

This textbook is based on a series of lectures and exercises that we gave at the University of Zurich for Master students in Statistics and Biostatistics. The first edition was a substantial extension of the German book "*Methoden der statistischen Inferenz: Likelihood und Bayes*", published by Spektrum Akademischer Verlag (Held 2008). This new second edition corrects previous errata, extends the material to time series analysis and includes several new examples and exercises.

Many people have helped in various ways. We would like to thank Eva and Reinhard Furrer, Torsten Hothorn, Andrea Riebler, Malgorzata Roos and Kaspar Rufibach for their support in the first edition. Special thanks go to Manuela Ott, who was instrumental in writing the second edition of the book. Last but not least, we are grateful to Eva Hiripi from Springer-Verlag Heidelberg for her continuing support and the Editors of Statistics in Biology and Health for welcoming us in their series.

Zürich, Switzerland Leonhard Held, Daniel Sabanés Bové
March 2019

Contents

Introduction

1

Contents

Statistics is a discipline with different branches. This book describes two central approaches to statistical inference, likelihood inference and Bayesian inference. Both concepts have in common that they use statistical models depending on unknown parameters to be estimated from the data. Moreover, both are constructive, i.e. provide precise procedures for obtaining the required results. A central role is played by the likelihood function, which is determined by the choice of a statistical model. While a likelihood approach bases inference only on the likelihood, the Bayesian approach combines the likelihood with prior information. Hybrid approaches also exist.

What do we want to learn from data using statistical inference? We can distinguish three major goals. Of central importance is to estimate the unknown parameters of a statistical model. This is the so-called *estimation problem*. However, how do we know that the chosen model is correct? We may have a number of statistical models and want to identify the one that describes the data best. This is the so-called *model selection problem*. And finally, we may want to predict future observations based on the observed ones. This is the *prediction problem*.

L. Held, D. Sabanés Bové, *Likelihood and Bayesian Inference*,
Statistics for Biology and Health, https://doi.org/10.1007/978-3-662-60792-3_1,

1.1 Examples

Several examples from biology and health will be considered throughout this book, many of them more than once viewed from different perspectives or tackled with different techniques. We will now give a brief overview.

1.1.1 Inference for a Proportion

One of the oldest statistical problems is the estimation of a probability based on an observed proportion. The underlying statistical model assumes that a certain event of interest occurs with probability π, say. For example, a possible event is the occurrence of a specific genetic defect, for example Klinefelter's syndrome, among male newborns in a *population* of interest. Suppose now that n male newborns are screened for that genetic defect and $x \in \{0, 1, \dots, n\}$ newborns do have it, i.e. $n - x$ newborns do not have this defect. The statistical task is now to estimate from this *sample* the underlying probability π for Klinefelter's syndrome for a randomly selected male newborn in that population.

The statistical model described above is called *binomial*. In that model, n is fixed, and x is a the realisation of a binomial random variable. However, the binomial model is not the only possible one: A proportion may in fact be the result of a different sampling scheme with the role of x and n being reversed. The resulting *negative binomial* model fixes x and checks all incoming newborns for the genetic defect considered until x newborns with the genetic defects are observed. Thus, in this model the total number n of newborns screened is random and follows a negative binomial distribution. See Appendix A.5.1 for properties of the binomial and negative binomial distributions. The observed proportion x/n of newborns with Klinefelter's syndrome is an estimate of the probability π in both cases, but the statistical model underlying the sampling process may still affect our inference for π. We will return to this issue in Sect. 2.5.2.

Probabilities are often transformed to *odds*, where the odds $\omega = \pi/(1 - \pi)$ are defined as the ratio of the probability π of the event considered and the probability $1 - \pi$ of the complementary event. For example, a probability $\pi = 0.5$ corresponds to 1 to 1 odds ($\omega = 1$), while odds of 9 to 1 ($\omega = 9$) are equivalent to a probability of $\pi = 0.9$. The corresponding estimate of ω is given by the empirical odds $x/(n - x)$. It is easy to show that any odds ω can be back-transformed to the corresponding probability π via $\pi = \omega/(1 + \omega)$.

1.1.2 Comparison of Proportions

Closely related to inference for a proportion is the comparison of two proportions. For example, a clinical study may be conducted to compare the risk of a certain disease in a treatment and a control group. We now have two unknown risk probabilities π_1 and π_2, with observations x_1 and x_2 and samples sizes n_1 and n_2 in

Table 1.1 Incidence of preeclampsia (observed proportion) in nine randomised placebo-controlled clinical trials of diuretics. Empirical odds ratios (OR) are also given. The studies are labelled with the name of the principal author

Trial	Treatment		Control		OR
Weseley	11 %	(14/131)	10 %	(14/136)	1.04
Flowers	5 %	(21/385)	13 %	(17/134)	0.40
Menzies	25 %	(14/57)	50 %	(24/48)	0.33
Fallis	16 %	(6/38)	45 %	(18/40)	0.23
Cuadros	1 %	(12/1011)	5 %	(35/760)	0.25
Landesman	10 %	(138/1370)	13 %	(175/1336)	0.74
Krans	3 %	(15/506)	4 %	(20/524)	0.77
Tervila	6 %	(6/108)	2 %	(2/103)	2.97
Campbell	42 %	(65/153)	39 %	(40/102)	1.14

the two groups. Different measures are now employed to compare the two groups, among which the *risk difference* $\pi_1 - \pi_2$ and the *risk ratio* π_1/π_2 are the most common ones. The *odds ratio*

$$\frac{\omega_1}{\omega_2} = \frac{\pi_1/(1-\pi_1)}{\pi_2/(1-\pi_2)},$$

the ratio of the odds ω_1 and ω_2, is also often used. Note that if the risk in the two groups is equal, i.e. $\pi_1 = \pi_2$, then the risk difference is zero, while both the risk ratio and the odds ratio is one. Statistical methods can now be employed to investigate if the simpler model with one parameter $\pi = \pi_1 = \pi_2$ can be preferred over the more complex one with different risk parameters π_1 and π_2. Such questions may also be of interest if more than two groups are considered.

A controlled clinical trial compares the effect of a certain treatment with a control group, where typically either a standard treatment or a placebo treatment is provided. Several randomised controlled clinical trials have investigated the use of diuretics in pregnancy to prevent preeclampsia. Preeclampsia is a medical condition characterised by high blood pressure and significant amounts of protein in the urine of a pregnant woman. It is a very dangerous complication of a pregnancy and may affect both the mother and fetus. In each trial women were randomly assigned to one of the two treatment groups. Randomisation is used to exclude possible subjective influence from the examiner and to ensure equal distribution of relevant risk factors in the two groups.

The results of nine such studies are reported in Table 1.1. For each trial the observed proportions x_i/n_i in the treatment and placebo control group ($i = 1, 2$) are given, as well as the corresponding *empirical odds ratio*

$$\frac{x_1/(n_1 - x_1)}{x_2/(n_2 - x_2)}.$$

One can see substantial variation of the empirical odds ratios reported in Table 1.1. This raises the question if this variation is only statistical in nature or if there is evidence for additional heterogeneity between the studies. In the latter case the true treatment effect differs from trial to trial due to different inclusion criteria, different underlying populations, or other reasons. Such questions are addressed in a *meta-analysis*, a combined analysis of results from different studies.

1.1.3 The Capture–Recapture Method

The capture–recapture method aims to estimate the size of a population of individuals, say the number N of fish in a lake. To do so, a sample of M fish is drawn from the lake, with all the fish marked and then thrown back into the lake. After a sufficient time, a second sample of size n is taken, and the number x of marked fish in that sample is recorded.

The goal is now to infer N from M, n and x. An ad-hoc estimate can be obtained by equating the proportion of marked fish in the lake with the corresponding proportion in the sample:

$$\frac{M}{N} \approx \frac{x}{n}.$$

This leads to the estimate $\hat{N} \approx M \cdot n/x$ for the number N of fish in the lake. As we will see in Example 2.2, there is a rigorous theoretical basis for this estimate. However, the estimate $\hat{N}$ has an obvious deficiency for $x = 0$, where $\hat{N}$ is infinite. Other estimates without this deficiency are available. Appropriate statistical techniques will enable us to quantify the uncertainty associated with the different estimates of N.

1.1.4 Hardy–Weinberg Equilibrium

The *Hardy–Weinberg equilibrium* (after Godfrey H. Hardy, 1879–1944, and Wilhelm Weinberg, 1862–1937) plays a central role in population genetics. Consider a population of diploid, sexually reproducing individuals and a specific locus on a chromosome with alleles A and a. If the allele frequencies of A and a in the population are υ and $1-\upsilon$, then the expected genotype frequencies of AA, Aa and aa are

$$\pi_1 = \upsilon^2, \qquad \pi_2 = 2\upsilon(1-\upsilon) \quad \text{and} \quad \pi_3 = (1-\upsilon)^2. \tag{1.1}$$

The Hardy–Weinberg equilibrium implies that the allele frequency υ determines the expected frequencies of the genotypes. If a population is not in Hardy–Weinberg equilibrium at a specific locus, then two parameters π_1 and π_2 are necessary to describe the distribution. The *de Finetti diagram* shown in Fig. 1.1 is a useful graphical visualisation of Hardy–Weinberg equilibrium.

It is often of interest to investigate whether a certain population is in Hardy–Weinberg equilibrium at a particular locus. For example, a random sample of $n = 747$ individuals has been taken in a study of MN blood group frequencies in Iceland. The MN blood group in humans is under the control of a pair of alleles,

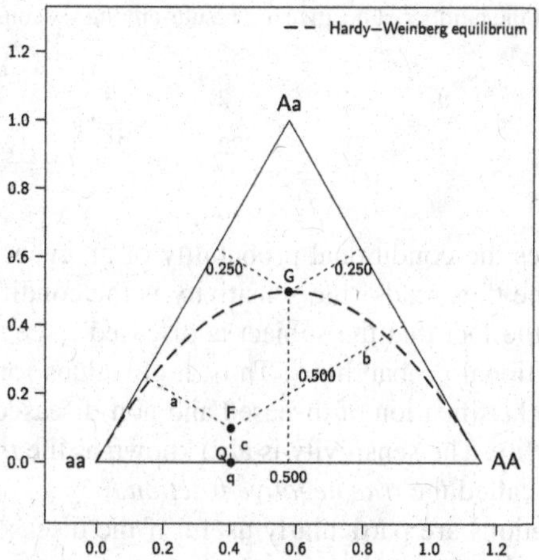

Fig. 1.1 The de Finetti diagram, named after the Italian statistician Bruno de Finetti (1906–1985), displays the expected relative genotype frequencies $\Pr(AA) = \pi_1$, $\Pr(aa) = \pi_3$ and $\Pr(Aa) = \pi_2$ in a bi-allelic, diploid population as the length of the perpendiculars a, b and c from the inner point F to the sides of an equilateral triangle. The ratio of the straight length $\overline{aa, Q}$ to the side length $\overline{aa, AA}$ is the relative allele frequency υ of A. Hardy–Weinberg equilibrium is represented by all points on the parabola $2\upsilon(1-\upsilon)$. For example, the point G represents such a population with $\upsilon = 0.5$, whereas population F has substantially less heterozygous Aa than expected under Hardy–Weinberg equilibrium

M and N. Most people in the Eskimo population are MM, while other populations tend to possess the opposite genotype NN. In the sample from Iceland, the frequencies of the underlying genotypes MM, MN and NN turned out to be $x_1 = 233$, $x_2 = 385$ and $x_3 = 129$. If we assume that the population is in Hardy–Weinberg equilibrium, then the statistical task is to estimate the unknown allele frequency υ from these data. Statistical methods can also address the question if the equilibrium assumption is supported by the data or not. This is a model selection problem, which can be addressed with a significance test or other techniques.

1.1.5 Estimation of Diagnostic Tests Characteristics

Screening of individuals is a popular public health approach to detect diseases in an early and hopefully curable stage. In order to screen a large population, it is imperative to use a fairly cheap diagnostic test, which typically makes errors in the disease classification of individuals. A useful diagnostic test will have high

$$\textit{sensitivity} = \Pr(\text{positive test} \mid \text{subject is diseased}) \quad \text{and}$$
$$\textit{specificity} = \Pr(\text{negative test} \mid \text{subject is healthy}).$$

Table 1.2 Distribution of the number of positive test results among six consecutive screening tests of 196 colon cancer cases

Number k of positive tests	0	1	2	3	4	5	6
Frequency Z_k	?	37	22	25	29	34	49

Here Pr(A | B) denotes the conditional probability of an event A, given the information B. The first line thus reads "the sensitivity is the conditional probability of a positive test, given the fact that the subject is diseased"; see Appendix A.1.1 for more details on conditional probabilities. Thus, high values for the sensitivity and specificity mean that classification of diseased and non-diseased individuals is correct with high probability. The sensitivity is also known as the *true positive fraction* whereas specificity is called the *true negative fraction*.

Screening examinations are particularly useful if the disease considered can be treated better in an earlier stage than in a later stage. For example, a diagnostic study in Australia involved 38 000 individuals, which have been screened for the presence of colon cancer repeatedly on six consecutive days with a simple diagnostic test. 3000 individuals had at least one positive test result, which was subsequently verified with a coloscopy. 196 cancer cases were eventually identified, and Table 1.2 reports the frequency of positive test results among those. Note that the number Z_0 of cancer patients that have never been positively tested is unavailable by design. The closely related *false negative fraction*

$$\Pr(\text{negative test} \mid \text{subject is diseased}),$$

which is 1 − sensitivity, is often of central public health interest. Statistical methods can be used to estimate this quantity and the number of undetected cancer cases Z_0. Similarly, the *false positive fraction*

$$\Pr(\text{positive test} \mid \text{subject is healthy})$$

is 1 − specificity.

1.1.6 Quantifying Disease Risk from Cancer Registry Data

Cancer registries collect incidence and mortality data on different cancer locations. For example, data on the incidence of lip cancer in Scotland have been collected between 1975 and 1980. The raw counts of cancer cases in 56 administrative districts of Scotland will vary a lot due to heterogeneity in the underlying population counts. Other possible reasons for variation include different age distributions or heterogeneity in underlying risk factors for lip cancer in the different districts.

A common approach to adjust for age heterogeneity is to calculate the expected number of cases using age standardisation. The *standardised incidence ratio* (SIR) of observed to expected number of cases is then often used to visually display ge-

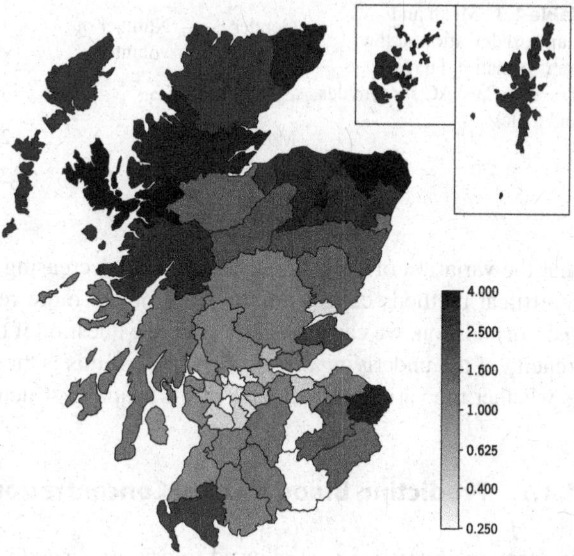

Fig. 1.2 The geographical distribution of standardised incidence ratios (SIRs) of lip cancer in Scotland, 1975–1980. Note that some SIRs are below or above the interval [0.25, 4] and are marked *white* and *black*, respectively

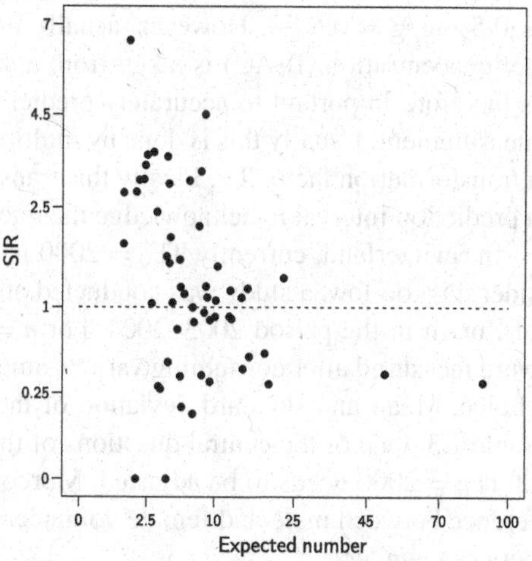

Fig. 1.3 Plot of the standardised incidence ratios (SIR) versus the expected number of lip cancer cases. Both variables are shown on a square-root scale to improve visibility. The *horizontal line* SIR = 1 represents equal observed and expected cases

ographical variation in disease risk. If the SIR is equal to 1, then the observed incidence is as expected. Figure 1.2 maps the corresponding SIRs for lip cancer in Scotland.

However, SIRs are unreliable indicators of disease incidence, in particular if the disease is rare. For example, a small district may have zero observed cases just by chance such that the SIR will be exactly zero. In Fig. 1.3, which plots the SIRs versus the number of expected cases, we can identify two such districts. More generally, the statistical variation of the SIRs will depend on the population counts, so more extreme SIRs will tend to occur in less populated areas, even if the underlying disease risk does not vary from district to district. Indeed, we can see from Fig. 1.3

Table 1.3 Mean and standard deviation of the transformation factor TF = BAC/BrAC for females and males

Gender	Number of volunteers	Transformation factor	
		Mean	Standard deviation
Female	33	2318.5	220.1
Male	152	2477.5	232.5
Total	185	2449.2	237.8

that the variation of the SIRs increases with decreasing number of expected cases. Statistical methods can be employed to obtain more reliable estimates of disease risk. In addition, we can also investigate the question if there is evidence for heterogeneity of the underlying disease risk at all. If this is the case, then another question is whether the variation in disease risk is spatially structured or not.

1.1.7 Predicting Blood Alcohol Concentration

In many countries it is not allowed to drive a car with a blood alcohol concentration (BAC) above a certain threshold. For example, in Switzerland this threshold is 0.5 mg/g = 0.5 ‰. However, usually only a measurement of the breath alcohol concentration (BrAC) is taken from a suspicious driver in the first instance. It is therefore important to accurately predict the BAC measurement from the BrAC measurement. Usually this is done by multiplication of the BrAC measurement with a transformation factor TF. Ideally this transformation should be accompanied with a prediction interval to acknowledge the uncertainty of the BAC prediction.

In Switzerland, currently $\text{TF}_0 = 2000$ is used in practice. As some experts consider this too low, a study was conducted at the Forensic Institute of the University of Zurich in the period 2003–2004. For $n = 185$ volunteers, both BrAC and BAC were measured after consuming various amounts of alcoholic beverages of personal choice. Mean and standard deviation of the ratio TF = BAC/BrAC are shown in Table 1.3. One of the central questions of the study was if the currently used factor of $\text{TF}_0 = 2000$ needs to be adjusted. Moreover, it is of interest if the empirical difference between male and female volunteers provides evidence of a true difference between genders.

1.1.8 Analysis of Survival Times

A randomised placebo-controlled trial of Azathioprine for primary biliary cirrhosis (PBC) was designed with patient survival as primary outcome. PBC is a chronic and eventually fatal liver disease, which affects mostly women. Table 1.4 gives the survival times (in days) of the 94 patients who have been treated with Azathioprine. The reported survival time is *censored* for 47 (50 %) of the patients. A censored survival time does not represent the time of death but the last time point when the patient was still known to be alive. It is not known whether, and if so when, a woman

Table 1.4 Survival times of 94 patients under Azathioprine treatment in days. Censored observations are marked with a plus sign

8+	9	38	96	144	167	177	191+	193	201
207	251	287+	335+	379+	421	425	464	498+	500
574+	582+	586	616	630	636	647	651+	688	743
754	769+	797	799+	804	828+	904+	932+	947	962+
974	1113+	1219	1247	1260	1268	1292+	1408	1436+	1499
1500	1522	1552	1554	1555+	1626+	1649+	1942	1975	1982+
1998+	2024+	2058+	2063+	2101+	2114+	2148	2209	2254+	2338+
2384+	2387+	2415+	2426	2436+	2470	2495+	2500	2522	2529+
2744+	2857	2929	3024	3056+	3247+	3299+	3414+	3456+	3703+
3906+	3912+	4108+	4253+						

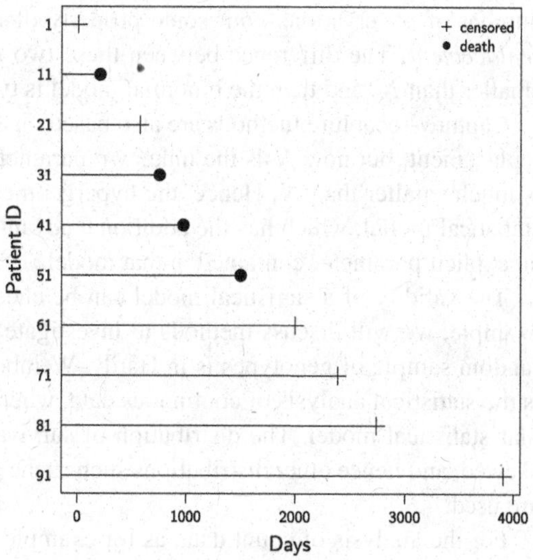

Fig. 1.4 Illustration of partially censored survival times using the first 10 observations of the first column in Table 1.4. A survival time marked with a plus sign is censored, whereas the other survival times are actual deaths

with censored survival time actually died of PBC. Possible reasons for censoring include drop-out of the study, e.g. due to moving away, or death by some other cause, e.g. due to a car accident. Figure 1.4 illustrates this type of data.

1.2 Statistical Models

The formulation of a suitable probabilistic model plays a central role in the statistical analysis of data. The terminology *statistical model* is also common. A statistical model will describe the probability distribution of the data as a function of an unknown *parameter*. If there is more than one unknown parameter, i.e. the unknown parameters form a *parameter vector*, then the model is a *multiparameter model*. In this book we will concentrate on *parametric models*, where the number of parameters is fixed, i.e. does not depend on the sample size. In contrast, in a *non-parametric model* the number of parameters grows with the sample size and may even be infinite.

Appropriate formulation of a statistical model is based on careful considerations on the origin and properties of the data at hand. Certain approximations may often be useful in order to simplify the model formulation. Often the observations are assumed to be a *random sample*, i.e. independent realisations from a known distribution. See Appendix A.5 for a comprehensive list of commonly used probability distributions.

For example, estimation of a proportion is often based on a random sample of size n drawn *without replacement* from some population with N individuals. The appropriate statistical model for the number of observations in the sample with the property of interest is the hypergeometric distribution. However, the hypergeometric distribution can be approximated by a binomial one, a statistical model for the number of observations with some property of interest in a random sample *with replacement*. The difference between these two models is negligible if n is much smaller than N, and then the binomial model is typically preferred.

Capture–recapture methods are also based on a random sample of size n without replacement, but now N is the unknown parameter of interest, so it is unclear if n is much smaller than N. Hence, the hypergeometric distribution is the appropriate statistical model, which has the additional advantage that the quantity of interest is an explicit parameter contained in that model.

The validity of a statistical model can be checked with statistical methods. For example, we will discuss methods to investigate if the underlying population of a random sample of genotypes is in Hardy–Weinberg equilibrium. Another example is the statistical analysis of continuous data, where the normal distribution is a popular statistical model. The distribution of survival times, for example, is typically skewed, and hence other distributions such as the gamma or the Weibull distribution are used.

For the analysis of count data, as for example the number of lip cancer cases in the administrative districts of Scotland from Example 1.1.6, a suitable distribution has to be chosen. A popular choice is the Poisson distribution, which is suitable if the mean and variance of the counts are approximately equal. However, in many cases there is *overdispersion*, i.e. the variance is larger than the mean. Then the Poisson-gamma distribution, a generalisation of the Poisson distribution, is a suitable choice.

Statistical models can become considerably more complex if necessary. For example, the statistical analysis of survival times needs to take into account that some of the observations are censored, so an additional model (or some simplifying assumption) for the censoring mechanism is typically needed. The formulation of a suitable statistical model for the data obtained in the diagnostic study described in Example 1.1.5 also requires careful thought since the study design does not deliver direct information on the number of patients with solely negative test results.

1.3 Contents and Notation of the Book

Chapter 2 introduces the central concept of a likelihood function and the maximum likelihood estimate. Basic elements of frequentist inference are summarised in Chap. 3. Frequentist inference based on the likelihood, as described in Chaps. 4 and 5, enables us to construct confidence intervals and significance tests for parameters of interest. Bayesian inference combines the likelihood with a prior distribution and is conceptually different from the frequentist approach. Chapter 6 describes the central aspects of this approach. Chapter 7 gives an introduction to model selection from both a likelihood and a Bayesian perspective, while Chap. 8 discusses the use of modern numerical methods for Bayesian inference and Bayesian model selection. In Chap. 9 we give an introduction to the construction and the assessment of probabilistic predictions. Finally, Chap. 10 describes methodology for time series analysis. Every chapter ends with exercises and some references to additional literature.

Modern statistical inference is unthinkable without the use of a computer. Numerous numerical techniques for optimisation and integration are employed to solve statistical problems. This book emphasises the role of the computer and gives many examples with explicit `R` code. Appendix C is devoted to the background of these numerical techniques. Modern statistical inference is also unthinkable without a solid background in mathematics, in particular probability, which is covered in Appendix A. A collection of the most common probability distributions and their properties is also given. Appendix B describes some central results from matrix algebra and calculus which are used in this book.

We finally describe some notational issues. Mathematical results are given in italic font and are often followed by a proof of the result, which ends with an open square (□). A filled square (■) denotes the end of an example. Definitions end with a diamond (♦). Vectorial parameters $\boldsymbol{\theta}$ are reproduced in boldface to distinguish them from scalar parameters θ. Similarly, independent univariate random variables X_i from a certain distribution contribute to a random sample $X_{1:n} = (X_1, \dots, X_n)$, whereas dependent univariate random variables $X_1, \dots, X_n$ form a general sample $X = (X_1, \dots, X_n)$. A random sample of n independent multivariate random variables $\boldsymbol{X}_i = (X_{i1}, \dots, X_{ik})^\top$ is denoted as $\boldsymbol{X}_{1:n} = (\boldsymbol{X}_1, \dots, \boldsymbol{X}_n)$. On page 389 we give a concise overview of the notation used in this book.

1.4 References

Estimation and comparison of proportions are discussed in detail in Connor and Imrey (2005). The data on preeclampsia trials is cited from Collins et al. (1985). Applications of capture–recapture techniques are described in Seber (1982). Details on the Hardy–Weinberg equilibrium can be found in Lange (2002), the data from Iceland are taken from Falconer and Mackay (1996). The colon cancer screening data is taken from Lloyd and Frommer (2004), while the data on lip cancer in Scotland is taken from Clayton and Bernardinelli (1992). The study on breath and blood alcohol concentration is described in Iten (2009) and Iten and Wüst (2009). Kirkwood and Sterne (2003) report data on the clinical study on the treatment of primary biliary cirrhosis with Azathioprine. Jones et al. (2014) is a recent book on statistical computing, which provides much of the background necessary to follow our numerical examples using R. For a solid but accessible treatment of probability theory, we recommend Grimmett and Stirzaker (2001, Chaps. 1–7).

2 Likelihood

Contents

The term *likelihood* has been introduced by Sir Ronald A. Fisher (1890–1962). The likelihood function forms the basis of likelihood-based statistical inference.

2.1 Likelihood and Log-Likelihood Function

Let $X = x$ denote a realisation of a random variable or vector X with probability mass or density function $f(x;\theta)$, cf. Appendix A.2. The function $f(x;\theta)$ depends on the realisation x and on typically unknown parameters θ, but is otherwise assumed to be known. It typically follows from the formulation of a suitable statistical model. Note that θ can be a scalar or a vector; in the latter case we will write the parameter vector $\boldsymbol{\theta}$ in boldface. The space $\mathcal{T}$ of all possible realisations of X is called *sample space*, whereas the parameter θ can take values in the *parameter space* Θ.

L. Held, D. Sabanés Bové, *Likelihood and Bayesian Inference*,
Statistics for Biology and Health, https://doi.org/10.1007/978-3-662-60792-3_2,

The function $f(x;\theta)$ describes the distribution of the random variable X for fixed parameter θ. The goal of statistical inference is to infer θ from the observed datum $X = x$. Playing a central role in this task is the *likelihood function* (or simply *likelihood*)

$$L(\theta; x) = f(x;\theta), \quad \theta \in \Theta,$$

viewed as a function of θ for fixed x. We will often write $L(\theta)$ for the likelihood if it is clear which observed datum x the likelihood refers to.

Definition 2.1 (Likelihood function) The *likelihood function* $L(\theta)$ is the probability mass or density function of the observed data x, viewed as a function of the unknown parameter θ. ♦

For discrete data, the likelihood function is the probability of the observed data viewed as a function of the unknown parameter θ. This definition is not directly transferable to continuous observations, where the probability of every exactly measured observed datum is strictly speaking zero. However, in reality continuous measurements are always rounded to a certain degree, and the probability of the observed datum x can therefore be written as $\Pr(x - \frac{\varepsilon}{2} \le X \le x + \frac{\varepsilon}{2})$ for some small rounding interval width $\varepsilon > 0$. Here X denotes the underlying true continuous measurement.

The above probability can be re-written as

$$\Pr\left(x - \frac{\varepsilon}{2} \le X \le x + \frac{\varepsilon}{2}\right) = \int_{x-\frac{\varepsilon}{2}}^{x+\frac{\varepsilon}{2}} f(y;\theta)\, dy \approx \varepsilon \cdot f(x;\theta),$$

so the probability of the observed datum x is approximately proportional to the density function $f(x;\theta)$ of X at x. As we will see later, the multiplicative constant ε can be ignored, and we therefore use the density function $f(x;\theta)$ as the likelihood function of a continuous datum x.

2.1.1 Maximum Likelihood Estimate

Plausible values of θ should have a relatively high likelihood. The most plausible value with maximum value of $L(\theta)$ is the *maximum likelihood estimate*.

Definition 2.2 (Maximum likelihood estimate) The *maximum likelihood estimate* (MLE) $\hat{\theta}_{\mathrm{ML}}$ of a parameter θ is obtained through maximising the likelihood function:

$$\hat{\theta}_{\mathrm{ML}} = \arg\max_{\theta \in \Theta} L(\theta).$$

♦

In order to compute the MLE, we can safely ignore multiplicative constants in $L(\theta)$, as they have no influence on $\hat{\theta}_{\mathrm{ML}}$. To simplify notation, we therefore often only report a likelihood function $L(\theta)$ without multiplicative constants, i.e. the *likelihood kernel*.

Definition 2.3 (Likelihood kernel) The likelihood kernel is obtained from a likelihood function by removing all multiplicative constants. We will use the symbol $L(\theta)$ both for likelihood functions and kernels. ♦

It is often numerically convenient to use the *log-likelihood function*

$$l(\theta) = \log L(\theta),$$

the natural logarithm of the likelihood function, for computation of the MLE. The logarithm is a strictly monotone function, and therefore

$$\hat{\theta}_{\mathrm{ML}} = \arg\max_{\theta \in \Theta} l(\theta).$$

Multiplicative constants in $L(\theta)$ turn to additive constants in $l(\theta)$, which again can often be ignored. A log-likelihood function without additive constants is called *log-likelihood kernel*. We will use the symbol $l(\theta)$ both for log-likelihood functions and kernels.

Example 2.1 (Inference for a proportion) Let $X \sim \mathrm{Bin}(n, \pi)$ denote a binomially distributed random variable. For example, $X = x$ may represent the observed number of babies with Klinefelter's syndrome among n male newborns. The number of male newborns n is hence known, while the true prevalence π of Klinefelter's syndrome among male newborns is unknown.

The corresponding likelihood function is

$$L(\pi) = \binom{n}{x} \pi^x (1-\pi)^{n-x} \quad \text{for } \pi \in (0, 1)$$

with unknown parameter $\pi \in (0, 1)$ and sample space $\mathcal{T} = \{0, 1, \dots, n\}$. The multiplicative term $\binom{n}{x}$ does not depend on π and can therefore be ignored, i.e. it is sufficient to consider the likelihood kernel $\pi^x (1-\pi)^{n-x}$. The likelihood function $L(\pi)$ is displayed in Fig. 2.1 for a sample size of $n = 10$ with $x = 2$ and $x = 0$ babies with Klinefelter's syndrome, respectively.

The log-likelihood kernel turns out to be

$$l(\pi) = x \log \pi + (n - x) \log(1 - \pi)$$

with derivative

$$\frac{dl(\pi)}{d\pi} = \frac{x}{\pi} - \frac{n-x}{1-\pi}.$$

Setting this derivative to zero gives the MLE $\hat{\pi}_{\mathrm{ML}} = x/n$, the relative frequency of Klinefelter's syndrome in the sample. The MLEs are marked with a vertical line in Fig. 2.1. ■

The uniqueness of the MLE is not guaranteed, and in certain examples there may exist at least two parameter values $\hat{\theta}_1 \neq \hat{\theta}_2$ with $L(\hat{\theta}_1) = L(\hat{\theta}_2) = \arg\max_{\theta \in \Theta} L(\theta)$. In other situations, the MLE may not exist at all. The following example illustrates

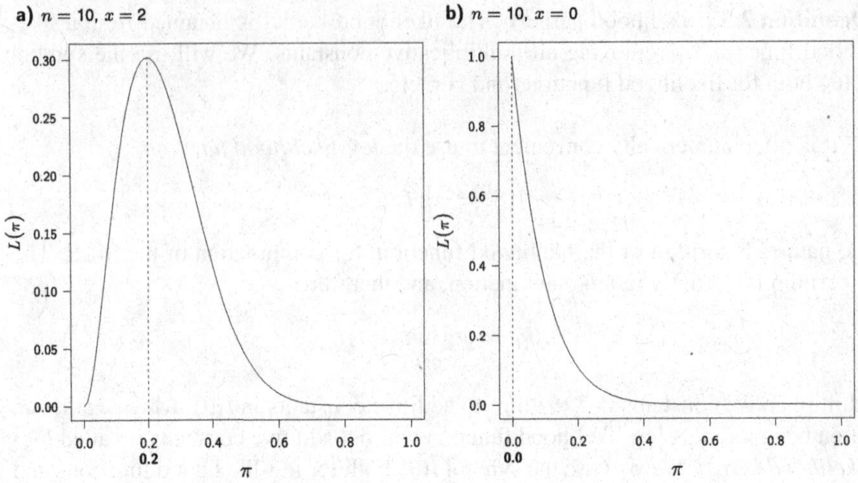

Fig. 2.1 Likelihood function for π in a binomial model. The MLEs are marked with a *vertical line*

that application of the capture–recapture method can result both in non-unique and non-existing MLEs.

Example 2.2 (Capture–recapture method) As described in Sect. 1.1.3, the goal of capture–recapture methods is to estimate the number N of individuals in a population. To achieve that goal, M individuals are marked and randomly mixed with the total population. A sample of size n without replacement is then drawn, and the number $X = x$ of marked individuals is determined. The suitable statistical model for X is therefore a hypergeometric distribution

$$X \sim \text{HypGeom}(n, N, M)$$

with probability mass function

$$\Pr(X = x) = f(x; \theta = N) = \frac{\binom{M}{x}\binom{N-M}{n-x}}{\binom{N}{n}}$$

for $x \in \mathcal{T} = \{\max\{0, n - (N - M)\}, \dots, \min(n, M)\}$. The likelihood function for N is therefore

$$L(N) = \frac{\binom{M}{x}\binom{N-M}{n-x}}{\binom{N}{n}}$$

for $N \in \Theta = \{\max(n, M + n - x), \max(n, M + n - x) + 1, \dots\}$, where we could have ignored the multiplicative constant $\binom{M}{x}\frac{n!}{(n-x)!}$. Figure 2.2 displays this likelihood function for certain values of x, n and M. Note that the unknown parameter

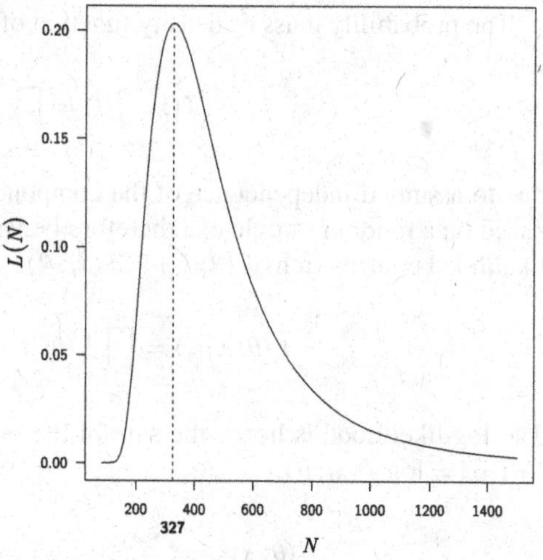

Fig. 2.2 Likelihood function for N in the capture–recapture experiment with $M = 26$, $n = 63$ and $x = 5$. The (unique) MLE is $\hat{N}_{\text{ML}} = 327$

$\theta = N$ can only take integer values and is not continuous, although the figure suggests the opposite.

It is possible to show (cf. Exercise 3) that the likelihood function is maximised at $\hat{N}_{\text{ML}} = \lfloor M \cdot n/x \rfloor$, where $\lfloor y \rfloor$ denotes the largest integer not greater than y. For example, for $M = 26$, $n = 63$ and $x = 5$ (cf. Fig. 2.2), we obtain $\hat{N}_{\text{ML}} = \lfloor 26 \cdot 63/5 \rfloor = \lfloor 327.6 \rfloor = 327$.

However, sometimes the MLE is not unique, and the likelihood function attains the same value at $\hat{N}_{\text{ML}} - 1$. For example, for $M = 13$, $n = 10$ and $x = 5$, we have $\hat{N}_{\text{ML}} = 13 \cdot 10/5 = 26$, but $\hat{N}_{\text{ML}} = 25$ also attains exactly the same value of $L(N)$. This can easily be verified empirically using the R-function dhyper, cf. Table A.1.

```
M <- 13
n <- 10
x <- 5
ml <- c(25, 26)
(dhyper(x=x, m=M, n=ml-M, k=n))
[1] 0.311832 0.311832
```

On the other hand, the MLE will not exist for $x = 0$ because the likelihood function $L(N)$ is then monotonically increasing. ■

We often have not only one observation x but a series $x_1, \ldots, x_n$ of n observations from $f(x; \theta)$, usually assumed to be independent. This leads to the concept of a *random sample*.

Definition 2.4 (Random sample) Data $x_{1:n} = (x_1, \ldots, x_n)$ are realisations of a *random sample* $X_{1:n} = (X_1, \ldots, X_n)$ of size n if the random variables $X_1, \ldots, X_n$ are independent and identically distributed from some distribution with probability mass or density function $f(x; \theta)$. The number n of observations is called the *sample size*. This may be denoted as $X_i \overset{\text{iid}}{\sim} f(x; \theta)$, $i = 1, \ldots, n$. ♦

The probability mass or density function of $X_{1:n}$ is

$$f(x_{1:n};\theta)=\prod_{i=1}^{n} f(x_i;\theta)$$

due to assumed independence of the components of $X_{1:n}$. The likelihood function based on a random sample can therefore be written as the product of the individual likelihood contributions $L(\theta;x_i)=f(x_i;\theta)$:

$$L(\theta;x_{1:n})=\prod_{i=1}^{n} L(\theta;x_i)=\prod_{i=1}^{n} f(x_i;\theta).$$

The log-likelihood is hence the sum of the individual log-likelihood contributions $l(\theta;x_i)=\log f(x_i;\theta)$:

$$l(\theta;x_{1:n})=\sum_{i=1}^{n} l(\theta;x_i)=\sum_{i=1}^{n}\log f(x_i;\theta). \tag{2.1}$$

Example 2.3 (Analysis of survival times) Let $X_{1:n}$ denote a random sample from an exponential distribution $\operatorname{Exp}(\lambda)$. Then

$$L(\lambda)=\prod_{i=1}^{n}\{\lambda\exp(-\lambda x_i)\}=\lambda^n\exp\left(-\lambda\sum_{i=1}^{n}x_i\right)$$

is the likelihood function of $\lambda\in\mathbb{R}^+$. The log-likelihood function is therefore

$$l(\lambda)=n\log\lambda-\lambda\sum_{i=1}^{n}x_i$$

with derivative

$$\frac{dl(\lambda)}{d\lambda}=\frac{n}{\lambda}-\sum_{i=1}^{n}x_i.$$

Setting the derivative to zero, we easily obtain the MLE $\hat{\lambda}_{\mathrm{ML}}=1/\bar{x}$ where $\bar{x}=\sum_{i=1}^{n}x_i/n$ is the mean observed survival time. If our interest is instead in the theoretical mean $\mu=1/\lambda$ of the exponential distribution, then the likelihood function takes the form

$$L(\mu)=\mu^{-n}\exp\left(-\frac{1}{\mu}\sum_{i=1}^{n}x_i\right),\quad \mu\in\mathbb{R}^+,$$

with MLE $\hat{\mu}_{\mathrm{ML}}=\bar{x}$.

For pure illustration, we now consider the $n=47$ non-censored PBC survival times from Example 1.1.8 and assume that they are exponentially distributed. We emphasise that this approach is in general not acceptable, as ignoring the censored observations will introduce bias if the distributions of censored and uncensored events differ. It is also less efficient, as a certain proportion of the available data

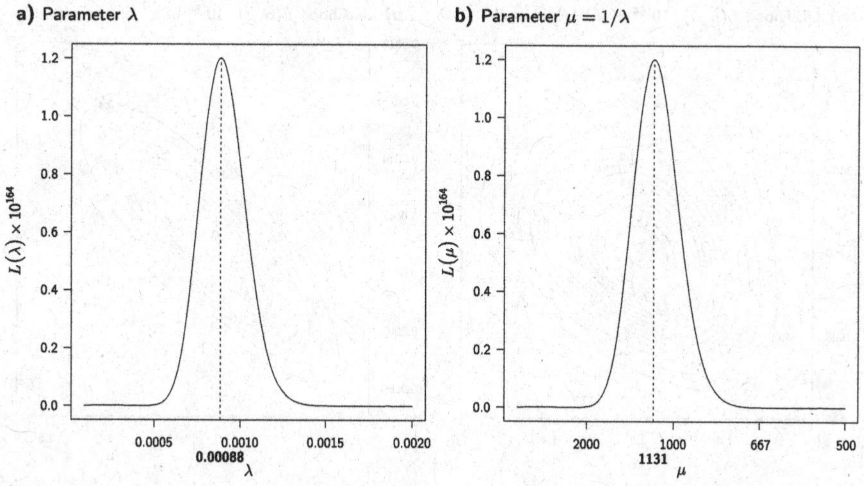

Fig. 2.3 Likelihood function for λ (*left*) and μ (*right*) assuming independent and exponentially distributed PBC-survival times. Only uncensored observations are taken into account

is ignored. In Example 2.8 we will therefore also take into account the censored observations.

The likelihood functions for the rate parameter λ and the mean survival time $\mu = 1/\lambda$ are shown in Fig. 2.3. Note that the actual values of the likelihood functions are identical, only the scale of the x-axis is transformed. This illustrates that the likelihood function and in particular the MLE are *invariant* with respect to one-to-one transformations of the parameter θ, see Sect. 2.1.3 for more details. It also shows that a likelihood function cannot be interpreted as a density function of a random variable. Indeed, assume that $L(\lambda)$ was an (unnormalised) density function; then the density of $\mu = 1/\lambda$ would be not equal to $L(1/\mu)$ because this change of variables would also involve the derivative of the inverse transformation, cf. Eq. (A.11) in Appendix A.2.3.

The assumption of exponentially distributed survival times may be unrealistic, and a more flexible statistical model may be warranted. Both the *gamma* and the *Weibull* distributions include the exponential distribution as a special case. The Weibull distribution $\text{Wb}(\mu, \alpha)$ is described in Appendix A.5.2 and depends on two parameters μ and α, which both are required to be positive. A random sample $X_{1:n}$ from a Weibull distribution has the density

$$f(x_{1:n}; \mu, \alpha) = \prod_{i=1}^{n} f(x_i; \mu, \alpha) = \prod_{i=1}^{n} \frac{\alpha}{\mu}\left(\frac{x_i}{\mu}\right)^{\alpha-1} \exp\left\{-\left(\frac{x_i}{\mu}\right)^{\alpha}\right\},$$

and the corresponding likelihood function can be written as

$$L(\mu, \alpha) = \frac{\alpha^n}{\mu^{n\alpha}}\left(\prod_{i=1}^{n} x_i\right)^{\alpha-1} \exp\left\{-\sum_{i=1}^{n}\left(\frac{x_i}{\mu}\right)^{\alpha}\right\}, \quad \mu, \alpha > 0.$$

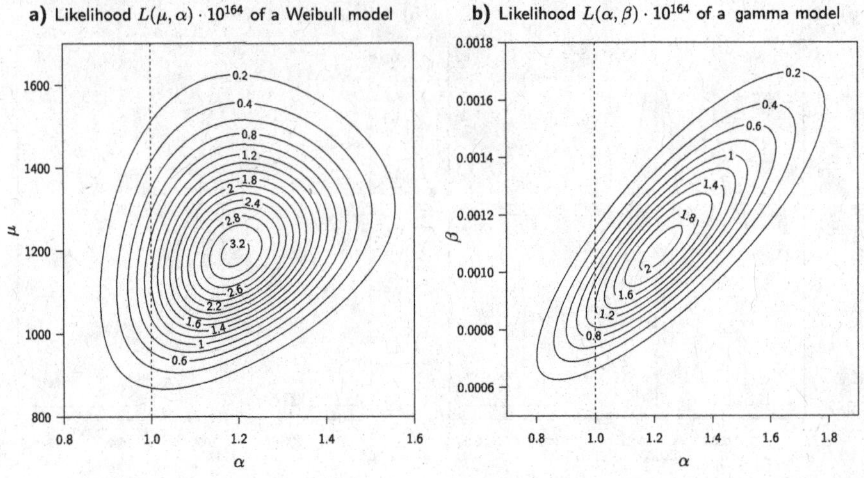

a) Likelihood $L(\mu, \alpha) \cdot 10^{164}$ of a Weibull model
b) Likelihood $L(\alpha, \beta) \cdot 10^{164}$ of a gamma model

Fig. 2.4 Flexible modelling of survival times is achieved by a Weibull or gamma model. The corresponding likelihood functions are displayed here. The *vertical line* at $\alpha = 1$ corresponds to the exponential model in both cases

For $\alpha = 1$, we obtain the exponential distribution with expectation $\mu = 1/\lambda$ as a special case.

A contour plot of the Weibull likelihood, a function of two parameters, is displayed in Fig. 2.4a. The likelihood function is maximised at $\alpha = 1.19$, $\mu = 1195$. The assumption of exponentially distributed survival times does not appear to be completely unrealistic, but the likelihood values for $\alpha = 1$ are somewhat lower. In Example 5.9 we will calculate a confidence interval for α, which can be used to quantify the plausibility of the exponential model.

If we assume that the random sample comes from a gamma distribution $G(\alpha, \beta)$, the likelihood is (cf. again Appendix A.5.2)

$$L(\alpha, \beta) = \prod_{i=1}^{n} \frac{\beta^\alpha}{\Gamma(\alpha)} x_i^{\alpha-1} \exp(-\beta x_i) = \left\{\frac{\beta^\alpha}{\Gamma(\alpha)}\right\}^n \left(\prod_{i=1}^{n} x_i\right)^{\alpha-1} \exp\left(-\beta \sum_{i=1}^{n} x_i\right).$$

The exponential distribution with parameter $\lambda = \beta$ corresponds to the special case $\alpha = 1$. Plausible values α and β of the gamma likelihood function tend to lie on the diagonal in Fig. 2.4b: for larger values of α, plausible values of β tend to be also larger. The sample is apparently informative about the mean $\mu = \alpha/\beta$, but not so informative about the components α and β of that ratio.

Alternatively, the gamma likelihood function can be *reparametrised*, and the parameters $\mu = \alpha/\beta$ and $\phi = 1/\beta$, say, could be used. The second parameter ϕ now represents the variance-to-mean ratio of the gamma distribution. Figure 2.5 displays the likelihood function using this new parametrisation. The dependence between the two parameters appears to be weaker than for the initial parametrisation shown in Fig. 2.4b. ■

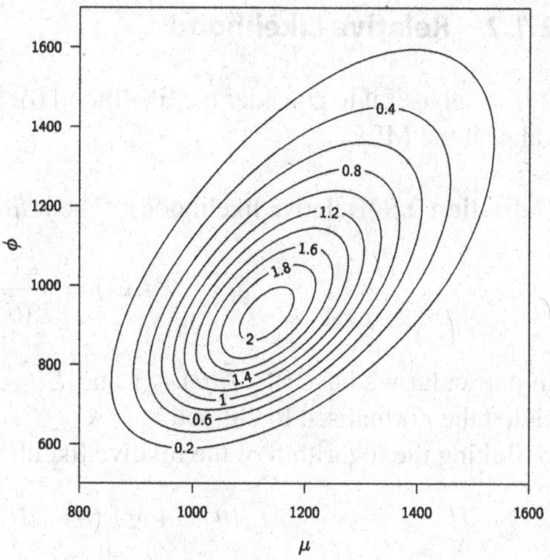

Fig. 2.5 Likelihood $L(\mu, \phi) \cdot 10^{164}$ of the reparametrised gamma model

A slightly less restrictive definition of a random sample still requires independence, but no longer that the components X_i do all have the same distribution. For example, they may still belong to the same distribution family, but with different parameters.

Example 2.4 (Poisson model) Consider Example 1.1.6 and denote the observed and expected number of cancer cases in the $n = 56$ regions of Scotland with x_i and e_i, respectively, $i = 1, \ldots, n$. The simplest model for such registry data assumes that the underlying relative risk λ is the same in all regions and that the observed counts x_i's constitute independent realisations from Poisson distributions with means $e_i \lambda$. The random variables X_i hence belong to the same distributional family but are not identically distributed since the mean parameter $e_i \lambda$ varies from observation to observation.

The log-likelihood kernel of the relative risk λ turns out to be

$$l(\lambda) = \sum_{i=1}^{n} x_i \log \lambda - \sum_{i=1}^{n} e_i \lambda,$$

and the MLE of λ is

$$\hat{\lambda}_{\text{ML}} = \sum_{i=1}^{n} x_i \Big/ \sum_{i=1}^{n} e_i = \bar{x}/\bar{e},$$

where $\bar{x} = \sum_{i=1}^{n} x_i/n$ and $\bar{e} = \sum_{i=1}^{n} e_i/n$ denote the mean observed and expected number of cases, respectively. ■

2.1.2 Relative Likelihood

It is often useful to consider the likelihood (or log-likelihood) function relative to its value at the MLE.

Definition 2.5 (Relative likelihood) The *relative likelihood* is

$$\tilde{L}(\theta) = \frac{L(\theta)}{L(\hat{\theta}_{\mathrm{ML}})}.$$

In particular we have $0 \le \tilde{L}(\theta) \le 1$ and $\tilde{L}(\hat{\theta}_{\mathrm{ML}}) = 1$. The relative likelihood is also called the normalised likelihood.

Taking the logarithm of the relative likelihood gives the *relative log-likelihood*

$$\tilde{l}(\theta) = \log \tilde{L}(\theta) = l(\theta) - l(\hat{\theta}_{\mathrm{ML}}),$$

where we have $-\infty < \tilde{l}(\theta) \le 0$ and $\tilde{l}(\hat{\theta}_{\mathrm{ML}}) = 0$. ♦

Example 2.5 (Inference for a proportion) All different likelihood functions are displayed for a binomial model (cf. Example 2.1) with sample size $n = 10$ and observation $x = 2$ in Fig. 2.6. Note that the change from an ordinary to a relative likelihood changes the scaling of the y-axis, but the shape of the likelihood function remains the same. This is also true for the log-likelihood function. ■

It is important to consider the entire likelihood function as the carrier of the information regarding θ provided by the data. This is far more informative than to consider only the MLE and to disregard the likelihood function itself. Using the values of the relative likelihood function gives us a method to derive a set of parameter values (usually an interval), which are supported by the data. For example, the following categorisation based on thresholding the relative likelihood function using the cutpoints $1/3$, $1/10$, $1/100$ and $1/1000$ has been proposed:

$$\begin{aligned}
1 \ge \tilde{L}(\theta) &> \frac{1}{3} && \theta \text{ very plausible,}\\
\frac{1}{3} \ge \tilde{L}(\theta) &> \frac{1}{10} && \theta \text{ plausible,}\\
\frac{1}{10} \ge \tilde{L}(\theta) &> \frac{1}{100} && \theta \text{ less plausible,}\\
\frac{1}{100} \ge \tilde{L}(\theta) &> \frac{1}{1000} && \theta \text{ barely plausible,}\\
\frac{1}{1000} \ge \tilde{L}(\theta) &\ge 0 && \theta \text{ not plausible.}
\end{aligned}$$

However, such a *pure likelihood* approach to inference has the disadvantage that the scale and the thresholds are somewhat arbitrarily chosen. Indeed, the likelihood on its own does not allow us to quantify the support for a certain set of parameter values

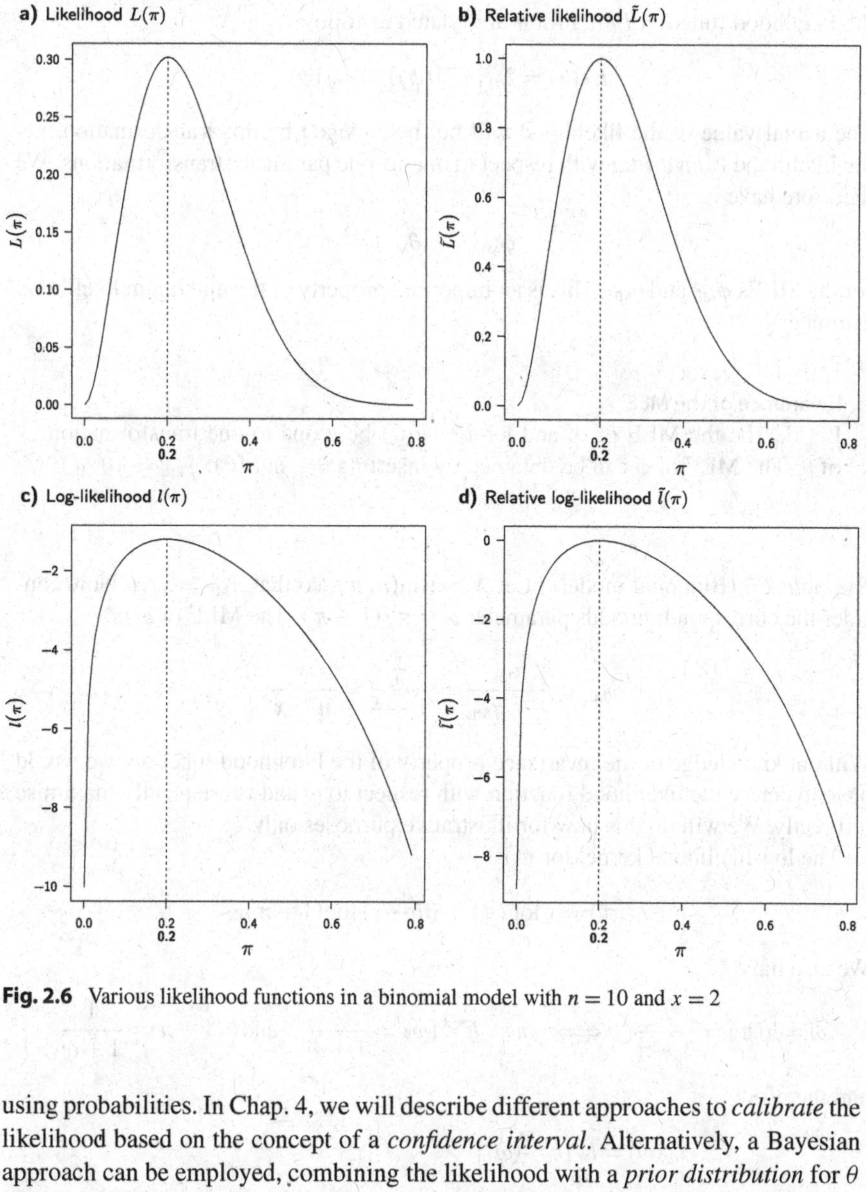

Fig. 2.6 Various likelihood functions in a binomial model with $n = 10$ and $x = 2$

using probabilities. In Chap. 4, we will describe different approaches to *calibrate* the likelihood based on the concept of a *confidence interval*. Alternatively, a Bayesian approach can be employed, combining the likelihood with a *prior distribution* for θ and using the concept of a *credible interval*. This approach is outlined in Chap. 6.

2.1.3 Invariance of the Likelihood

Suppose we parametrise the distribution of X not with respect to θ, but with respect to a one-to-one transformation $\phi = h(\theta)$. The likelihood function $L_\phi(\phi)$ for ϕ and the likelihood function $L_\theta(\theta)$ for θ are related as follows:

$$L_\theta(\theta) = L_\theta\{h^{-1}(\phi)\} = L_\phi(\phi).$$

The actual value of the likelihood will not be changed by this transformation, i.e. the likelihood is *invariant* with respect to one-to-one parameter transformations. We therefore have

$$\hat{\phi}_{\mathrm{ML}} = h(\hat{\theta}_{\mathrm{ML}})$$

for the MLEs $\hat{\phi}_{\mathrm{ML}}$ and $\hat{\theta}_{\mathrm{ML}}$. This is an important property of the maximum likelihood estimate:

Invariance of the MLE
Let $\hat{\theta}_{\mathrm{ML}}$ be the MLE of θ, and let $\phi = h(\theta)$ be a one-to-one transformation of θ. The MLE of ϕ can be obtained by inserting $\hat{\theta}_{\mathrm{ML}}$ in $h(\theta)$: $\hat{\phi}_{\mathrm{ML}} = h(\hat{\theta}_{\mathrm{ML}})$.

Example 2.6 (Binomial model) Let $X \sim \mathrm{Bin}(n, \pi)$, so that $\hat{\pi}_{\mathrm{ML}} = x/n$. Now consider the corresponding odds parameter $\omega = \pi/(1-\pi)$. The MLE of ω is

$$\hat{\omega}_{\mathrm{ML}} = \frac{\hat{\pi}_{\mathrm{ML}}}{1-\hat{\pi}_{\mathrm{ML}}} = \frac{\frac{x}{n}}{1-\frac{x}{n}} = \frac{x}{n-x}.$$

Without knowledge of the invariance property of the likelihood function, we would have to derive the likelihood function with respect to ω and subsequently maximise it directly. We will do this now for illustrative purposes only.

The log-likelihood kernel for π is

$$l_\pi(\pi) = x\log(\pi) + (n-x)\log(1-\pi).$$

We also have

$$\omega = h(\pi) = \frac{\pi}{1-\pi} \iff \pi = h^{-1}(\omega) = \frac{\omega}{1+\omega} \quad \text{and} \quad 1-\pi = \frac{1}{1+\omega}$$

and therefore

$$\begin{aligned} l_\omega(\omega) &= l_\pi\{h^{-1}(\omega)\} \\ &= x\log\left(\frac{\omega}{1+\omega}\right) + (n-x)\log\left(\frac{1}{1+\omega}\right) \\ &= x\log(\omega) - n\log(1+\omega). \end{aligned}$$

The derivative with respect to ω turns out to be

$$\frac{dl_\omega(\omega)}{d\omega} = \frac{x}{\omega} - \frac{n}{1+\omega},$$

so the root $\hat{\omega}_{ML}$ must fulfil $x(1+\hat{\omega}_{ML}) = n\,\hat{\omega}_{ML}$. We easily obtain

$$\hat{\omega}_{ML} = \frac{x}{n-x}.$$ ■

Example 2.7 (Hardy–Weinberg equilibrium) Let us now consider Example 1.1.4 and the observed frequencies $x_1 = 233$, $x_2 = 385$ and $x_3 = 129$ of the three genotypes MM, MN and NN. Assuming Hardy–Weinberg equilibrium, the multinomial log-likelihood kernel

$$l(\pi) = \sum_{i=1}^{3} x_i \log(\pi_i)$$

can be written with (1.1) as

$$\begin{aligned} l(\upsilon) &= x_1 \log(\upsilon^2) + x_2 \log\{2\upsilon(1-\upsilon)\} + x_3 \log\{(1-\upsilon)^2\} \\ &= 2x_1 \log(\upsilon) + \underbrace{x_2 \log(2)}_{=\text{const}} + x_2 \log(\upsilon) + x_2 \log(1-\upsilon) + 2x_3 \log(1-\upsilon) \\ &= (2x_1 + x_2)\log(\upsilon) + (x_2 + 2x_3)\log(1-\upsilon) + \text{const}. \end{aligned}$$

The log-likelihood kernel for the allele frequency υ is therefore $(2x_1 + x_2)\log(\upsilon) + (x_2 + 2x_3)\log(1-\upsilon)$, which can be identified as a binomial log-likelihood kernel for the success probability υ with $2x_1 + x_2$ successes and $x_2 + 2x_3$ failures.

The MLE of υ is therefore

$$\hat{\upsilon}_{ML} = \frac{2x_1 + x_2}{2x_1 + 2x_2 + 2x_3} = \frac{2x_1 + x_2}{2n} = \frac{x_1 + x_2/2}{n},$$

which is exactly the proportion of A alleles in the sample. For the data above, we obtain $\hat{\upsilon}_{ML} \approx 0.570$. The MLEs of π_1, π_2 and π_3 (assuming Hardy–Weinberg equilibrium) are therefore

$$\hat{\pi}_1 = \hat{\upsilon}_{ML}^2 \approx 0.324, \qquad \hat{\pi}_2 = 2\hat{\upsilon}_{ML}(1-\hat{\upsilon}_{ML}) \approx 0.490 \quad \text{and}$$

$$\hat{\pi}_3 = (1-\hat{\upsilon}_{ML})^2 \approx 0.185,$$

using the invariance property of the likelihood. ■

In the last example, the transformation to which the MLE is invariant is not really a one-to-one transformation. A more detailed view of the situation is the following: We have the more general multinomial model with two parameters π_1, π_2 (π_3 is determined by these) and the simpler Hardy–Weinberg model with one parameter υ. We can restrict the multinomial model to the Hardy–Weinberg model, which is hence a special case of the multinomial model. If we obtain an MLE for υ, we can hence calculate the resulting MLEs for π_1, π_2 and also π_3. However, in the other direction, i.e. by first calculating the unrestricted MLE $\hat{\pi}_{ML}$ in the multinomial model, we could not calculate a corresponding MLE $\hat{\upsilon}_{ML}$ in the simpler Hardy–Weinberg model.

2.1.4 Generalised Likelihood

Declaring probability mass and density functions as appropriate likelihood functions is not always sufficient. In some situations this definition must be suitably generalised. A typical example is the analysis of survival data with some observations being censored.

Assume that observed survival times $x_1, \dots, x_n$ are independent realisations from a distribution with density function $f(x; \theta)$ and corresponding distribution function $F(x; \theta) = \Pr(X \le x; \theta)$. The likelihood contribution of a non-censored observation x_i is then (as before) $f(x_i; \theta)$. However, a censored observation will contribute the term $1 - F(x_i; \theta) = \Pr(X_i > x_i; \theta)$ to the likelihood since in this case we only know that the actual (unobserved) survival time is larger than x_i.

Compact notation can be achieved using the *censoring indicator* $\delta_i, i = 1, \dots, n$, with $\delta_i = 0$ if the survival time x_i is censored and $\delta_i = 1$ if it is observed. Due to independence of the observations, the likelihood can be written as

$$L(\theta) = \prod_{i=1}^{n} f(x_i; \theta)^{\delta_i} \{1 - F(x_i; \theta)\}^{1-\delta_i}. \tag{2.2}$$

Example 2.8 (Analysis of survival times) A simple statistical model to describe survival times is to assume an exponential distribution with density and distribution function

$$f(x) = \lambda \exp(-\lambda x) \quad \text{and}$$
$$F(x) = 1 - \exp(-\lambda x),$$

respectively, so $1 - F(x) = \exp(-\lambda x)$. The likelihood function (2.2) now reduces to

$$L(\lambda) = \lambda^{n\bar{\delta}} \exp(-\lambda n \bar{x}),$$

where $\bar{\delta}$ is the observed proportion of uncensored observations, and $\bar{x}$ is the mean observed survival time of all (censored and uncensored) observations. The MLE is $\hat{\lambda}_{ML} = \bar{\delta}/\bar{x}$. Due to invariance of the MLE, the estimate for the mean $\mu = 1/\lambda$ is $\hat{\mu}_{ML} = \bar{x}/\bar{\delta}$.

Among the $n = 94$ observations from Example 1.1.8, there are $\sum_{i=1}^{n} \delta_i = 47$ uncensored, and the total follow-up time is $\sum_{i=1}^{n} x_i = 143192$ days. The estimated rate is $\hat{\lambda}_{ML} = 47/143\,192 = 32.82$ per 100 000 days, and the MLE of the expected survival time is $\hat{\mu}_{ML} = 3046.6$ days. This is substantially larger than in the analysis of the uncensored observations only (cf. Example 2.3), where we have obtained the estimate $\hat{\mu}_{ML} = 1130.8$ days. ■

2.2 Score Function and Fisher Information

The MLE of θ is obtained by maximising the (relative) likelihood function,

$$\hat{\theta}_{\mathrm{ML}} = \arg\max_{\theta \in \Theta} L(\theta) = \arg\max_{\theta \in \Theta} \tilde{L}(\theta).$$

For numerical reasons, it is often easier to maximise the log-likelihood $l(\theta) = \log L(\theta)$ or the relative log-likelihood $\tilde{l}(\theta) = l(\theta) - l(\hat{\theta}_{\mathrm{ML}})$ (cf. Sect. 2.1), which yields the same result since

$$\hat{\theta}_{\mathrm{ML}} = \arg\max_{\theta \in \Theta} l(\theta) = \arg\max_{\theta \in \Theta} \tilde{l}(\theta).$$

However, the log-likelihood function $l(\theta)$ has much larger importance, besides simplifying the computation of the MLE. Especially, its first and second derivatives are important and have their own names, which are introduced in the following. For simplicity, we assume that θ is a scalar.

Definition 2.6 (Score function) The first derivative of the log-likelihood function

$$S(\theta) = \frac{dl(\theta)}{d\theta}$$

is called the *score function*. ◆

Computation of the MLE is typically done by solving the *score equation* $S(\theta) = 0$.

The second derivative, the curvature, of the log-likelihood function is also of central importance and has its own name.

Definition 2.7 (Fisher information) The negative second derivative of the log-likelihood function

$$I(\theta) = -\frac{d^2 l(\theta)}{d\theta^2} = -\frac{dS(\theta)}{d\theta}$$

is called the *Fisher information*. The value of the Fisher information at the MLE $\hat{\theta}_{\mathrm{ML}}$, i.e. $I(\hat{\theta}_{\mathrm{ML}})$, is the *observed Fisher information*. ◆

Note that the MLE $\hat{\theta}_{\mathrm{ML}}$ is a function of the observed data, which explains the terminology "observed" Fisher information for $I(\hat{\theta}_{\mathrm{ML}})$.

Example 2.9 (Normal model) Suppose we have realisations $x_{1:n}$ of a random sample from a normal distribution $\mathrm{N}(\mu, \sigma^2)$ with unknown mean μ and known variance σ^2. The log-likelihood kernel and score function are then

$$l(\mu) = -\frac{1}{2\sigma^2}\sum_{i=1}^{n}(x_i - \mu)^2 \quad \text{and}$$

$$S(\mu) = \frac{1}{\sigma^2}\sum_{i=1}^{n}(x_i - \mu),$$

respectively. The solution of the score equation $S(\mu) = 0$ is the MLE $\hat{\mu}_{\text{ML}} = \bar{x}$. Taking another derivative gives the Fisher information

$$I(\mu) = \frac{n}{\sigma^2},$$

which does not depend on μ and so is equal to the observed Fisher information $I(\hat{\mu}_{\text{ML}})$, no matter what the actual value of $\hat{\mu}_{\text{ML}}$ is.

Suppose we switch the roles of the two parameters and treat μ as known and σ^2 as unknown. We now obtain

$$\hat{\sigma}^2_{\text{ML}} = \sum_{i=1}^{n}(x_i - \mu)^2/n$$

with Fisher information

$$I(\sigma^2) = \frac{1}{\sigma^6}\sum_{i=1}^{n}(x_i - \mu)^2 - \frac{n}{2\sigma^4}.$$

The Fisher information of σ^2 now really depends on its argument σ^2. The observed Fisher information turns out to be

$$I(\hat{\sigma}^2_{\text{ML}}) = \frac{n}{2\hat{\sigma}^4_{\text{ML}}}.$$ ■

It is instructive at this stage to adopt a *frequentist* point of view and to consider the MLE $\hat{\mu}_{\text{ML}} = \bar{x}$ from Example 2.9 as a random variable, i.e. $\hat{\mu}_{\text{ML}} = \bar{X}$ is now a function of the random sample $X_{1:n}$. We can then easily compute $\text{Var}(\hat{\mu}_{\text{ML}}) = \text{Var}(\bar{X}) = \sigma^2/n$ and note that

$$\text{Var}(\hat{\mu}_{\text{ML}}) = \frac{1}{I(\hat{\mu}_{\text{ML}})}$$

holds. In Sect. 4.2.3 we will see that this equality is approximately valid for other statistical models. Indeed, under certain regularity conditions, the variance $\text{Var}(\hat{\theta}_{\text{ML}})$ of the MLE turns out to be approximately equal to the inverse observed Fisher information $1/I(\hat{\theta}_{\text{ML}})$, and the accuracy of this approximation improves with increasing sample size n. Example 2.9 is a special case, where this equality holds exactly for any sample size.

Example 2.10 (Binomial model) The score function of a binomial observation $X = x$ with $X \sim \text{Bin}(n, \pi)$ is

$$S(\pi) = \frac{dl(\pi)}{d\pi} = \frac{x}{\pi} - \frac{n-x}{1-\pi}$$

and has been derived already in Example 2.1. Taking the derivative of $S(\pi)$ gives the Fisher information

$$\begin{aligned} I(\pi) &= -\frac{d^2 l(\pi)}{d\pi^2} = -\frac{dS(\pi)}{d\pi} \\ &= \frac{x}{\pi^2} + \frac{n-x}{(1-\pi)^2} \\ &= n\left\{\frac{x/n}{\pi^2} + \frac{(n-x)/n}{(1-\pi)^2}\right\}. \end{aligned}$$

Plugging in the MLE $\hat{\pi}_{\text{ML}} = x/n$, we finally obtain the observed Fisher information

$$I(\hat{\pi}_{\text{ML}}) = \frac{n}{\hat{\pi}_{\text{ML}}(1-\hat{\pi}_{\text{ML}})}.$$

This result is plausible if we take a frequentist point of view and consider the MLE as a random variable. Then

$$\text{Var}(\hat{\pi}_{\text{ML}}) = \text{Var}\left(\frac{X}{n}\right) = \frac{1}{n^2} \cdot \text{Var}(X) = \frac{1}{n^2} n\pi(1-\pi) = \frac{\pi(1-\pi)}{n},$$

so the variance of $\hat{\pi}_{\text{ML}}$ has the same form as the inverse observed Fisher information; the only difference is that the MLE $\hat{\pi}_{\text{ML}}$ is replaced by the true (and unknown) parameter π. The inverse observed Fisher information is hence an estimate of the variance of the MLE. ■

How does the observed Fisher information change if we reparametrise our statistical model? Here is the answer to this question.

Result 2.1 (Observed Fisher information after reparametrisation) *Let $I_\theta(\hat{\theta}_{\text{ML}})$ denote the observed Fisher information of a scalar parameter θ and suppose that $\phi = h(\theta)$ is a one-to-one transformation of θ. The observed Fisher information $I_\phi(\hat{\phi}_{\text{ML}})$ of ϕ is then*

$$I_\phi(\hat{\phi}_{\text{ML}}) = I_\theta(\hat{\theta}_{\text{ML}})\left\{\frac{dh^{-1}(\hat{\phi}_{\text{ML}})}{d\phi}\right\}^2 = I_\theta(\hat{\theta}_{\text{ML}})\left\{\frac{dh(\hat{\theta}_{\text{ML}})}{d\theta}\right\}^{-2}. \tag{2.3}$$

Proof The transformation h is assumed to be one-to-one, so $\theta = h^{-1}(\phi)$ and $l_\phi(\phi) = l_\theta\{h^{-1}(\phi)\}$. Application of the chain rule gives

$$
\begin{aligned}
S_\phi(\phi) &= \frac{dl_\phi(\phi)}{d\phi} = \frac{dl_\theta\{h^{-1}(\phi)\}}{d\phi} \\
&= \frac{dl_\theta(\theta)}{d\theta} \cdot \frac{dh^{-1}(\phi)}{d\phi} \\
&= S_\theta(\theta) \cdot \frac{dh^{-1}(\phi)}{d\phi}.
\end{aligned} \tag{2.4}
$$

The second derivative of $l_\phi(\phi)$ can be computed using the product and chain rules:

$$
\begin{aligned}
I_\phi(\phi) &= -\frac{dS_\phi(\phi)}{d\phi} = -\frac{d}{d\phi}\left\{S_\theta(\theta) \cdot \frac{dh^{-1}(\phi)}{d\phi}\right\} \\
&= -\frac{dS_\theta(\theta)}{d\phi} \cdot \frac{dh^{-1}(\phi)}{d\phi} - S_\theta(\theta) \cdot \frac{d^2h^{-1}(\phi)}{d\phi^2} \\
&= -\frac{dS_\theta(\theta)}{d\theta} \cdot \left\{\frac{dh^{-1}(\phi)}{d\phi}\right\}^2 - S_\theta(\theta) \cdot \frac{d^2h^{-1}(\phi)}{d\phi^2} \\
&= I_\theta(\theta)\left\{\frac{dh^{-1}(\phi)}{d\phi}\right\}^2 - S_\theta(\theta) \cdot \frac{d^2h^{-1}(\phi)}{d\phi^2}.
\end{aligned}
$$

Evaluating $I_\phi(\phi)$ at the MLE $\phi = \hat{\phi}_{\text{ML}}$ (so $\theta = \hat{\theta}_{\text{ML}}$) leads to the first equation in (2.3) (note that $S_\theta(\hat{\theta}_{\text{ML}}) = 0$). The second equation follows with

$$
\frac{dh^{-1}(\phi)}{d\phi} = \left\{\frac{dh(\theta)}{d\theta}\right\}^{-1} \quad \text{for } \frac{dh(\theta)}{d\theta} \neq 0. \tag{2.5}
$$

□

Example 2.11 (Binomial model) In Example 2.6 we saw that the MLE of the odds $\omega = \pi/(1-\pi)$ is $\hat{\omega}_{\text{ML}} = x/(n-x)$. What is the corresponding observed Fisher information? First, we compute the derivative of $h(\pi) = \pi/(1-\pi)$, which is

$$
\frac{dh(\pi)}{d\pi} = \frac{1}{(1-\pi)^2}.
$$

Using the observed Fisher information of π derived in Example 2.10, we obtain

$$
\begin{aligned}
I_\omega(\hat{\omega}_{\text{ML}}) &= I_\pi(\hat{\pi}_{\text{ML}})\left\{\frac{dh(\hat{\pi}_{\text{ML}})}{d\pi}\right\}^{-2} = \frac{n}{\hat{\pi}_{\text{ML}}(1-\hat{\pi}_{\text{ML}})} \cdot (1-\hat{\pi}_{\text{ML}})^4 \\
&= n \cdot \frac{(1-\hat{\pi}_{\text{ML}})^3}{\hat{\pi}_{\text{ML}}} = \frac{(n-x)^3}{nx}.
\end{aligned}
$$

As a function of x for fixed n, the observed Fisher information $I_\omega(\hat{\omega}_{\text{ML}})$ is monotonically decreasing (the numerator is monotonically decreasing, and the denominator is monotonically increasing). In other words, the observed Fisher information increases with decreasing MLE $\hat{\omega}_{\text{ML}}$.

The observed Fisher information of the log odds $\phi = \log(\omega)$ can be similarly computed, and we obtain

$$I_\phi(\hat{\phi}_{\text{ML}}) = I_\omega(\hat{\omega}_{\text{ML}})\left(\frac{1}{\hat{\omega}_{\text{ML}}}\right)^{-2} = \frac{(n-x)^3}{nx} \cdot \frac{x^2}{(n-x)^2} = \frac{x(n-x)}{n}.$$

Note that $I_\phi(\hat{\phi}_{\text{ML}})$ does not change if we redefine successes as failures and vice versa. This is also the case for the observed Fisher information $I_\pi(\hat{\pi}_{\text{ML}})$ but not for $I_\omega(\hat{\omega}_{\text{ML}})$. ■

2.3 Numerical Computation of the Maximum Likelihood Estimate

Explicit formulas for the MLE and the observed Fisher information can typically only be derived in simple models. In more complex models, numerical techniques have to be applied to compute maximum and curvature of the log-likelihood function. We first describe the application of general purpose optimisation algorithms to this setting and will discuss the Expectation-Maximisation (EM) algorithm in Sect. 2.3.2.

2.3.1 Numerical Optimisation

Application of the *Newton–Raphson algorithm* (cf. Appendix C.1.3) requires the first two derivatives of the function to be maximised, so for maximising the log-likelihood function, we need the score function and the Fisher information. Iterative application of the equation

$$\theta^{(t+1)} = \theta^{(t)} + \frac{S(\theta^{(t)})}{I(\theta^{(t)})}$$

gives after convergence (i.e. $\theta^{(t+1)} = \theta^{(t)}$) the MLE $\hat{\theta}_{\text{ML}}$. As a by-product, the observed Fisher information $I(\hat{\theta}_{\text{ML}})$ can also be extracted.

To apply the Newton–Raphson algorithm in `R`, the function `optim` can conveniently be used, see Appendix C.1.3 for details. We need to pass the log-likelihood function as an argument to `optim`. Explicitly passing the score function into `optim` typically accelerates convergence. If the derivative is not available, it can sometimes be computed symbolically using the `R` function `deriv`. Generally no derivatives need to be passed to `optim` because it can approximate them numerically. Particularly useful is the option `hessian = TRUE`, in which case `optim` will also return the negative observed Fisher information.

Example 2.12 (Screening for colon cancer) The goal of Example 1.1.5 is to estimate the false negative fraction of a screening test, which consists of six consecutive medical examinations. Let π denote the probability for a positive test result of the ith diseased individual and denote by X_i the number of positive test results among the six examinations. We start by assuming that individual test results are independent and that π does not vary from patient to patient (two rather unrealistic assumptions, as we will see later), so that X_i is binomially distributed: $X_i \sim \text{Bin}(N = 6, \pi)$. However, due to the study design, we will not observe a patient with $X_i = 0$ positive tests. We therefore need to use the *truncated binomial distribution* as the appropriate statistical model. The corresponding log-likelihood can be derived by considering

$$\Pr(X_i = k \mid X_i > 0) = \frac{\Pr(X_i = k)}{\Pr(X_i > 0)}, \quad k = 1, \dots, 6, \tag{2.6}$$

and turns out to be (cf. Example C.1 in Appendix C)

$$l(\pi) = \sum_{k=1}^{N} Z_k \{k \log(\pi) + (N - k) \log(1 - \pi)\} - n \log\{1 - (1 - \pi)^N\}. \tag{2.7}$$

Here Z_k denotes the number of patients with k positive test results, and $n = \sum_{k=1}^{N} Z_k = 196$ is the total number of diseased patients with at least one positive test result.

Computation of the MLE is now most conveniently done with numerical techniques. To do so, we write an R function `log.likelihood`, which returns the log-likelihood kernel of the unknown probability (`pi`) for a given vector of counts (`data`) and maximise, it with the `optim` function.

```
## Truncated binomial log-likelihood function
## pi: the parameter, the probability of a positive test result
## data: vector with counts Z_1, ..., Z_N
log.likelihood <- function(pi, data)
  {
      n <- sum(data)
      k <- length(data)
      vec <- seq_len(k)
      result <- sum(data * (vec * log(pi) + (k-vec) * log(1-pi))) -
               n * log(1 - (1-pi)^k)

  }
data <- c(37, 22, 25, 29, 34, 49)
eps <- 1e-10
result <- optim(0.5, log.likelihood, data = data,
                method = "L-BFGS-B", lower = eps, upper = 1-eps,
                control = list(fnscale = -1), hessian = TRUE)
(ml <- result$par)
[1] 0.6240838
```

The MLE turns out to be $\hat{\pi}_{\text{ML}} = 0.6241$, and the observed Fisher information $I(\hat{\pi}_{\text{ML}}) = 4885.3$ is computed via

```
(observed.fisher <- - result$hessian)
         [,1]
[1,] 4885.251
```

Invariance of the MLE immediately gives the MLE of the false negative fraction $\xi = \Pr(X_i = 0)$ via

$$\hat{\xi}_{ML} = (1 - \hat{\pi}_{ML})^N = 0.0028.$$

A naive estimate of the number of undetected cancer cases Z_0 can be obtained by solving $\hat{Z}_0/(196 + \hat{Z}_0) = \hat{\xi}_{ML}$ for $\hat{Z}_0$:

$$\hat{Z}_0 = 196 \cdot \frac{\hat{\xi}_{ML}}{1 - \hat{\xi}_{ML}} = 0.55. \tag{2.8}$$

It turns out that this estimate can be justified as a maximum likelihood estimate. To see this, note that the probability to detect a cancer case in one particular application of the six-stage screening test is $1 - \xi$. The number of samples until the first cancer case is detected therefore follows a geometric distribution with success probability $1 - \xi$, cf. Table A.1 in Appendix A.5.1. Thus, if n is the observed number of detected cancer cases, the total number of cancer cases $Z_0 + n$ follows a negative binomial distribution with parameters n and $1 - \xi$ (again cf. Appendix A.5.1):

$$Z_0 + n \sim \mathrm{NBin}(n, 1 - \xi),$$

so the expectation of $Z_0 + n$ is $\mathsf{E}(Z_0 + n) = n/(1 - \xi)$. Our interest is in the expectation of Z_0,

$$\mathsf{E}(Z_0) = \frac{n}{1 - \xi} - n = n \cdot \frac{\xi}{1 - \xi}. \tag{2.9}$$

The MLE (2.8) of $\mathsf{E}(Z_0)$ can now easily be obtained by replacing ξ with $\hat{\xi}_{ML}$.

A closer inspection of Table 1.2 makes 0.55 expected undetected cancer cases rather implausible and raises questions about the appropriateness of the binomial model. Indeed, a naive extrapolation of the observed frequencies Z_k, $k = 1, \dots, 6$, leads to a considerably larger values of Z_0. The fact that the binomial model does not fit the data well can also be seen from the frequencies in Table 1.2 with a "bathtub" shape with extreme values $k = 1$ and $k = 5, 6$ more likely than the intermediate values $k = 2, 3, 4$. Such a form is impossible with a (truncated) binomial distribution. This can also be seen from the expected frequencies under the binomial model with $\hat{\pi}_{ML} = 0.6241$, which are 5.5, 22.9, 50.8, 63.2, 42.0 and 11.6 for $k = 1, \dots, 6$. The difference to the observed frequencies Z_k is quite striking. In Example 5.18 we will apply a goodness-of-fit test, which quantifies the evidence against this model. Earlier in Chap. 5 we will introduce an alternative model with two parameters, which describes the observed frequencies much better. ■

2.3.2 The EM Algorithm

The *Expectation-Maximisation (EM) algorithm* is an iterative algorithm to compute MLEs. In certain situations it is particularly easy to apply, as the following example illustrates.

Example 2.13 (Screening for colon cancer) We reconsider Example 1.1.5, where the number Z_k of 196 cancer patients with $k \geq 1$ positive test results among six cancer colon screening tests has been recorded. Due to the design of the study, we have no information on the number Z_0 of patients with solely negative test results (cf. Table 1.2). Numerical techniques allow us to fit a truncated binomial distribution to the observed data $Z_1, \dots, Z_6$, cf. Example 2.12.

However, the EM algorithm could also be used to compute the MLEs. The idea is that an explicit and simple formula for the MLE of π would be available if the number Z_0 was known as well:

$$\hat{\pi} = \frac{\sum_{k=0}^{6} k \cdot Z_k}{6 \cdot \sum_{k=0}^{6} Z_k}. \tag{2.10}$$

Indeed, in this case we are back in the untruncated binomial case with $\sum_{k=0}^{6} k \cdot Z_k$ positive tests among $6 \cdot \sum_{k=0}^{6} Z_k$ tests. However, Z_0 is unknown, but if π and hence $\xi = (1 - \pi)^6$ are known, Z_0 can be estimated by the expectation of a negative binomial distribution (cf. Eq. (2.9) at the end of Example 2.12):

$$\hat{Z}_0 = \mathsf{E}(Z_0) = n \cdot \frac{\xi}{1 - \xi}, \tag{2.11}$$

where $n = 196$ and $\xi = (1 - \pi)^6$. The EM algorithm now iteratively computes (2.10) and (2.11) and replaces the terms Z_0 and π on the right-hand sides with their current estimates $\hat{Z}_0$ and $\hat{\pi}$, respectively. Implementation of the algorithm is shown in the following R-code:

```
## data set
fulldata <- c(NA, 37, 22, 25, 29, 34, 49)
k <- 0:6
n <- sum(fulldata[-1])
## impute start value for Z0 (first element)
## and initialise some different old value
fulldata[1] <- 10
Z0old <- 9
## the EM algorithm
while(abs(Z0old - fulldata[1]) >= 1e-7)
{
    Z0old <- fulldata[1]
    pi <- sum(fulldata * k) / sum(fulldata) / 6
    xi <- (1-pi)^6
    fulldata[1] <- n * xi / (1-xi)
}
```

This method quickly converges, as illustrated in Table 2.1, with starting value $Z_0 = 10$. Note that the estimate $\hat{\pi}$ from Table 2.1 is numerically equal to the MLE (cf. Example 2.12) already after 5 iterations. As a by-product, we also obtain the estimate $\hat{Z}_0 = 0.55$. ■

A general derivation of the EM algorithm distinguishes *observed* data X ($Z_1, \dots, Z_6$ in Example 2.13) and *unobserved* data U (Z_0 in Example 2.13). The joint probability mass or density function of the *complete* data X and U can always

Table 2.1 Values of the EM algorithm until convergence (Difference between old and new estimate $\hat{\pi}$ smaller than 10^{-7})

Iteration	$\hat{\pi}$	$\hat{Z}_0$
1	0.5954693	0.8627256
2	0.6231076	0.5633868
3	0.6240565	0.5549056
4	0.6240835	0.5546665
5	0.6240842	0.5546597
6	0.6240842	0.5546596

be written as (cf. Eq. (A.7) in Appendix A.2)

$$f(x,u) = f(u \mid x) f(x),$$

so the corresponding log-likelihood functions of θ obey the relationship

$$l(\theta; x, u) = l(\theta; u \mid x) + l(\theta; x). \tag{2.12}$$

Note that the log-likelihood functions cannot be written in the simple form $l(\theta)$ as they are based on different data: $l(\theta; x, u)$ is the complete data log-likelihood, while $l(\theta; x)$ is the observed data log-likelihood. Now u is unobserved, so we replace it in (2.12) by the random variable U:

$$l(\theta; x, U) = l(\theta; U \mid x) + l(\theta; x)$$

and consider the expectation of this equation with respect to $f(u \mid x; \theta')$ (it will soon become clear why we distinguish θ and θ'):

$$\underbrace{\mathsf{E}\{l(\theta; x, U); \theta'\}}_{=:Q(\theta;\theta')} = \underbrace{\mathsf{E}\{l(\theta; U \mid x); \theta'\}}_{=:C(\theta;\theta')} + \underbrace{l(\theta; x)}_{=l(\theta)}. \tag{2.13}$$

Note that the last term does not change, as it does not depend on U. So if

$$Q(\theta; \theta') \geq Q(\theta'; \theta'), \tag{2.14}$$

we have with (2.13):

$$l(\theta) - l(\theta') \geq C(\theta'; \theta') - C(\theta; \theta') \geq 0, \tag{2.15}$$

where the last inequality follows from the information inequality (cf. Appendix A.3.8).

This leads to the EM algorithm with starting value θ':

1. *Expectation (E-step):* Compute $Q(\theta; \theta')$.
2. *Maximisation (M-step):* Maximise $Q(\theta; \theta')$ with respect to θ to obtain θ''.
3. Now iterate Steps 1 and 2 (i.e. set $\theta' = \theta''$ and apply Step 1), until the values of θ converge. A possible stopping criterion is $|\theta' - \theta''| < \varepsilon$ for some small $\varepsilon > 0$.

Equation (2.15) implies that every iteration of the EM algorithm increases the log-likelihood. This follows from the fact that—through maximisation—$Q(\theta'';\theta')$ is larger than $Q(\theta;\theta')$ for all θ, so in particular (2.14) holds (with $\theta=\theta'$), and therefore $l(\theta'')\geq l(\theta')$. In contrast, the Newton–Raphson algorithm is not guaranteed to increase the log-likelihood in every iteration.

Example 2.14 (Example 2.13 continued) The joint probability mass function of observed data $X=(Z_1,\ldots,Z_6)$ and unobserved data $U=Z_0$ is multinomially distributed (cf. Appendix A.5.3) with size parameter equal to $n+Z_0$ and probability vector $\boldsymbol{p}$ with entries

$$p_k=\binom{6}{k}\pi^k(1-\pi)^{6-k},$$

$k=0,\ldots,6$, which we denote by

$$(Z_0,Z_1,\ldots,Z_6)^\top\sim \mathrm{M}_7(n+Z_0,\boldsymbol{p}).$$

The corresponding log-likelihood

$$l(\pi)=\sum_{k=0}^{6}Z_k\log(p_k) \tag{2.16}$$

is linear in the unobserved data Z_0. Therefore, the E-step of the EM algorithm is achieved if we replace Z_0 with its expectation conditional on the observed data $X=(Z_1,\ldots,Z_6)$ and on the unknown parameter π'. From Example 2.12 we know that this expectation is $\mathsf{E}(Z_0\mid Z_1=z_1,\ldots,Z_6=z_6;\pi')=n\cdot\xi'/(1-\xi')$ with $n=z_1+\cdots+z_6$ and $\xi'=(1-\pi')^6$. This completes the E-step.

The M-step now maximises (2.16) with $Z_0=n\cdot\xi'/(1-\xi')$ with respect to π. We have argued earlier that the complete data MLE is (2.10). This estimate can formally be derived from the complete data log-likelihood kernel

$$l(\pi)=\sum_{k=0}^{6}Z_k\bigl\{(6-k)\log(1-\pi)+k\log(\pi)\bigr\},$$

easily obtained from (2.16), which leads to the complete data score function

$$S(\pi)=\left(\sum_{k=0}^{6}k\cdot Z_k/\pi-6\sum_{k=0}^{6}Z_k\right)/(1-\pi).$$

It is easy to show that the root of this equation is (2.10). ■

One can also show that the EM algorithm always converges to a local or global maximum, or at least to a saddlepoint of the log-likelihood. However, the convergence can be quite slow; typically, more iterations are required than for the Newton–Raphson algorithm. Another disadvantage is that the algorithm does not automatically give the observed Fisher information. Of course, this can be calculated after convergence if the second derivative of the log-likelihood $l(\theta; x)$ of the observed data x is available.

2.4 Quadratic Approximation of the Log-Likelihood Function

An important approximation of the log-likelihood function is based on a quadratic function. To do so, we apply a Taylor approximation of second order (cf. Appendix B.2.3) around the MLE $\hat{\theta}_{\mathrm{ML}}$:

$$\begin{aligned} l(\theta) &\approx l(\hat{\theta}_{\mathrm{ML}}) + \frac{dl(\hat{\theta}_{\mathrm{ML}})}{d\theta}(\theta - \hat{\theta}_{\mathrm{ML}}) + \frac{1}{2}\frac{d^2 l(\hat{\theta}_{\mathrm{ML}})}{d\theta^2}(\theta - \hat{\theta}_{\mathrm{ML}})^2 \\ &= l(\hat{\theta}_{\mathrm{ML}}) + S(\hat{\theta}_{\mathrm{ML}})(\theta - \hat{\theta}_{\mathrm{ML}}) - \frac{1}{2} \cdot I(\hat{\theta}_{\mathrm{ML}})(\theta - \hat{\theta}_{\mathrm{ML}})^2. \end{aligned}$$

Due to $S(\hat{\theta}_{\mathrm{ML}}) = 0$, the quadratic approximation of the relative log-likelihood is

$$\tilde{l}(\theta) = l(\theta) - l(\hat{\theta}_{\mathrm{ML}}) \approx -\frac{1}{2} \cdot I(\hat{\theta}_{\mathrm{ML}})(\theta - \hat{\theta}_{\mathrm{ML}})^2. \tag{2.17}$$

Example 2.15 (Poisson model) Assume that we have one observation $x = 11$ from a Poisson distribution $\mathrm{Po}(e\lambda)$ with known offset $e = 3.04$ and unknown parameter λ. The MLE of λ is $\hat{\lambda}_{\mathrm{ML}} = x/e = 3.62$, cf. Example 2.4. The observed Fisher information turns out to be $I(\hat{\lambda}_{\mathrm{ML}}) = x/\hat{\lambda}^2_{\mathrm{ML}}$, so that the quadratic approximation of the relative log-likelihood is

$$\tilde{l}(\lambda) \approx -\frac{1}{2}\frac{x}{\hat{\lambda}^2_{\mathrm{ML}}}(\lambda - \hat{\lambda}_{\mathrm{ML}})^2.$$

Figure 2.7 displays $\tilde{l}(\lambda)$ and its quadratic approximation. ■

Example 2.16 (Normal model) Let $X_{1:n}$ denote a random sample from a normal distribution $\mathrm{N}(\mu, \sigma^2)$ with unknown mean μ and known variance σ^2. We know from Example 2.9 that

$$\begin{aligned} l(\mu) &= -\frac{1}{2\sigma^2}\sum_{i=1}^{n}(x_i - \mu)^2 \\ &= -\frac{1}{2\sigma^2}\left\{\sum_{i=1}^{n}(x_i - \bar{x})^2 + n(\bar{x} - \mu)^2\right\}, \\ l(\hat{\mu}_{\mathrm{ML}}) &= -\frac{1}{2\sigma^2}\sum_{i=1}^{n}(x_i - \bar{x})^2, \quad \text{and hence} \end{aligned}$$

$$\tilde{l}(\mu) = l(\mu) - l(\hat{\mu}_{\mathrm{ML}}) = -\frac{n}{2\sigma^2}(\bar{x} - \mu)^2,$$

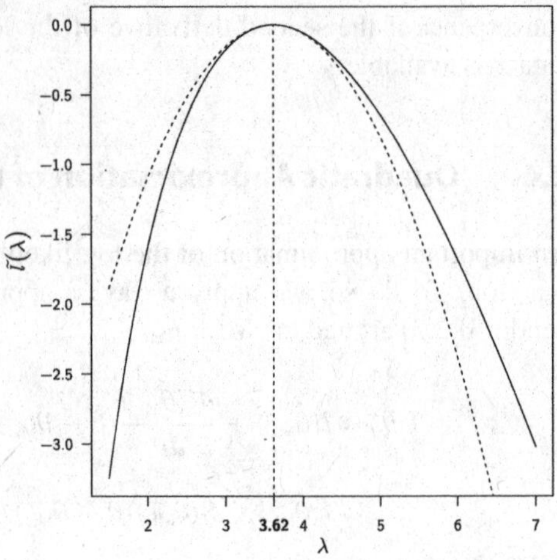

Fig. 2.7 Relative log-likelihood $\tilde{l}(\lambda)$ and its quadratic approximation (*dashed line*) for a single observation $x = 11$ from a Poisson distribution with mean $e\lambda$ and known offset $e = 3.04$

but we also have

$$-\frac{1}{2} \cdot I(\hat{\mu}_{\text{ML}})(\mu - \hat{\mu}_{\text{ML}})^2 = -\frac{n}{2\sigma^2}(\mu - \bar{x})^2.$$

Both sides of Eq. (2.17) are hence identical, so the quadratic approximation is here exact. ■

Under certain regularity conditions, which we will not discuss here, it can be shown that a quadratic approximation of the log-likelihood improves with increasing sample size. The following example illustrates this phenomenon in the binomial model.

Example 2.17 (Binomial model) Figure 2.8 displays the relative log-likelihood of the success probability π in a binomial model with sample size $n = 10, 50, 200, 1000$. The observed datum x has been fixed at $x = 8, 40, 160, 800$ such that the MLE of π is $\hat{\pi}_{\text{ML}} = 0.8$ in all four cases. We see that the quadratic approximation of the relative log-likelihood improves with increasing sample size n. The two functions are nearly indistinguishable for $n = 1000$. ■

The advantage of the quadratic approximation of the relative log-likelihood lies in the fact that we only need to know the MLE $\hat{\theta}_{\text{ML}}$ and the observed Fisher information $I(\hat{\theta}_{\text{ML}})$, no matter what the actual log-likelihood looks like. However, in certain pathological cases the approximation may remain poor even if the sample size is larger.

Example 2.18 (Uniform model) Let $X_{1:n}$ denote a random sample from a continuous uniform distribution $\text{U}(0, \theta)$ with unknown upper limit $\theta \in \mathbb{R}^+$. The density function of the uniform distribution is

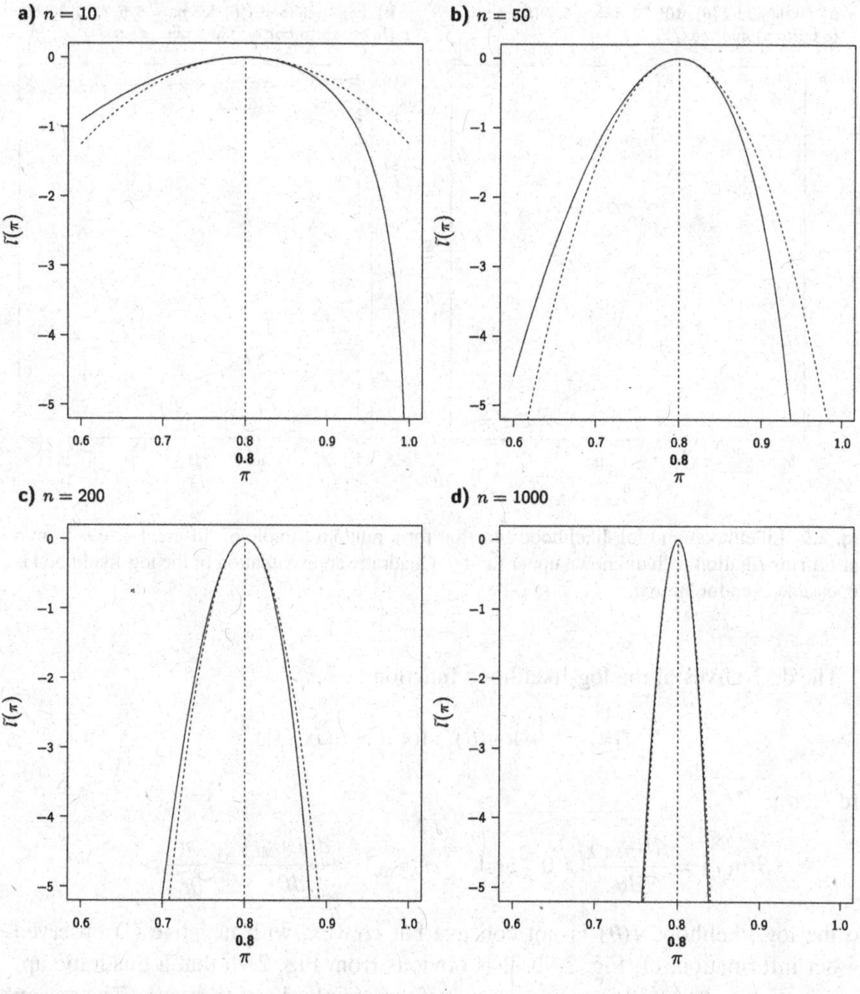

Fig. 2.8 Quadratic approximation (*dashed line*) of the relative log-likelihood (*solid line*) of the success probability π in a binomial model

$$f(x;\theta) = \frac{1}{\theta} \mathsf{I}_{[0,\theta)}(x)$$

with *indicator function* $\mathsf{I}_A(x)$ equal to one if $x \in A$ and zero otherwise. The likelihood function of θ is

$$L(\theta) = \begin{cases} \prod_{i=1}^n f(x_i;\theta) = \theta^{-n} & \text{for } \theta \geq \max_i(x_i), \\ 0 & \text{otherwise,} \end{cases}$$

with MLE $\hat{\theta}_{\mathrm{ML}} = \max_i(x_i)$, cf. Fig. 2.9a.

a) Likelihood $L(\theta) \cdot 10^5$ for $n = 5, 6, 7$ (from top to bottom) and $\max_i x_i = 7$

b) Log-likelihood $l(\theta)$ for $n = 5, 6, 7, 10, 30, 60$ (from top to bottom) and $\max_i x_i = 7$

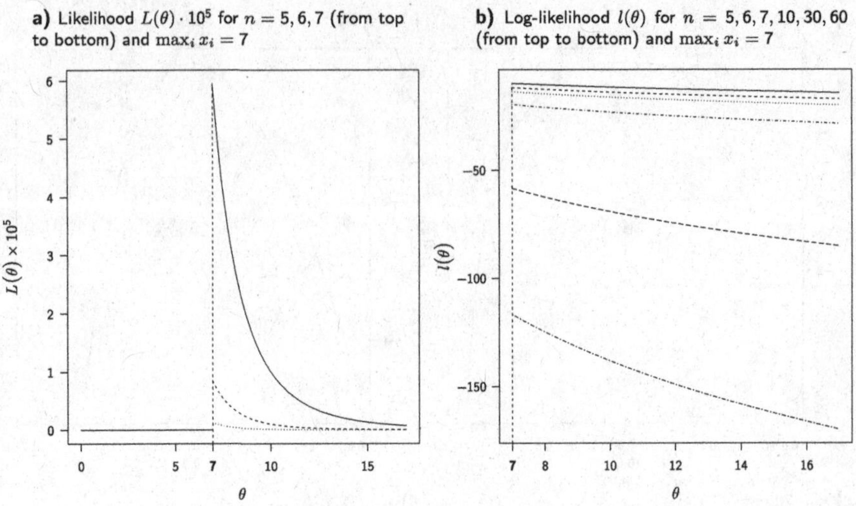

Fig. 2.9 Likelihood and log-likelihood function for a random sample of different size n from a uniform distribution with unknown upper limit θ. Quadratic approximation of the log-likelihood is impossible even for large n

The derivatives of the log-likelihood function

$$l(\theta) = -n \log(\theta) \quad \text{for } \theta > \max_i(x_i)$$

are

$$S(\hat{\theta}_{\mathrm{ML}}) = \frac{dl(\hat{\theta}_{\mathrm{ML}})}{d\theta} \neq 0 \quad \text{and} \quad -I(\hat{\theta}_{\mathrm{ML}}) = \frac{d^2 l(\hat{\theta}_{\mathrm{ML}})}{d\theta^2} = \frac{n}{\hat{\theta}_{\mathrm{ML}}^2} > 0,$$

so the log-likelihood $l(\theta)$ is not concave but convex, with negative (!) observed Fisher information, cf. Fig. 2.9b. It is obvious from Fig. 2.9b that a quadratic approximation to $l(\theta)$ will remain poor even if the sample size n increases. The reason for the irregular behaviour of the likelihood function is that the support of the uniform distribution depends on the unknown parameter θ. ∎

2.5   Sufficiency

Under certain regularity conditions, a likelihood function can be well characterised by the MLE and the observed Fisher information. However, Example 2.18 illustrates that this is not always the case. An alternative characterisation of likelihood functions is in terms of *sufficient statistics*, a concept which we will introduce in the following. We will restrict our attention to random samples, but the description could be easily generalised if required.

Definition 2.8 (Statistic) Let $x_{1:n}$ denote the realisation of a random sample $X_{1:n}$ from a distribution with probability mass or density function $f(x;\theta)$. Any function $T=h(X_{1:n})$ of $X_{1:n}$ with realisation $t=h(x_{1:n})$ is called a *statistic*. ♦

For example, the mean $\bar{X}=\sum_{i=1}^{n} X_i/n$ is a statistic. Also the maximum $\max_i(X_i)$ and the range $\max_i(X_i)-\min_i(X_i)$ are statistics.

Definition 2.9 (Sufficiency) A statistic $T=h(X_{1:n})$ is *sufficient* for θ if the conditional distribution of $X_{1:n}$ given $T=t$ is independent of θ, i.e. if

$$f(x_{1:n} \mid T=t)$$

does not depend on θ. ♦

Example 2.19 (Poisson model) Let $x_{1:n}$ denote the realisation of a random sample $X_{1:n}$ from a $\mathrm{Po}(\lambda)$ distribution with unknown rate parameter λ. The statistic $T=X_1+\cdots+X_n$ is sufficient for λ since the conditional distribution of $X_{1:n} \mid T=t$ is multinomial with parameters not depending on λ. Indeed, first note that $f(t \mid X_{1:n}=x_{1:n})=1$ if $t=x_1+\cdots+x_n$ and 0 elsewhere. We also know from Appendix A.5.1 that $T\sim \mathrm{Po}(n\lambda)$, and therefore we have

$$\begin{aligned} f(x_{1:n} \mid t) &= \frac{f(t \mid x_{1:n}) f(x_{1:n})}{f(t)} \\ &= \frac{f(x_{1:n})}{f(t)} \\ &= \frac{\prod_{i=1}^{n}\{\frac{\lambda^{x_i}}{x_i!}\exp(-\lambda)\}}{\frac{(n\lambda)^t}{t!}\exp(-n\lambda)} \\ &= \frac{t!}{\prod_{i=1}^{n} x_i!}\left(\frac{1}{n}\right)^t, \end{aligned}$$

which can easily be identified as the probability mass function of a multinomial distribution with size parameter $t=x_1+\cdots+x_n$ and all probabilities equal to $1/n$, compare Appendix A.5.3. ■

A sufficient statistic T contains all relevant information from the sample $X_{1:n}$ with respect to θ. To show that a certain statistic is sufficient, the following result is helpful.

Result 2.2 (Factorisation theorem) *Let $f(x_{1:n};\theta)$ denote the probability mass or density function of the random sample $X_{1:n}$. A statistic $T=h(X_{1:n})$ with realisation $t=h(x_{1:n})$ is sufficient for θ if and only if there exist functions $g_1(t;\theta)$ and $g_2(x_{1:n})$ such that for all possible realisations $x_{1:n}$ and all possible parameter values $\theta\in\Theta$,*

$$f(x_{1:n};\theta)=g_1(t;\theta)\cdot g_2(x_{1:n}). \tag{2.18}$$

Note that $g_1(t; \theta)$ depends on the argument $x_{1:n}$ only through $t = h(x_{1:n})$, but also depends on θ. The second term $g_2(x_{1:n})$ must not depend on θ.

A proof of this result can be found in Casella and Berger (2001, p. 276). As a function of θ, we can easily identify $g_1(t; \theta)$ as the likelihood kernel, cf. Definition 2.3. The second term $g_2(x_{1:n})$ is the corresponding multiplicative constant.

Example 2.20 (Poisson model) We already know from Example 2.19 that $T = h(X_{1:n}) = X_1 + \cdots + X_n$ is sufficient for λ, so the factorisation (2.18) must hold. This is indeed the case:

$$
\begin{aligned}
f(x_{1:n}; \theta) &= \prod_{i=1}^{n} f(x_i; \theta) \\
&= \prod_{i=1}^{n} \left\{ \frac{\lambda^{x_i}}{x_i!} \exp(-\lambda) \right\} \\
&= \underbrace{\lambda^t \exp(-n\lambda)}_{g_1(t;\lambda)} \underbrace{\prod_{i=1}^{n} \frac{1}{x_i!}}_{g_2(x_{1:n})}.
\end{aligned}
$$

■

Example 2.21 (Normal model) Let $x_{1:n}$ denote a realisation of a random sample from a normal distribution $\mathrm{N}(\mu, \sigma^2)$ with known variance σ^2. We now show that the *sample mean* $\bar{X} = \sum_{i=1}^{n} X_i / n$ is sufficient for μ. First, note that

$$
\begin{aligned}
f(x_{1:n}; \mu) &= \prod_{i=1}^{n} \left(2\pi\sigma^2\right)^{-\frac{1}{2}} \exp\left\{ -\frac{1}{2} \cdot \frac{(x_i - \mu)^2}{\sigma^2} \right\} \\
&= \left(2\pi\sigma^2\right)^{-\frac{n}{2}} \exp\left\{ -\frac{1}{2} \cdot \frac{\sum_{i=1}^{n}(x_i - \mu)^2}{\sigma^2} \right\}.
\end{aligned}
$$

Now

$$
\sum_{i=1}^{n} (x_i - \mu)^2 = \sum_{i=1}^{n} (x_i - \bar{x})^2 + n(\bar{x} - \mu)^2,
$$

and we can therefore factorise $f(x_{1:n}; \mu)$ as follows:

$$
f(x_{1:n}; \mu) = \underbrace{\left(2\pi\sigma^2\right)^{-\frac{n}{2}} \exp\left\{ -\frac{1}{2} \cdot \frac{\sum(x_i - \bar{x})^2}{\sigma^2} \right\}}_{g_2(x_{1:n})} \cdot \underbrace{\exp\left\{ -\frac{1}{2} \cdot \frac{n(\bar{x} - \mu)^2}{\sigma^2} \right\}}_{g_1(t;\mu) \text{ with } t=\bar{x}}.
$$

Result 2.2 now ensures that the sample mean $\bar{X}$ is sufficient for μ. Note that, for example, also $n\bar{X} = \sum_{i=1}^{n} X_i$ is sufficient for μ.

Suppose now that also σ^2 is unknown, i.e. $\boldsymbol{\theta} = (\mu, \sigma^2)$, and assume that $n \geq 2$. It is easy to show that now $T = (\bar{X}, S^2)$, where

$$S^2 = \frac{1}{n-1} \sum_{i=1}^{n} (X_i - \bar{X})^2$$

is the *sample variance*, is sufficient for $\boldsymbol{\theta}$. Another sufficient statistic for $\boldsymbol{\theta}$ is $\tilde{T} = (\sum_{i=1}^{n} X_i, \sum_{i=1}^{n} X_i^2)$. ■

Example 2.22 (Blood alcohol concentration) If we are prepared to assume that the transformation factor is normally distributed, knowledge of n, $\bar{x}$ and s^2 (or s) in each group (cf. Table 1.3) is sufficient to formulate the likelihood function. It is not necessary to know the actual observations. ■

Definition 2.10 (Likelihood ratio) The quantity

$$\Lambda_{x_{1:n}}(\theta_1, \theta_2) = \frac{L(\theta_1; x_{1:n})}{L(\theta_2; x_{1:n})} = \frac{\tilde{L}(\theta_1; x_{1:n})}{\tilde{L}(\theta_2; x_{1:n})}$$

is the *likelihood ratio* of one parameter value θ_1 relative to another parameter value θ_2 with respect to the realisation $x_{1:n}$ of a random sample $X_{1:n}$. ♦

Note that likelihood ratios between any two parameter values θ_1 and θ_2 can be calculated from the relative likelihood function $\tilde{L}(\theta; x_{1:n})$. Note also that

$$\tilde{L}(\theta; x_{1:n}) = \Lambda_{x_{1:n}}(\theta, \hat{\theta}_{\text{ML}})$$

because $\tilde{L}(\hat{\theta}_{\text{ML}}; x_{1:n}) = 1$, so the relative likelihood function can be recovered from the likelihood ratio.

Result 2.3 *A statistic $T = h(X_{1:n})$ is sufficient for θ if and only if for any pair $x_{1:n}$ and $\tilde{x}_{1:n}$ such that $h(x_{1:n}) = h(\tilde{x}_{1:n})$,*

$$\Lambda_{x_{1:n}}(\theta_1, \theta_2) = \Lambda_{\tilde{x}_{1:n}}(\theta_1, \theta_2) \tag{2.19}$$

for all $\theta_1, \theta_2 \in \Theta$.

Proof We show the equivalence of the factorisation (2.18) and Eq. (2.19). Suppose that (2.18) holds. Then

$$\Lambda_{x_{1:n}}(\theta_1, \theta_2) = \frac{g\{h(x_{1:n}); \theta_1\} \cdot h(x_{1:n})}{g\{h(x_{1:n}); \theta_2\} \cdot h(x_{1:n})} = \frac{g\{h(x_{1:n}); \theta_1\}}{g\{h(x_{1:n}); \theta_2\}},$$

so if $h(x_{1:n}) = h(\tilde{x}_{1:n})$, we have $\Lambda_{x_{1:n}}(\theta_1, \theta_2) = \Lambda_{\tilde{x}_{1:n}}(\theta_1, \theta_2)$ for all θ_1 and θ_2.

Conversely, suppose that (2.19) holds if $h(x_{1:n}) = h(\tilde{x}_{1:n})$, so $\Lambda_{x_{1:n}}(\theta_1, \theta_2)$ is a function (say g^*) of $h(x_{1:n})$, θ_1 and θ_2 only. Let us now fix $\theta_2 = \theta_0$. With $\theta = \theta_1$ we obtain

$$\frac{f(x_{1:n}; \theta)}{f(x_{1:n}; \theta_0)} = \Lambda_{x_{1:n}}(\theta, \theta_0) = g^*\{h(x_{1:n}), \theta, \theta_0\},$$

so (2.18) holds:

$$f(x_{1:n}; \theta) = \underbrace{g^*\{h(x_{1:n}), \theta, \theta_0\}}_{g_1\{h(x_{1:n}); \theta\}} \underbrace{f(x_{1:n}; \theta_0)}_{g_2(x_{1:n})}.$$

□

This result establishes an important relationship between a sufficient statistic T and the likelihood function: If $T = h(X_{1:n})$ is a sufficient statistic and $h(x_{1:n}) = h(\tilde{x}_{1:n})$, then $x_{1:n}$ and $\tilde{x}_{1:n}$ define the same likelihood ratio.

The following result establishes another important property of the likelihood function. We distinguish in the following the likelihood $L(\theta; x_{1:n})$ with respect to a realisation $x_{1:n}$ of a random sample $X_{1:n}$ and the likelihood $L(\theta; t)$ with respect to the corresponding realisation t of a sufficient statistic $T = h(X_{1:n})$ of the same random sample $X_{1:n}$.

Result 2.4 *Let $L(\theta; x_{1:n})$ denote the likelihood function with respect to a realisation $x_{1:n}$ of a random sample $X_{1:n}$. Let $L(\theta; t)$ denote the likelihood with respect to the realisation $t = h(x_{1:n})$ of a sufficient statistic $T = h(X_{1:n})$ for θ. For all possible realisations $x_{1:n}$, the ratio*

$$\frac{L(\theta; x_{1:n})}{L(\theta; t)}$$

will then not depend on θ, i.e. the two likelihood functions are (up to a proportionality constant) identical.

Proof To show Result 2.4, first note that $f(t \mid x_{1:n}) = 1$ if $t = h(x_{1:n})$ and 0 otherwise, so $f(x_{1:n}, t) = f(x_{1:n}) f(t \mid x_{1:n}) = f(x_{1:n})$ if $t = h(x_{1:n})$. For $t = h(x_{1:n})$, the likelihood function can therefore be written as

$$\begin{aligned} L(\theta; x_{1:n}) &= f(x_{1:n}; \theta) = f(x_{1:n}, t; \theta) = f(x_{1:n} \mid t; \theta) f(t; \theta) \\ &= f(x_{1:n} \mid t; \theta) L(\theta; t). \end{aligned}$$

Now T is sufficient for θ, so $f(x_{1:n} \mid t; \theta) = f(x_{1:n} \mid t)$ does not depend on θ. Therefore, $L(\theta; x_{1:n}) \propto L(\theta; t)$. The sign "$\propto$", in words "proportional to", means that there is a constant $C > 0$ (not depending on θ) such that $L(\theta; x_{1:n}) = C \cdot L(\theta; t)$. □

Example 2.23 (Binomial model) Let $X_{1:n}$ denote a random sample from a Bernoulli distribution $\mathrm{B}(\pi)$ with unknown parameter $\pi \in (0, 1)$. The likelihood function based on the realisation $x_{1:n}$ equals

$$L(\pi; x_{1:n}) = f(x_{1:n}; \pi) = \prod_{i=1}^{n} \pi^{x_i}(1-\pi)^{1-x_i} = \pi^t(1-\pi)^{n-t},$$

where $t = \sum_{i=1}^{n} x_i$. Obviously, $T = h(x_{1:n}) = \sum_{i=1}^{n} X_i$ is a sufficient statistic for π. Now T follows the binomial distribution $\mathrm{Bin}(n, \pi)$, so the likelihood function with respect to its realisation t is

$$L(\pi, t) = \binom{n}{t}\pi^t(1-\pi)^{n-t}.$$

As Result 2.4 states, the likelihood functions with respect to $x_{1:n}$ and t are identical up to the multiplicative constant $\binom{n}{t}$. ■

Example 2.23 has shown that regarding the information about the proportion π, the whole random sample $X_{1:n}$ can be compressed into the total number of successes $T = \sum_{i=1}^{n} X_i$ without any loss of information. This will be important in Chap. 4, where we consider asymptotic properties of ML estimation, i.e. properties of certain statistics for sample size $n \to \infty$. Then we can consider a single binomial random variable $X \sim \mathrm{Bin}(n, \pi)$ because it implicitly contains the whole information of n independent Bernoulli random variables with respect to π. We can also approximate the binomial distribution $\mathrm{Bin}(n, \pi)$ by a Poisson distribution $\mathrm{Po}(n\pi)$ when π is small compared to n. Therefore, we can consider a single Poisson random variable $\mathrm{Po}(e\lambda)$ and assume that the asymptotic properties of derived statistics are a good approximation of their finite sample properties. We will often use this Poisson model parametrisation with expected number of cases $e = n \cdot p$ and relative risk $\lambda = \pi/p$, using a reference probability p, see e.g. Example 2.4.

2.5.1 Minimal Sufficiency

We have seen in the previous section that sufficient statistics are not unique. In particular, the original sample $X_{1:n}$ is always sufficient due to Result 2.2:

$$f(x_{1:n}; \theta) = \underbrace{f(x_{1:n}; \theta)}_{=g_1\{h(x_{1:n})=x_{1:n};\theta\}} \cdot \underbrace{1}_{=g_2(x_{1:n})}.$$

The concept of minimal sufficiency ensures that a sufficient statistic cannot be reduced further.

Definition 2.11 (Minimal sufficiency) A sufficient statistic $T = h(X_{1:n})$ for θ is called *minimal sufficient*, if, for every possible realisation $x_{1:n}$ of $X_{1:n}$, $t = h(x_{1:n})$ can be written as a transformation of the realisation $\tilde{t}$ of any other sufficient statistic $\tilde{T} = \tilde{h}(X_{1:n})$. ♦

The following result describes the relationship between two minimal sufficient statistics.

Result 2.5 *If T and $\tilde{T}$ are minimal sufficient statistics, then there exists a one-to-one function g such that $\tilde{T} = g(T)$ and $T = g^{-1}(\tilde{T})$.*

Loosely speaking, a minimal sufficient statistic is unique up to any one-to-one transformation. For example, if T is minimal sufficient, then $T/2$ will also be minimal sufficient, but $|T|$ will not be minimal sufficient if T can take values that differ only in sign.

Result 2.6 *A necessary and sufficient criterion for a statistic $T(x_{1:n})$ to be minimal sufficient is that $h(x_{1:n}) = h(\tilde{x}_{1:n})$ if and only if*

$$\Lambda_{x_{1:n}}(\theta_1, \theta_2) = \Lambda_{\tilde{x}_{1:n}}(\theta_1, \theta_2)$$

for all θ_1, θ_2.

This is an extension of Result 2.3, where the sufficiency is characterised. For sufficiency, only the implication of equal likelihood ratios from equal statistics is needed. For minimal sufficiency, in addition, the implication of equal statistics from equal likelihood ratios is required. A proof of this result can be found in Young and Smith (2005, p. 92).

Result 2.6 means that any minimal sufficient statistic creates the same partition of the sample space as the likelihood ratio function. A more exact formulation is based on equivalence classes, where two different observations $x_{1:n}$ and $\tilde{x}_{1:n}$ are equivalent if they lead to the same likelihood ratio function. The likelihood ratio function partitions the sample space into the same equivalence classes as any minimal sufficient statistic. Therefore, the likelihood ratio is a one-to-one function of a minimal sufficient statistic and hence also minimal sufficient. Since we have described above that the likelihood ratio and the likelihood function can be recovered from each other, they are one-to-one transformations of each other. Hence, also the likelihood is minimal sufficient:

Minimal sufficiency of the likelihood
The likelihood function $L(\theta)$ is minimal sufficient for θ.

This implies that the likelihood function contains the whole information of the data with respect to θ. Any further reduction will result in information loss.

Example 2.24 (Normal model) Let $x_{1:n}$ denote a realisation of a random sample from a normal distribution $\mathrm{N}(\mu, \sigma^2)$ with known variance σ^2. The mean $h(X_{1:n}) = \bar{X}$ is minimal sufficient for μ, whereas $\tilde{T}(X_{1:n}) = (\bar{X}, S^2)$ is sufficient but not minimal sufficient for μ. ■

2.5.2 The Likelihood Principle

Are there general principles how to infer information from data? In the previous section we have seen that sufficient statistics contain the complete information of a sample with respect to an unknown parameter. It is thus natural to state the *sufficiency principle*:

Sufficiency principle
Consider a random sample $X_{1:n}$ from a distribution with probability mass or density function $f(x;\theta)$ and unknown parameter θ. Assume that $T = h(X_{1:n})$ is a sufficient statistic for θ. If $h(x_{1:n}) = h(\tilde{x}_{1:n})$ for two realisations of $X_{1:n}$, then inference for θ should be the same whether $x_{1:n}$ or $\tilde{x}_{1:n}$ has been observed.

The likelihood function is also sufficient, so we immediately have the *likelihood principle*:

Likelihood principle
If realisations $x_{1:n}$ and $\tilde{x}_{1:n}$ from a random sample $X_{1:n}$ with probability mass or density function $f(x;\theta)$ have proportional likelihood functions, i.e. $L(\theta; x_{1:n}) \propto L(\theta; \tilde{x}_{1:n})$ for all θ, then inference for θ should be the same, whether $x_{1:n}$ or $\tilde{x}_{1:n}$ is observed.

This principle is also called the *weak likelihood principle* to distinguish it from the *strong likelihood principle*:

Strong likelihood principle
Suppose $x_{1:n_1}$ is a realisation from a random sample $X_{1:n_1}$ with probability mass or density function $f_1(x;\theta)$. Let $\tilde{x}_{1:n_2}$ denote a realisation from a random sample $\tilde{X}_{1:n_2}$ with probability mass or density function $f_2(x;\theta)$, not necessarily identical to $f_1(x;\theta)$. If the corresponding two likelihood functions are proportional, i.e. $L_1(\theta; x_{1:n_1}) \propto L_2(\theta; \tilde{x}_{1:n_2})$, then inference for θ should be the same, whether $x_{1:n_1}$ from $f_1(x;\theta)$ or $\tilde{x}_{1:n_2}$ from $f_2(x;\theta)$ has been observed.

To illustrate the strong likelihood principle, we consider a classical example.

Example 2.25 (Binomial and negative binomial model) A binomial model $X \sim \text{Bin}(m, \pi)$ is appropriate if a fixed number m of independent and identical Bernoulli experiments is conducted. The random variable X denotes the number of successes of the event considered with success probability π. The likelihood function of $n_1 = 1$ realisation x is then

$$L_1(\pi) = \binom{m}{x} \pi^x (1-\pi)^{m-x}.$$

As discussed in Sect. 1.1.1, alternatively we can imagine a design where we conduct independent Bernoulli experiments until we have a total of x successes. In this inverse binomial model x is fixed, but now the number of required samples m is a realisation of a random variable M, say. Of interest is again the parameter π, the unknown probability of the event of interest, with likelihood function

$$L_2(\pi) = \binom{m-1}{x-1} \pi^x (1-\pi)^{m-x},$$

derived from the realisation m of a random sample of size $n_2 = 1$ from the negative binomial distribution $M \sim \text{NBin}(x, \pi)$. The likelihood functions $L_1(\theta)$ and $L_2(\theta)$ are (up to different multiplicative constants) identical if m and x are the same. The strong likelihood principle requires that statistical inference for θ must be the same, whether or not the data have arisen from the binomial or the negative binomial model. ■

2.6 Exercises

1. Examine the likelihood function in the following examples.
   (a) In a study of a fungus that infects wheat, 250 wheat seeds are disseminated after contaminating them with the fungus. The research question is how large the probability θ is that an infected seed can germinate. Due to technical problems, the exact number of germinated seeds cannot be evaluated, but we know only that less than 25 seeds have germinated. Write down the likelihood function for θ based on the information available from the experiment.
   (b) Let $X_{1:n}$ be a random sample from an $\text{N}(\theta, 1)$ distribution. However, only the largest value of the sample, $Y = \max(X_1, \dots, X_n)$, is known. Show that the density of Y is

   $$f(y) = n\{\Phi(y-\theta)\}^{n-1} \varphi(y-\theta), \quad y \in \mathbb{R},$$

   where $\Phi(\cdot)$ is the distribution function, and $\varphi(\cdot)$ is the density function of the standard normal distribution $\text{N}(0, 1)$. Derive the distribution function of Y and the likelihood function $L(\theta)$.
   (c) Let $X_{1:3}$ denote a random sample of size $n = 3$ from a Cauchy $\text{C}(\theta, 1)$ distribution, cf. Appendix A.5.2. Here $\theta \in \mathbb{R}$ denotes the location parameter of the Cauchy distribution with density

   $$f(x) = \frac{1}{\pi} \frac{1}{1+(x-\theta)^2}, \quad x \in \mathbb{R}.$$

   Derive the likelihood function for θ.

(d) Using R, produce a plot of the likelihood functions:
   i. $L(\theta)$ in 1(a).
   ii. $L(\theta)$ in 1(b) if the observed sample is $x = (1.5, 0.25, 3.75, 3.0, 2.5)$.
   iii. $L(\theta)$ in 1(c) if the observed sample is $x = (0, 5, 9)$.

2. A first-order autoregressive process $X_0, X_1, \ldots, X_n$ is specified by the conditional distribution

$$X_i \mid X_{i-1} = x_{i-1}, \ldots, X_0 = x_0 \sim \mathrm{N}(\alpha \cdot x_{i-1}, 1), \quad i = 1, 2, \ldots, n,$$

and some initial distribution for X_0. This is a popular model for time series data.

(a) Consider the observation $X_0 = x_0$ as fixed. Show that the log-likelihood kernel for a realisation $x_1, \ldots, x_n$ can be written as

$$l(\alpha) = -\frac{1}{2}\sum_{i=1}^{n}(x_i - \alpha x_{i-1})^2.$$

(b) Derive the score equation for α, compute $\hat{\alpha}_{\mathrm{ML}}$ and verify that it is really the maximum of $l(\alpha)$.
(c) Create a plot of $l(\alpha)$ and compute $\hat{\alpha}_{\mathrm{ML}}$ for the following sample:

$$(x_0, \ldots, x_6) = (-0.560, -0.510, 1.304, 0.722, 0.490, 1.960, 1.441).$$

3. Show that in Example 2.2 the likelihood function $L(N)$ is maximised at $\hat{N} = \lfloor M \cdot n/x \rfloor$, where $\lfloor x \rfloor$ is the largest integer that is smaller than x. To this end, analyse the monotonic behaviour of the ratio $L(N)/L(N-1)$. In which cases is the MLE not unique? Give a numeric example.
4. Derive the MLE of π for an observation x from a geometric $\mathrm{Geom}(\pi)$ distribution. What is the MLE of π based on a realisation $x_{1:n}$ of a random sample from this distribution?
5. A sample of 197 animals has been analysed regarding a specific phenotype. The number of animals with phenotypes AB, Ab, aB and ab, respectively, turned out to be

$$\boldsymbol{x} = (x_1, x_2, x_3, x_4)^\top = (125, 18, 20, 34)^\top.$$

A genetic model now assumes that the counts are realisations of a multinomially distributed multivariate random variable $\boldsymbol{X} \sim \mathrm{M}_4(n, \boldsymbol{\pi})$ with $n = 197$ and probabilities $\pi_1 = (2+\phi)/4$, $\pi_2 = \pi_3 = (1-\phi)/4$ and $\pi_4 = \phi/4$ (Rao 1973, p. 368).

(a) What is the parameter space of ϕ? See Table A.3 in Appendix A for details on the multinomial distribution and the parameter space of $\boldsymbol{\pi}$.
(b) Show that the likelihood kernel function for ϕ, based on the observation $\boldsymbol{x}$, has the form

$$L(\phi) = (2+\phi)^{m_1}(1-\phi)^{m_2}\phi^{m_3}$$

and derive expressions for m_1, m_2 and m_3 depending on $\boldsymbol{x}$.

(c) Derive an explicit formula for the MLE $\hat{\phi}_{\text{ML}}$, depending on m_1, m_2 and m_3. Compute the MLE given the data given above.

(d) What is the MLE of $\theta = \sqrt{\phi}$?

6. Show that $h(X) = \max_i(X_i)$ is sufficient for θ in Example 2.18.
7. (a) Let $(X_1, \ldots, X_n)$ be a sample from a distribution with density

$$f(x_i; \theta) = \begin{cases} \exp(i\theta - x_i), & x_i \geq i\theta, \\ 0, & x_i < i\theta, \end{cases}$$

for X_i, $i = 1, \ldots, n$. Show that $T = \min_i(X_i/i)$ is a sufficient statistic for θ.

(b) Let $X_{1:n}$ denote a random sample from a distribution with density

$$f(x; \theta) = \exp\{-(x - \theta)\}, \quad \theta < x < \infty, \ -\infty < \theta < \infty.$$

Derive a minimal sufficient statistic for θ.

8. Let $T = h(X_{1:n})$ be a sufficient statistic for θ, $g(\cdot)$ a one-to-one function, and $\tilde{T} = \tilde{h}(X_{1:n}) = g\{h(X_{1:n})\}$. Show that $\tilde{T}$ is sufficient for θ.
9. Let X_1 and X_2 denote two independent exponentially $\text{Exp}(\lambda)$ distributed random variables with parameter $\lambda > 0$. Show that $h(X_1, X_2) = X_1 + X_2$ is sufficient for λ.

2.7 References

A good introduction to likelihood methods is given by Pawitan (2001). More ambitious and rigorous is the presentation in Davison (2003). The pure likelihood approach is described in Edwards (1992) and Royall (1997).

3 Elements of Frequentist Inference

Contents

Maximum likelihood estimation has been introduced as an intuitive technique to derive the "most likely" parameter value θ for the observation x. But what properties does this estimate have? Is it good or perhaps even the best estimate in a certain sense? Are there other useful estimates? Can we derive an interval of plausible parameter values based on the likelihood, and can we quantify the associated certainty of the interval?

Before answering these questions in Chap. 4, we will first introduce some elementary concepts of frequentist inference in this chapter. Frequentist inference is based on hypothetical repetitions of the underlying sampling experiment. We will discuss frequentist properties of both point and interval estimates of an unknown parameter θ. Frequentist interval estimates are called confidence intervals to distinguish them from the Bayesian counterparts, so-called credible intervals as discussed in Chap. 6. We will touch upon basic concepts of significance tests and introduce the P-value as a quantitative measure of the evidence against a null hypothesis.

3.1 Unbiasedness and Consistency

Let $X_{1:n}$ denote a random sample from a distribution $f(x; \theta)$, which depends on the true (but unknown) parameter value θ. Our goal is to estimate θ based on the real-

L. Held, D. Sabanés Bové, *Likelihood and Bayesian Inference*,
Statistics for Biology and Health, https://doi.org/10.1007/978-3-662-60792-3_3,

isation $x_{1:n}$ of $X_{1:n}$. The MLE $\hat{\theta}_{\text{ML}}$ is one particular estimate of θ, but here we will consider any possible estimate $\hat{\theta}$. To investigate and compare frequentist properties of different estimates we first define the notion of an estimator based on a random sample.

Definition 3.1 (Estimator) Consider a real-valued statistic (cf. Definition 2.8)

$$T_n = h(X_{1:n}),$$

based on a random sample $X_{1:n}$ from a distribution with probability mass or density function $f(x; \theta)$ where θ is an unknown scalar parameter θ. If the random variable T_n is computed to make inference about θ, then it is called an *estimator*. We may simply write T rather than T_n if the sample size n is not important. The particular value $t = h(x_{1:n})$ that an estimator takes for a realisation $x_{1:n}$ of the random sample $X_{1:n}$ is called an *estimate*. ♦

What is a good estimator? At first glance it seems reasonable that a useful estimator is "on average" equal to the true value θ. This leads to the notion of unbiased estimators.

Definition 3.2 (Unbiasedness) An estimator T_n is called *unbiased* for θ if

$$\mathsf{E}(T_n) = \theta$$

for all $\theta \in \Theta$ and for all $n \in \mathbb{N}$. Otherwise, the estimator T_n is called *biased*. ♦

Example 3.1 (Sample variance) Let $X_{1:n}$ denote a random sample from a distribution with unknown mean μ and variance $\sigma^2 > 0$. It is easy to show that the sample mean $\bar{X} = n^{-1} \sum_{i=1}^{n} X_i$ has the following properties:

$$\mathsf{E}(\bar{X}) = \mu \quad \text{and} \quad \text{Var}(\bar{X}) = \frac{\sigma^2}{n}.$$

So $\hat{\mu} = \bar{X}$ is an unbiased estimator of the mean μ. The sample variance

$$S^2 = \frac{1}{n-1} \sum_{i=1}^{n} (X_i - \bar{X})^2 \tag{3.1}$$

is unbiased for the true variance σ^2. To see this, first note that

$$\begin{aligned} \sum_{i=1}^{n} (X_i - \bar{X})^2 &= \sum_{i=1}^{n} \{X_i - \mu - (\bar{X} - \mu)\}^2 \\ &= \sum_{i=1}^{n} (X_i - \mu)^2 - 2 \sum_{i=1}^{n} (X_i - \mu)(\bar{X} - \mu) + n(\bar{X} - \mu)^2 \end{aligned}$$

$$= \sum_{i=1}^{n} (X_i - \mu)^2 - 2n(\bar{X} - \mu)^2 + n(\bar{X} - \mu)^2$$

$$= \sum_{i=1}^{n} (X_i - \mu)^2 - n(\bar{X} - \mu)^2,$$

so

$$\begin{aligned} \mathsf{E}\{(n-1)S^2\} &= \mathsf{E}\left\{\sum_{i=1}^{n} (X_i - \bar{X})^2\right\} \\ &= n \cdot \mathsf{E}\{(X_i - \mu)^2\} - n \cdot \mathsf{E}\{(\bar{X} - \mu)^2\} \\ &= n \cdot \text{Var}(X_i) - n \cdot \text{Var}(\bar{X}) = n \cdot \sigma^2 - n \cdot \frac{\sigma^2}{n} \\ &= (n-1)\sigma^2, \end{aligned}$$

and therefore $\mathsf{E}(S^2) = \sigma^2$. ■

Note that unbiased estimators are not invariant under nonlinear transformations. For example, the *sample standard deviation* $S = \sqrt{S^2}$ is a biased estimator of the standard deviation $\sigma = \sqrt{\sigma^2}$. This can be shown using Jensen's inequality (cf. Appendix A.3.7):

$$\mathsf{E}(S) = \mathsf{E}(\sqrt{S^2}) < \sqrt{\sigma^2} = \sigma$$

since $g(x) = \sqrt{x}$ is a strictly concave function and S is (for $\sigma^2 > 0$) not degenerate. So S on average underestimates the standard deviation σ but is still commonly used to estimate σ.

The following example illustrates that an unbiased estimator is not necessarily better than a biased one.

Example 3.2 (Geometric model) Let x denote $n = 1$ realisation from a geometric $\text{Geom}(\pi)$ distribution with probability mass function

$$f(x; \pi) = \pi(1 - \pi)^{(x-1)}$$

and unknown parameter $\pi \in (0, 1)$, cf. Table A.1. We assume that the parameter space is an open interval, the limits $\pi = 0$ and $\pi = 1$ are not included. This is a common assumption since $f(x; \pi = 0)$ is not a proper distribution and X is for $\pi = 1$ deterministic, i.e. only the realisation $x = 1$ can occur.

Now there is only one unbiased estimate of π because the requirement

$$\mathsf{E}\{T(X)\} = \sum_{x=1}^{\infty} T(x)\pi(1 - \pi)^{(x-1)} = \pi \tag{3.2}$$

for all $\pi \in (0, 1)$ leads to the solution

$$\hat{\pi}(x) = T(x) = \begin{cases} 1 & \text{if } x = 1, \\ 0 & \text{else.} \end{cases} \tag{3.3}$$

This solution is unique, because the requirement (3.2) can be written as

$$\sum_{x=1}^{\infty} T(x)(1-\pi)^{(x-1)} = 1 = \sum_{x=1}^{\infty} \hat{\pi}(x)(1-\pi)^{(x-1)},$$

and the power series on the left and right sides are equal only if their coefficients series, $T(x)$ and $\hat{\pi}(x)$, are identical.

So (3.3) is the only unbiased estimator of π but appears to be not very sensible, as it can only take two distinct values on the border of the parameter space. In contrast, the ML estimator $\hat{\pi}_{\text{ML}} = 1/X$ (cf. Exercise 4 in Chap. 2) seems more sensible but is biased. Indeed, we have $\mathsf{E}(X) = 1/\pi$, and since $h(x) = 1/x$ is strictly convex on the positive real line and X is not degenerate, we have

$$\mathsf{E}(1/X) > 1/\mathsf{E}(X) = \pi$$

as a consequence of Jensen's inequality (cf. Appendix A.3.7). ■

Asymptotic unbiasedness is a weaker requirement of an estimator. An asymptotically unbiased estimator may be biased for any finite sample size n, but the bias should go to zero for $n \to \infty$.

Definition 3.3 (Asymptotic unbiasedness) An estimator T_n is called *asymptotically unbiased* for θ if

$$\lim_{n\to\infty} \mathsf{E}(T_n) = \theta$$

for all $\theta \in \Theta$. ♦

An unbiased estimator is also asymptotically unbiased, but not vice versa. Asymptotic unbiasedness has a strong connection to the notion of consistency.

Definition 3.4 (Consistency) An estimator T_n is called *consistent* if T_n converges as $n \to \infty$ in probability to θ, compare Appendix A.4.1. If T_n converges in mean square to θ (compare again Appendix A.4.1), then we have *consistency in mean square*. ♦

The underlying idea for defining consistency is that the estimator should be able to identify the true parameter value when the sample size increases. Appendix A.4.1 lists important relationships between the different modes of convergence. These properties translate to the relationships between the corresponding notions of consistency. In particular, a mean square consistent estimate is also consistent. Application of the continuous mapping theorem (see Appendix A.4.2) shows that any continuous function h of a consistent estimator for θ is consistent for $h(\theta)$. One can also establish that a mean square consistent estimator is asymptotically unbiased, while this is in general not true for a consistent estimator.

Consistency in mean square implies that

$$\mathrm{MSE} = \mathsf{E}\{(T_n - \theta)^2\} \tag{3.4}$$

goes to zero as $n \to \infty$. The quantity (3.4) is called the *mean squared error* (MSE). The mean squared error is of particular importance because the following decomposition holds:

$$\mathrm{MSE} = \mathrm{Var}(T_n) + \{\mathsf{E}(T_n) - \theta\}^2. \tag{3.5}$$

The quantity $\mathsf{E}(T_n) - \theta$ is called the *bias* of T_n. The mean squared error is therefore the sum of the variance of an estimator and its squared bias. Note that, of course, the bias of an asymptotically unbiased estimator goes to zero for $n \to \infty$. If the variance of the estimator goes also to zero as $n \to \infty$, then the estimator is, due to the decomposition (3.5), consistent in mean square and also consistent.

Example 3.3 (Sample variance) As we have seen in Example 3.1, the sample variance S^2 is unbiased for σ^2. An alternative estimator of σ^2 is

$$\tilde{S}^2 = \frac{1}{n}\sum_{i=1}^{n}(X_i - \bar{X})^2,$$

which has the expectation

$$\mathsf{E}(\tilde{S}^2) = \frac{n-1}{n}\mathsf{E}(S^2) = \frac{n-1}{n}\sigma^2.$$

So this estimator is biased but still asymptotically unbiased. Suppose that the fourth central moment $c_4 = \mathsf{E}[\{X - \mathsf{E}(X)\}^4]$ of X (cf. Sect. A.3.3) exists. Then

$$\mathrm{Var}(S^2) = \frac{1}{n}\left\{c_4 - \left(\frac{n-3}{n-1}\right)\sigma^4\right\},$$

cf. Exercise 4. Therefore, $\mathrm{Var}(S^2) \to 0$ as $n \to \infty$ and also $\mathrm{Var}(\tilde{S}^2) \to 0$ as $n \to \infty$. So both S^2 and $\tilde{S}^2$ are consistent in mean square. ■

3.2 Standard Error and Confidence Interval

An estimator T will be equal to the true parameter θ only in rare cases. However, it will often be close to θ in a certain sense if the estimator is useful. The *standard error* quantifies how much an estimator varies in hypothetical independent repetitions of the sampling experiment. If the estimator is unbiased or at least asymptotically unbiased, then the standard error is a consistent estimator of the standard deviation of T. To keep notation simple we will use the term *standard error* for both the estimator and its realisation.

Starting with a definition of the notion of the standard error, we will continue introducing *confidence intervals*, i.e. intervals which cover plausible values of the unknown parameter. Confidence intervals are *interval estimators* and consist of a lower and an upper limit. This is in contrast to the real-valued *(point) estimator*. We will use the term *confidence interval* for both the estimator and its realisation.

3.2.1 Standard Error

To quantify the statistical variability of an estimator T and to construct confidence intervals for the unknown parameter θ, the variance $\operatorname{Var}(T)$ or the standard deviation $\sqrt{\operatorname{Var}(T)}$ of T is needed. However, $\operatorname{Var}(T)$ is often just as unknown as the unknown parameter itself. In many cases we can nevertheless consistently estimate the standard deviation $\sqrt{\operatorname{Var}(T)}$ with the *standard error*:

Definition 3.5 (Standard error) Let $X_{1:n}$ denote a random sample, and $T_n = h(X_{1:n})$ an estimator of an unknown parameter θ. Suppose V is a consistent estimator of $\operatorname{Var}(T_n)$. The continuous mapping theorem (see Appendix A.4.2) then guarantees that the *standard error* $\sqrt{V}$ is a consistent estimator of the standard deviation $\sqrt{\operatorname{Var}(T)}$:

$$\operatorname{se}(T) = \sqrt{V}.$$ ♦

Example 3.4 (Sample variance) Let $X_{1:n}$ denote a random sample from a distribution with unknown mean μ and variance σ^2. Now $\hat{\mu} = \bar{X}$ is an unbiased estimator of μ, and S^2 is a consistent (and even unbiased) estimator of σ^2. We further have $\operatorname{Var}(\bar{X}) = \sigma^2/n$, so $V = S^2/n$ is a consistent estimator of $\operatorname{Var}(\bar{X})$, and we obtain the following standard error of $\hat{\mu} = \bar{X}$:

$$\operatorname{se}(\bar{X}) = \frac{S}{\sqrt{n}}.$$

Using Example 3.3, we see that also $\tilde{S}/\sqrt{n}$ is a standard error of $\bar{X}$, which illustrates that an estimator can have different standard errors. ■

3.2.2 Confidence Interval

We will now introduce the concept of a confidence interval.

Definition 3.6 (Confidence interval) For fixed $\gamma \in (0, 1)$, a $\gamma \cdot 100\,\%$ *confidence interval* for θ is defined by *two* statistics $T_l = h_l(X_{1:n})$ and $T_u = h_u(X_{1:n})$ based on a random sample $X_{1:n}$, which fulfil

$$\Pr(T_l \leq \theta \leq T_u) = \gamma$$

for all $\theta \in \Theta$. The statistics T_l and T_u are the *limits* of the confidence interval, and we assume $T_l \leq T_u$ throughout. The *confidence level* γ is also called *coverage probability*. ◆

The limits of a confidence interval are functions of the random sample $X_{1:n}$ and therefore also random. In contrast, the unknown parameter θ is fixed. If we imagine identical repetitions of the underlying statistical sampling experiment, then a $\gamma \cdot 100\ \%$ confidence interval will cover the unknown parameter θ in $\gamma \cdot 100\ \%$ of all cases.

Confidence interval
For repeated random samples from a distribution with unknown parameter θ, a $\gamma \cdot 100\ \%$ confidence interval will cover θ in $\gamma \cdot 100\ \%$ of all cases.

However, we are not allowed to say that the unknown parameter θ is within a $\gamma \cdot 100\ \%$ confidence interval with probability γ since θ is not a random variable. Such a Bayesian interpretation is possible exclusively for *credible intervals*, see Definition 6.3.

In the following we will concentrate on the commonly used *two-sided* confidence intervals. *One-sided* confidence intervals can be obtained using $T_l = -\infty$ or $T_u = \infty$. However, such intervals are rarely used in practice.

Example 3.5 (Normal model) Let $X_{1:n}$ denote a random sample from a normal distribution $\mathrm{N}(\mu, \sigma^2)$ with known variance σ^2. The interval $[T_l, T_u]$ with limits

$$T_l = \bar{X} - q \cdot \frac{\sigma}{\sqrt{n}} \quad \text{and}$$

$$T_u = \bar{X} + q \cdot \frac{\sigma}{\sqrt{n}}$$

defines a $\gamma \cdot 100\ \%$ confidence interval for μ, where $q = z_{(1+\gamma)/2}$ denotes the $(1+\gamma)/2$ quantile of the standard normal distribution. To prove this, we have to show that $[T_l, T_u]$ has coverage probability γ for all μ. Now

$$\bar{X} \sim \mathrm{N}\left(\mu, \frac{\sigma^2}{n}\right),$$

so

$$Z = \sqrt{n}\frac{\bar{X} - \mu}{\sigma} \tag{3.6}$$

has a standard normal distribution. Due to the symmetry of the standard normal distribution around zero, we have $z_{(1-\gamma)/2} = -q$, and therefore

$$\begin{aligned}\Pr(-q \le Z \le q) &= \Pr(Z \le q) - \Pr(Z < -q) \\ &= \frac{1+\gamma}{2} - \frac{1-\gamma}{2} \\ &= \gamma. \end{aligned} \tag{3.7}$$

Plugging the definition (3.6) into (3.7) and rearranging so that the unknown parameter μ is in the centre of the inequalities, we finally obtain the coverage probability

$$\Pr(T_l \le \mu \le T_u) = \gamma.$$

■

As for estimators, the question arises how we can characterise a "good" confidence interval. We will discuss this topic at a later stage in Sect. 4.6. As a cautionary note, the following example illustrates that properly defined confidence intervals may not be sensible at all from a likelihood perspective.

Example 3.6 (Uniform model) Let X_1, X_2 denote a random sample of size $n = 2$ from a continuous uniform distribution $\mathrm{U}(\theta - 0.5, \theta + 0.5)$ with unknown parameter $\theta \in \mathbb{R}$ (cf. Appendix A.5.2). It is straightforward to show (cf. Exercise 5) that the interval $[T_l, T_u]$ with limits

$$\begin{aligned} T_l &= \min(X_1, X_2) \quad \text{and} \\ T_u &= \max(X_1, X_2) \end{aligned}$$

is a 50 % confidence interval for θ.

Suppose t_l and t_u are realisations of T_l and T_u. Then the likelihood function is

$$L(\theta) = \mathsf{I}_{(t_u - 0.5, t_l + 0.5)}(\theta),$$

where $\mathsf{I}_A(\theta)$ denotes the indicator function of a set A (cf. Example 2.18). So only parameter values in the interval $(t_u - 0.5, t_l + 0.5)$ have a positive likelihood. Suppose now we observe a specific realisation x_1 and x_2 with $t_u - t_l = \max(x_1, x_2) - \min(x_1, x_2) \ge 0.5$. Then we have $t_l \le t_u - 0.5 \le \theta \le t_l + 0.5 \le t_u$, so *all* possible values of θ (those with positive likelihood) are within the interval $[t_l, t_u]$. This has to be contrasted with the confidence level of $[T_l, T_u]$, which is only 50 %.

This example illustrates that confidence intervals, which are not based on the likelihood function but are constructed solely based on frequentist properties of the two random variables T_l and T_u, can have rather odd properties in certain applications.

■

3.2.3 Pivots

In order to construct confidence intervals, the concept of a *pivot* is important.

Definition 3.7 (Pivot) A *pivot* is a function of the data X (viewed as random) and the true parameter θ, with distribution *not* depending on θ. The distribution of a pivot is called *pivotal distribution*. An *approximate pivot* is a pivot whose distribution does not asymptotically depend on the true parameter θ. ♦

So a pivot is a statistic, which also depends on the true parameter θ, having a distribution not depending on θ.

Using a pivot, we can construct confidence intervals which are valid for all possible values of θ. For illustration, in Example 3.5 the random variable Z defined in Eq. (3.6) is a pivot, which was used to construct a confidence interval for the mean μ of a normal random sample. The following example describes the construction of a confidence interval for the mean μ of an exponential random sample using a pivot.

Example 3.7 (Exponential model) Let $X_{1:n}$ denote a random sample from an exponential distribution $\text{Exp}(\lambda)$, i.e. $F(x) = \Pr(X_i \leq x) = 1 - \exp(-\lambda x)$, where the rate parameter λ is unknown. The distribution function of λX_i is therefore

$$\Pr(\lambda X_i \leq x) = \Pr\left(X_i \leq \frac{x}{\lambda}\right) = 1 - \exp(-x),$$

so $\lambda X_i \sim \text{Exp}(1) = \text{G}(1, 1)$ no matter what the actual value of λ is. The sum

$$\sum_{i=1}^{n} \lambda X_i = \lambda n \bar{X} \tag{3.8}$$

then follows a gamma $\text{G}(n, 1)$ distribution (cf. Appendix A.5.2), so is a pivot for λ.

For illustration, consider $n = 47$ non-censored survival times from Sect. 1.1.8 and assume that they follow an exponential distribution with unknown parameter λ. The total survival time in this sample is $n\bar{x} = 53\,146$ days. The 2.5 % and 97.5 % quantiles of the $\text{G}(47, 1)$ distribution are $q_{0.025} = 34.53$ and $q_{0.975} = 61.36$, respectively, so

$$\Pr(34.53 \leq \lambda n \bar{X} \leq 61.36) = 0.95.$$

A rearrangement gives

$$q_{0.025} \leq \lambda n \bar{X} \leq q_{0.975} \quad \Longleftrightarrow \quad q_{0.025}/(n\bar{X}) \leq \lambda \leq q_{0.975}/(n\bar{X}),$$

and we obtain a 95 % confidence interval for the rate λ with limits $6.5 \cdot 10^{-4}$ and $1.15 \cdot 10^{-3}$ events per day. The inverse values finally give a 95 % confidence interval for the expected survival time $\mathsf{E}(X_i) = 1/\lambda$ from 866.2 to 1539.0 days.

However, note that the assumption of an exponential distribution may not be realistic. In particular, this assumption implies that $\mathsf{E}(X_i) = \sqrt{\mathrm{Var}(X_i)}$. Also, ignoring uncensored observations is likely to lead to bias, compare Example 2.8, where we have taken the censored observations into account. ■

In applications we often have an additional unknown *nuisance parameter* η, which is not of primary interest, but must be taken into account in the statistical analysis. To construct a confidence interval for the parameter of interest θ in the presence of a nuisance parameter η, the distribution of a pivot for θ must not depend on θ nor on η; the pivot itself must not depend on η.

Example 3.8 (Normal model) Let $X_{1:n}$ denote a random sample from a normal distribution $\mathrm{N}(\mu, \sigma^2)$, where both mean μ and variance σ^2 are unknown. Suppose that the parameter of interest is μ while σ^2 is a nuisance parameter. The *t statistic*

$$T = \sqrt{n}\frac{\bar{X} - \mu}{S} \sim \mathrm{t}(n-1)$$

is a pivot for μ since the distribution of T is independent of μ and T does not depend on σ^2. It is well known that T follows a standard t distribution (cf. Appendix A.5.2) with $n-1$ degrees of freedom, thus we have

$$\Pr\{t_{(1-\gamma)/2}(n-1) \le T \le t_{(1+\gamma)/2}(n-1)\} = \gamma,$$

where $t_\alpha(k)$ denotes the α quantile of the standard t distribution with k degrees of freedom. Using the property $t_\alpha(n-1) = -t_{1-\alpha}(n-1)$, which follows from the symmetry of the standard-t distribution around zero, the interval with limits

$$\bar{X} \pm \frac{S}{\sqrt{n}} \cdot t_{(1+\gamma)/2}(n-1) \tag{3.9}$$

is a $\gamma \cdot 100\,\%$ confidence interval for μ.

If instead σ^2 is the parameter of interest, then

$$\frac{n-1}{\sigma^2} S^2 \sim \chi^2(n-1)$$

is a pivot for σ^2 with a chi-squared distribution with $n-1$ degrees of freedom as pivotal distribution. Using this pivot, we can easily construct confidence intervals for σ^2:

$$\begin{aligned}
\gamma &= \Pr\left\{\chi^2_{(1-\gamma)/2}(n-1) \le \frac{n-1}{\sigma^2} S^2 \le \chi^2_{(1+\gamma)/2}(n-1)\right\} \\
&= \Pr\left\{1/\chi^2_{(1-\gamma)/2}(n-1) \ge \frac{\sigma^2}{(n-1)S^2} \ge 1/\chi^2_{(1+\gamma)/2}(n-1)\right\} \\
&= \Pr\{(n-1)S^2/\chi^2_{(1+\gamma)/2}(n-1) \le \sigma^2 \le (n-1)S^2/\chi^2_{(1-\gamma)/2}(n-1)\}, \qquad (3.10)
\end{aligned}$$

where $\chi^2_\alpha(k)$ denotes the α quantile of the chi-squared distribution with k degrees of freedom. ■

Example 3.9 (Blood alcohol concentration) We want to illustrate the method with the study on blood alcohol concentration, cf. Sect. 1.1.7. Assuming that the transformation factors $X_{1:n}$ follow a normal distribution $\mathrm{N}(\mu, \sigma^2)$, we obtain the 95 % confidence interval (3.9) for μ from 2414.7 to 2483.7. The point estimate is the arithmetic mean $\bar{x} = 2449.2$. Furthermore, computing (3.10) and taking the square root of both limits yields the 95 % confidence interval from 215.8 to 264.8 for the standard deviation σ. Note that the bounds are not symmetric around the point estimate $s = 237.8$. ■

Construction of exact confidence intervals using exact pivots is often impossible. It is often easier to use approximate pivots to construct approximate confidence intervals. A particularly useful approximate pivot is the one described in the following:

Result 3.1 (*z*-statistic) *Let $X_{1:n}$ denote a random sample from some distribution with probability mass or density function $f(x; \theta)$. Let T_n denote a consistent estimator of the parameter θ with standard error* $\mathrm{se}(T_n)$. *Then the z*-statistic

$$Z(\theta) = \frac{T_n - \theta}{\mathrm{se}(T_n)} \tag{3.11}$$

is an approximate pivot, which follows under regularity conditions an asymptotic standard normal distribution, so

$$T_n \pm z_{\frac{1+\gamma}{2}} \, \mathrm{se}(T_n) \tag{3.12}$$

are the limits of an approximate $\gamma \cdot 100$ % confidence interval for θ. It is often called the Wald confidence interval.

The name of this approximate pivot derives from the quantiles of the standard normal distribution, which are often (also in this book) denoted by the symbol z. We now show why the z-statistic does indeed have an approximate standard normal distribution:

Proof The estimator T_n is consistent for θ, so it must be asymptotically unbiased, i.e. $\mathsf{E}(T_n) \to \theta$ as $n \to \infty$. Using the central limit theorem (cf. Appendix A.4.4), we have under regularity conditions

$$\frac{T_n - \theta}{\sqrt{\mathrm{Var}(T_n)}} \overset{\mathrm{a}}{\sim} \mathrm{N}(0, 1).$$

By definition the standard error $\mathrm{se}(T_n)$ is consistent for $\sqrt{\mathrm{Var}(T_n)}$, and therefore

$$\frac{\sqrt{\mathrm{Var}(T_n)}}{\mathrm{se}(T_n)} \xrightarrow{P} 1.$$

By Slutsky's theorem (cf. Appendix A.4.2) it follows that we can replace the unknown standard deviation of the estimator with the standard error, i.e.

$$\frac{\sqrt{\mathrm{Var}(T_n)}}{\mathrm{se}(T_n)} \cdot \frac{T_n - \theta}{\sqrt{\mathrm{Var}(T_n)}} = \frac{T_n - \theta}{\mathrm{se}(T_n)} \overset{a}{\sim} \mathrm{N}(0, 1).$$

□

Example 3.10 (Analysis of survival times) The standard deviation of the survival times is $s = 874.4$ days, so we obtain with (3.12) the limits of an approximate 95 % confidence interval for the mean survival time $\mathsf{E}(X_i)$:

$$\bar{x} \pm 1.96 \cdot \frac{874.4}{\sqrt{47}} = 1130.8 \pm 250.0 = 880.8 \text{ and } 1380.8 \text{ days}.$$

Here we have used the 97.5 % quantile $z_{0.975} \approx 1.96$ of the standard normal distribution in the calculation. This number is worth remembering because it appears very often in formulas for approximate 95 % confidence intervals.

This confidence interval is different from the one derived under the assumption of an exponential distribution, compare Example 3.7. This can be explained by the fact that the empirical standard deviation of the observed survival times $s = 874.4$ is smaller than the empirical mean $\bar{x} = 1130.8$. By contrast, under the assumption of an exponential distribution, the standard deviation must equal the mean. Therefore, the above confidence interval is smaller. ■

Example 3.11 (Inference for a proportion) We now consider the problem sketched in Sect. 1.1.1 and aim to construct a confidence interval for the unknown success probability π of a binomial sample $X \sim \mathrm{Bin}(n, \pi)$, where X denotes the number of successes, and n is the (known) number of trials. A commonly used estimator of π is $\hat{\pi} = X/n$ with variance

$$\mathrm{Var}(\hat{\pi}) = \frac{\pi(1-\pi)}{n},$$

cf. Example 2.10. Because $\hat{\pi}$ is consistent for π,

$$\mathrm{se}(\hat{\pi}) = \sqrt{\hat{\pi}(1-\hat{\pi})/n}$$

can be identified as a standard error of $\hat{\pi}$. The $\gamma \cdot 100$ % Wald confidence interval for π therefore has limits

$$\hat{\pi} \pm z_{(1+\gamma)/2} \sqrt{\frac{\hat{\pi}(1-\hat{\pi})}{n}}. \tag{3.13}$$

■

It is possible that there exists more than one approximate pivot for a particular parameter θ. For finite sample size, we typically have different confidence intervals, and we need criteria to compare their properties. One particularly useful criterion is the *actual* coverage probability of a confidence interval with *nominal* coverage probability γ. Complications will arise as the actual coverage probability may depend on the unknown parameter θ. Another criterion is the width of the confidence interval because smaller intervals are to be preferred for a given coverage probability. In Example 4.22 we will describe a case study where we empirically compare various confidence intervals for proportions.

3.2.4 The Delta Method

Suppose $\hat{\theta}$ is a consistent estimator of θ with standard error $\mathrm{se}(\hat{\theta})$. The *delta method* (cf. Appendix A.4.5) allows us to compute a standard error of $h(\hat{\theta})$ where $h(\theta)$ is some transformation of θ with $dh(\theta)/d\theta \neq 0$.

The delta method

A standard error of $h(\hat{\theta})$ can be obtained by multiplying the standard error of $\hat{\theta}$ with the absolute value of the derivative $dh(\theta)/d\theta$ evaluated at $\hat{\theta}$:

$$\mathrm{se}\{h(\hat{\theta})\} = \mathrm{se}(\hat{\theta}) \cdot \left| \frac{dh(\hat{\theta})}{d\theta} \right|.$$

Assuming that $dh(\theta)/d\theta \neq 0$ at the true value θ, the delta method allows us to compute a Wald confidence interval for $h(\theta)$ with limits

$$h(\hat{\theta}) \pm z_{(1+\gamma)/2}\,\mathrm{se}\{h(\hat{\theta})\}.$$

If h is assumed to be one-to-one, then we can apply the inverse function h^{-1} to the above limits and obtain another approximate confidence interval for θ. In general, if h is a strictly increasing function, then

$$T_l \leq h(\theta) \leq T_u$$

is equivalent to

$$h^{-1}(T_l) \leq \theta \leq h^{-1}(T_u).$$

Hence, the coverage probability of the back-transformed confidence interval $[h^{-1}(T_l), h^{-1}(T_u)]$ for θ will be the same as for the confidence interval $[T_l, T_u]$ for $h(\theta)$. For the Wald confidence intervals, the limits of the back-transformed confidence interval will not equal the limits

$$\hat{\theta} \pm z_{(1+\gamma)/2}\,\mathrm{se}(\hat{\theta})$$

of the original $\gamma \cdot 100\,\%$ Wald confidence interval for θ, whenever h is a nonlinear transformation. In particular, the new limits will no longer be symmetric around $\hat{\theta}$.

The question arises which transformation $h(\theta)$ should be selected to compute a Wald confidence interval for θ. One argument against certain transformations is the requirement that a confidence interval should be *boundary-respecting*, i.e. not cover impossible values outside the parameter space. The following example gives a confidence interval that is not boundary-respecting and uses a transformation to fix that problem.

Example 3.12 (Inference for a proportion) Let $X \sim \text{Bin}(n, \pi)$ and suppose that $n = 100$, $x = 2$, i.e. $\hat{\pi} = 0.02$ with $\text{se}(\hat{\pi}) = \sqrt{\hat{\pi}(1-\hat{\pi})/n} = 0.014$. In Example 3.11 we derived the 95 % Wald confidence interval (3.13) for π with limits

$$0.02 \pm 1.96 \cdot 0.014 = -0.007 \text{ and } 0.047,$$

so the lower limit is negative and thus outside the parameter space $(0, 1)$.

Alternatively, one can first apply the logit function

$$\phi = h(\pi) = \text{logit}(\pi) = \log\left(\frac{\pi}{1-\pi}\right),$$

which transforms a probability $\pi \in (0, 1)$ to the log odds $\phi \in \mathbb{R}$. We now need to compute the MLE of ϕ and its standard error. Invariance of the MLE gives

$$\hat{\phi}_{\text{ML}} = \text{logit}(\hat{\pi}_{\text{ML}}) = \log\left(\frac{\hat{\pi}_{\text{ML}}}{1-\hat{\pi}_{\text{ML}}}\right) = \log\left(\frac{x}{n-x}\right).$$

The standard error of $\hat{\phi}_{\text{ML}}$ can be computed with the delta method:

$$\text{se}(\hat{\phi}_{\text{ML}}) = \text{se}(\hat{\pi}_{\text{ML}}) \cdot \left|\frac{dh(\hat{\pi}_{\text{ML}})}{d\pi}\right|,$$

where

$$\frac{dh(\hat{\pi}_{\text{ML}})}{d\pi} = \frac{1-\pi}{\pi} \cdot \frac{1-\pi+\pi}{(1-\pi)^2} = \frac{1}{\pi(1-\pi)},$$

so we obtain

$$\begin{aligned}
\text{se}(\hat{\phi}_{\text{ML}}) &= \sqrt{\frac{\hat{\pi}_{\text{ML}}(1-\hat{\pi}_{\text{ML}})}{n}} \cdot \frac{1}{\hat{\pi}_{\text{ML}}(1-\hat{\pi}_{\text{ML}})} \\
&= \frac{1}{\sqrt{n \cdot \hat{\pi}_{\text{ML}}(1-\hat{\pi}_{\text{ML}})}} = \sqrt{\frac{n}{x(n-x)}} \\
&= \sqrt{\frac{1}{x} + \frac{1}{n-x}}.
\end{aligned}$$

For the above data, we obtain $\text{se}(\hat{\phi}_{\text{ML}}) \approx 0.714$, and the 95 % Wald confidence interval for $\phi = \text{logit}(\pi)$ has limits

$$\text{logit}(0.02) \pm 1.96 \cdot 0.714 = -5.29 \text{ and } -2.49.$$

Back-transformation using the inverse logit function $h^{-1}(\phi) = \exp(\phi)/\{1+\exp(\phi)\}$ gives the interval $[0.01, 0.076]$ for π. By construction this interval can only cover values in the unit interval, so it appears to be more useful. However, note that for $x = 0$ (and also for $x = n$), the standard error $\text{se}(\hat{\phi}_{\text{ML}})$ is infinite, so it is impossible to compute a useful confidence interval. Also the original confidence interval (3.13) fails in this case with standard error $\text{se}(\hat{\pi}) = 0$. We will discuss an alternative confidence interval that produces sensible results also in this case in Example 4.10. ■

A different approach to construct Wald confidence intervals is to select the function h such that $\text{Var}\{h(\hat{\theta})\}$ does not (at least approximately) depend on θ. We will discuss this approach in Sect. 4.3.

3.2.5 The Bootstrap

Finally, we describe a *Monte Carlo* technique, which can also be used for constructing confidence intervals. Instead of analytically computing confidence intervals, the help of a computer is needed here. Details on Monte Carlo methods for Bayesian computations are given in Chap. 8.

We are interested in a model parameter θ and would like to estimate it with the corresponding statistic $\hat{\theta}$ of the realisation $x_{1:n}$ of a random sample. Standard errors and confidence intervals allow us to account for the sampling variability of the underlying random variables $X_i \stackrel{\text{iid}}{\sim} f(x;\theta)$, $i = 1, \dots, n$. These can often be calculated from $x_{1:n}$ by the use of mathematical derivations, as discussed previously in this section.

Of course, it would be easier when many samples $x_{1:n}^{(1)}, \dots, x_{1:n}^{(B)}$ from the population would be available. Then we could directly estimate the distribution of the parameter estimate $\hat{\theta}(X_{1:n})$ implied by the distribution of the random sample $X_{1:n}$. The idea of the *bootstrap*, which was invented by Bradley Efron (born 1938) in 1979, is simple: We use the given sample $x_{1:n}$ to obtain an estimate $\hat{f}(x)$ of the unknown probability mass or density function $f(x)$. Then we draw bootstrap samples $x_{1:n}^{(1)}, \dots, x_{1:n}^{(B)}$ by sampling from $\hat{f}(x)$ instead from $f(x)$, so, each bootstrap sample is a realisation of $X_i \stackrel{\text{iid}}{\sim} \hat{f}(x)$, $i = 1, \dots, n$. Afterwards, we can directly estimate quantities of the distribution of $\hat{\theta}(X_{1:n})$ from the bootstrap samples $\hat{\theta}(x_{1:n}^{(1)}), \dots, \hat{\theta}(x_{1:n}^{(B)})$. For example, we can compute the bootstrap standard error of $\hat{\theta}$ by estimating the standard deviation of the bootstrap samples. Analogously, we can compute the 2.5 % and 97.5 % quantiles of the bootstrap samples to obtain a 95 % bootstrap confidence interval for θ. With this approach, we can directly estimate the uncertainty attached to our estimate $\hat{\theta} = \hat{\theta}(x_{1:n})$ in the original data set $x_{1:n}$.

The most straightforward estimate of $f(x)$ is the empirical distribution $\hat{f}_n(x)$, which puts weight $1/n$ on each realisation $x_1, \dots, x_n$ of the sample. Random sampling from $\hat{f}_n(x)$ reduces to drawing data points with replacement from the original sample $x_{1:n}$. The name of the bootstrap method traces back to this simple trick, which is "pulling oneself up by one's bootstraps". This procedure does not make

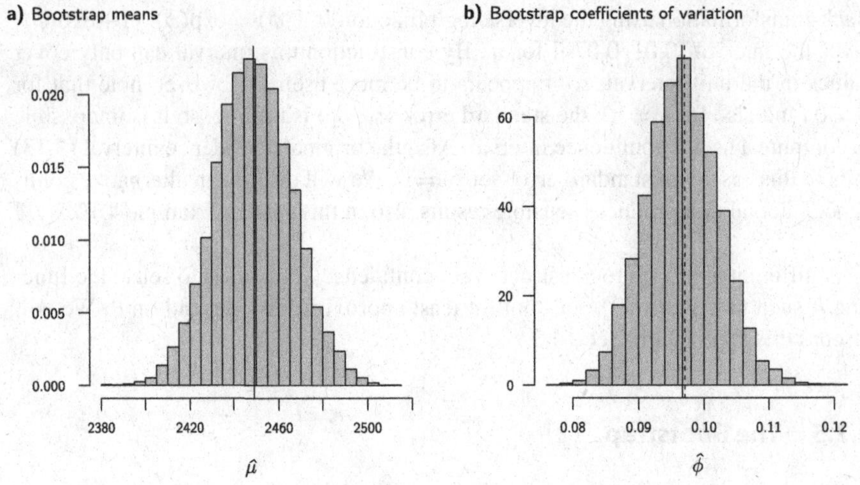

Fig. 3.1 Histogram of bootstrap means and coefficients of variation for the transformation factor. The means of the bootstrap samples $\hat{\mu}(x_{1:n}^{(b)})$ and $\hat{\phi}(x_{1:n}^{(b)})$ are marked by *continuous vertical lines*, and the estimates $\hat{\mu}$ and $\hat{\phi}$ in the original sample are marked by *dashed vertical lines*

any parametric assumptions about $f(x)$ and is therefore known as the *nonparametric bootstrap*. Note that there are finitely many different bootstrap samples (actually n^n ordered samples if there are no duplicate observations) that can be drawn from $\hat{f}_n(x)$. Since the nonparametric bootstrap distribution of $\hat{\theta}(X_{1:n})$ is discrete, we could in principle avoid the sampling at all and work with the theoretical distribution instead to obtain uncertainty estimates. However, this is only feasible for very small sample sizes, say $n < 10$. Otherwise, we have to proceed with Monte Carlo sampling from the nonparametric bootstrap distribution, which reduces the statistical accuracy only marginally in comparison with the sampling error of $X_{1:n}$ when enough (e.g. $B > 10\,000$) bootstrap samples are used.

Example 3.13 (Blood alcohol concentration) In Example 3.9 we computed the mean transformation factor of $\hat{\mu} = \bar{x} = 2449.2$ with 95 % Wald confidence interval from 2414.7 to 2483.7.

The 95 % bootstrap confidence interval can now be obtained as the 2.5 % and 97.5 % quantiles of the bootstrap means, which are shown in Fig. 3.1a. With $B = 10\,000$ bootstrap samples, we obtain the 95 % confidence interval from 2415.0 to 2483.4. This bootstrap confidence interval is very close to the Wald confidence interval. This good agreement between both methods is due to the symmetry of the distribution of the bootstrap means. The fact that both methods yield almost the same intervals is reassuring for the results obtained with standard methods. When both methods yield clearly different results, we are warned, and the results should be interpreted with caution. In general, the bootstrap confidence interval is to be preferred because it makes less assumptions than the standard Wald confidence interval. Of course, this comes at the cost of more extensive calculations and additional Monte Carlo error through stochastic simulation. ■

The great advantage of the bootstrap is that it can easily be applied to more complicated statistics. If analytical methods for the computation of confidence intervals are missing or rely on strong assumptions, we can always resort to bootstrap confidence intervals.

Example 3.14 (Blood alcohol concentration) Besides the mean transformation factor μ (see Example 3.13), we may also be interested in a bootstrap confidence interval for the *coefficient of variation* $\phi = \sigma/\mu$. The corresponding *sample coefficient of variation* $\hat{\phi} = s/\bar{x}$ can be computed for every bootstrap sample; the corresponding histogram is shown in Fig. 3.1b. Based on the quantiles of the bootstrap distribution, we obtain the 95 % bootstrap confidence interval from 0.086 to 0.108, with point estimate of 0.097. In the same manner, a 95 % bootstrap confidence interval can be constructed e.g. for the median transformation factor, with the result [2411, 2479]. ■

We can also estimate the bias of an estimator using the bootstrap. The bias of $\hat{\theta}$ was defined in Sect. 3.1 as

$$\mathsf{E}_f\{\hat{\theta}(X_{1:n})\} - \theta,$$

where we use the subscript f to emphasise that the expectation is with respect to the distribution $f(x)$. Since the population parameter θ is linked to the distribution $f(x)$, plugging in $\hat{f}(x)$ for $f(x)$ yields the bootstrap bias estimate

$$\mathsf{E}_{\hat{f}}\{\hat{\theta}(X_{1:n})\} - \hat{\theta} \approx \frac{1}{B}\sum_{b=1}^{B}\hat{\theta}\big(x_{1:n}^{(b)}\big) - \hat{\theta},$$

where $\hat{\theta}$ is the estimate in the original sample. A bias-corrected estimate of θ is thus given by

$$\tilde{\theta} = \hat{\theta} - \big[\mathsf{E}_{\hat{f}}\{\hat{\theta}(X_{1:n})\} - \hat{\theta}\big] = 2\hat{\theta} - \mathsf{E}_{\hat{f}}\{\hat{\theta}(X_{1:n})\} \approx 2\hat{\theta} - \frac{1}{B}\sum_{b=1}^{B}\hat{\theta}\big(x_{1:n}^{(b)}\big).$$

The bootstrap approach described above for computing confidence intervals is called the *percentile method.* If the bias of the estimator is relatively large, then more sophisticated methods, which take this into account, should be used. We will now sketch the *bias-corrected and accelerated* (BCa) method. Consider a bootstrap estimator $\hat{\theta}$ for θ. Assume that there exists a one-to-one transformation $\phi = g(\theta)$ of the original parameter with estimate $\hat{\phi} = g(\hat{\theta})$ and associated standard error $\operatorname{se}(\hat{\phi}) = \tau(1 + a\phi)$, such that

$$\frac{\hat{\phi} - \phi}{\operatorname{se}(\hat{\phi})} \overset{a}{\sim} \mathrm{N}(-c, 1). \tag{3.14}$$

This represents a generalisation of the approximate normal distribution in Result 3.1, where c determines the remaining bias of the transformed bootstrap estimator. The "acceleration constant" a steers the dependence of the standard error of $\hat{\phi}$ on the true value ϕ, while τ is a constant. With (3.14) it is straightforward to construct a confidence interval for ϕ. As in Sect. 3.2.4, the limits of this confidence interval can then be transformed back to the θ-scale by the inverse transformation g^{-1}.

However, the BCa method does not require us to specify the normalising transformation g or the standard error constant τ. Given a set of B bootstrap samples having empirical distribution function $\hat{G}_B$ for the bootstrap estimates $\hat{\theta}(x_{1:n}^{(b)})$, the BCa confidence interval with confidence level γ has limits

$$\hat{G}_B^{-1}\{\Phi(q_{(1-\gamma)/2})\} \quad \text{and} \quad \hat{G}_B^{-1}\{\Phi(q_{(1+\gamma)/2})\}, \tag{3.15}$$

where $\hat{G}_B^{-1}(\alpha)$ is the α quantile of the bootstrap estimates, Φ is the standard normal distribution function, and

$$q_\alpha = c + \frac{c + z_\alpha}{1 - a(c + z_\alpha)}.$$

Note that if the bias constant $c = 0$ and the acceleration constant $a = 0$, then $q_\alpha = z_\alpha$ where $z_\alpha = \Phi^{-1}(\alpha)$ is the α quantile of the standard normal distribution, and hence (3.15) reduces to the simple percentile interval with limits $\hat{G}_B^{-1}\{(1-\gamma)/2\}$ and $\hat{G}_B^{-1}\{(1+\gamma)/2\}$. In general, these two constants can be estimated as

$$\hat{c} = \Phi^{-1}\{\hat{G}_B(\hat{\theta})\}$$

and

$$\hat{a} = \frac{1}{6}\frac{\sum_{i=1}^{n}\{\hat{\theta}^* - \hat{\theta}(x_{-i})\}^3}{[\sum_{i=1}^{n}\{\hat{\theta}^* - \hat{\theta}(x_{-i})\}^2]^{3/2}},$$

where $\hat{\theta}(x_{-i})$ is the estimate of θ obtained from the original sample $x_{1:n}$ excluding observation x_i, and $\hat{\theta}^* = \frac{1}{n}\sum_{i=1}^{n}\hat{\theta}(x_{-i})$ is their average.

Example 3.15 (Blood alcohol concentration) We see in Fig. 3.1 that while, for the mean parameter μ (see Example 3.13), the average of the bootstrap estimates is very close to the original estimate $\hat{\mu}$ (bias estimate divided by standard error: -0.006), for the coefficient of variation ϕ (see Example 3.14), the relative size of the bias estimate is larger (bias estimate divided by standard error: -0.070). Because the bias estimates are negative, the estimates are expected to be too small, and the bias correction shifts them upwards: $\hat{\mu} = 2449.176$ is corrected to $\tilde{\mu} = 2449.278$, and $\hat{\phi} = 0.09708$ is corrected to $\tilde{\phi} = 0.09747$.

The 95 % BCa confidence intervals are computed from the estimated constants $\hat{a} = 0.002$ and $\hat{c} = 0.018$ for the mean estimator $\hat{\mu}$ and $\hat{a} = 0.048$ and $\hat{c} = 0.087$ for the coefficient of variation estimator $\hat{\phi}$. Both constants are further away from zero for $\hat{\phi}$, indicating a greater need to move from the simple percentile method to the BCa method. For μ, we obtain the 95 % BCa confidence interval $[2416, 2484]$, which is practically identical to the percentile interval computed before. For ϕ, we obtain $[0.088, 0.110]$, which has a notably larger upper bound than the percentile interval. Indeed, not the 2.5 % and 97.5 % quantiles of the bootstrap estimates are used here, but the 5.1 % and 99.1 % quantiles. ■

For practical implementation of bootstrap methods, we recommend the R-package `boot`, which provides BCa and other types of improved bootstrap confidence intervals via the function `boot.ci`. It also helps in implementing the *parametric bootstrap*, which estimates the unknown probability mass or density function $f(x)$ by taking samples from $f(x; \hat{\theta})$, a known distribution parametrised by θ, which is estimated from the original sample x as $\hat{\theta}$. This approach can also be used in the presence of nuisance parameters.

Sometimes, especially in irregular models like in the following example, both the nonparametric bootstrap and the parametric bootstrap fail.

Example 3.16 (Uniform model) As in Example 2.18, we consider again a random sample $X_{1:n}$ from a continuous uniform distribution $\mathrm{U}(0, \theta)$. We are interested in the parameter θ, which has MLE $\hat{\theta} = \max_i(x_i)$, and want to generate a confidence interval for it with bootstrap methods.

If we use the nonparametric bootstrap, we have probability $1 - 1/n$ that $\max_i(x_i)$ is not drawn, at each of the n draws necessary to generate a new bootstrap sample $x^{(b)}$. So the probability that $\max_i(x_i)$ is not part of a new bootstrap sample is $(1 - 1/n)^n$, or conversely, the probability that it is contained in a new bootstrap sample $x^{(b)}$ is

$$1 - \left(1 - \frac{1}{n}\right)^n > 1 - \exp(-1) \approx 0.632.$$

This lower bound is the limit as $n \to \infty$. Therefore, the bootstrap distribution always puts more than 0.632 probability mass on $\hat{\theta}(X) = \hat{\theta}$, the MLE in the original sample. That means that even if the sample size n increases, we do not get more information on the distribution of the MLE by using the nonparametric bootstrap. The reason is that the empirical distribution $\hat{f}_n(x)$ is not a good estimator in the tails of the distribution $f(x)$.

If we apply a parametric bootstrap instead, we would draw each observation of a new bootstrap sample independently from $\mathrm{U}(0, \hat{\theta})$, which leads to a continuous bootstrap distribution on $\hat{\theta}(X)$.

However, with both bootstrap approaches, there is a positive probability that the maximum $\max_i(x_i)$ is not included in a new bootstrap sample. Even worse, the confidence intervals computed from the bootstrap samples will always contain values smaller than the maximum. But these values are impossible because then the maximum value could not have been observed at all. Hence, a sensible confidence interval will have a *lower* bound not smaller than the observed maximum, see Example 4.20. ■

3.3 Significance Tests and P-Values

It is often of interest to quantify the evidence against a specific *null hypothesis* H_0 given the observed data. Such a null hypothesis can often be represented by a particular value θ_0 of the unknown parameter, i.e. $H_0 : \theta = \theta_0$. For example, in clinical studies a typical null hypothesis is that there is no difference in the effect of two treatments. If the outcome is binary, then this null hypothesis corresponds to an odds ratio $\theta_0 = 1$.

A *significance test* quantifies the evidence against H_0 using the *P-value*, defined as the conditional probability of the observed or more extreme data given H_0.

P-Value
The P-value is the probability, under the assumption of the null hypothesis H_0, of obtaining a result equal to or more extreme than what was actually observed.

A P-value is a frequentist concept based on a large number of repetitions of the study under the null hypothesis (for example, with odds ratio equal to 1). The P-value can now be viewed as the proportion of these studies that provide equal or less support for the null hypothesis than the observed study. The smaller the P-value, the more evidence there is against the null hypothesis.

Calculation of a P-value depends on what we mean by "more extreme". This is usually accomplished by explicit specification of an *alternative hypothesis* H_1. Usually, the alternative hypothesis is represented by all parameter values except θ_0, i.e. $H_1 : \theta \neq \theta_0$. This corresponds to the so-called *two-sided* significance test, in contrast to the *one-sided* significance test, where the alternative hypothesis is either $H_1 : \theta < \theta_0$ or $H_1 : \theta > \theta_0$. It turns out that a two-sided P-value is usually twice as large as the corresponding one-sided P-value where the alternative hypothesis is in the direction of the observed effect. At first sight it seems that one-sided P-values should be preferred if the alternative hypothesis has been specified a priori, for example in a study protocol. However, in biomedical research one-sided P-values are often not well received. For example, in the absence of a study protocol there is always the possibility that the alternative hypothesis has been selected *after* having observed the data supporting it, in order to obtain a P-value just half the size of the two-sided one.

There is another argument for using two-sided P-values. In certain situations we may investigate a series of null hypotheses, not just one. For example, in a genomic study we may want to identify genes that are associated with a specific phenotype of interest. It turns out that two-sided P-values follow under the null hypothesis (approximately) a standard uniform distribution. It is then convenient to investigate if the empirical distribution of such P-values follows a uniform distribution or not.

We will interpret P-values as an informal and continuous measure of the evidence against a certain null hypothesis. A rough but useful categorisation of P-values uses integer thresholds on the $\log_{10}$-scale and is given in Fig. 3.2.

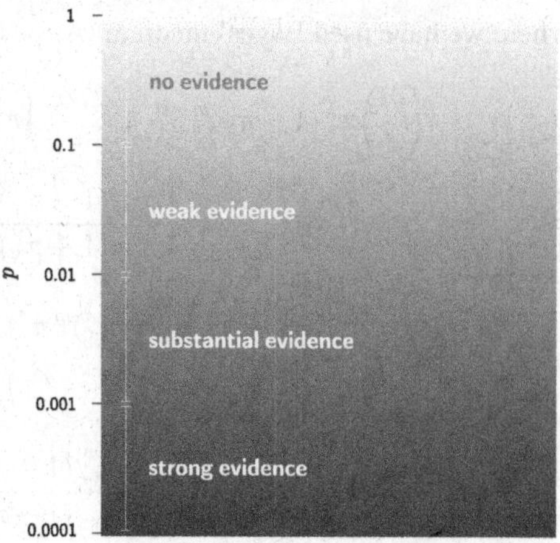

Fig. 3.2 Evidence against the null hypothesis H_0 for different P-values p

Table 3.1 Incidence of preeclampsia in a randomised placebo-controlled clinical trial of diuretics

		Treatment	
		Diuretics	Placebo
Preeclampsia	Yes	$x = 6$	$y = 2$
	No	$m - x = 102$	$n - y = 101$
		$m = 108$	$n = 103$

Example 3.17 (Fisher's exact test) Let θ denote the odds ratio and suppose we want to test the null hypothesis $H_0 : \theta = 1$ against the alternative $H_1 : \theta > 1$. Let $\hat{\theta}_{\text{ML}} = \hat{\theta}_{\text{ML}}(x)$ denote the observed odds ratio, assumed to be larger than 1. The one-sided P-value for the alternative $H_1 : \theta > 1$ is then

$$P\text{-value} = \Pr\{\hat{\theta}_{\text{ML}}(X) \geq \hat{\theta}_{\text{ML}} \mid H_0 : \theta = 1\},$$

where $\hat{\theta}_{\text{ML}}(X)$ denotes the MLE of θ viewed as a function of the data X.

For illustration, consider the data from the clinical study in Table 1.1 labelled as "Tervila". Table 3.1 summarises the data in a 2×2 table. The observed odds ratio is $6 \cdot 101/(2 \cdot 102) \approx 2.97$. We will show in the following that if we fix both margins of the 2×2 table and if we assume that true odds ratio equals 1, i.e. $\theta = 1$, the distribution of each entry of the table follows a hypergeometric distribution with all parameters determined by the margins. This result can be used to calculate a P-value for $H_0 : \theta = 1$. Note that it is sufficient to consider one entry of the table, for example x_1; the values of the other entries directly follow from the fixed margins.

Using the notation given in Table 3.1, we assume that $X \sim \text{Bin}(m, \pi_x)$ and likewise $Y \sim \text{Bin}(n, \pi_y)$, independent of X. Now let $Z = X + Y$. Then our interest is in

$$\Pr(X = x \mid Z = z) = \frac{\Pr(X = x) \cdot \Pr(Z = z \mid X = x)}{\Pr(Z = z)}, \tag{3.16}$$

where we have used Bayes' theorem (A.8). The numerator in (3.16) is

$$\begin{aligned}
&\binom{m}{x}\pi_x^x(1-\pi_x)^{m-x}\cdot\binom{n}{z-x}\pi_y^{z-x}(1-\pi_y)^{n-z+x}\\
&=\binom{m}{x}\binom{n}{z-x}\left\{\frac{\pi_x/(1-\pi_x)}{\pi_y/(1-\pi_y)}\right\}^x(1-\pi_x)^m\pi_y^z(1-\pi_y)^{n-z}\\
&=\binom{m}{x}\binom{n}{z-x}(1-\pi_x)^m\pi_y^z(1-\pi_y)^{n-z}
\end{aligned}$$

since we assume that the odds ratio

$$\theta=\frac{\pi_x/(1-\pi_x)}{\pi_y/(1-\pi_y)}=1.$$

The denominator in (3.16) can be written—using the law of total probability (A.9)—as

$$\mathsf{Pr}(Z=z)=\sum_{s=0}^{z}\mathsf{Pr}(X_1=s)\cdot\mathsf{Pr}(Z=z\,|\,X_1=s),$$

so we finally obtain

$$\mathsf{Pr}(X=x\,|\,Z=z)=\frac{\binom{m}{x}\binom{n}{z-x}}{\sum_{s=0}^{z}\binom{m}{s}\binom{n}{z-s}},$$

i.e. $X\,|\,Z=z\sim\mathrm{HypGeom}(z,m+n,m)$.

For the data shown in Table 3.1, we have

$$X\,|\,Z=z\sim\mathrm{HypGeom}(z=8,m+n=211,m=108),$$

so the one-sided P-value can be calculated as the sum of the hypergeometric probabilities to observe $x=6$, 7 or 8 entries:

$$\frac{\binom{108}{6}\binom{103}{2}}{\binom{211}{8}}+\frac{\binom{108}{7}\binom{103}{1}}{\binom{211}{8}}+\frac{\binom{108}{8}\binom{103}{0}}{\binom{211}{8}}=0.118+0.034+0.004=0.156.$$

Calculation of a two-sided P-value is also possible in this scenario. A common approach is to add all hypergeometric probabilities that are equal to or less than the probability of the observed table. For the data considered, this corresponds to adding the probabilities for $x=0$, 1 or 2 to the one-sided P-value, and we obtain the two-sided P-value 0.281. Both P-values do not provide any evidence against the null hypothesis that the true odds ratio is 1. ■

Calculation of P-values is often based on the realisation $t=T(x)$ of a *test statistic* T, which follows (at least approximately) a known distribution under the assumption of the null hypothesis. A pivot, fixing the parameter value θ at the null hypothesis value θ_0, is the obvious choice for a test statistic.

Example 3.18 (Analysis of survival times) Suppose that survival times are exponentially distributed with rate λ as in Example 3.7 and we wish to test the null hypothesis $H_0 : \lambda_0 = 1/1000$, i.e. that the mean survival time θ is $1/\lambda_0 = 1000$ days. Using the pivot (3.8) with $\lambda = \lambda_0 = 1/1000$, we obtain the test statistic $T = n\bar{X}/1000$ with realisation

$$t = \frac{n\bar{x}}{1000} = 53.146.$$

The distribution of this test statistic is under the null hypothesis $G(n = 47, 1)$, so the one-sided P-value (using the alternative $H_1 : \lambda < 1/1000$) can be easily calculated using the function pgamma in R:

```
t
[1] 53.146
n
[1] 47
pgamma(t, shape=n, rate=1, lower.tail=FALSE)
[1] 0.1818647
```

The one-sided P-value turns out to be 0.18, so under the assumption of exponentially distributed survival times, there is no evidence against the null hypothesis of a mean survival time equal to 1000 days. ■

Many pivots follow asymptotically a standard normal distribution, in which case the P-value can easily be calculated based on the standard normal distribution function. If t is the *observed test statistic*, then the two-sided P-value is $2\Pr(T \geq |t|) = 2\Phi(-|t|)$, where T denotes a standard normal random variable, and $\Phi(x)$ its distribution function.

Example 3.19 (Analysis of survival times) Using the approximate pivot

$$Z(\theta_0) = \frac{T_n - \theta_0}{\text{se}(T_n)}, \tag{3.17}$$

from (3.11) we can test the null hypothesis that the mean survival time is $\theta_0 = 1000$ days, but now without assuming exponentially distributed survival times, similarly as with the confidence interval in Example 3.10. Here T_n denotes a consistent estimator of the parameter θ with standard error $\text{se}(T_n)$.

The realisation of the test statistic (3.17) turns out to be

$$z = \frac{1130.8 - 1000}{874.4/\sqrt{47}} = 1.03$$

for the PBC data. The one-sided P-value can now be calculated using the standard normal distribution function as $\Phi\{-|z|\}$ and turns out to be 0.15. The P-value is fairly similar to the one based on the exponential model and provides no evidence against the null hypothesis. A two-sided P-value can be easily obtained as twice the one-sided one.

Note that we have used the sample standard deviation $s = 874.4$ in the calculation of the denominator of (3.17), based on the formula

$$s^2 = \frac{1}{n-1}\sum_{i=1}^{n}(x_i - \bar{x})^2, \tag{3.18}$$

where $\bar{x}$ is the empirical mean survival time. The P-value is calculated assuming that the null hypothesis is true, so it can be argued that $\bar{x}$ in (3.18) should be replaced by the null hypothesis value 1000. In this case, $n-1$ can be replaced by n to ensure that, under the null hypothesis, s^2 is unbiased for σ^2. This leads to a slightly smaller value of the test statistic and consequently to the slightly larger P-value 0.16. ■

In practice the null hypothesis value is often the null value $\theta_0 = 0$. For example, we might want to test the null hypothesis that the risk difference is zero. Similarly, if we are interested to test the null hypothesis that the odds ratio is one, then this corresponds to a log odds ratio of zero. For such null hypotheses, the test statistic (3.17) takes a particularly simple form as the estimate T_n of θ divided by its standard error:

$$Z = \frac{T_n}{\mathrm{se}(T_n)}.$$

A realisation of Z is called the *Z-value*.

It is important to realise that the P-value is a conditional probability under the assumption that the null hypothesis is true. The P-value is not the probability of the null hypothesis given the data, which is a common misinterpretation. The posterior probability of a certain statistical model is a Bayesian concept (see Sect. 7.2.1), which makes sense if prior probabilities are assigned to the null hypothesis and its counterpart, the alternative hypothesis. However, from a frequentist perspective a null hypothesis can only be true or false. As a consequence, the P-value is commonly viewed as an informal measure of the evidence against a null hypothesis. Note also that a large P-value cannot be viewed as evidence *for* the null hypothesis; a large P-value represents absence of evidence against the null hypothesis, and "absence of evidence is not evidence of absence".

The Neyman–Pearson approach to statistical inference rejects the P-value as an informal measure of the evidence against the null hypothesis. Instead, this approach postulates that there are only two possible "decisions" that can be reached after having observed data: either "rejecting" or "not rejecting" the null hypothesis. This theory then introduces the probability of the *Type-I error* α, defined as the conditional probability of rejecting the null hypothesis although it is true in a series of hypothetical repetitions of the study considered. It can now be easily shown that the resulting *hypothesis test* will have a Type-I error probability equal to α, if the null hypothesis is rejected whenever the P-value is smaller than α. Note that this construction requires the Type-I error probability α to be specified before the study is conducted. Indeed, in a clinical study the probability of the Type-I error (usually 5 %) will be fixed already in the study protocol. However, in observational studies the P-value is commonly misinterpreted as a *post-hoc* Type-I error probability. For example, suppose that a P-value of 0.029 has been observed. This misconception would suggest that the probability of rejecting the null hypothesis although it is true in a series of imaginative repetitions of the study is 0.029. This interpretation of the P-value is not correct, as it mixes a truly frequentist (unconditional) concept (the probability of the Type-I error) with the P-value, a measure of the evidence of the observed data against the null hypothesis, i.e. an (at least partly) conditional concept.

In this book we will mostly use significance rather than hypothesis tests and interpret P-values as a continuous measure of the evidence against the null hypothesis, see Fig. 3.2. However, there is need to emphasise the *duality of hypothesis tests and confidence intervals*. Indeed, the result of a two-sided hypothesis test of the null hypothesis $H_0 : \theta = \theta_0$ at Type-I error probability α can be read off from the corresponding $(1 - \alpha) \cdot 100\ \%$ confidence interval for θ: If and only if θ_0 is within the confidence interval, then the Neyman–Pearson test would not reject the null hypothesis.

Duality of hypothesis tests and confidence intervals
The set of values θ_0 for which a certain hypothesis test does not reject the null hypothesis $H_0 : \theta = \theta_0$ at Type-I error probability α is a $(1 - \alpha) \cdot 100\ \%$ confidence interval for θ.

So confidence intervals can be built based on *inverting* a certain hypothesis test.

P-values and significance tests are probably the most commonly used statistical tools for routine investigation of scientific hypotheses. However, because of the widespread misinterpretation of P-values and significance tests, there have been attempts to replace or at least accompany P-values by confidence intervals. Indeed, confidence intervals are richer in the sense that we can always calculate (at least approximately) a P-value for a certain null hypothesis from a confidence interval at level 95 %, say, but the reverse step requires additional knowledge of $\hat{\theta}_{\text{ML}}$. In addition, confidence intervals give a range of possible values for an effect size, so they inform not only about statistical significance but also about the practical relevance of a certain parameter estimate.

3.4 Exercises

1. Sketch why the MLE

$$\hat{N}_{\text{ML}} = \left\lfloor \frac{M \cdot n}{x} \right\rfloor$$

in the capture–recapture experiment (cf. Example 2.2) cannot be unbiased. Show that the alternative estimator

$$\hat{N} = \frac{(M+1) \cdot (n+1)}{(x+1)} - 1$$

is unbiased if $N \leq M + n$.

2. Let $X_{1:n}$ be a random sample from a distribution with mean μ and variance $\sigma^2 > 0$. Show that

$$\mathsf{E}(\bar{X}) = \mu \quad \text{and} \quad \text{Var}(\bar{X}) = \frac{\sigma^2}{n}.$$

3. Let $X_{1:n}$ be a random sample from a normal distribution with mean μ and variance $\sigma^2 > 0$. Show that the estimator

$$\hat{\sigma} = \sqrt{\frac{n-1}{2}} \frac{\Gamma(\frac{n-1}{2})}{\Gamma(\frac{n}{2})} S$$

is unbiased for σ, where S is the square root of the sample variance S^2 in (3.1).

4. Show that the sample variance S^2 can be written as

$$S^2 = \frac{1}{2n(n-1)} \sum_{i,j=1}^{n} (X_i - X_j)^2.$$

Use this representation to show that

$$\text{Var}(S^2) = \frac{1}{n}\left\{c_4 - \left(\frac{n-3}{n-1}\right)\sigma^4\right\},$$

where $c_4 = \mathsf{E}[\{X - \mathsf{E}(X)\}^4]$ is the fourth central moment of X.

5. Show that the confidence interval defined in Example 3.6 indeed has coverage probability 50 % for all values $\theta \in \Theta$.

6. Consider a random sample $X_{1:n}$ from the uniform model $\text{U}(0, \theta)$, cf. Example 2.18. Let $Y = \max\{X_1, \dots, X_n\}$ denote the maximum of the random sample $X_{1:n}$. Show that the confidence interval for θ with limits

$$Y \text{ and } (1-\gamma)^{-1/n} Y$$

has coverage γ.

7. Consider a population with mean μ and variance σ^2. Let $X_1, \dots, X_5$ be independent draws from this population. Consider the following estimators for μ:

$$T_1 = \frac{1}{5}(X_1 + X_2 + X_3 + X_4 + X_5),$$

$$T_2 = \frac{1}{3}(X_1 + X_2 + X_3),$$

$$T_3 = \frac{1}{8}(X_1 + X_2 + X_3 + X_4) + \frac{1}{2}X_5,$$

$$T_4 = X_1 + X_2 \quad \text{and}$$

$$T_5 = X_1.$$

   (a) Which estimators are unbiased for μ?
   (b) Compute the MSE of each estimator.

8. The distribution of a multivariate random variable $\boldsymbol{X}$ belongs to an exponential family of order p if the logarithm of its probability mass or density function can be written as

$$\log\{f(\boldsymbol{x};\boldsymbol{\tau})\} = \sum_{i=1}^{p} \eta_i(\boldsymbol{\tau}) T_i(\boldsymbol{x}) - B(\boldsymbol{\tau}) + c(\boldsymbol{x}). \tag{3.19}$$

Here $\boldsymbol{\tau}$ is the p-dimensional parameter vector, and T_i, η_i, B and c are real-valued functions. It is assumed that the set $\{1, \eta_1(\boldsymbol{\tau}), \ldots, \eta_p(\boldsymbol{\tau})\}$ is linearly independent. Then we define the canonical parameters $\theta_1 = \eta_1(\tau_1), \ldots, \theta_p = \eta_p(\tau_p)$. With $\boldsymbol{\theta} = (\theta_1, \ldots, \theta_p)^\top$ and $\boldsymbol{T}(\boldsymbol{x}) = (T_1(\boldsymbol{x}), \ldots, T_p(\boldsymbol{x}))^\top$ we can write the log density in canonical form:

$$\log\{f(\boldsymbol{x};\boldsymbol{\theta})\} = \boldsymbol{\theta}^\top \boldsymbol{T}(\boldsymbol{x}) - A(\boldsymbol{\theta}) + c(\boldsymbol{x}). \tag{3.20}$$

Exponential families are interesting because most of the commonly used distributions, such as the Poisson, geometric, binomial, normal and gamma distribution, are exponential families. Therefore, it is worthwhile to derive general results for exponential families, which can then be applied to many distributions at once. For example, two very useful results for the exponential family of order one in canonical form are $\mathsf{E}\{T(X)\} = dA/d\theta(\theta)$ and $\text{Var}\{T(X)\} = d^2A/d\theta^2(\theta)$.

(a) Show that $\boldsymbol{T}(\boldsymbol{X})$ is minimal sufficient for $\boldsymbol{\theta}$.

(b) Show that the density of the Poisson distribution $\text{Po}(\lambda)$ can be written in the forms (3.19) and (3.20), respectively. Thus, derive the expectation and variance of $X \sim \text{Po}(\lambda)$.

(c) Show that the density of the normal distribution $\text{N}(\mu, \sigma^2)$ can be written in the forms (3.19) and (3.20), respectively, where $\boldsymbol{\tau} = (\mu, \sigma^2)^\top$. Hence, derive a minimal sufficient statistic for $\boldsymbol{\tau}$.

(d) Show that for an exponential family of order one, $I(\hat{\tau}_{\text{ML}}) = J(\hat{\tau}_{\text{ML}})$. Verify this result for the Poisson distribution.

(e) Show that for an exponential family of order one in canonical form, $I(\theta) = J(\theta)$. Verify this result for the Poisson distribution.

(f) Suppose $X_{1:n}$ is a random sample from a one-parameter exponential family with canonical parameter θ. Derive an expression for the log-likelihood $l(\theta)$.

9. Assume that survival times $X_{1:n}$ form a random sample from a gamma distribution $\text{G}(\alpha, \alpha/\mu)$ with mean $\mathsf{E}(X_i) = \mu$ and shape parameter α.

(a) Show that $\bar{X} = n^{-1}\sum_{i=1}^{n} X_i$ is a consistent estimator of the mean survival time μ.

(b) Show that $X_i/\mu \sim \text{G}(\alpha, \alpha)$.

(c) Define the approximate pivot from Result 3.1,

$$Z = \frac{\bar{X} - \mu}{S/\sqrt{n}},$$

where $S^2 = (n-1)^{-1} \sum_{i=1}^{n} (X_i - \bar{X})^2$. Using the result from above, show that the distribution of Z does not depend on μ.

(d) For $n = 10$ and $\alpha \in \{1, 2, 5, 10\}$, simulate 100 000 samples from Z and compare the resulting 2.5 % and 97.5 % quantiles with those from the asymptotic standard normal distribution. Is Z a good approximate pivot?

(e) Show that $\bar{X}/\mu \sim \mathrm{G}(n\alpha, n\alpha)$. If α was known, how could you use this quantity to derive a confidence interval for μ?

(f) Suppose α is unknown; how could you derive a confidence interval for μ?

10. All beds in a hospital are numbered consecutively from 1 to $N > 1$. In one room a doctor sees $n \leq N$ beds, which are a random subset of all beds, with (ordered) numbers $X_1 < \cdots < X_n$. The doctor now wants to estimate the total number of beds N in the hospital.

(a) Show that the joint probability mass function of $X = (X_1, \ldots, X_n)$ is

$$f(x; N) = \binom{N}{n}^{-1} \mathsf{I}_{\{n,\ldots,N\}}(x_n).$$

(b) Show that X_n is minimal sufficient for N.

(c) Confirm that the probability mass function of X_n is

$$f_{X_n}(x_n; N) = \frac{\binom{x_n - 1}{n-1}}{\binom{N}{n}} \mathsf{I}_{\{n,\ldots,N\}}(x_n).$$

(d) Show that

$$\hat{N} = \frac{n+1}{n} X_n - 1$$

is an unbiased estimator of N.

(e) Study the ratio $L(N+1)/L(N)$ and derive the ML estimator of N. Compare it with $\hat{N}$.

3.5 References

The methods discussed in this chapter can be found in many books on statistical inference, for example in Lehmann and Casella (1998), Casella and Berger (2001) or Young and Smith (2005). The section on the bootstrap has only touched the surface of a wealth of so-called *resampling methods* for frequentist statistical inference. More details can be found e.g. in Davison and Hinkley (1997) and Chihara and Hesterberg (2019).

Frequentist Properties of the Likelihood

4

Contents

In Chap. 2 we have considered the likelihood and related quantities such as the log-likelihood, the score function, the MLE and the (observed) Fisher information for a fixed observation $X = x$ from a distribution with probability mass or density function $f(x; \theta)$. For example, in a binomial model with known sample size n and unknown probability π we have

$$\begin{aligned} \text{log-likelihood} \quad & l(\pi; x) = x \log \pi + (n - x) \log(1 - \pi), \\ \text{score function} \quad & S(\pi; x) = \frac{x}{\pi} - \frac{n - x}{1 - \pi}, \\ \text{MLE} \quad & \hat{\pi}_{\mathrm{ML}}(x) = \frac{x}{n}, \end{aligned}$$

L. Held, D. Sabanés Bové, *Likelihood and Bayesian Inference*,
Statistics for Biology and Health, https://doi.org/10.1007/978-3-662-60792-3_4,

and Fisher information $I(\pi; x) = \frac{x}{\pi^2} + \frac{n-x}{(1-\pi)^2}$.

Now we take a different point of view and apply the concepts of frequentist inference as outlined in Chap. 3. To this end, we consider $S(\pi)$, $\hat{\pi}_{\mathrm{ML}}$ and $I(\pi)$ as *random variables*, with distribution derived from the random variable $X \sim \mathrm{Bin}(n, \pi)$. The above equations now read

$$S(\pi; X) = \frac{X}{\pi} - \frac{n-X}{1-\pi},$$

$$\hat{\pi}_{\mathrm{ML}}(X) = \frac{X}{n}, \quad \text{and} \quad I(\pi; X) = \frac{X}{\pi^2} + \frac{n-X}{(1-\pi)^2}$$

where $X \sim \mathrm{Bin}(n, \pi)$ is an identical replication of the experiment underlying our statistical model. The parameter π is now fixed and denotes the true (unknown) parameter value. To ease notation, we will often not explicitly state the dependence of the random variables $S(\pi)$, $\hat{\pi}_{\mathrm{ML}}$ and $I(\pi)$ on the random variable X.

The results we will describe in the following sections are valid under a standard set of regularity conditions, often called *Fisher regularity conditions*.

Definition 4.1 (Fisher regularity conditions) Consider a distribution with probability mass or density function $f(x; \theta)$ with unknown parameter $\theta \in \Theta$. *Fisher regularity conditions* hold if

1. the parameter space Θ is an open interval, i.e. θ must not be at the boundary of the parameter space,
2. the support of $f(x; \theta)$ does not depend on θ,
3. the probability mass or density functions $f(x; \theta)$ indexed by θ are distinct, i.e.

$$f(x; \theta_1) \neq f(x; \theta_2) \quad \text{whenever } \theta_1 \neq \theta_2, \tag{4.1}$$

4. the likelihood $L(\theta) = f(x; \theta)$ is twice continuously differentiable with respect to θ,
5. the integral $\int f(x; \theta)\, dx$ can be twice differentiated under the integral sign. ♦

This chapter will introduce three important test statistics based on the likelihood: the score statistic, the Wald statistic and the likelihood ratio statistic. Many of the results derived are *asymptotic*, i.e. are valid only for a random sample $X_{1:n}$ with relatively large sample size n. A case study on different confidence intervals for proportions completes this chapter.

4.1 The Expected Fisher Information and the Score Statistic

In this section we will derive frequentist properties of the score function and the Fisher information. We will introduce the score statistic, which is useful to derive likelihood-based significance tests and confidence intervals.

4.1.1 The Expected Fisher Information

The Fisher information $I(\theta; x)$ of a parameter θ, the negative second derivative of the log-likelihood (cf. Sect. 2.2), depends in many cases not only on θ, but also on the observed data $X = x$. To free oneself from this dependence, it appears natural to consider the *expected Fisher information*, i.e. the expectation of the Fisher information $I(\theta; X)$,

$$J(\theta) = \mathsf{E}\{I(\theta; X)\},$$

where $I(\theta; X)$ is viewed as a function of the random variable X. Note that taking the expectation with respect to the distribution $f(x; \theta)$ of X implies that θ is the true (unknown) parameter.

Definition 4.2 (Expected Fisher information) The expectation of the Fisher information $I(\theta; X)$, viewed as a function of the data X with distribution $f(x; \theta)$, is the *expected Fisher information* $J(\theta)$. ♦

We will usually assume that the expected Fisher information $J(\theta)$ is positive and bounded, i.e. $0 < J(\theta) < \infty$.

Example 4.1 (Binomial model) If the data X follow a binomial distribution, $X \sim \text{Bin}(n, \pi)$, then we know from Example 2.10 that the Fisher information of π equals

$$I(\pi; x) = \frac{x}{\pi^2} + \frac{n - x}{(1 - \pi)^2}.$$

Using $\mathsf{E}(X) = n\pi$, we obtain the expected Fisher information

$$\begin{aligned} J(\pi) &= \mathsf{E}\{I(\pi; X)\} \\ &= \mathsf{E}\left\{\frac{X}{\pi^2}\right\} + \mathsf{E}\left\{\frac{n - X}{(1 - \pi)^2}\right\} \\ &= \frac{\mathsf{E}(X)}{\pi^2} + \frac{n - \mathsf{E}(X)}{(1 - \pi)^2} \\ &= \frac{n\pi}{\pi^2} + \frac{n - n\pi}{(1 - \pi)^2} \\ &= \frac{n}{\pi} + \frac{n}{1 - \pi} = \frac{n}{\pi(1 - \pi)}. \end{aligned}$$

Note that the only difference to the observed Fisher information $I(\hat{\pi}_{\text{ML}}; x)$ derived in Example 2.10 is the replacement of the MLE $\hat{\pi}_{\text{ML}}$ with the true value π. ■

The expected Fisher information can also be described as the variance of the score function. Before showing this general result, we first study a specific example.

Example 4.2 (Binomial model) For a binomial distributed random variable $X \sim \text{Bin}(n, \pi)$, the score function of the parameter π has been computed in Example 2.1:

$$S(\pi; x) = \frac{d}{d\pi} l(\pi) = \frac{x}{\pi} - \frac{n-x}{1-\pi}.$$

We now replace x with X and calculate the expectation and variance of the score function $S(\pi; X)$, viewed as a function of the random variable $X \sim \text{Bin}(n, \pi)$, where π denotes the true success probability.

With $\mathsf{E}(X) = n\pi$ and $\text{Var}(X) = n\pi(1-\pi)$ we obtain

$$\mathsf{E}\{S(\pi; X)\} = \frac{n\pi}{\pi} - \frac{n - n\pi}{1-\pi} = n - n = 0$$

and

$$\begin{aligned}
\text{Var}\{S(\pi; X)\} &= \text{Var}\left(\frac{X}{\pi} - \frac{n-X}{1-\pi}\right) \\
&= \text{Var}\left\{X\left(\frac{1}{\pi} + \frac{1}{1-\pi}\right) - \frac{n}{1-\pi}\right\} \\
&= \text{Var}\left(\frac{X}{\pi(1-\pi)}\right) = \frac{n\pi(1-\pi)}{\pi^2(1-\pi)^2} = \frac{n}{\pi(1-\pi)}.
\end{aligned}$$

So the expectation of the score function at the true parameter value θ is zero, and the variance is equal to the expected Fisher information:

$$\text{Var}\{S(\pi; X)\} = J(\pi) = \mathsf{E}\{I(\pi; X)\}.$$ ■

Under the Fisher regularity conditions, this result holds in general:

Result 4.1 (Expectation and variance of the score function) *Under the Fisher regularity conditions, which ensure that the order of differentiation and integration can be changed, we have*:

$$\begin{aligned}
\mathsf{E}\{S(\theta; X)\} &= 0, \\
\text{Var}\{S(\theta; X)\} &= J(\theta).
\end{aligned}$$

Proof To prove Result 4.1, we assume without loss of generality that $L(\theta) = f(x; \theta)$, i.e. all multiplicative constants in $f(x; \theta)$ are included in the likelihood function. First, note that for continuous X we have

$$\begin{aligned}
\mathsf{E}\{S(\theta; X)\} &= \int S(\theta; x) f(x; \theta)\, dx \\
&= \int \left\{\frac{d}{d\theta} \log L(\theta)\right\} f(x; \theta)\, dx \\
&= \int \frac{\frac{d}{d\theta} L(\theta)}{L(\theta)} f(x; \theta)\, dx, \quad \text{with the chain rule,}
\end{aligned}$$

$$
\begin{aligned}
&= \int \frac{d}{d\theta} L(\theta)\,dx, \quad \text{due to } L(\theta) = f(x;\theta),\\
&= \frac{d}{d\theta}\int L(\theta)\,dx, \quad \text{under the above regularity condition,}\\
&= \frac{d}{d\theta} 1 = 0.
\end{aligned}
$$

So we have $\mathsf{E}\{S(\theta; X)\} = 0$, and therefore $\mathrm{Var}\{S(\theta; X)\} = \mathsf{E}\{S(\theta; X)^2\}$. It is therefore sufficient to show that $J(\theta) = \mathsf{E}\{S(\theta; X)^2\}$ holds:

$$
\begin{aligned}
J(\theta) &= \mathsf{E}\left\{-\frac{d^2}{d\theta^2}\log L(\theta)\right\}\\
&= \mathsf{E}\left\{-\frac{d}{d\theta}\frac{\frac{d}{d\theta}L(\theta)}{L(\theta)}\right\}, \quad \text{with the chain rule,}\\
&= \mathsf{E}\left\{-\frac{\frac{d^2}{d\theta^2}L(\theta)\cdot L(\theta) - \{\frac{d}{d\theta}L(\theta)\}^2}{L(\theta)^2}\right\}, \quad \text{with the quotient rule,}\\
&= -\mathsf{E}\left\{\frac{\frac{d^2}{d\theta^2}L(\theta)}{L(\theta)}\right\} + \mathsf{E}\left\{\frac{(\frac{d}{d\theta}L(\theta))^2}{L(\theta)^2}\right\}\\
&= -\int \frac{\frac{d^2}{d\theta^2}L(\theta)}{L(\theta)} f(x;\theta)\,dx + \int \frac{\{\frac{d}{d\theta}L(\theta)\}^2}{L(\theta)^2} f(x;\theta)\,dx,
\end{aligned}
$$

using the above regularity condition

$$
\begin{aligned}
&= -\frac{d^2}{d\theta^2}\underbrace{\int L(\theta)\,dx}_{=1} + \underbrace{\int \left\{\frac{d}{d\theta}\log L(\theta)\right\}^2 f(x;\theta)\,dx}_{=S(\theta)^2}\\
&= \mathsf{E}\{S(\theta; X)^2\}.
\end{aligned}
$$

For discrete X the result is still valid with integration replaced by summation. □

Result 4.1 says that the score function at the true parameter value θ is on average zero. This suggests that the MLE, which has a score function value of zero, is on average equal to the true value. However, this is in general not true. It is true, though, asymptotically, as we will see in Result 4.9.

The variance of the score function is due to Result 4.1 equal to the expected Fisher information. How can we interpret this result? The expected Fisher information is the average negative curvature of the log-likelihood at the true value. If the log-likelihood function is steep and has a lot of curvature, then the information with respect to θ is large, and the score function at the true value θ will vary a lot. Conversely, if the log likelihood function is flat, then the score function will not vary much at the true value θ, so does not have much information with respect to θ.

4.1.2 Properties of the Expected Fisher Information

We now study properties of the expected Fisher information.

Result 4.2 (Expected Fisher information from a random sample) *Let $X_{1:n}$ denote a random sample from a distribution with probability mass or density function $f(x;\theta)$. Let $J(\theta)$ denote the* expected unit Fisher information, *i.e. the expected Fisher information from one observation X_i from $f(x;\theta)$. The expected Fisher information $J_{1:n}(\theta)$ from the whole random sample $X_{1:n}$ is then*

$$J_{1:n}(\theta) = n \cdot J(\theta).$$

This property follows directly from the additivity of the log-likelihood function for random samples, see Eq. (2.1).

Example 4.3 (Normal model) Consider a random sample $X_{1:n}$ from a normal distribution $\mathrm{N}(\mu, \sigma^2)$, where our interest lies in the expectation μ, and we assume that the variance σ^2 is known. We know from Example 2.9 that the unit Fisher information is

$$I(\mu; x_i) = \frac{1}{\sigma^2}.$$

Now $I(\mu; x_i)$ does not depend on x_i and thus equals the expected unit Fisher information $J(\mu)$, and the expected Fisher information from the whole random sample $X_{1:n}$ is $J_{1:n}(\mu) = nJ(\mu) = nI(\mu) = n/\sigma^2$.

Suppose now that we are interested in the expected Fisher information of the unknown variance σ^2 and treat the mean μ as known. We know from Example 2.9 that the unit Fisher information of σ^2 is

$$I(\sigma^2; x_i) = \frac{1}{\sigma^6}(x_i - \mu)^2 - \frac{1}{2\sigma^4}.$$

Using $\mathsf{E}\{(X_i - \mu)^2\} = \mathrm{Var}(X_i) = \sigma^2$, we easily obtain the expected unit Fisher information of σ^2:

$$J(\sigma^2) = \frac{1}{\sigma^4} - \frac{1}{2\sigma^4} = \frac{1}{2\sigma^4}.$$

The expected Fisher information from the whole random sample $X_{1:n}$ is therefore $J_{1:n}(\sigma^2) = nJ(\sigma^2) = n/(2\sigma^4)$. ■

The following result establishes that the property described in Result 2.1 for the observed Fisher information also holds for the expected Fisher information.

Result 4.3 (Expected Fisher information of a transformation) *Let $J_\theta(\theta)$ denote the expected Fisher information of a scalar parameter θ, and $\phi = h(\theta)$ a one-to-one transformation of θ. The expected Fisher information $J_\phi(\phi)$ of ϕ can then be* *calculated as follows*:

$$J_\phi(\phi) = J_\theta(\theta)\left\{\frac{dh^{-1}(\phi)}{d\phi}\right\}^2 = J_\theta(\theta)\left\{\frac{dh(\theta)}{d\theta}\right\}^{-2}.$$

Proof Using (2.4), we have

$$S_\phi(\phi; X) = S_\theta(\theta; X) \cdot \frac{dh^{-1}(\phi)}{d\phi}$$

and with Result 4.1 we have

$$\begin{aligned} J_\phi(\phi) &= \operatorname{Var}\{S_\phi(\phi; X)\} \\ &= \operatorname{Var}\{S_\theta(\theta; X)\} \cdot \left\{\frac{dh^{-1}(\phi)}{d\phi}\right\}^2 \\ &= J_\theta(\theta) \cdot \left\{\frac{dh^{-1}(\phi)}{d\phi}\right\}^2. \end{aligned}$$

The second equation follows from (2.5). □

Example 4.4 (Binomial model) The expected Fisher information of the success probability π in a binomial experiment is

$$J_\pi(\pi) = \frac{n}{\pi(1-\pi)},$$

compare Example 4.1. The expected Fisher information of the log odds $\phi = h(\pi) = \log\{\pi/(1-\pi)\}$ therefore is

$$J_\phi(\phi) = J_\pi(\pi)\left\{\frac{1}{\pi(1-\pi)}\right\}^{-2} = n\pi(1-\pi)$$

due to

$$\frac{dh(\pi)}{d\pi} = \frac{1}{\pi(1-\pi)}.$$

This corresponds to the observed Fisher information, as $J_\phi(\hat{\phi}_{\mathrm{ML}}) = I_\phi(\hat{\phi}_{\mathrm{ML}})$, cf. Example 2.11. ■

Example 4.5 (Normal model) In Example 4.3 we showed that the expected Fisher information of the variance σ^2 is $n/(2\sigma^4)$. Applying Result 4.3 to $\theta = \sigma^2$ and $\sigma = h(\theta) = \sqrt{\theta}$, we obtain the expected Fisher information of σ:

$$J_\sigma(\sigma) = J_{\sigma^2}(\sigma^2) \cdot |2\sigma|^2 = \frac{n}{2\sigma^4} \cdot 4\sigma^2 = \frac{2n}{\sigma^2},$$

using $dh^{-1}(\sigma)/d\sigma = d\sigma^2/d\sigma = 2\sigma$. ■

Location and scale parameters are commonly encountered in statistics. Here is a definition.

Definition 4.3 (Location and scale parameters) Let X denote a random variable with probability mass or density function $f_X(x) = f(x;\theta)$, depending one scalar parameter θ. If the probability mass or density function $f_Y(y)$ of $Y = X + c$, where $c \in \mathbb{R}$ is a constant, has the form $f_X(y;\theta + c)$, then θ is called a *location parameter*. If the probability mass or density function $f_Y(y)$ of $Y = cX$, where $c \in \mathbb{R}^+$ is a positive constant, has the form $f_X(y; c\theta)$ then θ is called a *scale parameter*. ♦

Example 4.6 (Normal model) Consider a normal random variable $X \sim \mathrm{N}(\mu, \sigma^2)$ with density

$$f_X(x) = \frac{1}{\sqrt{2\pi\sigma^2}} \exp\left\{-\frac{1}{2}\frac{(x-\mu)^2}{\sigma^2}\right\}.$$

We first view $f_X(x) = f_X(x;\mu)$ as a function of the mean μ. The density of $Y = X + c$ now is

$$\begin{aligned} f_Y(y) &= \frac{1}{\sqrt{2\pi\sigma^2}} \exp\left\{-\frac{1}{2}\frac{\{(y-c)-\mu\}^2}{\sigma^2}\right\} \\ &= \frac{1}{\sqrt{2\pi\sigma^2}} \exp\left\{-\frac{1}{2}\frac{(y-(\mu+c))^2}{\sigma^2}\right\} = f_X(y;\mu+c), \end{aligned}$$

using the change of variables formula (Appendix A.2.3), so μ can be identified as a location parameter.

Suppose now that $\mu = 0$ and consider $f_X(x) = f(x;\sigma)$ as a function of the standard deviation σ. The density of $Y = cX$ now is

$$\begin{aligned} f_Y(y) &= \frac{1}{c}\frac{1}{\sqrt{2\pi\sigma^2}} \exp\left\{-\frac{1}{2}\frac{(y/c)^2}{\sigma^2}\right\} \\ &= \frac{1}{\sqrt{2\pi(c\sigma)^2}} \exp\left\{-\frac{1}{2}\frac{y^2}{(c\sigma)^2}\right\} = f_X(y;c\sigma), \end{aligned}$$

so σ can be identified as a scale parameter. ■

The expected Fisher information of location or scale parameters has specific properties:

Result 4.4 (Expected Fisher information of location and scale parameters)

1. *The expected Fisher information $J(\theta)$ of a location parameter θ does not depend on θ, i.e. is constant.*
2. *The expected Fisher information $J(\theta)$ of a scale parameter θ has the form*

$$J(\theta) \propto \theta^{-2}.$$

Example 4.7 (Normal model) Consider a random sample $X_{1:n}$ from a normal distribution $\mathrm{N}(\mu, \sigma^2)$. The expected Fisher information of the location parameter μ is $J(\mu) = n/\sigma^2$ and indeed independent of μ (cf. Example 4.3). In Example 4.5 we have also shown that $J(\sigma) = 2n/\sigma^2$, which is in line with Result 4.4, which states that the expected Fisher information of the scale parameter σ must be proportional to σ^{-2}. ■

4.1.3 The Score Statistic

Result 4.1 has established formulas for the expectation and variance of the score function. We can also make an asymptotic statement about the whole distribution of the score function for a random sample:

Result 4.5 (Score statistic) *Consider a random sample $X_{1:n}$ from a distribution with probability mass or density function $f(x;\theta)$. Under the Fisher regularity conditions, the following holds*:

$$\frac{S(\theta; X_{1:n})}{\sqrt{J_{1:n}(\theta)}} \overset{\mathrm{a}}{\sim} \mathrm{N}(0, 1). \tag{4.2}$$

This result identifies the *score statistic* (4.2) as an approximate pivot for θ, with the standard normal pivotal distribution, cf. Definition 3.7. We note that the symbol $\overset{\mathrm{a}}{\sim}$ is to be understood as convergence in distribution as $n \to \infty$, cf. Appendix A.4.1. The alternative formulation

$$S(\theta; X_{1:n}) \overset{\mathrm{a}}{\sim} \mathrm{N}\bigl(0, J_{1:n}(\theta)\bigr)$$

is mathematically less precise but makes it more explicit that the asymptotic variance of the score function $S(\theta; X_{1:n})$ is equal to the expected Fisher information $J_{1:n}(\theta)$, a result which is commonly used in practice. However, Eq. (4.2) with a limit distribution not depending on n is more rigid.

Proof To prove (4.2), consider

$$Y_i = \frac{d}{d\theta} \log\bigl\{f(X_i; \theta)\bigr\}$$

so $S(\theta; X_{1:n}) = \sum_{i=1}^{n} Y_i$. We know from Result 4.1 that $\mathsf{E}(Y_i) = 0$ and $\mathrm{Var}(Y_i) = J(\theta)$, where $J(\theta)$ denotes the expected unit Fisher information. The expected Fisher information $J_{1:n}(\theta)$ with respect to the random sample $X_{1:n}$ is therefore $n \cdot J(\theta)$, see Result 4.2. We can now apply the central limit theorem (cf. Appendix A.4.4) and obtain

$$\frac{1}{\sqrt{n}} S(\theta; X_{1:n}) = \frac{1}{\sqrt{n}} \sum_{i=1}^{n} Y_i \xrightarrow{D} \mathrm{N}\bigl(0, J(\theta)\bigr), \quad \text{so} \quad \frac{S(\theta; X_{1:n})}{\sqrt{J_{1:n}(\theta)}} \overset{\mathrm{a}}{\sim} \mathrm{N}(0, 1).$$ □

The next result shows that we can replace the expected Fisher information $J_{1:n}(\theta)$ with the ordinary Fisher information $I(\theta, X_{1:n})$ in Eq. (4.2).

Result 4.6 *We can replace in Result 4.5 the expected Fisher information $J_{1:n}(\theta)$ by the ordinary Fisher information $I(\theta; X_{1:n})$, i.e.*

$$\frac{S(\theta; X_{1:n})}{\sqrt{I(\theta; X_{1:n})}} \overset{\text{a}}{\sim} \text{N}(0, 1). \tag{4.3}$$

Proof Due to

$$I(\theta; X_i) = -\frac{d^2}{d\theta^2} \log\{f(X_i; \theta)\} \quad \text{and}$$
$$\mathsf{E}\{I(\theta; X_i)\} = J(\theta),$$

the law of large numbers (cf. Appendix A.4.3) gives

$$\frac{1}{n} I(\theta; X_{1:n}) = \frac{1}{n} \sum_{i=1}^{n} I(\theta; X_i) \xrightarrow{P} J(\theta), \quad \text{and therefore}$$
$$\frac{I(\theta; X_{1:n})}{n J(\theta)} = \frac{I(\theta; X_{1:n})}{J_{1:n}(\theta)} \xrightarrow{P} 1.$$

The continuous mapping theorem (cf. Appendix A.4.2) ensures that also the inverse square root converges to 1, i.e.

$$\frac{\sqrt{J_{1:n}(\theta)}}{\sqrt{I(\theta; X_{1:n})}} \xrightarrow{P} 1.$$

By Slutsky's theorem (cf. Appendix A.4.2) and Eq. (4.2) we finally obtain

$$\frac{\sqrt{J_{1:n}(\theta)}}{\sqrt{I(\theta; X_{1:n})}} \cdot \frac{S(\theta; X_{1:n})}{\sqrt{J_{1:n}(\theta)}} \xrightarrow{D} 1 \cdot \text{N}(0, 1),$$

i.e. we have established (4.3). □

There are two further variants of the score statistic. We can replace the true parameter value θ in the denominator $\sqrt{J_{1:n}(\theta)}$ of Eq. (4.2) with $\hat{\theta}_{\text{ML}} = \hat{\theta}_{\text{ML}}(X_{1:n})$. Likewise, we can replace θ in the denominator $\sqrt{I_{1:n}(\theta; X_{1:n})}$ of Eq. (4.3) with $\hat{\theta}_{\text{ML}} = \hat{\theta}_{\text{ML}}(X_{1:n})$. This will be justified later on page 98. In applications this requires the calculation of the MLE $\hat{\theta}_{\text{ML}}$.

The score statistic in Eq. (4.2) forms the basis of the corresponding significance test, which we describe now.

4.1.4 The Score Test

The score statistic can be used to construct significance tests and confidence intervals. The significance test based on the score statistic is called *score test*. In the following we still assume that the data has arisen through an appropriate random sample of sufficiently large sample size n.

Suppose we are interested in the null hypothesis $H_0: \theta = \theta_0$ with two-sided alternative $H_1: \theta \neq \theta_0$. Both $S(\theta_0; x_{1:n})/\sqrt{J(\theta_0)}$ and $S(\theta_0; x_{1:n})/\sqrt{I(\theta_0; x_{1:n})}$ can now be used to calculate a P-value, as illustrated in the following example.

Example 4.8 (Scottish lip cancer) Consider Sect. 1.1.6, where we have observed (x_i) and expected (e_i) counts of lip cancer in $i = 1, \dots, 56$ regions of Scotland. We assume that x_i is a realisation from a Poisson distribution $\text{Po}(e_i \lambda_i)$ with known offset $e_i > 0$ and unknown rate parameter λ_i.

In contrast to Example 2.4, we assume here that the rate parameters λ_i differ between regions. Note that calculation of the expected counts e_i has been done under the constraint that the sum of the expected equals the sum of the observed counts in Scotland. The value $\lambda_0 = 1$ is hence a natural reference value for the rate parameters because it corresponds to the overall relative risk $\sum x_i / \sum e_i$.

Consider a specific region i and omit the subscript i for ease of notation. Note that the number of observed lip cancer cases x in a specific region can be viewed as the sum of a very large number of binary observations (lip cancer yes/no) based on the whole population in that region. So although we just have one Poisson observation, the usual asymptotics still apply, if the population size is large enough.

We want to test the null hypothesis $H_0: \lambda = \lambda_0$. It is easy to show that

$$
\begin{aligned}
S(\lambda; x) &= \frac{x}{\lambda} - e, \\
I(\lambda; x) &= x/\lambda^2, \\
\hat{\lambda}_{\text{ML}} &= x/e \quad \text{and} \\
J(\lambda) &= e/\lambda.
\end{aligned}
$$

The observed test statistic of the score test using the expected Fisher information in (4.2) is therefore

$$
T_1(x) = \frac{S(\lambda_0; x)}{\sqrt{J(\lambda_0)}} = \frac{x - e\lambda_0}{\sqrt{e\lambda_0}}.
$$

Using the ordinary Fisher information instead, we obtain the observed test statistic from (4.3)

$$
T_2(x) = \frac{S(\lambda_0; x)}{\sqrt{I(\lambda_0; x)}} = \frac{x - e\lambda_0}{\sqrt{x}}.
$$

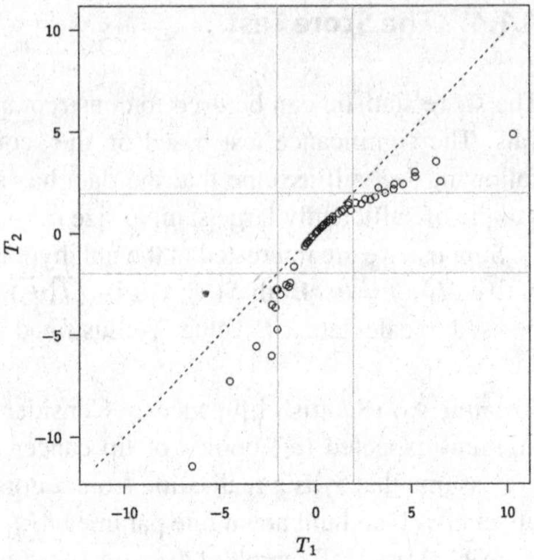

Fig. 4.1 Scatter plot of the values of the different score test statistics T_1 and T_2 for a unity relative risk of lip cancer in the 56 regions of Scotland. The *grey lines* mark the 2.5 % and 97.5 % quantiles of the standard normal distribution, the *diagonal line* corresponds to the equality $T_1 = T_2$

Note that

$$\frac{T_1(x)}{T_2(x)} = \sqrt{\frac{x}{e\lambda_0}}$$

so if $x > e\lambda_0$, both T_1 and T_2 are positive, and T_1 will have a larger value than T_2. Conversely, if $x < e\lambda_0$, both T_1 and T_2 are negative, and T_2 will have a larger absolute value than T_1. Therefore, $T_1 \geq T_2$ always holds.

We now consider all 56 regions separately in order to test the null hypothesis $H_0 : \lambda_i = \lambda_0 = 1$ for each region i. Figure 4.1 plots the values of the two test statistics T_1 and T_2 against each other.

Note that the test statistic T_2 is infinite for the two observations with $x_i = 0$. The alternative test statistic T_1 gives here still sensible values, namely -1.33 and -2.04. Of course, it has to be noted that the assumption of an approximate normal distribution is questionable for small x_i. This can be also seen from the discrepancies between T_1 and T_2, two test statistics which should be asymptotically equivalent.

It is interesting that 24 of the 56 regions have absolute values of T_1 larger than the critical value 1.96. This corresponds to 43 %, considerably more than the 5 % to be expected under the null hypothesis. This can also be seen in a histogram of the corresponding two-sided P-values, shown in Fig. 4.2a. We observe far more small P-values (< 0.1) than we would expect under the null hypothesis, where the P-values are (asymptotically) uniformly distributed.

This suggests that there is heterogeneity in relative lip cancer risk between the different regions despite the somewhat questionable asymptotic regime. Very similar results can be obtained with the test statistic T_2, excluding the two regions with zero observed counts, cf. Fig. 4.2b. ■

A remarkable feature of the score test is that it does not require calculation of the MLE $\hat{\theta}_{ML}$. This can make the application of the score test simpler than alternative methods, which often require knowledge of the MLE.

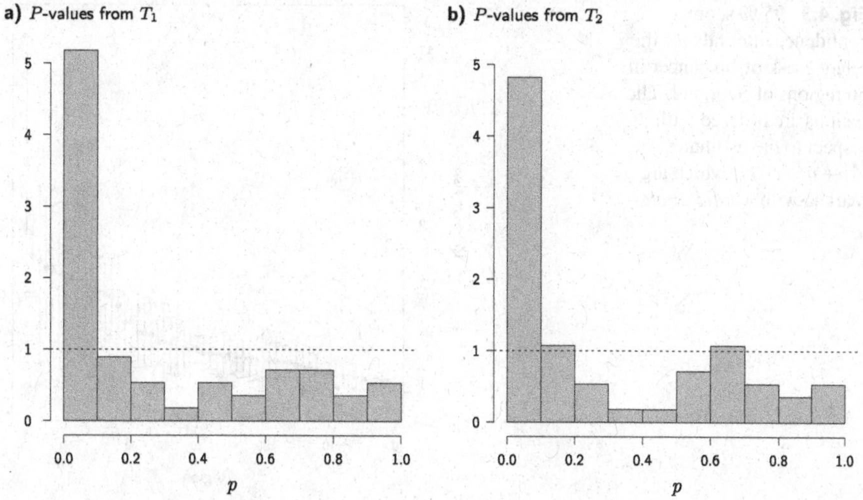

Fig. 4.2 Histograms of the P-values based on the two score test statistics T_1 and T_2 for a unity relative risk of lip cancer in the 56 regions of Scotland

4.1.5 Score Confidence Intervals

The approximate pivots (4.2) and (4.3) can also be used to compute approximate *score confidence intervals* for θ. Consider the score test statistic (4.2); then the duality of hypothesis tests and confidence intervals implies that all values θ_0 fulfilling

$$z_{(1-\gamma)/2} \leq \frac{S(\theta_0; x_{1:n})}{\sqrt{J_{1:n}(\theta_0)}} \leq z_{(1+\gamma)/2}$$

form a $\gamma \cdot 100\ \%$ score confidence interval for θ. Due to $z_{(1-\gamma)/2} = -z_{(1+\gamma)/2}$, this condition can also be written as

$$\frac{S(\theta_0; x_{1:n})^2}{J_{1:n}(\theta_0)} \leq z^2_{(1+\gamma)/2}.$$

The same holds for the score test statistic (4.3), which uses the observed Fisher information. In summary we have the two approximate $\gamma \cdot 100\ \%$ score confidence intervals

$$\left\{\theta : \frac{S(\theta; x_{1:n})^2}{J_{1:n}(\theta)} \leq q^2\right\} \quad \text{and} \quad \left\{\theta : \frac{S(\theta; x_{1:n})^2}{I(\theta; x_{1:n})} \leq q^2\right\}, \tag{4.4}$$

where $q = z_{(1+\gamma)/2}$ denotes the $(1+\gamma)/2$ quantile of the standard normal distribution. Computation is in general not straightforward, but explicit formulas can be derived in some special cases, as illustrated by the following examples.

Example 4.9 (Scottish lip cancer) We now want to compute a score confidence interval for the relative risk λ in one specific region.

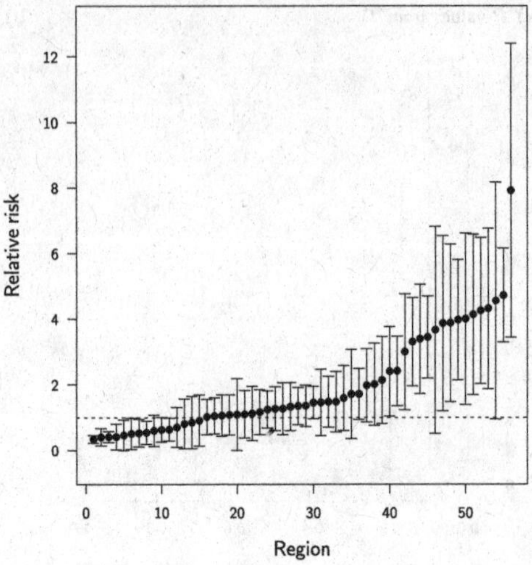

Fig. 4.3 95 % score confidence intervals for the relative risk of lip cancer in 56 regions of Scotland. The regions are ordered with respect to the estimates $(x_i + q^2/2)/e_i$, which are marked with a *solid circle*

With $S(\lambda; x)$ and $J(\lambda)$ from Example 4.8, we have

$$\frac{\{S(\lambda; x)\}^2}{J(\lambda)} = \frac{(x - \lambda e)^2}{\lambda e} \leq q^2.$$

We obtain the limits of this confidence interval by solving the quadratic equation

$$e^2\lambda^2 - (2x + q^2)e\lambda + x^2 = 0$$

for λ. The two solutions are

$$\lambda_{1/2} = \frac{x + q^2/2}{e} \pm \frac{q}{2e}\sqrt{4x + q^2}.$$

Note that the interval is boundary-respecting since the lower limit is always non-negative and equal to zero for $x = 0$. The interval limits are symmetric around $(x + q^2/2)/e$, but not around the MLE $\hat{\lambda}_{\mathrm{ML}} = x/e$. Nevertheless, it is easy to show that the MLE is always inside the confidence interval for any confidence level γ. For the data on lip cancer in Scotland, we obtain the 95 % score confidence intervals for the relative risk in each region, as displayed in Fig. 4.3. ■

Another famous example of a confidence interval based on the score statistic is the Wilson confidence interval for a binomial success probability π.

Example 4.10 (Wilson confidence interval) Edwin B. Wilson (1879–1964) suggested in 1927 an approximate $\gamma \cdot 100$ % confidence interval with limits

$$\frac{x + q^2/2}{n + q^2} \pm \frac{q\sqrt{n}}{n + q^2}\sqrt{\hat{\pi}_{\mathrm{ML}}(1 - \hat{\pi}_{\mathrm{ML}}) + \frac{q^2}{4n}} \tag{4.5}$$

for the success probability π in the binomial model $X \sim \text{Bin}(n, \pi)$; here $\hat{\pi}_{\text{ML}} = x/n$. We will now show that this confidence interval is a score confidence interval based on (4.4) using the expected Fisher information. First note that

$$S(\pi; x) = \frac{x}{\pi} - \frac{n-x}{1-\pi} = \frac{x - n\pi}{\pi(1-\pi)} \tag{4.6}$$

is the score function in the binomial model and

$$J(\pi) = \frac{n}{\pi(1-\pi)}$$

is the expected Fisher information. Now

$$\left\{\pi : \frac{|S(\pi; x)|}{\sqrt{J(\pi)}} \le q\right\} = \left\{\pi : \frac{S(\pi; x)^2}{J(\pi)} \le q^2\right\},$$

so the limits of this approximate $\gamma \cdot 100\ \%$ confidence interval for π are the solutions of the equation

$$\frac{\pi(1-\pi)}{n}\left\{\frac{x - n\pi}{\pi(1-\pi)}\right\}^2 = \frac{(x - n\pi)^2}{n\pi(1-\pi)} = q^2,$$

which is equivalent to solving the quadratic equation

$$\pi^2(n^2 + nq^2) + \pi(-2nx - nq^2) + x^2 = 0$$

for π. The two solutions are

$$\begin{aligned}
\pi_{1/2} &= \frac{2nx + nq^2 \pm \sqrt{n^2(2x + q^2)^2 - 4(n^2 + nq^2)x^2}}{2(n^2 + nq^2)} \\
&= \frac{q^2 + 2x}{2(n + q^2)} \pm \frac{\sqrt{(q^2 + 2x)^2 - 4(1 + q^2/n)x^2}}{2(n + q^2)} \\
&= \frac{x + q^2/2}{n + q^2} \pm \frac{q\sqrt{n}}{n + q^2}\sqrt{\frac{(q^2/2 + x)^2 - (1 + q^2/n)x^2}{q^2 n}}.
\end{aligned}$$

Rewriting the term under the square root as

$$\begin{aligned}
\frac{(q^2/2 + x)^2 - (1 + q^2/n)x^2}{q^2 n} &= \frac{q^4/4 + q^2 x + x^2 - x^2 - (q^2 x^2)/n}{q^2 n} \\
&= \frac{x - x^2/n}{n} + \frac{q^2}{4n}
\end{aligned}$$

$$= \frac{x}{n}\left(1 - \frac{x}{n}\right) + \frac{q^2}{4n}$$

$$= \hat{\pi}_{\mathrm{ML}}(1 - \hat{\pi}_{\mathrm{ML}}) + \frac{q^2}{4n},$$

we finally obtain Eq. (4.5). ■

An interesting property of the Wilson confidence interval is that the limits are always within the unit interval, i.e. the Wilson interval is boundary-respecting. For example, if there are $x = 0$ successes in n trials, the Wilson confidence interval has limits 0 and $q^2/(q^2 + n)$; for $x = n$ the limits are $n/(q^2 + n)$ and 1. The Wald interval does not have this property; see Example 4.22 for a thorough comparison of different confidence intervals for proportions.

The underlying reason of this property of the Wilson confidence interval is that the score confidence intervals based on the expected Fisher information $J(\theta)$ are invariant with respect to one-to-one transformations of the parameter θ. For example, suppose we would parametrise the binomial likelihood in terms of the log odds $\phi = \operatorname{logit}(\pi)$ instead of the success probability π. The limits of the score confidence interval for ϕ are then simply the logit-transformed limits of the original Wilson confidence interval for π. This is also true in general.

Result 4.7 (Invariance of score confidence intervals) *Let $\phi = h(\theta)$ denote a one-to-one transformation of θ and use subscripts to differentiate between the score function and expected Fisher information of the old (θ) and new parameter (ϕ). Then we have*

$$\left\{h(\theta) : \frac{S_\theta(\theta; x)^2}{J_\theta(\theta)} \le q^2\right\} = \left\{\phi : \frac{S_\phi(\phi; x)^2}{J_\phi(\phi)} \le q^2\right\}.$$

Proof This property follows immediately from Result 4.3 and Eq. (2.4):

$$\frac{S_\phi(\phi; x)^2}{J_\phi(\phi)} = \frac{\left\{S_\theta(\theta; x)\frac{dh^{-1}(\phi)}{d\phi}\right\}^2}{J_\theta(\theta)\left\{\frac{dh^{-1}(\phi)}{d\phi}\right\}^2} = \frac{S_\theta(\theta; x)^2}{J_\theta(\theta)}.$$ □

We can therefore transform the limits of a score interval for θ and do not need to re-calculate the score function of ϕ. This is an important and attractive property of the score confidence interval based on the expected Fisher information. As we will see in the next section, the more commonly used Wald confidence interval does not have this property.

4.2 The Distribution of the ML Estimator and the Wald Statistic

In this section we will first discuss some properties of the ML estimator. We then derive its asymptotic distribution, which leads to the Wald statistic, the most commonly used statistic for significance tests and confidence intervals.

4.2.1 Cramér–Rao Lower Bound

The variance of an unbiased estimator is commonly calculated to assess its precision: The smaller the variance the better it is. The Cramér–Rao lower bound forms a universal lower bound for the variance of all unbiased estimators. If an estimator has variance equal to the Cramér–Rao lower bound, it is *optimal* because there is no other unbiased estimator with smaller variance.

Result 4.8 *Let $T = h(X)$ denote an unbiased estimator of $g(\theta)$ based on some data X from a distribution with probability mass or density function $f(x;\theta)$, i.e.*

$$\mathsf{E}(T) = g(\theta) \quad \text{for all } \theta \in \Theta.$$

Let $J(\theta)$ denote the expected Fisher information of θ with respect to X. Under the Fisher regularity conditions, we then have the following property:

$$\operatorname{Var}(T) \geq \frac{g'(\theta)^2}{J(\theta)}. \tag{4.7}$$

In particular, if $g(\theta) = \theta$, we have

$$\operatorname{Var}(T) \geq \frac{1}{J(\theta)}. \tag{4.8}$$

The right-hand sides of (4.7) *and* (4.8) *are called* Cramér–Rao lower bounds.

Proof To prove (4.7), consider two arbitrary random variables S and T. Then the squared correlation between S and T is

$$\rho(S,T)^2 = \frac{\operatorname{Cov}(S,T)^2}{\operatorname{Var}(S) \cdot \operatorname{Var}(T)} \leq 1,$$

compare Appendix A.3.6. Suppose now that $S = S(\theta; X)$, i.e. S is the score function. We know that $\operatorname{Var}(S) = J(\theta)$, so we have

$$\operatorname{Var}(T) \geq \frac{\operatorname{Cov}(S,T)^2}{J(\theta)}.$$

It remains to show that $\operatorname{Cov}(S,T) = g'(\theta)$. Due to $\mathsf{E}(S) = 0$, we have

$$\begin{aligned}
\operatorname{Cov}(S,T) &= \mathsf{E}(S \cdot T) - \underbrace{\mathsf{E}(S)}_{=0}\mathsf{E}(T) \\
&= \int S(\theta; x) T(x) f(x;\theta)\, dx \\
&= \int \frac{\frac{d}{d\theta} f(x;\theta)}{f(x;\theta)} T(x) f(x;\theta)\, dx
\end{aligned}$$

$$= \frac{d}{d\theta} \int T(x) f(x;\theta)\, dx$$

$$= \frac{d}{d\theta} \mathsf{E}\{T(X)\} = \frac{d}{d\theta} g(\theta)$$

$$= g'(\theta).$$

For discrete X the result is still valid with integration replaced by summation. □

Example 4.11 (Poisson model) Suppose X follows a Poisson distribution $\text{Po}(e\lambda)$ with known offset $e > 0$ and unknown parameter λ. We know that $J(\lambda) = e/\lambda$. The ML estimator $\hat{\lambda}_{\text{ML}} = X/e$ is unbiased for λ and has variance $\text{Var}(\hat{\lambda}_{\text{ML}}) = \text{Var}(X)/e^2 = \lambda/e = 1/J(\lambda)$. Therefore, $\hat{\lambda}_{\text{ML}}$ attains the Cramér–Rao lower bound and is optimal, i.e. it has minimal variance under among all unbiased estimators. ■

The ML estimator in the last example is optimal. However, in general, optimal estimators are extremely rare. We will soon establish a more general but also slightly weaker result, which states that the ML estimator is in general *asymptotically optimal* because it is asymptotically unbiased with asymptotic variance equal to the inverse expected Fisher information. Estimators with this property are also called *efficient*.

Efficient estimator
An estimator that is asymptotically unbiased and asymptotically attains the Cramér–Rao lower bound is called efficient.

4.2.2 Consistency of the ML Estimator

We will now consider the ML estimator and show that it is consistent if the Fisher regularity conditions hold. In particular, we need to assume that the support of the statistical model $f(x;\theta)$ does not depend on θ, that $L(\theta)$ is continuous in θ and that θ is (as always in this section) a scalar.

Result 4.9 (Consistency of the ML estimator) *Let $X_{1:n}$ denote a random sample from $f(x;\theta_0)$ where θ_0 denotes the true parameter value. For $n \to \infty$, there is a consistent sequence of MLEs $\hat{\theta}_n$ (defined as the local maximum of the likelihood function).*

Proof We have to show that for any $\varepsilon > 0$ (for $n \to \infty$), there is a (possibly local) maximum $\hat{\theta}$ in the interval $(\theta_0 - \varepsilon, \theta_0 + \varepsilon)$. Now $L(\theta)$ is assumed to be continuous, so it is sufficient to show that the probability of

$$L(\theta_0) > L(\theta_0 - \varepsilon) \tag{4.9}$$

and

$$L(\theta_0) > L(\theta_0 + \varepsilon) \tag{4.10}$$

goes to 1 as $n \to \infty$. First, consider (4.9), which can be rewritten as

$$\frac{1}{n}\log\left\{\frac{L(\theta_0)}{L(\theta_0 - \varepsilon)}\right\} > 0.$$

The law of large numbers now ensures that

$$\frac{1}{n}\log\left\{\frac{L(\theta_0)}{L(\theta_0 - \varepsilon)}\right\} = \frac{1}{n}\sum_{i=1}^{n}\log\left\{\frac{f(x_i;\theta_0)}{f(x_i;\theta_0 - \varepsilon)}\right\} \xrightarrow{P} \mathsf{E}\left[\log\left\{\frac{f(X;\theta_0)}{f(X;\theta_0 - \varepsilon)}\right\}\right],$$

where X has density $f(x;\theta_0)$. Application of the information inequality (cf. Appendix A.3.8) gives with assumption (4.1):

$$\mathsf{E}\left[\log\left\{\frac{f(X;\theta_0)}{f(X;\theta_0 - \varepsilon)}\right\}\right] > 0.$$

For Eq. (4.10), we can argue similarly. □

Note that in this proof the MLE is defined as a *local* maximiser of the likelihood, so the uniqueness of the MLE is not shown. A proof with the MLE defined as the *global* maximiser requires additional assumptions, and the proof becomes more involved.

4.2.3 The Distribution of the ML Estimator

The following result, which establishes the asymptotic normality of the MLE, is one of the most important results of likelihood theory.

Result 4.10 *Let $X_{1:n}$ denote a random sample from $f(x;\theta_0)$ and suppose that $\hat{\theta}_{\mathrm{ML}} = \hat{\theta}_{\mathrm{ML}}(X)$ is consistent for θ_0. Assuming that the Fisher regularity conditions hold, we then have:*

$$\sqrt{n}(\hat{\theta}_{\mathrm{ML}} - \theta_0) \xrightarrow{D} \mathrm{N}\big(0, J(\theta_0)^{-1}\big),$$

where $J(\theta_0)$ denotes the expected unit Fisher information of one observation X_i from $f(x;\theta_0)$. The expected Fisher information of the full random sample is therefore $J_{1:n}(\theta_0) = n \cdot J(\theta_0)$, and we have

$$\sqrt{J_{1:n}(\theta_0)}(\hat{\theta}_{\mathrm{ML}} - \theta_0) \overset{a}{\sim} \mathrm{N}(0, 1). \tag{4.11}$$

This result identifies (4.11) as an approximate pivot for θ_0. Informally, we may say that

$$\hat{\theta}_{\mathrm{ML}} \overset{a}{\sim} \mathrm{N}\big(\theta_0, J_{1:n}(\theta_0)^{-1}\big).$$

As a by-product, the result establishes two further properties of MLEs:

Properties of the ML estimator

1. The ML estimator is asymptotically unbiased.
2. The variance of the ML estimator asymptotically attains the Cramér–Rao lower bound, so the ML estimator is efficient.

Proof To show Result 4.10, consider a first-order Taylor expansion of the score function around θ_0,

$$S(\theta; x_{1:n}) \approx S(\theta_0; x_{1:n}) - I(\theta_0; x_{1:n})(\theta - \theta_0).$$

For $\theta = \hat{\theta}_{\mathrm{ML}}$ with $S(\hat{\theta}_{\mathrm{ML}}; x_{1:n}) = 0$, we then have:

$$I(\theta_0; x_{1:n})(\hat{\theta}_{\mathrm{ML}} - \theta_0) \approx S(\theta_0; x_{1:n}) \quad \text{respectively}$$

$$\sqrt{n}(\hat{\theta}_{\mathrm{ML}} - \theta_0) \approx \sqrt{n} \cdot \frac{S(\theta_0; x_{1:n})}{I(\theta_0; x_{1:n})} = \left\{\frac{I(\theta_0; x_{1:n})}{n}\right\}^{-1} \cdot \frac{S(\theta_0; x_{1:n})}{\sqrt{n}}.$$

Result 4.5 now ensures that

$$\frac{S(\theta_0; X_{1:n})}{\sqrt{n}} \xrightarrow{D} \mathrm{N}\big(0, J(\theta_0)\big),$$

and we also have (using the proof of Result 4.6) that

$$\frac{I(\theta_0; X_{1:n})}{n} \xrightarrow{P} J(\theta_0).$$

The continuous mapping theorem (cf. Appendix A.4.2) gives

$$\left\{\frac{I(\theta_0; X_{1:n})}{n}\right\}^{-1} \xrightarrow{P} J(\theta_0)^{-1},$$

and application of Slutsky's theorem (cf. Appendix A.4.2) finally gives

$$\sqrt{n}(\hat{\theta}_{\mathrm{ML}} - \theta_0) \xrightarrow{D} J(\theta_0)^{-1} \cdot \mathrm{N}\big(0, J(\theta_0)\big) = \mathrm{N}\big(0, J(\theta_0)^{-1}\big). \qquad \square$$

We can now replace the expected Fisher information $J_{1:n}(\theta_0)$ in (4.11) with the ordinary Fisher information $I(\theta_0; X_{1:n})$, just as for the score statistic. Similarly, we can evaluate both the expected and the ordinary Fisher information not at the true parameter value θ_0, but at the MLE $\hat{\theta}_{\mathrm{ML}}$. In total, we have the following three variants of Result 4.10:

$$\hat{\theta}_{ML} \overset{a}{\sim} N\big(\theta_0, J_{1:n}(\hat{\theta}_{ML})^{-1}\big),$$
$$\hat{\theta}_{ML} \overset{a}{\sim} N\big(\theta_0, I(\theta_0; X_{1:n})^{-1}\big),$$
$$\hat{\theta}_{ML} \overset{a}{\sim} N\big(\theta_0, I(\hat{\theta}_{ML}; X_{1:n})^{-1}\big).$$

This illustrates that we can use both $1/\sqrt{I(\hat{\theta}_{ML}; x_{1:n})}$ and $1/\sqrt{J_{1:n}(\hat{\theta}_{ML})}$ as a standard error of the MLE $\hat{\theta}_{ML}$. This is an important result, as it justifies the usage of a standard error based on $I(\hat{\theta}_{ML}; x_{1:n})$, i.e. the negative curvature of the log-likelihood, evaluated at the MLE $\hat{\theta}_{ML}$.

Standard error of the ML estimator

The reciprocal square root of the observed Fisher information is a standard error of $\hat{\theta}_{ML}$:

$$\mathrm{se}(\hat{\theta}_{ML}) = 1/\sqrt{I(\hat{\theta}_{ML}; x_{1:n})}.$$

4.2.4 The Wald Statistic

To test the null hypothesis $H_0 : \theta = \theta_0$, we can use one of the two test statistics

$$\sqrt{I(\hat{\theta}_{ML})}(\hat{\theta}_{ML} - \theta_0) \overset{a}{\sim} N(0, 1) \tag{4.12}$$

and

$$\sqrt{J(\hat{\theta}_{ML})}(\hat{\theta}_{ML} - \theta_0) \overset{a}{\sim} N(0, 1), \tag{4.13}$$

which are both asymptotically standard normally distributed under the null hypothesis H_0. These two statistics are therefore approximate pivots for θ and are usually denoted as *Wald statistic*. The corresponding statistical test is called *Wald test* and has been developed by Abraham Wald (1902–1950) in 1939.

Example 4.12 (Scottish lip cancer) Consider again a specific region of Scotland, where we want to test whether the lip cancer risk is higher than on average. To test the corresponding null hypothesis $H_0 : \lambda = \lambda_0$ with the Wald test, we know from Example 4.8 that

$$I(\lambda) = x/\lambda^2, \qquad \hat{\lambda}_{ML} = x/e \quad \text{and} \quad J(\lambda) = e/\lambda.$$

It then follows that

$$J(\hat{\lambda}_{\mathrm{ML}}) = I(\hat{\lambda}_{\mathrm{ML}}) = e^2/x$$

and the observed Wald test statistic (4.12) is

$$T_3(x) = \sqrt{I(\hat{\lambda}_{\mathrm{ML}})}(\hat{\lambda}_{\mathrm{ML}} - \lambda_0) = \sqrt{e^2/x}(\hat{\lambda}_{\mathrm{ML}} - \lambda_0) = \frac{x - e\lambda_0}{\sqrt{x}}.$$

We see that the Wald test statistic T_3 is in this example equal to the score test statistic using the ordinary Fisher information T_2. ■

Using the duality of hypothesis tests and confidence intervals, *Wald confidence intervals* can be computed by inverting the Wald test based on the test statistics (4.12) or (4.13), respectively. Using (4.12), all values θ_0 that fulfil

$$z_{\frac{1-\gamma}{2}} \leq \sqrt{I(\hat{\theta}_{\mathrm{ML}})}(\hat{\theta}_{\mathrm{ML}} - \theta_0) \leq z_{\frac{1+\gamma}{2}}$$

would not lead to a rejection of the null hypothesis $H_0 : \theta = \theta_0$ when the significance level is $1 - \gamma$. These values form the $\gamma \cdot 100\,\%$ Wald confidence interval for θ with limits

$$\hat{\theta}_{\mathrm{ML}} \pm z_{\frac{1+\gamma}{2}} / \sqrt{I(\hat{\theta}_{\mathrm{ML}})}.$$

Because $1/\sqrt{I(\hat{\theta}_{\mathrm{ML}})}$ is commonly referred to as standard error of the MLE, we obtain the limits

$$\hat{\theta}_{\mathrm{ML}} \pm z_{\frac{1+\gamma}{2}} \operatorname{se}(\hat{\theta}_{\mathrm{ML}}),$$

of a $\gamma \cdot 100\,\%$ Wald confidence interval, already introduced in Result 3.1.

Example 4.13 (Hardy–Weinberg equilibrium) The likelihood function of υ is of binomial form with success probability υ, $2x_1 + x_2$ successes and $x_2 + 2x_3$ failures among $2n$ trials, where n is the sample size (cf. Example 2.7). The MLE of υ is therefore

$$\hat{\upsilon}_{\mathrm{ML}} = \frac{2x_1 + x_2}{2x_1 + 2x_2 + 2x_3} = \frac{2x_1 + x_2}{2n} = \frac{x_1 + \frac{1}{2}x_2}{n}.$$

Also by arguing with the binomial form of the likelihood, we obtain the standard error

$$\operatorname{se}(\hat{\upsilon}_{\mathrm{ML}}) = \sqrt{\hat{\upsilon}_{\mathrm{ML}}(1 - \hat{\upsilon}_{\mathrm{ML}})/(2n)}$$

of the MLE.

For the data from Example 2.7, we have $n = 747$ and obtain $\hat{\upsilon}_{\mathrm{ML}} = 0.570$ and

$$\operatorname{se}(\hat{\upsilon}_{\mathrm{ML}}) \approx 0.013,$$

so we can easily compute the 95 % Wald confidence interval [0.545, 0.595]. ■

As discussed in Sect. 3.2.4, Wald confidence intervals are not invariant to nonlinear transformations, and the choice of transformation may be guided by the requirement that the confidence interval respects the boundaries of the parameter space. We will now describe another approach for determining a suitable parameter transformation.

4.3 Variance-Stabilising Transformations

An alternative approach to constructing Wald confidence intervals is to search for a transformation $\phi = h(\theta)$ of the parameter θ such that the expected Fisher information $J_\phi(\phi)$ does not depend on ϕ. Such a transformation is called *variance-stabilising* since Result 4.10 ensures that the asymptotic variance of $\hat{\phi}_{\text{ML}}$ does not depend on ϕ. If the expected Fisher information $J_\theta(\theta)$ is already independent of θ, then no additional transformation of θ is necessary. For example, this is always the case if θ is a location parameter, cf. Result 4.4. If $J_\theta(\theta)$ is not independent of θ, the following result shows how to find the variance-stabilising transformation.

Result 4.11 (Calculation of the variance-stabilising transformation) *Let $X_{1:n}$ denote a random sample from a distribution with probability mass or density function $f(x;\theta)$ and associated expected Fisher information $J_\theta(\theta)$. The transformation*

$$\phi = h(\theta) \propto \int^\theta J_\theta(u)^{\frac{1}{2}}\, du \tag{4.14}$$

is then a variance-stabilising transformation of θ with $J_\phi(\phi)$ not depending on ϕ. Here $\int^\theta g(u)\, du$ denotes the value of the anti-derivative of a function g, evaluated at θ.

Proof To prove Result 4.11, note that

$$\frac{dh(\theta)}{d\theta} \propto J_\theta(\theta)^{\frac{1}{2}},$$

so with Result 4.3 we have

$$J_\phi(\phi) = J_\theta(\theta)\left\{\frac{dh(\theta)}{d\theta}\right\}^{-2} \propto J_\theta(\theta) J_\theta(\theta)^{-1} = 1. \qquad \square$$

Example 4.14 (Scottish lip cancer) In Example 4.9 we computed a score confidence interval for the relative lip cancer risk λ in a specific region of Scotland, cf. Sect. 1.1.6. The used data is the realisation x from a Poisson distribution $\text{Po}(e\lambda)$ with known offset e. The expected Fisher information of λ is $J_\lambda(\lambda) = e/\lambda \propto \lambda^{-1}$, cf. Example 4.8. Therefore,

$$\int^\lambda J_\lambda(u)^{1/2}\, du \propto \int^\lambda u^{-1/2}\, du = 2\sqrt{\lambda}.$$

We can ignore the constant factor 2, so we can identify the square root transformation $\phi = h(\lambda) = \sqrt{\lambda}$ as the variance-stabilising transformation of the relative rate λ of a Poisson distribution. The MLE of λ is $\hat{\lambda}_{\mathrm{ML}} = x/e$, so using the invariance of the MLE, we immediately obtain $\hat{\phi}_{\mathrm{ML}} = \sqrt{x/e}$.

We know from Result 4.11 that the expected Fisher information $J_\phi(\phi)$ does not depend on ϕ, but what is its value? Using Result 4.3, we can easily compute

$$J_\phi(\phi) = J_\lambda(\lambda)\left\{\frac{dh(\lambda)}{d\lambda}\right\}^{-2} = \frac{e}{\lambda}\left|\frac{1}{2}\lambda^{-\frac{1}{2}}\right|^{-2} = 4e,$$

so the asymptotic variance of $\hat{\phi}_{\mathrm{ML}}$ is $1/(4e)$.

We now use the variance-stabilising transformation to compute confidence intervals for all 56 regions in Scotland. As before, we construct confidence intervals for each of the regions separately. For example, in a particular region of Scotland there have been $x = 11$ lip cancer cases, but only $e = 3.04$ have been expected, so $\hat{\lambda}_{\mathrm{ML}} = x/e = 3.62$ and $\mathrm{se}(\hat{\lambda}_{\mathrm{ML}}) = 1/\sqrt{J_\lambda(\hat{\lambda}_{\mathrm{ML}})} = \sqrt{\hat{\lambda}_{\mathrm{ML}}/e} = 1.09$. This gives the 95 % Wald confidence interval for λ with limits

$$3.62 \pm 1.96 \cdot 1.09 = 1.48 \text{ and } 5.76.$$

Alternatively, we can compute the limits of a 95 % Wald confidence interval for $\phi = \sqrt{\lambda}$ using $\mathrm{se}(\hat{\phi}_{\mathrm{ML}}) = 1/\sqrt{J_\phi(\hat{\phi}_{\mathrm{ML}})} = \sqrt{1/(4e)}$:

$$\sqrt{3.62} \pm 1.96 \cdot \sqrt{1/(4 \cdot 3.04)} = 1.34 \text{ and } 2.46,$$

and back-transform those to the 95 % confidence interval [1.800, 6.07] for λ. Note that this interval is no longer symmetric around $\hat{\lambda}_{\mathrm{ML}} = 3.62$ but slightly shifted to the right.

Figure 4.4 displays the variance-stabilised Wald confidence intervals for the 56 regions of Scotland. The intervals look similar to the score intervals shown in Fig. 4.3. ■

Example 4.15 (Inference for a proportion) We now want to derive the variance-stabilising transformation in the binomial model. The ML estimator of the success probability π is $\hat{\pi}_{\mathrm{ML}} = \bar{X} = X/n$ with expected Fisher information

$$J(\pi) = \frac{n}{\pi(1-\pi)} \propto \left\{\pi(1-\pi)\right\}^{-1},$$

so we have to compute $\int^\pi \{u(1-u)\}^{-\frac{1}{2}}\, du$. Using the substitution

$$u = \sin(p)^2 \iff p = \arcsin(\sqrt{u}),$$

we have $1 - u = \cos(p)^2$, $du = 2\sin(p)\cos(p)dp$, and hence

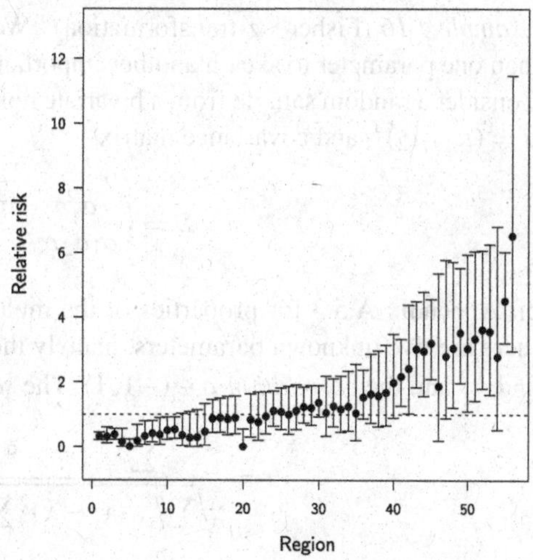

Fig. 4.4 95 % Wald confidence intervals for the relative risk of lip cancer incidence in 56 regions of Scotland based on the variance-stabilising transformation. The MLEs x_i/e_i are given as *solid circles*. The regions are ordered as in Fig. 4.3

$$\begin{aligned}\int^{\pi} \{u(1-u)\}^{-\frac{1}{2}}\,du &= \int^{\arcsin(\sqrt{\pi})} \frac{2\sin(p)\cos(p)}{|\sin(p)\cos(p)|}\,dp \\ &\propto \int^{\arcsin(\sqrt{\pi})} 2\,dp \\ &= 2\arcsin(\sqrt{\pi}).\end{aligned}$$

So $h(\pi) = \arcsin(\sqrt{\pi})$ is the variance-stabilising transformation with approximate variance $1/(4n)$, as can be shown easily. As intended, the approximate variance does not depend on π.

Suppose $n = 100$ and $x = 2$, i.e. $\hat{\pi}_{\text{ML}} = 0.02$ and $h(\hat{\pi}_{\text{ML}}) = \arcsin(\sqrt{\hat{\pi}_{\text{ML}}}) = 0.142$. The approximate 95 % confidence interval for $\phi = h(\pi) = \arcsin(\sqrt{\pi})$ now has the limits

$$0.142 \pm 1.96 \cdot 1/\sqrt{400} = 0.044 \text{ and } 0.240.$$

Back-transformation using $h^{-1}(\phi) = \sin(\phi)^2$ finally gives the 95 % confidence interval $[0.0019, 0.0565]$ for π. It should be noted that application of the back-transformation requires the limits of the confidence interval for ϕ to lie in the interval $[0, \pi/2 \approx 1.5708]$ (here π denotes the circle constant). In extreme cases, for example for $x = 0$, this may not be the case. ■

Finally, we want to derive the variance-stabilising transformation of a scale parameter θ. With Result 4.4, we have $J(\theta) \propto \theta^{-2}$, so

$$\int^{\theta} J(u)^{1/2}\,du \propto \int^{\theta} u^{-1}\,du = \log(\theta).$$

The variance-stabilising transformation of a scale parameter is therefore always the logarithmic transformation.

Example 4.16 (Fisher's z-transformation) We now consider a problem with more than one parameter to sketch another important variance-stabilising transformation. Consider a random sample from a bivariate normal distribution $N_2(\boldsymbol{\mu}, \boldsymbol{\Sigma})$ with mean $\boldsymbol{\mu} = (\mu_1, \mu_2)^\top$ and covariance matrix

$$\boldsymbol{\Sigma} = \begin{pmatrix} \sigma_1^2 & \sigma_1\sigma_2\rho \\ \sigma_1\sigma_2\rho & \sigma_2^2 \end{pmatrix},$$

cf. Appendix A.5.3 for properties of the multivariate normal distribution. In total, there are five unknown parameters, namely the means μ_1 and μ_2, the variances σ_1^2 and σ_2^2 and the *correlation* $\rho \in (-1, 1)$. The *sample correlation*

$$r = \frac{\sum_{i=1}^n (x_i - \bar{x})(y_i - \bar{y})}{\sqrt{\sum_{i=1}^n (x_i - \bar{x})^2 \sum_{i=1}^n (y_i - \bar{y})^2}}$$

is the MLE of ρ.

It can be shown (with the central limit theorem and the delta method) that r has an asymptotic normal distribution with mean ρ and variance $V(\rho) = (1-\rho^2)^2/n$ (compare also Exercise 2 in Chap. 5). The asymptotic variance depends on ρ, so we would like to find a transformation that removes this dependence.

Using (4.14) with the Fisher information replaced by the inverse variance, we obtain

$$\begin{aligned} h(\rho) &= \int^\rho V(u)^{-1/2}\, du \\ &\propto \int^\rho \frac{1}{1-u^2}\, du \\ &= \int^\rho \frac{1}{2(1+u)}\, du + \int^\rho \frac{1}{2(1-u)}\, du \\ &= \frac{1}{2}\log(1+\rho) - \frac{1}{2}\log(1-\rho) \\ &= \frac{1}{2}\log\left(\frac{1+\rho}{1-\rho}\right). \end{aligned}$$

This variance-stabilising transformation can also be written as $\zeta = h(\rho) = \tanh^{-1}(\rho)$ and has been discovered by R.A. Fisher. It is therefore known as *Fisher's z-transformation*. The transformed correlation $z = \tanh^{-1}(r)$ is approximately normally distributed with mean ζ and constant variance $1/n$ (note that $1/(n-3)$ is actually a slightly more accurate approximation of the variance). A confidence interval for ζ can now be back-transformed using the inverse Fisher's z-transformation

$$\rho = \tanh(\zeta) = \frac{\exp(2\zeta) - 1}{\exp(2\zeta) + 1}$$

to obtain a confidence interval for the correlation ρ. Note that this transformation ensures that the limits of the confidence interval for ρ are inside the parameter space $(-1, 1)$ of ρ. ■

Example 4.17 (Blood alcohol concentration) We want to illustrate the method with the study on blood alcohol concentration, cf. Sect. 1.1.7. We denote the BrAC measurement from the ith proband as x_i and the BAC value as y_i, $i = 1, \dots, n = 185$, and assume that they are realisations from a bivariate normal random sample. We obtain the means $\bar{x} = 1927.5$ and $\bar{y} = 0.7832$ and the sample correlation $r = 0.9725$. If we use the asymptotic normality of r, we obtain the standard error $\sqrt{V(r)} = 0.003994$ for r and the 95 % Wald confidence interval $[0.9646, 0.9803]$ for ρ.

The estimated transformed correlation is $z = \tanh(r) = 2.1357$, with 95 % Wald confidence interval $z \pm 1.96/\sqrt{n} = 1.9916$ and 2.2798 for ζ. Transforming this to the correlation scale gives the 95 % confidence interval $[0.9634, 0.9793]$ for ρ. This is similar to the original Wald confidence interval above. The reason is that the sample size n is quite large in this example, so that the asymptotics are effective for both scales. If we had just $n = 10$ observations giving the same estimate, we would get an upper bound for the original Wald confidence interval that is larger than 1. ■

4.4 The Likelihood Ratio Statistic

A second-order Taylor expansion of $l(\theta)$ at $\hat{\theta}_{\mathrm{ML}}$ as in (2.17) gives

$$l(\theta) \approx l(\hat{\theta}_{\mathrm{ML}}) - \frac{1}{2} \cdot I(\hat{\theta}_{\mathrm{ML}})(\theta - \hat{\theta}_{\mathrm{ML}})^2$$

so

$$2 \log\left\{\frac{L(\hat{\theta}_{\mathrm{ML}})}{L(\theta)}\right\} = 2\{l(\hat{\theta}_{\mathrm{ML}}) - l(\theta)\} \approx I(\hat{\theta}_{\mathrm{ML}})(\theta - \hat{\theta}_{\mathrm{ML}})^2.$$

The left-hand side

$$W = 2 \log\left\{\frac{L(\hat{\theta}_{\mathrm{ML}})}{L(\theta)}\right\} = -2\tilde{l}(\theta) \tag{4.15}$$

is called the *likelihood ratio statistic*. Here $W = W(X_{1:n})$ is a function of the random sample $X_{1:n}$ because the likelihood $L(\theta; X_{1:n})$ and the ML estimator $\hat{\theta}_{\mathrm{ML}} = \hat{\theta}_{\mathrm{ML}}(X_{1:n})$ depend on the data.

If θ denotes the true parameter value, then (4.12) implies that

$$I(\hat{\theta}_{\mathrm{ML}})(\hat{\theta}_{\mathrm{ML}} - \theta)^2 \overset{a}{\sim} \chi^2(1),$$

due to the fact that the squared standard normal random variable has a chi-squared distribution with one degree of freedom, cf. Appendix A.5.2. So the likelihood ratio statistic (4.15) follows a chi-squared distribution with one degree of freedom,

$$W \overset{a}{\sim} \chi^2(1),$$

and is an approximate pivot. It can be used both for significance testing and the calculation of confidence intervals.

4.4.1 The Likelihood Ratio Test

The *likelihood ratio test* to test the null hypothesis $H_0 : \theta = \theta_0$ is based on the likelihood ratio statistic (4.15) with θ replaced by θ_0. An equivalent formulation is based on the *signed likelihood ratio statistic*

$$\operatorname{sign}(\hat{\theta}_{\mathrm{ML}} - \theta_0) \cdot \sqrt{W}, \tag{4.16}$$

which under H_0 asymptotically follows the standard normal distribution.

Example 4.18 (Scottish lip cancer) We now want to compute the likelihood ratio test statistic for one specific region in Scotland with the null hypothesis $H_0 : \lambda = \lambda_0$. The log-likelihood equals

$$l(\lambda) = x \log(\lambda) - e\lambda,$$

so

$$W(x) = \begin{cases} 2[x\{\log(x) - \log(e\lambda_0) - 1\} + e\lambda_0] & \text{for } x > 0, \\ 2e\lambda_0 & \text{for } x = 0, \end{cases}$$

and we easily obtain the signed likelihood ratio statistic (4.16), cf. Fig. 4.5. If positive, the signed likelihood ratio statistic T_4 is more conservative than the score test in this example. If negative, it is less conservative due to $T_4 \le T_1$. Figure 4.6 gives a histogram of the corresponding P-values. The pattern is very similar to the one based on the score statistic shown in Fig. 4.2. Note that 25 out of 56 regions have a P-value smaller than 5 %, so many more than we would expect under the assumption that the null hypothesis holds in all regions. ■

4.4.2 Likelihood Ratio Confidence Intervals

Likelihood ratio confidence intervals can be obtained by inverting the likelihood ratio test. More specifically, for a given MLE $\hat{\theta}_{\mathrm{ML}}$, all parameter values θ with $W = 2\log\{L(\hat{\theta}_{\mathrm{ML}})/L(\theta)\} \le \chi^2_\gamma(1)$ form a $\gamma \cdot 100$ % likelihood ratio confidence interval. In exceptional cases (if the likelihood function is multimodal) this set will not be an interval.

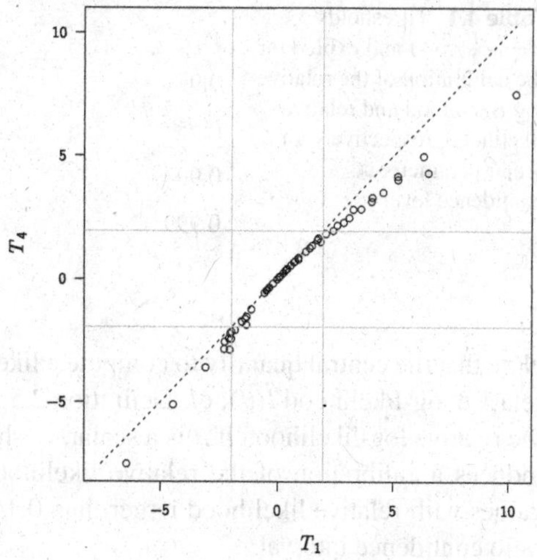

Fig. 4.5 Scatter plot of the score statistic T_1 versus the signed likelihood ratio statistic T_4 for the 56 regions of Scotland

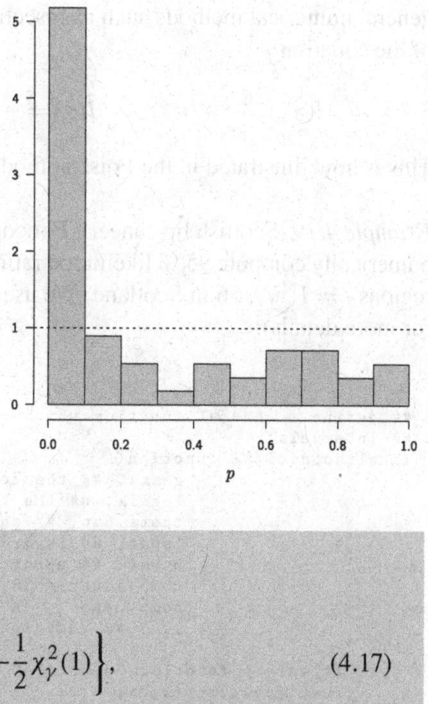

Fig. 4.6 Histogram of P-values obtained from the likelihood ratio test statistic T_4 for a unity relative risk of lip cancer in the 56 regions of Scotland

Likelihood ratio confidence intervals

The set

$$\left\{\theta : \tilde{l}(\theta) \geq -\frac{1}{2}\chi^2_\gamma(1)\right\}, \tag{4.17}$$

which can also be represented as

$$\left\{\theta : \tilde{L}(\theta) \geq \exp\left[-\frac{1}{2}\chi^2_\gamma(1)\right]\right\}, \tag{4.18}$$

forms an approximate $\gamma \cdot 100\,\%$ confidence interval for θ.

Table 4.1 Thresholds $c = -\frac{1}{2}\chi^2_\gamma(1)$ and $\exp(c)$ for the calibration of the relative log-likelihood and relative likelihood, respectively, of a scalar parameter at confidence level γ

γ	c	$\exp(c)$
0.9	−1.35	0.259
0.95	−1.92	0.147
0.99	−3.32	0.036
0.999	−5.41	0.004

Note that the central quantity to compute a likelihood ratio confidence interval is the relative log-likelihood $\tilde{l}(\theta)$, cf. Definition 2.5. We are now in a position to calibrate the relative log-likelihood if θ is a scalar, as shown in Table 4.1. Of course, this also induces a calibration of the relative likelihood $\tilde{L}(\theta)$. For example, all parameter values with relative likelihood larger than 0.147 will be within the 95 % likelihood ratio confidence interval.

Computation of the limits of a likelihood ratio confidence interval requires in general numerical methods such as bisection (see Appendix C.1.2) to find the roots of the equation

$$\tilde{l}(\theta) = -\frac{1}{2}\chi^2_\gamma(1).$$

This is now illustrated in the Poisson model.

Example 4.19 (Scottish lip cancer) For comparison with Examples 4.9 and 4.14, we numerically compute 95 % likelihood ratio confidence intervals for λ_i in each of the regions $i = 1, \dots, 56$ in Scotland. We use the R-function `uniroot` (cf. Appendix C for more details):

```
## define a general function which computes likelihood confidence
## intervals
likelihood.ci <- function(
                        gamma, ## the confidence level
                        loglik, ## the log-likelihood function
                        theta.hat, ## the MLE
                        lower, ## lower bound of parameter space
                        upper, ## upper bound of parameter space
                        comp.lower = TRUE, ## compute lower bound of CI?
                        comp.upper = TRUE, ## compute upper bound of CI?
                        ...) ## additional arguments for the log-likelihood
  {
      ## target function, such that f(theta)=0 gives CI limits
      f <- function(theta, ...)
      {
          loglik(theta, ...) - loglik(theta.hat, ...) + 1/2 *
                   qchisq(gamma, df=1)
      }

      ## compute lower and upper bounds of CI
      ret <- c()
      if(comp.lower)
      {
          hl.lower <- uniroot(f,
```

```
                                   interval = c(lower, theta.hat),
                                   ...)$root
            ret <- c(ret, hl.lower)
        }
        if(comp.upper)
        {
            hl.upper <- uniroot(f,
                                   interval = c(theta.hat, upper),
                                   ...)$root
            ret <- c(ret, hl.upper)
        }
        return(ret)
  }
## the log-likelihood of lambda in the Poisson model
loglik.poisson <- function(lambda, x, e, log = TRUE)
  {
        dpois(x = x, lambda = lambda * e, log = log)
  }
## get the data
x <- scotlandData[, 1]
e <- scotlandData[, 2]
## here we will save the bounds of the CIs
likCI <- matrix(nrow=length(x), ncol=2)
## confidence level
confLevel <- 0.95
## small positive value
eps <- sqrt(.Machine$double.eps)
## now process all regions
for(i in seq_along(x))
  {
        res <- likelihood.ci(gamma = confLevel,
                             loglik = loglik.poisson,
                             theta.hat = x[i] / e[i],
                             lower = eps, upper = 1/eps,
                             x = x[i], e = e[i],
                             comp.lower = x[i] > 0)
        if (x[i] == 0)
        {
            res <- c (0, res)
        }
        likCI[i, ] <- res
  }
```

Note that we are careful in the R-code when $x_i = 0$ because then the mode of the likelihood is at zero, which is the left boundary of the parameter space $\mathbb{R}_0^+$. Therefore, we only compute the upper bound numerically in that case. Moreover, we do not search for the lower bound very close to zero, but only a small $\varepsilon > 0$ away from that (here it is $\varepsilon \approx 1.5 \cdot 10^{-8}$). Figure 4.7 displays the resulting intervals. They look very similar to the ones based on the score statistic in Fig. 4.3. ■

It is interesting that Wald confidence intervals can be viewed as approximate likelihood ratio confidence intervals. To see this, we replace in (4.17) the relative log-likelihood $\tilde{l}(\theta)$ with its quadratic approximation (2.17),

$$\tilde{l}(\theta) \approx -\frac{1}{2} I(\hat{\theta}_{\mathrm{ML}})(\theta - \hat{\theta}_{\mathrm{ML}})^2,$$

to obtain

$$\left\{\theta : I(\hat{\theta}_{\mathrm{ML}})(\theta - \hat{\theta}_{\mathrm{ML}})^2 \le \chi^2_\gamma(1)\right\},$$

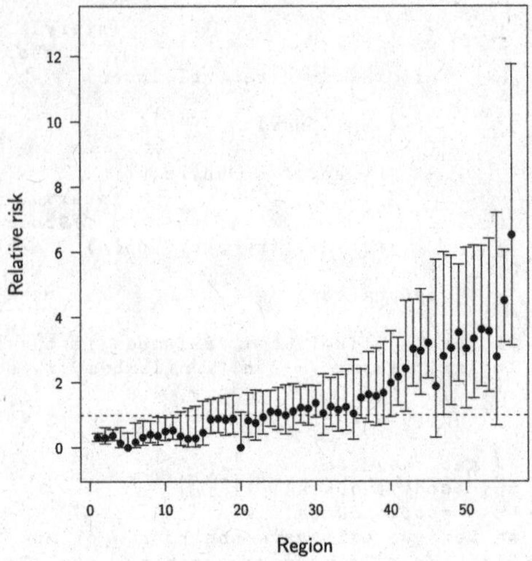

Fig. 4.7 95 % likelihood ratio confidence intervals for the relative risk in the 56 regions of Scotland. The MLEs x_i/e_i are given as *solid circles*. The regions are ordered as in Fig. 4.3

which corresponds to

$$\left\{\theta : \sqrt{I(\hat{\theta}_{\text{ML}})}|\theta - \hat{\theta}_{\text{ML}}| \leq z_{\frac{1+\gamma}{2}}\right\},$$

i.e. the Wald confidence interval. This is because $(z_{(1+\gamma)/2})^2 = \chi^2_\gamma(1)$ due to the relation between the standard normal and the chi-squared distribution. Figure 4.8 illustrates this for $X \sim \text{Po}(e\lambda)$ with known offset $e = 3.04$ and observation $x = 11$. This comparison suggests that likelihood ratio confidence intervals are more accurate than Wald confidence intervals, as they avoid the quadratic approximation of the log-likelihood. Various theoretical results support this intuitive finding. In particular, likelihood ratio intervals are invariant to one-to-one transformations, as we have shown (implicitly) already in Sect. 2.1.3.

Invariance of likelihood ratio confidence intervals
Likelihood ratio confidence intervals are invariant with respect to one-to-one transformations of the parameter.

Likelihood ratio confidence intervals can also be computed in cases where the quadratic approximation of the log-likelihood fails, as in the following example.

Example 4.20 (Uniform model) In Example 2.18 we considered the uniform model $\text{U}(0, \theta)$ with unknown upper bound θ as a counter-example, where quadratic approximation of the log-likelihood is not possible. However, we still can derive a likelihood ratio confidence interval. Because the likelihood from the realisation $x_{1:n}$ of a random sample is

$$L(\theta) = \mathsf{I}_{[0,\theta]}\Big(\max_i(x_i)\Big)\theta^{-n}$$

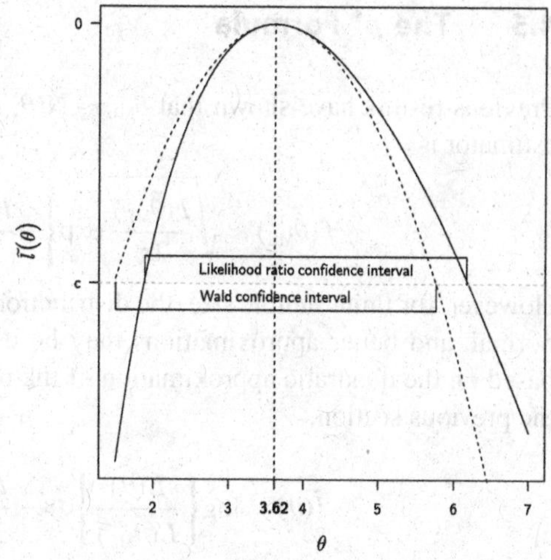

Fig. 4.8 Comparison of the 95 % likelihood ratio confidence interval for λ with the corresponding Wald confidence interval if $X \sim \text{Po}(e\lambda)$, $e = 3.04$ is known and $x = 11$ has been observed

and thus zero for values $\theta < \max_i(x_i) = \hat{\theta}_{\text{ML}}$, we only need to compute the upper limit of the interval. The relative likelihood is (in the range $\theta \geq \hat{\theta}_{\text{ML}}$)

$$\tilde{L}(\theta) = \frac{L(\theta)}{L(\hat{\theta}_{\text{ML}})} = \left(\frac{\hat{\theta}_{\text{ML}}}{\theta}\right)^n,$$

so from (4.18) we obtain the upper bound

$$\theta = \frac{\max_i(x_i)}{\exp\{-\frac{1}{2}\chi^2_\gamma(1)\}^{\frac{1}{n}}}$$

of the $\gamma \cdot 100$ % likelihood ratio confidence interval for θ.

Validity of this confidence interval is, however, questionable, as the Fisher regularity conditions (4.1) are not fulfilled. Indeed, we can show (Exercise 6 in Chap. 3) that the upper bound should be

$$\theta = \frac{\max_i(x_i)}{(1-\gamma)^{\frac{1}{n}}}$$

to achieve a confidence level of γ. Since $1 - \gamma < \exp\{-\frac{1}{2}\chi^2_\gamma(1)\}$ (see Table 4.1), this upper bound is larger than the one obtained from the above likelihood calculation, and the likelihood ratio confidence interval will have lower than nominal coverage. ■

4.5 The p^* Formula

Previous results have shown that $\hat{\theta}_{\text{ML}} \overset{\text{a}}{\sim} \text{N}(\theta, I(\hat{\theta}_{\text{ML}})^{-1})$, so the density of the ML estimator is

$$f(\hat{\theta}_{\text{ML}}) \approx \sqrt{\frac{I(\hat{\theta}_{\text{ML}})}{2\pi}} \exp\left\{-\frac{I(\hat{\theta}_{\text{ML}})}{2}(\hat{\theta}_{\text{ML}} - \theta)^2\right\}. \tag{4.19}$$

However, for finite sample size, the distribution of the ML estimator can be far from normal, and better approximations may be useful. One way to improve (4.19) is based on the quadratic approximation of the relative log-likelihood as described in the previous section,

$$\tilde{l}(\theta) = \log\left\{\frac{L(\theta)}{L(\hat{\theta}_{\text{ML}})}\right\} \approx -\frac{I(\hat{\theta}_{\text{ML}})}{2}(\hat{\theta}_{\text{ML}} - \theta)^2. \tag{4.20}$$

Plugging the left-hand side of (4.20) into the exponential function in (4.19), we obtain the following (approximate) formula for the density of the ML estimator $\hat{\theta}_{\text{ML}}$:

$$f(\hat{\theta}_{\text{ML}}) \approx p^*(\hat{\theta}_{\text{ML}}) = \sqrt{\frac{I(\hat{\theta}_{\text{ML}})}{2\pi}} \frac{L(\theta)}{L(\hat{\theta}_{\text{ML}})}.$$

This is the p^* *formula*, an approximation of the distribution of the ML estimator, which is often very accurate. Note that $p^*(\hat{\theta}_{\text{ML}})$ does not necessarily integrate to one, so an additional normalisation step may be necessary.

Example 4.21 (Normal model) Let $X_{1:n}$ denote a random sample from a normal distribution with unknown mean μ and known variance σ^2. The ML estimator of μ is $\hat{\mu}_{\text{ML}} = \bar{x} = \sum_{i=1}^n x_i/n$ with observed Fisher information $I(\hat{\mu}_{\text{ML}}) = n/\sigma^2$. Now $\bar{X}$ is sufficient for μ, so the likelihood function of μ is (up to a multiplicative constant)

$$L(\mu) = \exp\left\{-\frac{n}{2\sigma^2}(\bar{x} - \mu)^2\right\}$$

and $L(\hat{\mu}_{\text{ML}}) = 1$. With $\hat{\mu}_{\text{ML}} = \bar{x}$, we therefore obtain

$$p^*(\hat{\mu}_{\text{ML}}) = \sqrt{\frac{n/\sigma^2}{2\pi}} \exp\left\{-\frac{n}{2\sigma^2}(\hat{\mu}_{\text{ML}} - \mu)^2\right\},$$

the density of a normal distribution with mean μ and variance σ^2/n. So here the p^* formula gives the density of the exact distribution of the ML estimator. However, this result could also have been obtained using the Wald statistic.

Suppose now that μ is known but σ^2 is unknown. Then $\hat{\sigma}^2_{\text{ML}} = \sum_{i=1}^n (x_i - \mu)^2/n$, $I(\hat{\sigma}^2_{\text{ML}}) = n/(2\hat{\sigma}^4_{\text{ML}})$ and $J(\sigma^2) = n/(2\sigma^4)$. The usual normal approximation of the distribution of $\hat{\sigma}^2_{\text{ML}}$ has therefore mean σ^2 and variance $1/J(\sigma^2) = 2\sigma^4/n$.

Now the likelihood function of σ^2 is

$$L(\sigma^2) = (\sigma^2)^{-n/2} \exp\left(-\frac{n}{2\sigma^2}\hat{\sigma}^2_{\mathrm{ML}}\right),$$

so

$$L(\hat{\sigma}^2_{\mathrm{ML}}) = (\hat{\sigma}^2_{\mathrm{ML}})^{-n/2} \exp\left(-\frac{n}{2}\right).$$

We obtain the p^* formula

$$\begin{aligned}
p^*(\hat{\sigma}^2_{\mathrm{ML}}) &= \sqrt{\frac{I(\hat{\sigma}^2_{\mathrm{ML}})}{2\pi}} \frac{L(\sigma^2)}{L(\hat{\sigma}^2_{\mathrm{ML}})} \\
&= \sqrt{\frac{n/(2\hat{\sigma}^4_{\mathrm{ML}})}{2\pi}} \left(\frac{\sigma^2}{\hat{\sigma}^2_{\mathrm{ML}}}\right)^{-\frac{n}{2}} \frac{\exp(-\frac{n}{2\sigma^2}\hat{\sigma}^2_{\mathrm{ML}})}{\exp(-\frac{n}{2})} \\
&= \frac{1}{\sqrt{2\pi}}\sqrt{n/2}\exp(n/2)(\sigma^2)^{-n/2}(\hat{\sigma}^2_{\mathrm{ML}})^{n/2-1} \exp\left(-\frac{n}{2\sigma^2}\hat{\sigma}^2_{\mathrm{ML}}\right) \\
&\propto (\hat{\sigma}^2_{\mathrm{ML}})^{n/2-1} \exp\left(-\frac{n}{2\sigma^2}\hat{\sigma}^2_{\mathrm{ML}}\right),
\end{aligned}$$

which can be identified as the kernel of a $\mathrm{G}(n/2, n/(2\sigma^2))$ distribution, cf. Appendix A.5.2. After suitable normalisation, the p^* formula hence gives $\hat{\sigma}^2_{\mathrm{ML}} \sim \mathrm{G}(n/2, n/(2\sigma^2))$. Interestingly, this is the exact distribution of $\hat{\sigma}^2_{\mathrm{ML}}$, since we know from Example 3.8 that the pivot $n\hat{\sigma}^2_{\mathrm{ML}}/\sigma^2$ has a $\chi^2(n)$ distribution, i.e. a $\mathrm{G}(n/2, 1/2)$ distribution, so $\hat{\sigma}^2_{\mathrm{ML}} \sim \mathrm{G}(n/2, n/(2\sigma^2))$. Note that this distribution has mean σ^2 and variance $(2\sigma^4)/n$, which matches the mean and variance of the ordinary normal approximation. ■

4.6 A Comparison of Likelihood-Based Confidence Intervals

Using likelihood theory, we have discussed the following three approximate pivots to construct confidence intervals and significance tests:

Three approximate pivots for likelihood inference

$$S(\theta; X_{1:n})/\sqrt{J_{1:n}(\theta)} \overset{a}{\sim} \mathrm{N}(0, 1), \tag{4.21}$$

$$\sqrt{J_{1:n}(\theta)}(\hat{\theta}_{\mathrm{ML}} - \theta) \overset{a}{\sim} \mathrm{N}(0, 1), \tag{4.22}$$

$$2\log\frac{L(\hat{\theta}_{\mathrm{ML}})}{L(\theta)} \overset{a}{\sim} \chi^2(1). \tag{4.23}$$

The asymptotic pivotal distribution in (4.21) and (4.22) still holds if we replace $J_{1:n}(\theta)$ by $J_{1:n}(\hat{\theta}_{\text{ML}})$, $I(\theta; X_{1:n})$ or $I(\hat{\theta}_{\text{ML}}; X_{1:n})$. Regarding the choice of the Fisher information, it is typically recommended to use $I(\hat{\theta}_{\text{ML}}; X_{1:n})$. However, note that in exponential families, $I(\hat{\theta}_{\text{ML}}; X_{1:n}) = J_{1:n}(\hat{\theta}_{\text{ML}})$, cf. Exercise 8 in Chap. 3.

The large number of possible statistics, that are all asymptotically equivalent but give different results for finite samples is confusing. Which of the different pivots should we use in practice? The score statistic is applied only in special cases such as in the case of the Wilson confidence interval for a proportion (cf. Example 4.10). More commonly used are *Wald confidence intervals* with limits

$$\hat{\theta}_{\text{ML}} \pm z_{(1+\gamma)/2} \cdot \text{se}(\hat{\theta}_{\text{ML}}),$$

with $\text{se}(\hat{\theta}_{\text{ML}}) = \{I(\hat{\theta}_{\text{ML}})\}^{-\frac{1}{2}}$ or $\{J(\hat{\theta}_{\text{ML}})\}^{-\frac{1}{2}}$. However, Wald confidence intervals are not very accurate and may not be boundary-respecting. In contrast, likelihood ratio confidence intervals will by construction cover only parameter values with non-zero likelihood, a feature that is shared by score confidence intervals using the expected Fisher information. The score test has certain optimality features (it is the "locally most powerful test") and does neither require computation of the MLE nor of the observed Fisher information.

From that perspective it is surprising that Wald confidence intervals can be found so often in applications. One reason is that for a Wald confidence interval, only the MLE and the curvature of the log-likelihood at the MLE are required. Both quantities are computed easily using numerical optimisation techniques. The numerical computation of a likelihood ratio or a score confidence interval is typically more involved. Moreover, for large sample size, the different confidence intervals become more and more similar.

In the following case study we will empirically compare different confidence intervals for a proportion.

Example 4.22 (Inference for a proportion) Consider the binomial model $X \sim \text{Bin}(n, \pi)$ and suppose that we are interested in the calculation of a confidence interval for the success probability π (for known sample size n).

1. With $\hat{\pi}_{\text{ML}} = x/n$ and

$$\text{se}(\hat{\pi}_{\text{ML}}) = \{I(\hat{\pi}_{\text{ML}})\}^{-\frac{1}{2}} = \left\{\frac{n}{\hat{\pi}_{\text{ML}}(1-\hat{\pi}_{\text{ML}})}\right\}^{-\frac{1}{2}} = \sqrt{\frac{\hat{\pi}_{\text{ML}}(1-\hat{\pi}_{\text{ML}})}{n}},$$

we can easily compute the limits of the $\gamma \cdot 100\,\%$ Wald confidence interval:

$$\hat{\pi}_{\text{ML}} \pm q \cdot \text{se}(\hat{\pi}_{\text{ML}}),$$

where $q = z_{(1+\gamma)/2}$ denotes the $(1+\gamma)/2$ quantile of the standard normal distribution. However, the standard error will be zero for $x = 0$ and $x = n$, so in these cases we calculate the standard error based on $x = 0.5$ or $x = n - 0.5$ successes, respectively, and use it to calculate a one-sided confidence interval for π with lower limit 0 and upper limit 1, respectively. If, for the other cases, the limits of the Wald confidence interval lie outside the unit interval, they are rounded to 0 and 1, respectively.

2. The Wald confidence interval for $\phi = \operatorname{logit}(\pi)$ avoids this problem using the logit transformation. First note that (due to invariance of the MLE)

$$\hat{\phi}_{\mathrm{ML}} = \operatorname{logit}(\hat{\pi}_{\mathrm{ML}}) = \log\left(\frac{\hat{\pi}_{\mathrm{ML}}}{1-\hat{\pi}_{\mathrm{ML}}}\right) = \log\left(\frac{x}{n-x}\right).$$

The standard error of $\hat{\phi}_{\mathrm{ML}}$ can be easily computed with the delta method (see Example 3.12):

$$\operatorname{se}(\hat{\phi}_{\mathrm{ML}}) = \sqrt{\frac{1}{x} + \frac{1}{n-x}}.$$

The limits of the Wald confidence interval for $\phi = \operatorname{logit}(\pi)$,

$$\hat{\phi}_{\mathrm{ML}} \pm q \cdot \operatorname{se}(\hat{\phi}_{\mathrm{ML}}),$$

can finally be back-transformed using the inverse logit function

$$\pi(\phi) = \frac{\exp(\phi)}{1+\exp(\phi)} = \frac{1}{1+\exp(-\phi)}.$$

However, problems occur if $x = 0$ or $x = n$, where both $\hat{\phi}_{\mathrm{ML}}$ and $\operatorname{se}(\hat{\phi}_{\mathrm{ML}})$ are infinite. In these cases we calculate the MLE and its standard error using $x = 0.5$ or $x = n - 0.5$ successes, respectively, and use it to calculate a one-sided confidence interval with lower limit 0 and upper limit 1, respectively.

3. We can also calculate a Wald confidence interval for the variance-stabilising parameter transformation $\phi = \arcsin(\sqrt{\pi})$ (cf. Example 4.15), which has approximate variance equal to $1/(4n)$. This confidence interval has limits

$$\arcsin(\sqrt{\hat{\pi}_{\mathrm{ML}}}) \pm q \cdot (4n)^{-\frac{1}{2}},$$

which we back-transform to the unit interval using the inverse function $\pi = \sin(\phi)^2$. Note that for the extreme cases $x = 0$ or $x = n$, we calculate as before a one-sided confidence interval with lower limit equal to zero if $x = 0$ and upper limit equal to one if $x = n$.

4. The Wilson confidence interval described in Example 4.10 with limits

$$\frac{x+q^2/2}{n+q^2} \pm \frac{q\sqrt{n}}{n+q^2}\sqrt{\hat{\pi}_{\mathrm{ML}}(1-\hat{\pi}_{\mathrm{ML}}) + \frac{q^2}{4n}}$$

is another possibility. It is not centred around $\hat{\pi}_{\mathrm{ML}}$, but around

$$\frac{x+q^2/2}{n+q^2},$$

Table 4.2 Comparison of different confidence intervals for a binomial probability at level 95 % for various values of sample size n and number of successes x

(a) Different Wald confidence intervals

n	x	(1) Wald for π	(2) Wald for logit(π)	(3) Wald for arcsin($\sqrt{\pi}$)
10	0	0.000 to 0.113	0.000 to 0.364	0.000 to 0.066
10	1	0.000 to 0.286	0.014 to 0.467	0.000 to 0.349
10	5	0.190 to 0.810	0.225 to 0.775	0.210 to 0.790
100	0	0.000 to 0.012	0.000 to 0.049	0.000 to 0.007
100	10	0.041 to 0.159	0.055 to 0.176	0.049 to 0.166
100	50	0.402 to 0.598	0.403 to 0.597	0.403 to 0.597

(b) Score, likelihood ratio and Clopper–Pearson confidence intervals

n	x	(4) Wilson	(5) Likelihood	(6) Clopper–Pearson
10	0	0.000 to 0.278	0.000 to 0.175	0.000 to 0.308
10	1	0.018 to 0.404	0.006 to 0.372	0.003 to 0.445
10	5	0.237 to 0.763	0.218 to 0.782	0.187 to 0.813
100	0	0.000 to 0.037	0.000 to 0.019	0.000 to 0.036
100	10	0.055 to 0.174	0.051 to 0.169	0.049 to 0.176
100	50	0.404 to 0.596	0.403 to 0.597	0.398 to 0.602

the relative proportion in the sample after the addition of $q^2/2$ successes and $q^2/2$ non-successes. For illustration, if $\gamma = 0.95$, then $q^2/2 = 1.96^2/2$, so we will add slightly less than two successes and non-successes.

5. Numerical methods (cf. Appendix C) are required to compute the limits of the $\gamma \cdot 100$ % likelihood ratio confidence interval

$$\{\theta : \tilde{l}(\theta) \geq -\chi^2_\gamma(1)/2\}.$$

6. The "exact" *Clopper–Pearson confidence interval* (Clopper and Pearson 1934) has limits

$$b_{(1-\gamma)/2}(x, n-x+1) \quad \text{and} \quad b_{(1+\gamma)/2}(x+1, n-x)$$

if $x \notin \{0, n\}$, where $b_\alpha(\alpha, \beta)$ denotes the α quantile of the beta distribution with parameters α and β, see Appendix A.5.2. If $x = 0$, we set the lower limit to zero, and if $x = n$, we set the upper limit to one.

Table 4.2 gives the limits of the different 95 % confidence intervals for selected values of the sample size n and the number of successes x. There are substantial differences for small x and n, but the intervals become more similar for larger sample sizes and are quite close for $n = 100$ and $x = 50$.

What are the actual coverage probabilities of the different confidence intervals? For example, for $n = 50$, there are 51 different confidence intervals $\mathrm{CI}_\gamma(x)$ depending on the number of successes $x \in \mathcal{T} = \{0, 1, \ldots, 50\}$. We can now compute the

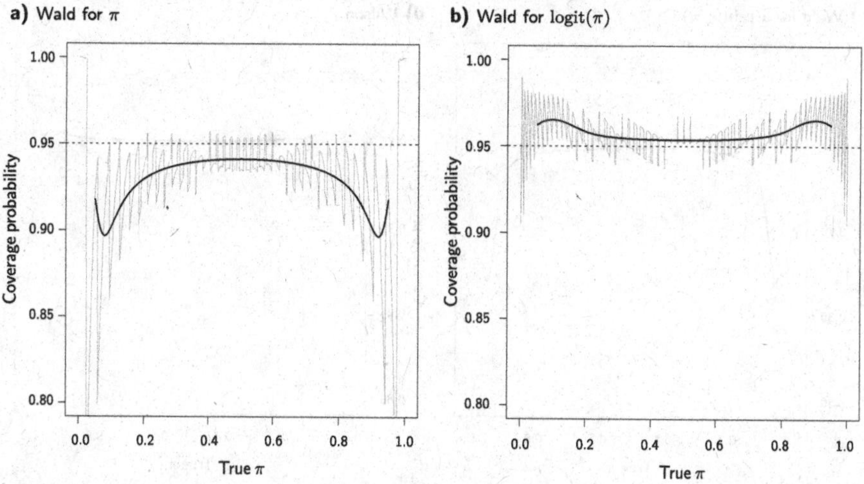

Fig. 4.9 Coverage probabilities of 95 % confidence intervals in a binomial experiment with $n = 50$ trials. The exact values are shown in *grey* for $\pi \in \{0.001, 0.002, \ldots, 0.999\}$, locally smoothed values are given in *black*

coverage probability

$$\Pr\{\pi \in \mathrm{CI}_\gamma(X)\} = \sum_{x \in \mathcal{T}} f(x; \pi, n) \mathsf{I}_{\mathrm{CI}_\gamma(x)}(\pi)$$

for every true parameter value π based on the binomial probability mass function $f(x; \pi, n)$.

Ideally, $\Pr\{\pi \in \mathrm{CI}_\gamma(X)\}$ should be equal to the nominal level γ for every sample size n and every true parameter value π. However, the binomial distribution is discrete, so all confidence intervals will only approximately have the nominal coverage level. Figure 4.9 illustrates that the true coverage of the various confidence intervals actually differs a lot. Shown are the actual coverage probabilities and smoothed ones, which give a better impression of the locally averaged coverage probability.

Figure 4.9 shows that the Wald confidence interval has nearly always smaller coverage than the nominal confidence level, sometimes considerably smaller. The variance-stabilised Wald confidence interval for $\arcsin(\sqrt{\pi})$ behaves somewhat better. The Wald confidence intervals for $\mathrm{logit}(\pi)$ tends to have larger coverage than the nominal level. The best locally averaged coverage is achieved by the Wilson confidence interval, followed by the likelihood ratio confidence interval, which behaves similarly for medium values of π and has slightly lower coverage than the Wilson confidence interval. Of particular interest is the behaviour of the "exact" Clopper–Pearson interval: its coverage is always larger than the nominal level, so this confidence interval appears to be anything else but "exact", at least in terms of coverage. Only in rare applications such a conservative confidence interval may be warranted. However, the Clopper–Pearson interval is widely used in practice, presumably due to the misleading specification "exact".

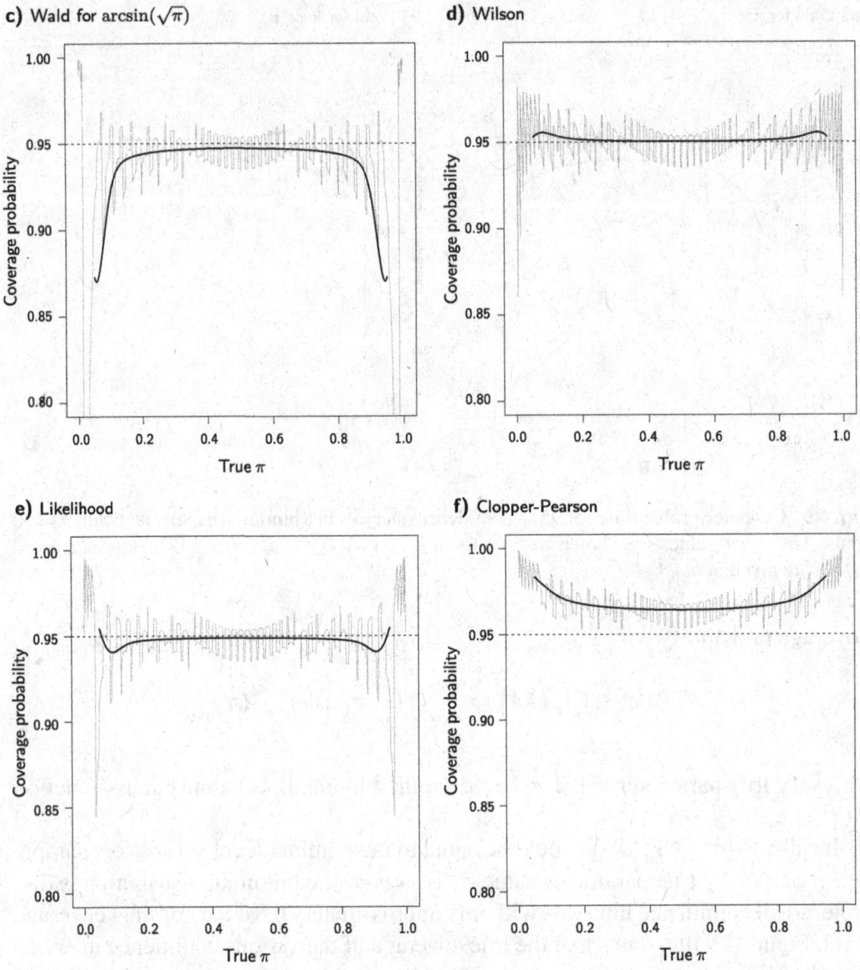

Fig. 4.9 (Continued)

Figure 4.10 displays the widths of the confidence intervals, which is an alternative quality criterion: If several confidence intervals attain the same nominal level, then the one with smaller width should be preferred. It is good to see from Fig. 4.10, which displays the widths for values of x in the range between 0 and 25, that the conservative Clopper–Pearson confidence interval has the largest width for $x \geq 4$. The Wilson confidence interval has the smallest width for $x > 10$, while for $x \leq 10$, the Wald for π and $\arcsin(\sqrt{\pi})$ and the likelihood ratio confidence interval have the smallest width. The Wald confidence interval for $\text{logit}(\pi)$ has a quite large width for $x \leq 10$.

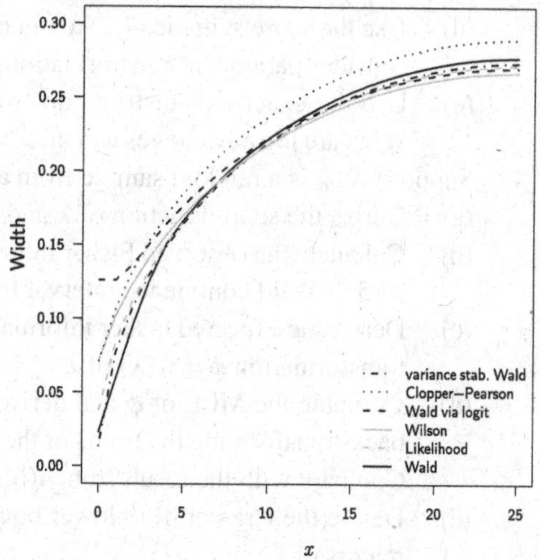

Fig. 4.10 Widths of the 95 % confidence intervals for π with a sample size of $n = 50$ in a binomial experiment, depending on the number of successes $X = x$. The values are symmetric around $x = 25.5$. The width of the Wald confidence interval on the logit scale is 1 for $x = 0$ and $x = 1$

To summarise, this empirical comparison identifies the Wilson confidence interval as the one with the best properties. The likelihood ratio confidence interval is not much worse than Wilson, but computation is more demanding. The different Wald confidence intervals do not have so good operational characteristics, in particular the "standard" interval for π has poor properties. The Clopper–Pearson confidence interval is a special case, as its coverage is always larger than (or equal to) the nominal level. Hence, it tends to have a larger width than the competing ones. ∎

4.7 Exercises

1. Compute an approximate 95 % confidence interval for the true correlation ρ based on the MLE $r = 0.7$, a sample of size of $n = 20$ and Fisher's z-transformation.
2. Derive a general formula for the score confidence interval in the Poisson model based on the Fisher information, cf. Example 4.9.
3. A study is conducted to quantify the evidence against the null hypothesis that less than 80 percent of the Swiss population have antibodies against the human herpesvirus. Among a total of 117 persons investigated, 105 had antibodies.
   (a) Formulate an appropriate statistical model and the null and alternative hypotheses. Which sort of P-value should be used to quantify the evidence against the null hypothesis?
   (b) Use the Wald statistic (4.12) and its approximate normal distribution to obtain a P-value.
   (c) Use the logit-transformation (compare Example 3.13) and the corresponding Wald statistic to obtain a P-value.

(d) Use the score statistic (4.2) to obtain a P-value. Why do we not need to consider parameter transformations when using this statistic?
(e) Use the exact null distribution from your model to obtain a P-value. What are the advantages and disadvantages of this procedure in general?

4. Suppose $X_{1:n}$ is a random sample from an $\text{Exp}(\lambda)$ distribution.
(a) Derive the score function of λ and solve the score equation to get $\hat{\lambda}_{\text{ML}}$.
(b) Calculate the observed Fisher information, the standard error of $\hat{\lambda}_{\text{ML}}$ and a 95 % Wald confidence interval for λ.
(c) Derive the expected Fisher information $J(\lambda)$ and the variance-stabilising transformation $\phi = h(\lambda)$ of λ.
(d) Compute the MLE of ϕ and derive a 95 % confidence interval for λ by back-transforming the limits of the 95 % Wald confidence interval for ϕ. Compare with the result from 4(b).
(e) Derive the Cramér–Rao lower bound for the variance of unbiased estimators of λ.
(f) Compute the expectation of $\hat{\lambda}_{\text{ML}}$ and use this result to construct an unbiased estimator of λ. Compute its variance and compare it to the Cramér–Rao lower bound.

5. An alternative parametrisation of the exponential distribution is

$$f_X(x) = \frac{1}{\theta}\exp\left(-\frac{x}{\theta}\right)\mathbb{I}_{\mathbb{R}^+}(x), \quad \theta > 0.$$

Let $X_{1:n}$ denote a random sample from this density. We want to test the null hypothesis $H_0: \theta = \theta_0$ against the alternative hypothesis $H_1: \theta \neq \theta_0$.
(a) Calculate both variants T_1 and T_2 of the score test statistic.
(b) A sample of size $n = 100$ gave $\bar{x} = 0.26142$. Quantify the evidence against $H_0: \theta_0 = 0.25$ using a suitable significance test.

6. In a study assessing the sensitivity π of a low-budget diagnostic test for asthma, each of n asthma patients is tested repeatedly until the first positive test result is obtained. Let X_i be the number of the first positive test for patient i. All patients and individual tests are independent, and the sensitivity π is equal for all patients and tests.
(a) Derive the probability mass function $f(x; \pi)$ of X_i.
(b) Write down the log-likelihood function for the random sample $X_{1:n}$ and compute the MLE $\hat{\pi}_{\text{ML}}$.
(c) Derive the standard error $\text{se}(\hat{\pi}_{\text{ML}})$ of the MLE.
(d) Give a general formula for an approximate 95 % confidence interval for π. What could be the problem of this interval?
(e) Now we consider the parametrisation with $\phi = \text{logit}(\pi) = \log\{\pi/(1-\pi)\}$. Derive the corresponding MLE $\hat{\phi}_{\text{ML}}$, its standard error and associated approximate 95 % confidence interval. What is the advantage of this interval?

(f) $n = 9$ patients did undergo the trial, and the observed numbers were $x = (3, 5, 2, 6, 9, 1, 2, 2, 3)$. Calculate the MLEs $\hat{\pi}_{\text{ML}}$ and $\hat{\phi}_{\text{ML}}$, the confidence intervals from 6(d) and 6(e) and compare them by transforming the latter back to the π-scale.

(g) Produce a plot of the relative log-likelihood function $\tilde{l}(\pi)$ and two approximations in the range $\pi \in (0.01, 0.5)$: The first approximation is based on the direct quadratic approximation of $\tilde{l}_\pi(\pi) \approx q_\pi(\pi)$, and the second approximation is based on the quadratic approximation of $\tilde{l}_\phi(\phi) \approx q_\phi(\phi)$, i.e. $q_\phi\{\text{logit}(\pi)\}$ values are plotted. Comment the result.

7. A simple model for the drug concentration in plasma over time after a single intravenous injection is $c(t) = \theta_2 \exp(-\theta_1 t)$ with $\theta_1, \theta_2 > 0$. For simplicity, we assume here that $\theta_2 = 1$.

(a) Assume that n probands had their concentrations c_i, $i = 1, \dots, n$, measured at the same single time-point t, and assume that the model $c_i \overset{\text{iid}}{\sim} \text{N}(c(t), \sigma^2)$ is appropriate for the data. Calculate the MLE of θ_1.

(b) Calculate the asymptotic variance of the MLE.

(c) In pharmacokinetic studies one is often interested in the area under the concentration curve, $\alpha = \int_0^\infty \exp(-\theta_1 t)\, dt$. Calculate the MLE for α and its variance estimate using the delta theorem.

(d) We now would like to determine the optimal time point for measuring the concentrations c_i. Minimise the asymptotic variance of the MLE with respect to t, when θ_1 is assumed to be known, to obtain an optimal time point t_{opt}.

8. Assume the gamma model $\text{G}(\alpha, \alpha/\mu)$ for the random sample $X_{1:n}$ with mean $\mathsf{E}(X_i) = \mu > 0$ and shape parameter $\alpha > 0$.

(a) First assume that α is known. Derive the MLE $\hat{\mu}_{\text{ML}}$ and the observed Fisher information $I(\hat{\mu}_{\text{ML}})$.

(b) Use the p^* formula to derive an asymptotic density of $\hat{\mu}_{\text{ML}}$ depending on the true parameter μ. Show that the kernel of this approximate density is exact in this case, i.e. it equals the kernel of the exact density known from Exercise 9 from Chap. 4.

(c) Stirling's approximation of the gamma function is

$$\Gamma(x) \approx \sqrt{\frac{2\pi}{x}} \frac{x^x}{\exp(x)}. \tag{4.24}$$

Show that approximating the normalising constant of the exact density with (4.24) gives the normalising constant of the approximate p^* formula density.

(d) Now assume that μ is known. Derive the log-likelihood, score function and Fisher information of α. Use the digamma function $\psi(x) = \frac{d}{dx}\log\{\Gamma(x)\}$ and the trigamma function $\psi'(x) = \frac{d}{dx}\psi(x)$.

(e) Show, by rewriting the score equation, that the MLE $\hat{\alpha}_{\text{ML}}$ fulfils

$$-n\psi(\hat{\alpha}_{\text{ML}}) + n\log(\hat{\alpha}_{\text{ML}}) + n = -\sum_{i=1}^{n}\log(x_i) + \frac{1}{\mu}\sum_{i=1}^{n} x_i + n\log(\mu). \tag{4.25}$$

Hence, show that the log-likelihood kernel can be written as

$$l(\alpha) = n\big[\alpha\log(\alpha) - \alpha - \log\{\Gamma(\alpha)\} + \alpha\psi(\hat{\alpha}_{\text{ML}}) - \alpha\log(\hat{\alpha}_{\text{ML}})\big].$$

(f) Implement an R-function of the p^* formula, taking as arguments the MLE value(s) $\hat{\alpha}_{\text{ML}}$, at which to evaluate the density, and the true parameter α. For numerical reasons, first compute the approximate log-density

$$\log f^*(\hat{\alpha}_{\text{ML}}) = -\frac{1}{2}\log(2\pi) + \frac{1}{2}\log\{I(\hat{\alpha}_{\text{ML}})\} + l(\alpha) - l(\hat{\alpha}_{\text{ML}})$$

and then exponentiate it. The R-functions `digamma`, `trigamma` and `lgamma` can be used to calculate $\psi(x)$, $\psi'(x)$ and $\log\{\Gamma(x)\}$, respectively.

(g) In order to illustrate the quality of this approximation, we consider the case with $\alpha = 2$ and $\mu = 3$. Simulate 10 000 data sets of size $n = 10$ and compute the MLE $\hat{\alpha}_{\text{ML}}$ for each of them by numerically solving (4.25) using the R-function `uniroot` (cf. Appendix C.1.2). Plot a histogram of the resulting 10 000 MLE samples (using `hist` with option `prob=TRUE`). Add the approximate density derived above to compare.

4.8 References

The methods discussed in this chapter are found in many books on likelihood inference, for example in Pawitan (2001) or Davison (2003), see also Millar (2011). Further details on Fisher regularity conditions can be found in Lehmann and Casella (1998, Chap. 6). Different confidence intervals for a binomial proportion are thoroughly discussed in Brown et al. (2001) and Connor and Imrey (2005), see also Newcombe (2013). The smoothed coverage probabilities have been computed using a specific kernel function as described in Bayarri and Berger (2004).

Likelihood Inference in Multiparameter Models

5

Contents

In the previous chapters we have discussed likelihood inference if the statistical model depends on only one unknown parameter. However, in many applications the statistical model contains more than one unknown parameter. This chapter describes how the concepts of likelihood inference for scalar parameters can be extended to vectorial parameters. We will print an unknown parameter vector $\boldsymbol{\theta}$ in bold, to distinguish it from a single scalar parameter θ. We start with two classical examples.

Example 5.1 (Normal model) The normal model has two unknown parameters, the expectation μ and the variance σ^2. Suppose that both are unknown and let $X_{1:n}$ denote a random sample from an $\mathrm{N}(\mu, \sigma^2)$ distribution, so the unknown parameter vector is $\boldsymbol{\theta} = (\mu, \sigma^2)^\top$. The log-likelihood

$$
\begin{aligned}
l(\boldsymbol{\theta}) = l(\mu, \sigma^2) &= -\frac{n}{2}\log(\sigma^2) - \frac{1}{2\sigma^2}\sum_{i=1}^{n}(x_i - \mu)^2 \\
&= -\frac{n}{2}\log(\sigma^2) - \frac{1}{2\sigma^2}\left\{(n-1)s^2 + n(\bar{x} - \mu)^2\right\}
\end{aligned}
$$

L. Held, D. Sabanés Bové, *Likelihood and Bayesian Inference*,
Statistics for Biology and Health, https://doi.org/10.1007/978-3-662-60792-3_5,

Fig. 5.1 Contour plot of the relative likelihood in the normal model

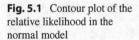

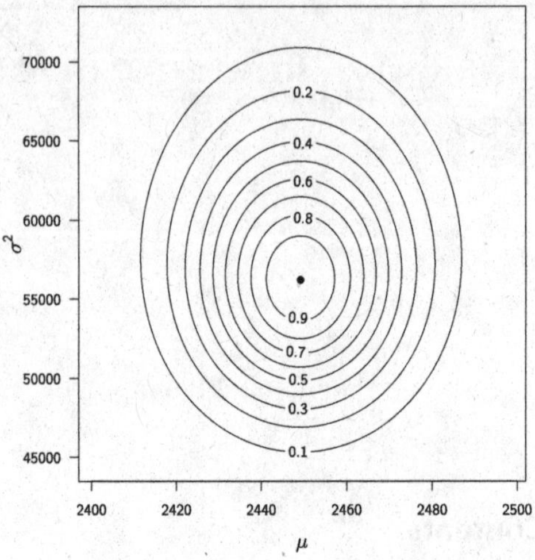

is hence a function of μ and σ^2. Note that the data $x_{1:n}$ enter the likelihood function through the sufficient statistics $\bar{x}$ and s^2, compare Example 2.21.

For illustration, consider the study on alcohol concentration from Sect. 1.1.7. Assume that there is no difference between genders, so that we can look at the overall $n = 185$ volunteers with empirical mean and variance of the transformation factors given in Table 1.3. The resulting relative likelihood is shown in Fig. 5.1. ■

Example 5.2 (Multinomial model) Let $\boldsymbol{X} \sim \mathrm{M}_k(n, \boldsymbol{\pi})$ denote a sample from a multinomial distribution with unknown parameter vector $\boldsymbol{\theta} = \boldsymbol{\pi} = (\pi_1, \dots, \pi_k)^\top$. The log-likelihood

$$l(\boldsymbol{\theta}) = l(\boldsymbol{\pi}) = l(\pi_1, \dots, \pi_k) = \sum_{i=1}^{k} x_i \log(\pi_i)$$

is a function of the k unknown probabilities $\pi_1, \dots, \pi_k$. Since they have to sum to one, there are actually only $k-1$ free parameters. To compute the MLE, this restriction has to be taken into account using the Lagrange method, cf. Example 5.4. For illustration, we consider the frequencies of the genotypes MM, MN and NN from Sect. 1.1.4, $x_1 = 233$, $x_2 = 385$ and $x_3 = 129$. Here $k = 3$, so we have a trinomial model with $k-1 = 2$ free parameters. Figure 5.2 displays the relative log-likelihood with respect to π_1 and π_2 (π_3 can be calculated via $\pi_3 = 1 - \pi_1 - \pi_2$). ■

5.1 Score Vector and Fisher Information Matrix

Definition 5.1 (Score vector) If the unknown parameter $\boldsymbol{\theta} = (\theta_1, \dots, \theta_p)^\top$ is a vector, the gradient of the log-likelihood

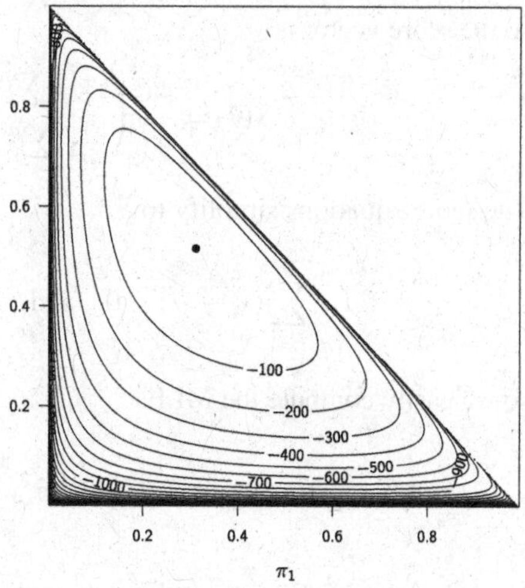

Fig. 5.2 Contour plot of the relative log-likelihood in the trinomial model

$$S(\boldsymbol{\theta}) = \frac{\partial}{\partial \boldsymbol{\theta}} l(\boldsymbol{\theta}) = \left(\frac{\partial}{\partial \theta_1} l(\boldsymbol{\theta}), \dots, \frac{\partial}{\partial \theta_p} l(\boldsymbol{\theta}) \right)^\top$$

is called the *score vector* or *score function*. ♦

Definition 5.2 (Score equations) The system of p equations

$$S(\boldsymbol{\theta}) = \mathbf{0}$$

are called the *score equations*. ♦

The MLE $\hat{\boldsymbol{\theta}}_{\text{ML}} = (\hat{\theta}_1, \dots, \hat{\theta}_p)^\top$ is now a vector and usually obtained by solving the score equations.

Definition 5.3 (Fisher information matrix) The (symmetric) *Fisher information matrix* $I(\boldsymbol{\theta})$ is the $p \times p$ matrix with entries

$$-\frac{\partial^2}{\partial \theta_i \partial \theta_j} l(\boldsymbol{\theta}), \quad 1 \le i, j \le p.$$

The *observed Fisher information matrix* is $I(\hat{\boldsymbol{\theta}}_{\text{ML}})$. ♦

Example 5.3 (Normal model) The partial derivatives of the log-likelihood are

$$\frac{\partial l(\boldsymbol{\theta})}{\partial \mu} = \frac{1}{\sigma^2} \sum_{i=1}^n (x_i - \mu) \quad \text{and}$$

$$\frac{\partial l(\boldsymbol{\theta})}{\partial \sigma^2} = -\frac{n}{2\sigma^2} + \frac{1}{2\sigma^4} \sum_{i=1}^n (x_i - \mu)^2,$$

so the score vector is

$$\boldsymbol{S}(\boldsymbol{\theta}) = \frac{1}{\sigma^2}\begin{pmatrix} \sum_{i=1}^n (x_i - \mu) \\ \frac{1}{2\sigma^2}\{\sum_{i=1}^n (x_i - \mu)^2 - n\} \end{pmatrix}.$$

The score equations simplify to

$$\sum_{i=1}^n (x_i - \mu) = 0 \quad \text{and} \quad \frac{1}{\sigma^2}\sum_{i=1}^n (x_i - \mu)^2 = n,$$

so we easily compute the MLEs

$$\hat{\mu}_{\text{ML}} = \bar{x} \quad \text{and} \quad \hat{\sigma}^2_{\text{ML}} = \frac{1}{n}\sum_{i=1}^n (x_i - \bar{x})^2.$$

Note that the MLE of σ^2 is biased, as discussed previously in Example 3.1. The Fisher information $\boldsymbol{I}(\boldsymbol{\theta})$, the matrix containing the negative second partial derivatives, turns out to be

$$\begin{aligned} \boldsymbol{I}(\boldsymbol{\theta}) &= -\begin{pmatrix} \frac{\partial^2 l(\boldsymbol{\theta})}{\partial \mu^2} & \frac{\partial^2 l(\boldsymbol{\theta})}{\partial \mu \partial \sigma^2} \\ \frac{\partial^2 l(\boldsymbol{\theta})}{\partial \sigma^2 \partial \mu} & \frac{\partial^2 l(\boldsymbol{\theta})}{\partial (\sigma^2)^2} \end{pmatrix} \\ &= \frac{1}{\sigma^2}\begin{pmatrix} n & \frac{1}{\sigma^2}\sum_{i=1}^n (x_i - \mu) \\ \frac{1}{\sigma^2}\sum_{i=1}^n (x_i - \mu) & \frac{1}{\sigma^4}\sum_{i=1}^n (x_i - \mu)^2 - \frac{n}{2\sigma^2} \end{pmatrix}. \end{aligned} \tag{5.1}$$

The observed Fisher information, obtained by replacing μ and σ^2 with $\hat{\mu}_{\text{ML}}$ and $\hat{\sigma}^2_{\text{ML}}$, respectively, turns out to be

$$\boldsymbol{I}(\hat{\boldsymbol{\theta}}_{\text{ML}}) = \begin{pmatrix} \frac{n}{\hat{\sigma}^2_{\text{ML}}} & 0 \\ 0 & \frac{n}{2\hat{\sigma}^4_{\text{ML}}} \end{pmatrix}$$

and so is a diagonal matrix. ■

Example 5.4 (Multinomial model) Maximisation of the log-likelihood in the multinomial model from Example 5.2,

$$l(\boldsymbol{\pi}) = \sum_{i=1}^k x_i \log(\pi_i),$$

under the constraint $g(\boldsymbol{\pi}) = \sum_{i=1}^k (\pi_i) - 1 = 0$ can be done using the Lagrange method (cf. Appendix B.2.5). The equation to solve is

$$\frac{\partial}{\partial \boldsymbol{\pi}} l(\boldsymbol{\pi}) = \lambda \cdot \frac{\partial}{\partial \boldsymbol{\pi}} g(\boldsymbol{\pi}),$$

and plugging in the gradients gives

$$\left(\frac{x_1}{\pi_1}, \dots, \frac{x_k}{\pi_k}\right)^\top = \lambda \cdot (1, \dots, 1)^\top.$$

The number of trials is n, so

$$n = \sum_{i=1}^{k} x_i = \sum_{i=1}^{k} \lambda \pi_i = \lambda \sum_{i=1}^{k} \pi_i = \lambda,$$

and therefore

$$\hat{\pi}_{\text{ML}} = (\hat{\pi}_1, \dots, \hat{\pi}_k)^\top,$$

where $\hat{\pi}_i = x_i / n$. So the MLEs of the multinomial probabilities are just the corresponding relative frequencies.

An alternative approach explicitly replaces π_k by $1 - \sum_{i=1}^{k-1} \pi_i$ in the probability mass function of the multinomial distribution. The corresponding log-likelihood of the trimmed parameter vector $\tilde{\pi}$ is then

$$l(\tilde{\pi}) = \sum_{i=1}^{k-1} x_i \log(\pi_i) + x_k \log\left(1 - \sum_{i=1}^{k-1} \pi_i\right),$$

which directly leads to the score vector

$$S(\tilde{\pi}) = \frac{\partial}{\partial \tilde{\pi}} l(\tilde{\pi}) = \left(\frac{x_1}{\pi_1} - \frac{x_k}{1 - \sum_{i=1}^{k-1} \pi_i}, \dots, \frac{x_{k-1}}{\pi_{k-1}} - \frac{x_k}{1 - \sum_{i=1}^{k-1} \pi_i}\right)^\top.$$

This defines a set of score equations, which will eventually lead to the same MLEs as before.

The Fisher information matrix of $\tilde{\pi}$ has dimension $(k-1) \times (k-1)$ and is

$$I(\tilde{\pi}) = -\frac{\partial}{\partial \tilde{\pi}} S(\tilde{\pi}) = \text{diag}\left\{\frac{x_i}{\pi_i^2}\right\}_{i=1}^{k-1} + \frac{x_k}{(1 - \sum_{i=1}^{k-1} \pi_i)^2} \cdot \mathbf{1}\mathbf{1}^\top, \qquad (5.2)$$

where $\mathbf{1}$ is the unity vector of length $k-1$ such that the matrix $\mathbf{1}\mathbf{1}^\top$ contains only ones. If we replace in (5.2) $\tilde{\pi}$ by $\hat{\tilde{\pi}}_{\text{ML}}$, we obtain the observed Fisher information matrix

$$I(\hat{\tilde{\pi}}_{\text{ML}}) = n\left\{\text{diag}(\hat{\tilde{\pi}}_{\text{ML}})^{-1} + \left(1 - \sum_{i=1}^{k-1} \hat{\pi}_i\right)^{-1} \mathbf{1}\mathbf{1}^\top\right\}.$$

Applying a useful formula for matrix inversion (cf. Appendix B.1.4), we obtain the inverse observed Fisher information matrix

$$I(\hat{\tilde{\pi}}_{\text{ML}})^{-1} = n^{-1}\left\{\text{diag}(\hat{\tilde{\pi}}_{\text{ML}}) - \hat{\tilde{\pi}}_{\text{ML}} \hat{\tilde{\pi}}_{\text{ML}}^\top\right\}. \qquad (5.3)$$

■

A quadratic approximation of the log-likelihood can be obtained with a multidimensional Taylor expansion: Suppose that $\boldsymbol{\theta} = (\theta_1, \dots, \theta_p)^\top$ is the unknown true p-dimensional parameter vector. Applying Eq. (B.6) from Appendix B.2.3, we obtain the quadratic approximation

$$\tilde{l}(\boldsymbol{\theta}) \approx -\frac{1}{2}(\boldsymbol{\theta} - \hat{\boldsymbol{\theta}}_{\text{ML}})^\top \boldsymbol{I}(\hat{\boldsymbol{\theta}}_{\text{ML}})(\boldsymbol{\theta} - \hat{\boldsymbol{\theta}}_{\text{ML}}) \tag{5.4}$$

of the relative log-likelihood, which only requires the MLE $\hat{\boldsymbol{\theta}}_{\text{ML}}$ and the observed Fisher information matrix $\boldsymbol{I}(\hat{\boldsymbol{\theta}}_{\text{ML}})$. It is instructive to compare this approximation with the corresponding one for a scalar parameter:

$$\tilde{l}(\theta) \approx -\frac{1}{2} I(\hat{\theta}_{\text{ML}})(\theta - \hat{\theta}_{\text{ML}})^2$$

as discussed in Sect. 2.4. As in the previous Chap. 4, many of the subsequent results are based on the validity of the quadratic approximation (5.4).

5.2 Standard Error and Wald Confidence Interval

The Wald statistic $(\hat{\theta}_{\text{ML}} - \theta)/\text{se}(\hat{\theta}_{\text{ML}}) \overset{\text{a}}{\sim} \text{N}(0, 1)$ for a scalar parameter θ can be generalised easily to multiparameter models, where we have a parameter vector $\boldsymbol{\theta} = (\theta_1, \dots, \theta_p)^\top$. Here we have

$$\frac{\hat{\theta}_i - \theta_i}{\text{se}(\hat{\theta}_i)} \overset{\text{a}}{\sim} \text{N}(0, 1), \quad i = 1, \dots, p,$$

where the standard error $\text{se}(\hat{\theta}_i)$ is defined as the square root of the ith diagonal entry of the inverse observed Fisher information matrix.

Standard error of the ML estimator

The square root of the ith diagonal element of the inverse observed Fisher information matrix is a standard error of the ith component $\hat{\theta}_i$ of the MLE $\hat{\boldsymbol{\theta}}_{\text{ML}}$:

$$\text{se}(\hat{\theta}_i) = \sqrt{\left[\boldsymbol{I}(\hat{\boldsymbol{\theta}}_{\text{ML}})^{-1}\right]_{ii}}.$$

This result can be used to calculate the limits of a $\gamma \cdot 100\,\%$ Wald confidence interval for θ_i:

$$\hat{\theta}_i \pm z_{(1+\gamma)/2}\,\text{se}(\hat{\theta}_i). \tag{5.5}$$

A justification of this procedure is given in Result 5.1; see also Sect. 5.4.2.

Example 5.5 (Normal model) In Example 5.3 we have derived the observed Fisher information matrix in the normal model. This matrix is diagonal, so its inverse can easily be obtained by inverting the diagonal elements:

$$\boldsymbol{I}(\hat{\boldsymbol{\theta}}_{\mathrm{ML}})^{-1} = \begin{pmatrix} \frac{\hat{\sigma}^2_{\mathrm{ML}}}{n} & 0 \\ 0 & \frac{2\hat{\sigma}^4_{\mathrm{ML}}}{n} \end{pmatrix}.$$

The standard errors are therefore $\mathrm{se}(\hat{\mu}_{\mathrm{ML}}) = \hat{\sigma}_{\mathrm{ML}}/\sqrt{n}$ and $\mathrm{se}(\hat{\sigma}^2_{\mathrm{ML}}) = \hat{\sigma}^2_{\mathrm{ML}}\sqrt{2/n}$, and we obtain the $\gamma \cdot 100$ % Wald confidence intervals with limits

$$\bar{x} \pm z_{(1+\gamma)/2}\frac{\hat{\sigma}_{\mathrm{ML}}}{\sqrt{n}} \quad \text{and}$$

$$\hat{\sigma}^2_{\mathrm{ML}} \pm z_{(1+\gamma)/2}\sqrt{\frac{2}{n}}\hat{\sigma}^2_{\mathrm{ML}}$$

for μ and σ^2, respectively. Note that these intervals are only asymptotically valid. Indeed, we know from Example 3.8 that the interval with limits

$$\bar{x} \pm t_{(1+\gamma)/2}(n-1) \cdot \frac{s}{\sqrt{n}},$$

where $s^2 = \sum_{i=1}^n (x_i - \bar{x})^2/(n-1)$, forms an exact $\gamma \cdot 100$ % confidence interval for μ, which is wider because $t_{(1+\gamma)/2}(n-1) > z_{(1+\gamma)/2}$ and $s^2 > \hat{\sigma}^2_{\mathrm{ML}}$. However, asymptotically the two intervals are equivalent because, as $n \to \infty$, both $(\hat{\sigma}^2_{\mathrm{ML}} - s^2) \to 0$ and $t_{(1+\gamma)/2}(n-1) \to z_{(1+\gamma)/2}$.

The lower limit of the confidence interval for σ^2 is $\hat{\sigma}^2_{\mathrm{ML}}(1 - z_{(1+\gamma)/2}\sqrt{2/n})$, which may be negative, for example for $\gamma = 95$ % and $n = 5$. This once again illustrates that Wald confidence intervals may not be boundary-respecting. The profile likelihood confidence intervals, which will be introduced in the next section, are boundary-respecting and should thus in general be preferred. ■

Example 5.6 (Hardy–Weinberg equilibrium) The MLEs of the probabilities in the trinomial model are

$$\hat{\pi}_1 = \frac{233}{747} \approx 0.312, \qquad \hat{\pi}_2 = \frac{385}{747} \approx 0.515 \quad \text{and} \quad \hat{\pi}_3 = \frac{129}{747} \approx 0.173.$$

These values are slightly different from the corresponding estimates assuming Hardy–Weinberg equilibrium:

$$\hat{\pi}_1 \approx 0.324, \qquad \hat{\pi}_2 \approx 0.490 \quad \text{and} \quad \hat{\pi}_3 \approx 0.185,$$

cf. Example 2.7. The standard errors of the MLEs in the multinomial model can be easily obtained from (5.3):

$$\mathrm{se}(\hat{\pi}_i) = \sqrt{\frac{\hat{\pi}_i(1-\hat{\pi}_i)}{n}},$$

which is exactly the same formula as in the binomial model. In our example we obtain $\mathrm{se}(\hat{\pi}_1) = 0.017$, $\mathrm{se}(\hat{\pi}_2) = 0.018$ and $\mathrm{se}(\hat{\pi}_3) = 0.014$. Note that the difference between the MLEs under Hardy–Weinberg equilibrium and the MLEs in the trinomial model is less than two standard errors for all three parameters. This suggests that the Hardy–Weinberg model fits these data quite well, although this interpretation ignores the uncertainty of the estimates in the Hardy–Weinberg model. We will describe more rigorous approaches to decide between these two models in Sect. 7.1. ■

5.3 Profile Likelihood

Visualisation of a multidimensional likelihood function of $\boldsymbol{\theta}$ is in general difficult, so instead one might want to look at one-dimensional likelihood functions of the components θ_i. The problem is how to eliminate the remaining *nuisance parameters* θ_j $(j \neq i)$ from the joint likelihood. This is done using the *profile likelihood*.

To ease notation, in the following the parameter vector of interest is denoted $\boldsymbol{\theta}$, and the vector of nuisance parameters is denoted $\boldsymbol{\eta}$.

Definition 5.4 (Profile likelihood) Let $L(\boldsymbol{\theta}, \boldsymbol{\eta})$ be the joint likelihood function of the parameter of interest $\boldsymbol{\theta}$ and the nuisance parameter $\boldsymbol{\eta}$. Let $\hat{\boldsymbol{\eta}}_{\mathrm{ML}}(\boldsymbol{\theta})$ denote the MLE with respect to $L(\boldsymbol{\theta}, \boldsymbol{\eta})$ for fixed $\boldsymbol{\theta}$. Then

$$L_p(\boldsymbol{\theta}) = \max_{\boldsymbol{\eta}} L(\boldsymbol{\theta}, \boldsymbol{\eta}) = L\{\boldsymbol{\theta}, \hat{\boldsymbol{\eta}}_{\mathrm{ML}}(\boldsymbol{\theta})\}$$

is called the *profile likelihood* of $\boldsymbol{\theta}$. The value of the profile likelihood at a particular parameter value $\boldsymbol{\theta}$ is obtained through maximising the joint likelihood $L(\boldsymbol{\theta}, \boldsymbol{\eta})$ with respect to the nuisance parameter $\boldsymbol{\eta}$. ♦

We often need numerical techniques to maximise the joint likelihood with respect to the nuisance parameter $\boldsymbol{\eta}$. To avoid this, it is tempting to just plug-in the MLE of the nuisance parameter $\boldsymbol{\eta}$ in the joint likelihood $L(\boldsymbol{\theta}, \boldsymbol{\eta})$, which leads to the following definition.

Definition 5.5 (Estimated likelihood) Let $(\hat{\boldsymbol{\theta}}_{\mathrm{ML}}, \hat{\boldsymbol{\eta}}_{\mathrm{ML}})$ denote the MLE of $(\boldsymbol{\theta}, \boldsymbol{\eta})$ based on the joint likelihood $L(\boldsymbol{\theta}, \boldsymbol{\eta})$. The function

$$L_e(\boldsymbol{\theta}) = L(\boldsymbol{\theta}, \hat{\boldsymbol{\eta}}_{\mathrm{ML}})$$

is called the *estimated likelihood* of $\boldsymbol{\theta}$. ♦

The estimated likelihood ignores the uncertainty of the MLE $\hat{\boldsymbol{\eta}}_{\mathrm{ML}}$: it assumes that the true value of the nuisance parameter $\boldsymbol{\eta}$ is equal to its estimate $\hat{\boldsymbol{\eta}}_{\mathrm{ML}}$. As a consequence, the estimated likelihood $L_e(\boldsymbol{\theta})$ will in general not correctly quantify the uncertainty with respect to $\boldsymbol{\theta}$.

The logarithm of the profile likelihood L_p defines the *profile log-likelihood* l_p. We can also easily define the *relative profile likelihood* $\tilde{L}_p$ and *relative profile log-likelihood* $\tilde{l}_p$, compare Definition 2.5. Equivalent definitions are possible for the estimated likelihood.

If θ is a scalar, then we can calculate confidence intervals based on the relative profile likelihood using the standard thresholds 0.147 and 0.036 for $\gamma = 95\ \%$ and $\gamma = 99\ \%$, compare Table 4.1. In complete analogy to Eq. (4.18), the set

$$\left\{\theta : \tilde{L}_p(\theta) \geq \exp\left[-\frac{1}{2}\chi^2_\gamma(1)\right]\right\}$$

is then an approximate $\gamma \cdot 100\ \%$ *profile likelihood confidence interval* for θ. As in Eq. (4.17), we can also derive this interval using the relative profile log-likelihood $\tilde{l}_p(\theta)$. A theoretical justification of this approach will be discussed in Sect. 5.5.

Example 5.7 (Normal model) Let $X_{1:n}$ denote a random sample from an $\mathrm{N}(\mu, \sigma^2)$ distribution with both parameters unknown. The MLE of σ^2 for fixed μ is

$$\hat{\sigma}^2_{\mathrm{ML}}(\mu) = \frac{1}{n}\sum_{i=1}^n (x_i - \mu)^2.$$

Plugging this in the joint likelihood of μ and σ^2,

$$L(\mu, \sigma^2) = (\sigma^2)^{-\frac{n}{2}} \exp\left\{-\frac{1}{2\sigma^2}\sum_{i=1}^n (x_i - \mu)^2\right\}, \tag{5.6}$$

gives the profile likelihood of μ (ignoring multiplicative constants):

$$\begin{aligned} L_p(\mu) &= L\{\mu, \hat{\sigma}^2_{\mathrm{ML}}(\mu)\} = \{\hat{\sigma}^2_{\mathrm{ML}}(\mu)\}^{-\frac{n}{2}} \\ &= \left\{\frac{1}{n}\sum_{i=1}^n (x_i - \mu)^2\right\}^{-\frac{n}{2}} \\ &= \left\{(\bar{x} - \mu)^2 + \frac{1}{n}\sum_{i=1}^n (x_i - \bar{x})^2\right\}^{-\frac{n}{2}} \\ &= \left\{(\bar{x} - \mu)^2 + \frac{n-1}{n}s^2\right\}^{-\frac{n}{2}}. \end{aligned}$$

The estimated likelihood of μ turns out to be

$$L_e(\mu) = L(\mu, \hat{\sigma}^2_{\mathrm{ML}}) = (\hat{\sigma}^2_{\mathrm{ML}})^{-\frac{n}{2}} \exp\left\{-\frac{1}{2\hat{\sigma}^2_{\mathrm{ML}}}\sum_{i=1}^n (x_i - \mu)^2\right\},$$

where $\hat{\sigma}^2_{\mathrm{ML}} = \sum_{i=1}^n (x_i - \bar{x})^2/n$.

a) Contour plot of the relative likelihood $\tilde{L}(\mu, \sigma^2)$ from (5.6) with lines for $\{\mu, \hat{\sigma}^2_{\text{ML}}(\mu)\}$ and $(\mu, \hat{\sigma}^2_{\text{ML}})$

b) Profile and estimated relative likelihood of μ: the threshold at 0.147 defines an approximate 95% profile likelihood confidence interval. Note that the corresponding interval based on the estimated likelihood is, if only slightly, smaller.

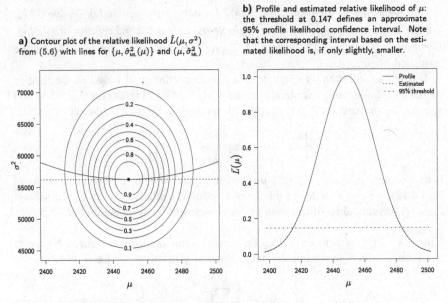

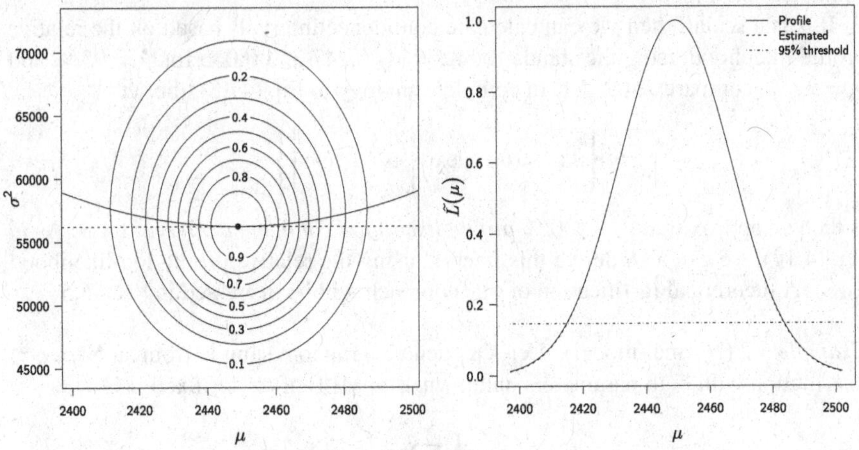

Fig. 5.3 Comparison of estimated and profile likelihood in the normal model

Both the profile and the estimated relative likelihood for the alcohol concentration data set from Sect. 1.1.7 are shown in Fig. 5.3b. Here the estimated likelihood is only slightly narrower than the profile likelihood. The difference is better to see in Fig. 5.3a, which marks the points of the joint relative likelihood that define the profile and estimated likelihood. The horizontal line $\sigma^2 = \hat{\sigma}^2_{\text{ML}}$ corresponds to the estimated likelihood, while the profile likelihood is obtained through maximisation of the joint likelihood with respect to the nuisance parameter. The values of the profile likelihood are marked in Fig. 5.3a using a solid line. The horizontal line at $\exp\{-\frac{1}{2}\chi^2_{0.95}(1)\} = 0.147$ in Fig. 5.3b allows the calculation of the 95 % profile likelihood confidence interval for μ, which turns out to be $[2414.83, 2483.53]$. For comparison, the exact 95 % confidence interval from Example 3.8 is $[2414.92, 2483.44]$.

Finally, we can also calculate the profile likelihood of σ^2 using $\hat{\mu}_{\text{ML}}(\sigma^2) = \bar{x}$:

$$L_p(\sigma^2) = (\sigma^2)^{-\frac{n}{2}} \exp\left\{-\frac{1}{2\sigma^2} \sum_{i=1}^{n} (x_i - \bar{x})^2\right\}.$$

Now $\hat{\mu}_{\text{ML}}(\sigma^2) = \bar{x}$ does not depend on σ^2, so the profile likelihood $L_p(\sigma^2)$ and the estimated likelihood $L_e(\sigma^2)$ are identical in this case. ■

Wald confidence intervals as defined in (5.5) can be justified by a quadratic approximation of the profile log-likelihood. The curvature of the profile log-likelihood at its maximum plays a central role in the argument. The following result derives this curvature for any dimension of the parameter of interest $\boldsymbol{\theta}$ and the nuisance parameter $\boldsymbol{\eta}$.

Result 5.1 (Curvature of the profile likelihood) *Let* $\boldsymbol{\gamma} = (\boldsymbol{\theta}^\top, \boldsymbol{\eta}^\top)^\top$ *(with* $\dim(\boldsymbol{\theta}) = p$ *and* $\dim(\boldsymbol{\eta}) = q$*) denote the unknown parameter vector with likelihood* $L(\boldsymbol{\theta}, \boldsymbol{\eta})$ *and observed Fisher information*

$$\boldsymbol{I}(\hat{\boldsymbol{\theta}}_{\mathrm{ML}}, \hat{\boldsymbol{\eta}}_{\mathrm{ML}}) = \begin{pmatrix} \boldsymbol{I}_{11} & \boldsymbol{I}_{12} \\ \boldsymbol{I}_{21} & \boldsymbol{I}_{22} \end{pmatrix},$$

a symmetric $(p+q) \times (p+q)$ *matrix (block dimensions for* $\boldsymbol{I}_{11}$: $p \times p$, $\boldsymbol{I}_{12}$: $p \times q$, $\boldsymbol{I}_{21}$: $q \times p$, $\boldsymbol{I}_{22}$: $q \times q$*). Now partition the inverse of* $\boldsymbol{I}(\hat{\boldsymbol{\theta}}_{\mathrm{ML}}, \hat{\boldsymbol{\eta}}_{\mathrm{ML}})$ *accordingly:*

$$\boldsymbol{I}(\hat{\boldsymbol{\theta}}_{\mathrm{ML}}, \hat{\boldsymbol{\eta}}_{\mathrm{ML}})^{-1} = \begin{pmatrix} \boldsymbol{I}^{11} & \boldsymbol{I}^{12} \\ \boldsymbol{I}^{21} & \boldsymbol{I}^{22} \end{pmatrix}.$$

The negative curvature of the profile log-likelihood of $\boldsymbol{\theta}$ *at* $\hat{\boldsymbol{\theta}}_{\mathrm{ML}}$ *is then equal to* $(\boldsymbol{I}^{11})^{-1}$.

Proof To prove this result, recall that the profile log-likelihood $l_p(\boldsymbol{\theta}) = l\{\boldsymbol{\theta}, \hat{\boldsymbol{\eta}}_{\mathrm{ML}}(\boldsymbol{\theta})\}$ is defined through the conditional MLE $\hat{\boldsymbol{\eta}}_{\mathrm{ML}}(\boldsymbol{\theta})$ for fixed $\boldsymbol{\theta}$, which is found by solving $\frac{\partial}{\partial \boldsymbol{\eta}} l(\boldsymbol{\theta}, \boldsymbol{\eta}) = 0$. Therefore, we know that

$$\frac{\partial}{\partial \boldsymbol{\eta}} l\{\boldsymbol{\theta}, \hat{\boldsymbol{\eta}}_{\mathrm{ML}}(\boldsymbol{\theta})\} = 0 \tag{5.7}$$

for all $\boldsymbol{\theta}$. Hence, the derivative of this function with respect to $\boldsymbol{\theta}$ is also zero:

$$\frac{\partial}{\partial \boldsymbol{\theta}} \left[\frac{\partial}{\partial \boldsymbol{\eta}^\top} l\{\boldsymbol{\theta}, \hat{\boldsymbol{\eta}}_{\mathrm{ML}}(\boldsymbol{\theta})\} \right] = 0.$$

Note that this derivative is an $r \times q$ matrix. Also note the order: First, derivatives of the log-likelihood is with respect to $\boldsymbol{\eta}$ are taken, then the conditional MLE $\hat{\boldsymbol{\eta}}_{\mathrm{ML}}(\boldsymbol{\theta})$ is plugged in for $\boldsymbol{\eta}$, and afterwards this function is differentiated. Since we have the convention that the function to be differentiated gives column values, we use the transpose symbol in the inner derivative here, which means that we transpose the resulting row vector after taking derivatives. Using the multivariate chain rule for differentiation, cf. Appendix B.2.2, we can rewrite this matrix as

$$\frac{\partial}{\partial \boldsymbol{\theta}} \left[\frac{\partial}{\partial \boldsymbol{\eta}^\top} l\{\boldsymbol{\theta}, \hat{\boldsymbol{\eta}}_{\mathrm{ML}}(\boldsymbol{\theta})\} \right] = \frac{\partial}{\partial \boldsymbol{\theta}} \boldsymbol{h}\{\boldsymbol{g}(\boldsymbol{\theta})\}$$

with $\boldsymbol{h}(\boldsymbol{\gamma}) = \frac{\partial}{\partial \boldsymbol{\eta}^\top} l(\boldsymbol{\gamma})$ and

$$\begin{aligned}
\boldsymbol{g}(\boldsymbol{\theta}) &= \left(\boldsymbol{\theta}, \hat{\boldsymbol{\eta}}_{\mathrm{ML}}(\boldsymbol{\theta})\right)^\top \\
&= \frac{\partial}{\partial \boldsymbol{\gamma}} \boldsymbol{h}\{\boldsymbol{g}(\boldsymbol{\theta})\} \cdot \frac{\partial}{\partial \boldsymbol{\theta}} \boldsymbol{g}(\boldsymbol{\theta}) \\
&= \begin{pmatrix} \frac{\partial}{\partial \boldsymbol{\theta}} \boldsymbol{h}\{\boldsymbol{g}(\boldsymbol{\theta})\} & \frac{\partial}{\partial \boldsymbol{\eta}} \boldsymbol{h}\{\boldsymbol{g}(\boldsymbol{\theta})\} \end{pmatrix} \begin{pmatrix} \mathbf{I}_q \\ \frac{\partial}{\partial \boldsymbol{\theta}} \hat{\boldsymbol{\eta}}_{\mathrm{ML}}(\boldsymbol{\theta}) \end{pmatrix}
\end{aligned}$$

$$
\begin{aligned}
&= \frac{\partial}{\partial \boldsymbol{\theta}} \boldsymbol{h}\{\boldsymbol{g}(\boldsymbol{\theta})\} + \frac{\partial}{\partial \boldsymbol{\eta}} \boldsymbol{h}\{\boldsymbol{g}(\boldsymbol{\theta})\} \cdot \frac{\partial}{\partial \boldsymbol{\theta}} \hat{\boldsymbol{\eta}}_{\mathrm{ML}}(\boldsymbol{\theta}) \\
&= \underbrace{\frac{\partial}{\partial \boldsymbol{\theta}} \frac{\partial}{\partial \boldsymbol{\eta}^\top} l\{\boldsymbol{\theta}, \hat{\boldsymbol{\eta}}_{\mathrm{ML}}(\boldsymbol{\theta})\}}_{r \times q} + \underbrace{\frac{\partial}{\partial \boldsymbol{\eta}} \frac{\partial}{\partial \boldsymbol{\eta}^\top} l\{\boldsymbol{\theta}, \hat{\boldsymbol{\eta}}_{\mathrm{ML}}(\boldsymbol{\theta})\}}_{r \times r} \cdot \underbrace{\frac{\partial}{\partial \boldsymbol{\theta}} \hat{\boldsymbol{\eta}}_{\mathrm{ML}}(\boldsymbol{\theta})}_{r \times q}.
\end{aligned}
$$

Note that in the first two matrices of the last line, the log-likelihood is differentiated twice each, and afterwards the conditional MLE $\hat{\boldsymbol{\eta}}_{\mathrm{ML}}(\boldsymbol{\theta})$ is plugged in. From above we know that $\boldsymbol{g}(\boldsymbol{\theta}) = 0$, so we can solve the equation for $\frac{\partial}{\partial \boldsymbol{\theta}} \hat{\boldsymbol{\eta}}_{\mathrm{ML}}(\boldsymbol{\theta})$ and obtain

$$
\frac{\partial}{\partial \boldsymbol{\theta}} \hat{\boldsymbol{\eta}}_{\mathrm{ML}}(\boldsymbol{\theta}) = -\left[\frac{\partial}{\partial \boldsymbol{\eta}} \frac{\partial}{\partial \boldsymbol{\eta}^\top} l\{\boldsymbol{\theta}, \hat{\boldsymbol{\eta}}_{\mathrm{ML}}(\boldsymbol{\theta})\}\right]^{-1} \frac{\partial}{\partial \boldsymbol{\theta}} \frac{\partial}{\partial \boldsymbol{\eta}^\top} l\{\boldsymbol{\theta}, \hat{\boldsymbol{\eta}}_{\mathrm{ML}}(\boldsymbol{\theta})\}. \tag{5.8}
$$

We can also use the multivariate chain rule to differentiate the profile log-likelihood $l_p(\boldsymbol{\theta})$. In order to be consistent with the notation in Appendix B.2.2, we use here a $1 \times q$ row vector for this derivative, instead of the otherwise used convention that the score vector is a column vector:

$$
\begin{aligned}
\frac{\partial}{\partial \boldsymbol{\theta}} l_p(\boldsymbol{\theta}) &= \frac{\partial}{\partial \boldsymbol{\theta}} l\{\boldsymbol{g}(\boldsymbol{\theta})\} \\
&= \frac{\partial}{\partial \boldsymbol{\gamma}} l\{\boldsymbol{g}(\boldsymbol{\theta})\} \cdot \frac{\partial}{\partial \boldsymbol{\theta}} \boldsymbol{g}(\boldsymbol{\theta}) \\
&= \frac{\partial}{\partial \boldsymbol{\theta}} l\{\boldsymbol{\theta}, \hat{\boldsymbol{\eta}}_{\mathrm{ML}}(\boldsymbol{\theta})\} + \frac{\partial}{\partial \boldsymbol{\eta}} l\{\boldsymbol{\theta}, \hat{\boldsymbol{\eta}}_{\mathrm{ML}}(\boldsymbol{\theta})\} \cdot \frac{\partial}{\partial \boldsymbol{\theta}} \hat{\boldsymbol{\eta}}_{\mathrm{ML}}(\boldsymbol{\theta}) \\
&= \frac{\partial}{\partial \boldsymbol{\theta}} l\{\boldsymbol{\theta}, \hat{\boldsymbol{\eta}}_{\mathrm{ML}}(\boldsymbol{\theta})\},
\end{aligned}
$$

using (5.7) for the last line. Differentiating this again, we obtain the curvature of the profile log-likelihood:

$$
\begin{aligned}
\frac{\partial}{\partial \boldsymbol{\theta}} \frac{\partial}{\partial \boldsymbol{\theta}^\top} l_p(\boldsymbol{\theta}) &= \frac{\partial}{\partial \boldsymbol{\theta}} \left[\frac{\partial}{\partial \boldsymbol{\theta}^\top} \{\boldsymbol{\theta}, \hat{\boldsymbol{\eta}}_{\mathrm{ML}}(\boldsymbol{\theta})\}\right] \\
&= \frac{\partial}{\partial \boldsymbol{\theta}} \frac{\partial}{\partial \boldsymbol{\theta}^\top} l\{\boldsymbol{\theta}, \hat{\boldsymbol{\eta}}_{\mathrm{ML}}(\boldsymbol{\theta})\} + \frac{\partial}{\partial \boldsymbol{\eta}} \frac{\partial}{\partial \boldsymbol{\theta}^\top} l\{\boldsymbol{\theta}, \hat{\boldsymbol{\eta}}_{\mathrm{ML}}(\boldsymbol{\theta})\} \cdot \frac{\partial}{\partial \boldsymbol{\theta}} \hat{\boldsymbol{\eta}}_{\mathrm{ML}}(\boldsymbol{\theta}).
\end{aligned} \tag{5.9}
$$

Substituting (5.8) into (5.9), we finally obtain

$$
\begin{aligned}
\frac{\partial}{\partial \boldsymbol{\theta}} \frac{\partial}{\partial \boldsymbol{\theta}^\top} l_p(\boldsymbol{\theta}) &= \frac{\partial}{\partial \boldsymbol{\theta}} \frac{\partial}{\partial \boldsymbol{\theta}^\top} l\{\boldsymbol{\theta}, \hat{\boldsymbol{\eta}}_{\mathrm{ML}}(\boldsymbol{\theta})\} \\
&\quad - \frac{\partial}{\partial \boldsymbol{\eta}} \frac{\partial}{\partial \boldsymbol{\theta}^\top} l\{\boldsymbol{\theta}, \hat{\boldsymbol{\eta}}_{\mathrm{ML}}(\boldsymbol{\theta})\} \left[\frac{\partial}{\partial \boldsymbol{\eta}} \frac{\partial}{\partial \boldsymbol{\eta}^\top} l\{\boldsymbol{\theta}, \hat{\boldsymbol{\eta}}_{\mathrm{ML}}(\boldsymbol{\theta})\}\right]^{-1} \\
&\quad \times \frac{\partial}{\partial \boldsymbol{\theta}} \frac{\partial}{\partial \boldsymbol{\eta}^\top} l\{\boldsymbol{\theta}, \hat{\boldsymbol{\eta}}_{\mathrm{ML}}(\boldsymbol{\theta})\}.
\end{aligned}
$$

Note that this holds for all $\boldsymbol{\theta}$. However, we are mainly interested in the negative curvature at the MLE $\hat{\boldsymbol{\theta}}_{\mathrm{ML}}$, i.e.

$$\begin{aligned} \boldsymbol{I}_p(\hat{\boldsymbol{\theta}}_{\mathrm{ML}}) &= -\frac{\partial}{\partial \boldsymbol{\theta}} \frac{\partial}{\partial \boldsymbol{\theta}^\top} l_p(\hat{\boldsymbol{\theta}}_{\mathrm{ML}}) \\ &= \boldsymbol{I}_{11} - \boldsymbol{I}_{12}(\boldsymbol{I}_{22})^{-1}\boldsymbol{I}_{21}. \end{aligned}$$

Application of the formula for the inversion of block matrices (cf. Appendix B.1.3) gives

$$\left(\boldsymbol{I}^{11}\right)^{-1} = \boldsymbol{I}_{11} - \boldsymbol{I}_{12}(\boldsymbol{I}_{22})^{-1}\boldsymbol{I}_{21}, \tag{5.10}$$

so we finally obtain $\boldsymbol{I}_p(\hat{\boldsymbol{\theta}}_{\mathrm{ML}}) = (\boldsymbol{I}^{11})^{-1}$. □

Note that with (5.10) and the symmetry of the Fisher information matrix we have

$$\begin{aligned} \left(\boldsymbol{I}^{11}\right)^{-1} &= \boldsymbol{I}_{11} - \underbrace{\boldsymbol{I}_{12}(\boldsymbol{I}_{22})^{-1}\boldsymbol{I}_{12}^\top}_{\geq 0}, \quad \text{so} \\ \left(\boldsymbol{I}^{11}\right)^{-1} &\leq \boldsymbol{I}_{11}, \end{aligned}$$

where this inequality denotes that the difference of the two matrices is (zero or) positive definite (cf. Appendix B.1.2). In particular, this implies that the diagonal elements of the left-hand side matrix are smaller than (or equal to) the corresponding diagonal elements of the right-hand side matrix. Now the Fisher information of the estimated likelihood at its maximum is $\boldsymbol{I}_{11}$. This inequality therefore shows that the observed Fisher information of the profile likelihood $(\boldsymbol{I}^{11})^{-1}$ is smaller than or equal to the observed Fisher information of the estimated likelihood $\boldsymbol{I}_{11}$. This once again illustrates that the estimated likelihood ignores the uncertainty in the estimation of the nuisance parameter $\boldsymbol{\eta}$. If $\boldsymbol{I}_{12} = \boldsymbol{I}_{21} = 0$, then $\boldsymbol{\theta}$ and $\boldsymbol{\eta}$ are *orthogonal*. With Eq. (5.10), we then have

$$\left(\boldsymbol{I}^{11}\right)^{-1} = \boldsymbol{I}_{11},$$

i.e. the observed Fisher information of profile and estimated likelihood are identical. As we have seen in Example 5.3, this is the case in the normal model.

With the above result, we can approximate the profile log-likelihood $\tilde{l}_p(\theta_i)$ of a scalar parameter θ_i by a quadratic function

$$\tilde{l}_p(\theta_i) \approx -\frac{1}{2} \cdot \left(I^{11}\right)^{-1}(\theta_i - \hat{\theta}_i)^2,$$

where I^{ii} denotes the ith diagonal entry of $\boldsymbol{I}(\hat{\boldsymbol{\theta}}_{\mathrm{ML}})^{-1}$. This result corresponds to our definition of Wald confidence intervals in Sect. 5.2.

The following result considers the special case of a difference between two parameters coming from independent data sets.

Result 5.2 (Standard error of a difference) *Suppose that the data can be split in two independent parts (denoted by* 0 *and* 1) *and the corresponding likelihood functions are parametrised by* α *and* β, *respectively. The standard error of* $\hat{\delta}_{\text{ML}} = \hat{\beta}_{\text{ML}} - \hat{\alpha}_{\text{ML}}$ *then has the form*

$$\operatorname{se}(\hat{\delta}_{\text{ML}}) = \sqrt{I_0(\hat{\alpha}_{\text{ML}})^{-1} + I_1(\hat{\beta}_{\text{ML}})^{-1}}.$$

Proof To show Result 5.2, we start with the log-likelihood decomposition

$$l(\alpha, \beta) = l_0(\alpha) + l_1(\beta),$$

which is due to the independence of the two data sets. Now $\beta = \alpha + \delta$, so the joint log-likelihood of α and δ is

$$l(\alpha, \delta) = l_0(\alpha) + l_1(\alpha + \delta).$$

Furthermore, the score functions for α and δ are

$$\frac{\partial}{\partial \alpha} l(\alpha, \delta) = S_0(\alpha) + S_1(\alpha + \delta) \quad \text{and}$$
$$\frac{\partial}{\partial \delta} l(\alpha, \delta) = S_1(\alpha + \delta),$$

where S_0 and S_1 are the score functions corresponding to l_0 and l_1, respectively. The Fisher information matrix is hence

$$\boldsymbol{I}(\alpha, \delta) = \begin{pmatrix} I_0(\alpha) + I_1(\alpha + \delta) & I_1(\alpha + \delta) \\ I_1(\alpha + \delta) & I_1(\alpha + \delta) \end{pmatrix},$$

where I_0 and I_1 denote the Fisher information corresponding to l_0 and l_1, respectively.

We would like to calculate the Fisher information (negative curvature) of the profile log-likelihood of δ at the MLE $\hat{\delta}_{\text{ML}}$. From Result 5.1 we know that we need to compute the inverse of $\boldsymbol{I}(\hat{\alpha}_{\text{ML}}, \hat{\delta}_{\text{ML}})$ and then take the inverse of its lower right element. Using Appendix B.1.1 and the notation from Result 5.1, we know that this lower right element is

$$\begin{aligned}
\frac{1}{I_p(\hat{\delta}_{\text{ML}})} &= I^{22} \\
&= \frac{1}{I_{11} I_{22} - I_{12} I_{21}} I_{11} \\
&= \frac{I_{11}}{I_{11} I_{22} - I_{22} I_{22}} \\
&= \frac{I_{11}}{(I_{11} - I_{22}) I_{22}}
\end{aligned}$$

$$= \frac{I_0(\hat{\alpha}_{\text{ML}}) + I_1(\hat{\alpha}_{\text{ML}} + \hat{\delta}_{\text{ML}})}{I_0(\hat{\alpha}_{\text{ML}}) I_1(\hat{\alpha}_{\text{ML}} + \hat{\delta}_{\text{ML}})}$$

$$= \frac{1}{I_0(\hat{\alpha}_{\text{ML}})} + \frac{1}{I_1(\hat{\alpha}_{\text{ML}} + \hat{\delta}_{\text{ML}})}, \tag{5.11}$$

which gives the result. See Exercise 3 for an alternative derivation of this formula. □

Example 5.8 (Comparison of proportions) Suppose that

$$X_1 \sim \text{Bin}(n_1, \pi_1) \quad \text{and} \quad X_2 \sim \text{Bin}(n_2, \pi_2)$$

are independent binomial random variables. To quantify any particular difference between the probabilities π_1 and π_2 in the two groups, the *odds ratio*

$$\theta = \frac{\pi_1/(1-\pi_1)}{\pi_2/(1-\pi_2)}$$

or the *log odds ratio* $\psi = \log(\theta)$ is often used. The equality $\pi_1 = \pi_2$, which is often used as a null hypothesis, corresponds to $H_0 : \theta = 1$ and $H_0 : \psi = 0$, respectively. Now

$$\psi = \log(\theta) = \log\left(\frac{\pi_1}{1-\pi_1}\right) - \log\left(\frac{\pi_2}{1-\pi_2}\right),$$

so with $\hat{\pi}_i = x_i/n_i$, $i = 1, 2$, and invariance of the MLE, we immediately obtain

$$\hat{\psi}_{\text{ML}} = \log\left(\frac{x_1}{n_1 - x_1}\right) - \log\left(\frac{x_2}{n_2 - x_2}\right) = \log\left\{\frac{x_1/(n_1 - x_1)}{x_2/(n_2 - x_2)}\right\}$$

as the MLE of ψ.

For illustration, consider Table 3.1 summarising the result from the clinical study in Table 1.1 labelled as "Tervila". Here we obtain

$$\hat{\psi}_{\text{ML}} = \log\left(\frac{6/102}{2/101}\right) \approx 1.089.$$

Now define the log odds $\phi_i = \log\{\pi_i/(1-\pi_i)\}$ for group $i = 1, 2$. We know from Example 2.11 that the observed Fisher information of the MLE $\hat{\phi}_i$ of ϕ_i is

$$I_i(\hat{\phi}_i) = \frac{x_i(n_i - x_i)}{n_i}.$$

With (5.11), the observed Fisher information of $\hat{\psi}_{\text{ML}}$ is therefore

$$I(\hat{\psi}_{\text{ML}}) = \left\{I_1(\hat{\phi}_1)^{-1} + I_2(\hat{\phi}_2)^{-1}\right\}^{-1}$$

$$= \left\{\frac{n_1}{x_1(n_1 - x_1)} + \frac{n_2}{x_2(n_2 - x_2)}\right\}^{-1}$$

$$= \left(\frac{1}{x_1} + \frac{1}{n_1 - x_1} + \frac{1}{x_2} + \frac{1}{n_2 - x_2} \right)^{-1},$$

so the standard error of $\hat{\psi}_{\text{ML}}$ is

$$\text{se}(\hat{\psi}_{\text{ML}}) = \sqrt{\frac{1}{x_1} + \frac{1}{n_1 - x_1} + \frac{1}{x_2} + \frac{1}{n_2 - x_2}}.$$

In our example we obtain

$$\text{se}(\hat{\psi}_{\text{ML}}) = \sqrt{\frac{1}{6} + \frac{1}{102} + \frac{1}{2} + \frac{1}{101}} \approx 0.828,$$

so the 95 % Wald confidence interval for ψ has limits

$$\hat{\psi}_{\text{ML}} \pm 1.96 \cdot \text{se}(\hat{\psi}_{\text{ML}}) = 1.089 \pm 1.96 \cdot 0.828 = -0.54 \text{ and } 2.71.$$

The reference value $\psi = 0$, which corresponds to the null hypothesis of no difference between the two groups, is inside this interval, so the corresponding P-value must be larger than 0.05.

Calculation of the profile likelihood of ψ is more difficult. The joint likelihood of the two binomial probabilities π_1 and π_2 is

$$L(\pi_1, \pi_2) = \pi_1^{x_1}(1 - \pi_1)^{n_1 - x_1} \pi_2^{x_2}(1 - \pi_2)^{n_2 - x_2}.$$

We now reparametrise the likelihood in terms of

$$\psi = \log\left\{\frac{\pi_1/(1 - \pi_1)}{\pi_2/(1 - \pi_2)}\right\} \quad \text{and} \quad \eta = \log\left(\frac{\pi_2}{1 - \pi_2}\right),$$

from which we can directly calculate the profile likelihood of the log odds ratio ψ. Note that we could have also chosen a different nuisance parameter, and for example let η be the log odds in group 1 instead. Invariance of the likelihood ensures that the profile likelihood of ψ would remain unchanged. The above reparametrisation implies

$$\pi_1 = \frac{\exp(\eta + \psi)}{1 + \exp(\eta + \psi)} \quad \text{and} \quad \pi_2 = \frac{\exp(\eta)}{1 + \exp(\eta)},$$

and we obtain the joint likelihood

$$\begin{aligned} L(\psi, \eta) &= \{\exp(\eta + \psi)\}^{x_1}\{1 + \exp(\eta + \psi)\}^{-n_1} \cdot \{\exp(\eta)\}^{x_2}\{1 + \exp(\eta)\}^{-n_2} \\ &= \exp\{\eta(x_1 + x_2)\} \exp(\psi x_1)\{1 + \exp(\eta + \psi)\}^{-n_1}\{1 + \exp(\eta)\}^{-n_2}, \end{aligned}$$

shown in Fig. 5.4a as relative log-likelihood.

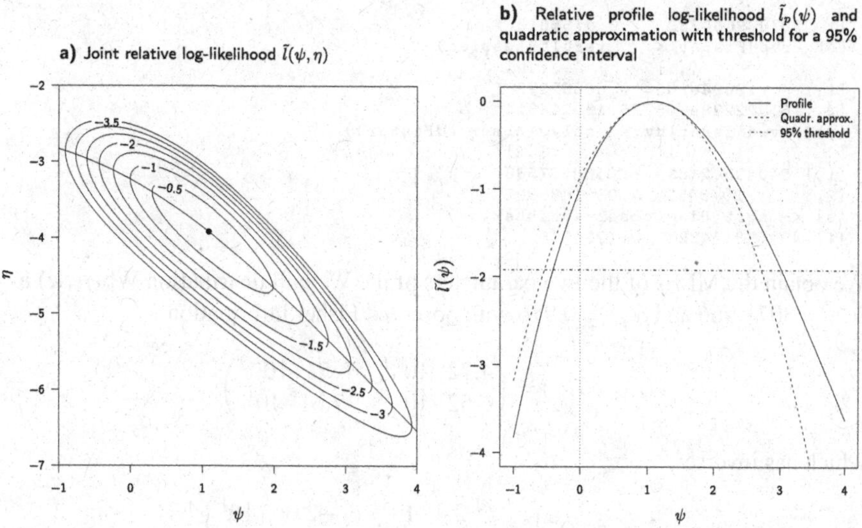
a) Joint relative log-likelihood $\tilde{l}(\psi, \eta)$
b) Relative profile log-likelihood $\tilde{l}_p(\psi)$ and quadratic approximation with threshold for a 95% confidence interval
Profile
Quadr. approx.
95% threshold

Fig. 5.4 Log-likelihood functions for the odds ratio ψ in the binomial model with the data from Table 3.1

Calculation of the profile likelihood $L_p(\psi) = \max_\eta L(\psi, \eta)$ can be done numerically, and we obtain the 95 % profile likelihood confidence interval $[-0.41, 3.02]$. Compared with the Wald confidence interval, the profile likelihood confidence interval is slightly shifted to the right but still covers the null hypothesis value $\psi = 0$, compare Fig. 5.4b. ■

Example 5.9 (Analysis of survival times) We now reconsider a Weibull model $\mathrm{Wb}(\mu, \alpha)$ for the survival data described in Example 2.3 but want to incorporate the censored observations as well in our analysis, cf. Sect. 2.1.4. The data `pbcTreat` have the form ($\texttt{time}_i$, $\texttt{d}_i$). MLEs and observed Fisher information can be calculated with the following R program (the functions `dweibull()`, `pweibull()` are explained in Appendix A.5.2):

```
weibullLik <- function(mualpha, log = TRUE)
{
    mu <- mualpha[1]
    alpha <- mualpha[2]
    loglik <- with(pbcTreat,
                   sum(d * dweibull (time, alpha, mu, log = TRUE) +
                       (1 - d) * pweibull(time, alpha, mu,
                                          lower.tail = FALSE,
                                          log.p = TRUE)))
    if(log)
        return(loglik)
    else
        return(exp(loglik))
}
start <- c(1000, 1)
result <- optim(start, weibullLik, control=list(fnscale=-1),
                hessian=TRUE)
(ml <- result$par)
```

```
[1] 3078.9660333    0.9756013
(observedFisher <- - result$hessian)
             [,1]         [,2]
[1,] 4.718004e-06  0.006522228
[2,] 6.522228e-03 76.139214514
(observedFisherInv <- solve(observedFisher))
             [,1]         [,2]
[1,] 240425.21837 -20.59527568
[2,]    -20.59528   0.01489807
(se <- sqrt(diag(observedFisherInv)))
[1] 490.3317432   0.1220576
```

We obtain the MLEs of the two parameters of the Weibull distribution $\text{Wb}(\mu, \alpha)$ as $\hat{\mu}_{\text{ML}} = 3078.966$ and $\hat{\alpha}_{\text{ML}} = 0.976$ with observed Fisher information

$$\boldsymbol{I}(\hat{\mu}_{\text{ML}}, \hat{\alpha}_{\text{ML}}) = \begin{pmatrix} 4.72 \cdot 10^{-6} & 6.52 \cdot 10^{-3} \\ 6.52 \cdot 10^{-3} & 7.61 \cdot 10^{1} \end{pmatrix},$$

which has inverse

$$\boldsymbol{I}(\hat{\mu}_{\text{ML}}, \hat{\alpha}_{\text{ML}})^{-1} = \begin{pmatrix} 2.4 \cdot 10^{5} & -2.06 \cdot 10^{1} \\ -2.06 \cdot 10^{1} & 1.49 \cdot 10^{-2} \end{pmatrix}.$$

We therefore have $\text{se}(\hat{\mu}_{\text{ML}}) = \sqrt{2.4 \cdot 10^5} = 490.3$ and $\text{se}(\hat{\alpha}_{\text{ML}}) = \sqrt{1.49 \cdot 10^{-2}} = 0.122$.

We can now calculate, for example, the limits of a 95 % Wald confidence interval for α:

$$0.976 \pm 1.96 \cdot 0.122 = 0.736 \text{ and } 1.215.$$

Note that the value $\alpha = 1$, which corresponds to a simple exponential model, is within this interval. The relative profile log-likelihood of α can be calculated numerically and is shown in Fig. 5.5. The 95 % profile likelihood confidence interval (using the threshold -1.92, cf. Table 4.1) turns out to be $[0.753, 1.231]$ and is slightly shifted to the right in comparison with the Wald confidence interval.

If we instead assume a gamma model, we obtain (using the parametrisation from Fig. 2.5) $\hat{\mu}_{\text{ML}} = 3093.8$ and $\hat{\phi}_{\text{ML}} = 3196.9$ with observed Fisher information

$$\boldsymbol{I}(\hat{\mu}_{\text{ML}}, \hat{\phi}_{\text{ML}}) = \begin{pmatrix} 1.4 \cdot 10^{-5} & -6.9 \cdot 10^{-6} \\ -6.9 \cdot 10^{-6} & 4.6 \cdot 10^{-6} \end{pmatrix}$$

and standard errors $\text{se}(\hat{\mu}_{\text{ML}}) = 520.2$ and $\text{se}(\hat{\phi}_{\text{ML}}) = 910.3$.

What would we have obtained in the original parametrisation with parameters $\alpha = \mu/\phi$ and $\beta = 1/\phi$? Invariance of the likelihood immediately gives

$$\hat{\alpha}_{\text{ML}} = \frac{\hat{\mu}_{\text{ML}}}{\hat{\phi}_{\text{ML}}} = 0.968 \quad \text{and} \quad \hat{\beta}_{\text{ML}} = \frac{1}{\hat{\phi}_{\text{ML}}} = 3.1 \cdot 10^{-4}.$$

In Example 5.13 we will apply the multivariate delta method to compute the standard errors of $\hat{\alpha}_{\text{ML}}$ and $\hat{\beta}_{\text{ML}}$ without directly maximising the log-likelihood of α and β. ■

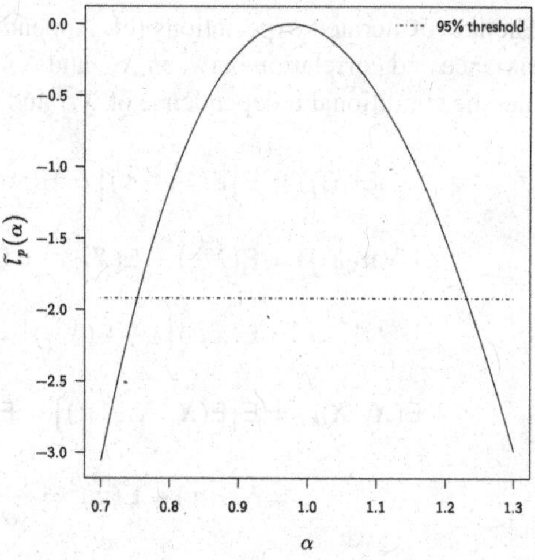

Fig. 5.5 Relative profile log-likelihood of α

Example 5.10 (Screening for colon cancer) A more general than the binomial model from Example 2.12 is based on the *beta-binomial distribution*, which allows for heterogeneity of the success probability π across individuals. Assume that the numbers of positive test results $X_{1:n}$, $n = 196$, are a random sample from a beta-binomial $\operatorname{BeB}(N, \alpha, \beta)$ distribution (cf. Appendix A.5.1). The probability for k positive test results among $N = 6$ tests is therefore

$$\Pr(X_i = k) = \binom{N}{k} \frac{\mathrm{B}(\alpha + k, \beta + N - k)}{\mathrm{B}(\alpha, \beta)}$$

for $k = 0, \ldots, N$ and $\alpha, \beta > 0$, where $\mathrm{B}(x, y)$ denotes the beta function (cf. Appendix B.2.1).

As before, we need to truncate this distribution using Eq. (2.6) to obtain the log-likelihood

$$l(\alpha, \beta) = \sum_{k=1}^{N} Z_k \log\left(\frac{\mathrm{B}(\alpha + k, \beta + N - k)}{\mathrm{B}(\alpha, \beta)}\right) - n \log\left(1 - \frac{\mathrm{B}(\alpha, \beta + N)}{\mathrm{B}(\alpha, \beta)}\right).$$

However, because the parameters α and β are difficult to interpret, instead the reparametrisation

$$\mu = \frac{\alpha}{\alpha + \beta} \quad \text{and} \quad \rho = \frac{1}{\alpha + \beta + 1}$$

is often used. Here $N\mu$ is the mean of X_i, and ρ is the correlation between two binary observations from the same individual. This can be easily seen from the definition of the beta-binomial distribution as marginal distribution of X_i when $X_i \mid \pi \sim \operatorname{Bin}(N, \pi)$ and $\pi \sim \operatorname{Be}(\alpha, \beta)$. Therefore, X_i is a sum of binary dependent random variables $X_{ij} \in \{0, 1\}$, $j = 1, \ldots, N$:

$$X_i = X_{i1} + \cdots + X_{iN}.$$

The law of iterated expectations (cf. Appendix A.3.4) can be used to calculate covariance and correlation between X_{ij} and X_{ik} for $j \neq k$ (cf. Appendix A.3.6). Note that the conditional independence of X_{ij} and X_{ik} given π is exploited:

$$\begin{aligned}
\mathsf{E}(X_{ij}) &= \mathsf{E}\{\mathsf{E}(X_{ij} \mid \pi)\} = \mathsf{E}(\pi) = \frac{\alpha}{\alpha+\beta}, \\
\operatorname{Var}(X_{ij}) &= \mathsf{E}(X_{ij}^2) - \mathsf{E}(X_{ij})^2 = \mathsf{E}(X_{ij}) - \mathsf{E}(X_{ij})^2 \\
&= \mathsf{E}(X_{ij})\{1 - \mathsf{E}(X_{ij})\} = \frac{\alpha\beta}{(\alpha+\beta)^2}, \\
\mathsf{E}(X_{ij}X_{ik}) &= \mathsf{E}\{\mathsf{E}(X_{ij}X_{ik} \mid \pi)\} = \mathsf{E}\{\mathsf{E}(X_{ij} \mid \pi)\,\mathsf{E}(X_{ik} \mid \pi)\} = \mathsf{E}(\pi^2) \\
&= \operatorname{Var}(\pi) + \mathsf{E}(\pi)^2 = \frac{\alpha\beta}{(\alpha+\beta)^2(\alpha+\beta+1)} + \left(\frac{\alpha}{\alpha+\beta}\right)^2, \\
\operatorname{Cov}(X_{ij}, X_{ik}) &= \mathsf{E}(X_{ij}X_{ik}) - \mathsf{E}(X_{ij})\,\mathsf{E}(X_{ik}) = \frac{\alpha\beta}{(\alpha+\beta)^2(\alpha+\beta+1)} \quad \text{and} \\
\rho = \operatorname{Corr}(X_{ij}, X_{ik}) &= \frac{\operatorname{Cov}(X_{ij}, X_{ik})}{\sqrt{\operatorname{Var}(X_{ij})\operatorname{Var}(X_{ik})}} = \frac{1}{\alpha+\beta+1}.
\end{aligned}$$

Therefore, $\rho = 1/(\alpha+\beta+1)$ can be interpreted as correlation between X_{ij} and X_{ik}.

Because our primary interest lies in the false negative fraction

$$\xi = \Pr(X_i = 0) = \frac{\mathrm{B}(\alpha, \beta+N)}{\mathrm{B}(\alpha, \beta)},$$

we actually prefer a third reparametrisation using ξ and ρ as seen in Fig. 5.6a. Based on this, we obtain the profile log-likelihood of ξ shown in Fig. 5.6b with MLE $\hat{\xi}_{\mathrm{ML}} = 0.238$ and 95 % profile likelihood confidence interval [0.113, 0.554]. Note that profile and estimated log-likelihood are substantially different.

The naive estimate of Z_0 is now $\hat{Z}_0 = 196 \cdot 0.238/(1-0.238) \approx 61.22$, much larger than in the binomial model ($\hat{Z}_0 \approx 0.55$), compare Example 2.12. The 95 % profile likelihood confidence interval [24.97, 243.46] for Z_0 can be calculated using the transformation $Z_0 = 196 \cdot \xi/(1-\xi)$ of the limits of the 95 % profile likelihood confidence interval for ξ. ■

5.4 Frequentist Properties of the Multiparameter Likelihood

As in Chap. 4, we now consider the score vector and the Fisher information matrix as random, i.e. both the score vector $\boldsymbol{S}(\boldsymbol{\theta}; X)$ and the Fisher information matrix $\boldsymbol{I}(\boldsymbol{\theta}; X)$ are functions of the data X. We define the *expected Fisher information matrix*, shortly *expected Fisher information*, $\boldsymbol{J}(\boldsymbol{\theta})$ as

$$\boldsymbol{J}(\boldsymbol{\theta}) = \mathsf{E}\{\boldsymbol{I}(\boldsymbol{\theta}; X)\}.$$

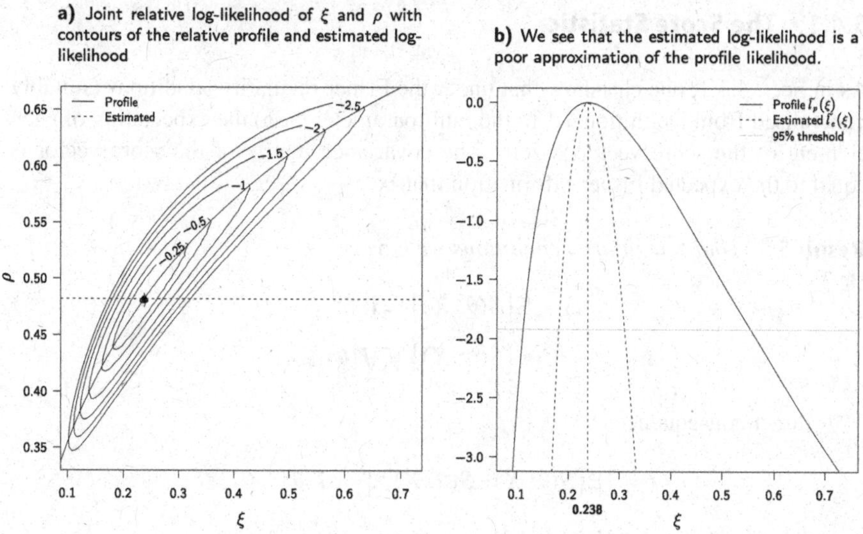
a) Joint relative log-likelihood of ξ and ρ with contours of the relative profile and estimated log-likelihood
b) We see that the estimated log-likelihood is a poor approximation of the profile likelihood.
Profile
Estimated
95% threshold
ρ
ξ
0.238

Fig. 5.6 Comparison of (relative) profile and estimated likelihood of the false negative fraction ξ in the beta-binomial model

Example 5.11 (Normal model) Elementwise computation of the expectation of (5.1) from Example 5.3 gives the expected Fisher information of a normal random sample

$$J(\boldsymbol{\theta}) = \begin{pmatrix} \frac{n}{\sigma^2} & 0 \\ 0 & \frac{n}{2\sigma^4} \end{pmatrix}.$$ ■

Example 5.12 (Multinomial model) Elementwise computation of the expectation of (5.2) from Example 5.4 gives the expected Fisher information in the multinomial model as

$$J(\tilde{\boldsymbol{\pi}}) = n \left\{ \operatorname{diag}(\tilde{\boldsymbol{\pi}})^{-1} + \left(1 - \sum_{i=1}^{k-1} \pi_i\right)^{-1} \mathbf{1}\mathbf{1}^\top \right\}.$$

Applying the Sherman–Morrison formula (cf. Appendix B.1.4), we can easily compute the inverse of the expected Fisher information:

$$J(\tilde{\boldsymbol{\pi}})^{-1} = n^{-1} \left\{ \operatorname{diag}(\tilde{\boldsymbol{\pi}}) - \tilde{\boldsymbol{\pi}} \tilde{\boldsymbol{\pi}}^\top \right\}.$$ ■

In complete analogy to the case of a scalar parameter, we now define three approximate pivots: the score, the Wald and the likelihood ratio statistic. The important properties of these statistics are sketched in the following.

5.4.1 The Score Statistic

As in Sect. 4.1.1, one can show that under the Fisher regularity conditions (suitably generalised from Definition 4.1 to the multiparameter case) the expectation of each element of the score vector is zero. The covariance matrix of the score vector is equal to the expected Fisher information matrix.

Result 5.3 *Under regularity conditions we have:*

$$\begin{aligned}\mathsf{E}\{\boldsymbol{S}(\boldsymbol{\theta}; X)\} &= \mathbf{0},\\ \operatorname{Cov}\{\boldsymbol{S}(\boldsymbol{\theta}; X)\} &= \boldsymbol{J}(\boldsymbol{\theta}).\end{aligned}$$

A direct consequence is

$$\mathsf{E}\{\boldsymbol{S}(\boldsymbol{\theta}; X) \cdot \boldsymbol{S}(\boldsymbol{\theta}; X)^\top\} = \boldsymbol{J}(\boldsymbol{\theta}).$$

A multivariate version of Result 4.5 is given by

$$\boldsymbol{J}_{1:n}(\boldsymbol{\theta})^{-\frac{1}{2}} \boldsymbol{S}(\boldsymbol{\theta}; X_{1:n}) \overset{a}{\sim} \mathrm{N}_p(\mathbf{0}, \mathbf{I}_p),$$

where $\mathbf{I}_p$ denotes the $p \times p$ identity matrix, and $\boldsymbol{A}^{-\frac{1}{2}}$ is the *Cholesky square root* of $\boldsymbol{A}^{-1}$ (cf. Appendix B.1.2), i.e.

$$\left(\boldsymbol{A}^{-\frac{1}{2}}\right)^\top \boldsymbol{A}^{-\frac{1}{2}} = \boldsymbol{A}^{-1}.$$

If $\boldsymbol{X} \sim \mathrm{N}_p(0, \mathbf{I}_p)$, then $\boldsymbol{X}^\top \boldsymbol{X} \sim \chi^2(p)$, so we also have

$$\begin{aligned}&\left\{\boldsymbol{J}_{1:n}(\boldsymbol{\theta})^{-\frac{1}{2}} \boldsymbol{S}(\boldsymbol{\theta}; X_{1:n})\right\}^\top \left\{\boldsymbol{J}_{1:n}(\boldsymbol{\theta})^{-\frac{1}{2}} \boldsymbol{S}(\boldsymbol{\theta}; X_{1:n})\right\}\\ &\quad= \boldsymbol{S}(\boldsymbol{\theta}; X_{1:n})^\top \left\{\boldsymbol{J}_{1:n}(\boldsymbol{\theta})^{-\frac{1}{2}}\right\}^\top \boldsymbol{J}_{1:n}(\boldsymbol{\theta})^{-\frac{1}{2}} \boldsymbol{S}(\boldsymbol{\theta}; X_{1:n})\\ &\quad= \boldsymbol{S}(\boldsymbol{\theta}; X_{1:n})^\top \boldsymbol{J}_{1:n}(\boldsymbol{\theta})^{-1} \boldsymbol{S}(\boldsymbol{\theta}; X_{1:n})\\ &\quad\overset{a}{\sim} \chi^2(p).\end{aligned}$$

We can replace as in Sect. 4.1.3 the expected Fisher information matrix $\boldsymbol{J}_{1:n}(\boldsymbol{\theta})$ by the Fisher information matrix $\boldsymbol{I}(\boldsymbol{\theta}; X_{1:n})$. We can also replace the true value $\boldsymbol{\theta}$ by the MLE $\hat{\boldsymbol{\theta}}_{\mathrm{ML}} = \hat{\boldsymbol{\theta}}_{\mathrm{ML}}(X_{1:n})$.

5.4.2 The Wald Statistic

The following results can be established, generalizing Result 4.10:

$$\boldsymbol{J}_{1:n}(\hat{\boldsymbol{\theta}}_{\mathrm{ML}})^{\frac{1}{2}} (\hat{\boldsymbol{\theta}}_{\mathrm{ML}} - \boldsymbol{\theta}) \overset{a}{\sim} \mathrm{N}_p(\mathbf{0}, \mathbf{I}_p), \tag{5.12}$$

$$\boldsymbol{I}(\hat{\boldsymbol{\theta}}_{\mathrm{ML}}; X_{1:n})^{\frac{1}{2}}(\hat{\boldsymbol{\theta}}_{\mathrm{ML}} - \boldsymbol{\theta}) \overset{\mathrm{a}}{\sim} \mathrm{N}_p(\mathbf{0}, \mathbf{I}_p). \tag{5.13}$$

So in a less rigorous formulation we can say that the MLE $\hat{\boldsymbol{\theta}}_{\mathrm{ML}}$ is asymptotically normal with mean equal to the true value $\boldsymbol{\theta}$ and covariance matrix equal to the inverse observed Fisher information:

$$\hat{\boldsymbol{\theta}}_{\mathrm{ML}} \overset{\mathrm{a}}{\sim} \mathrm{N}_p\big(\boldsymbol{\theta}, \boldsymbol{I}(\hat{\boldsymbol{\theta}}_{\mathrm{ML}})^{-1}\big). \tag{5.14}$$

A direct consequence of (5.14) is that the asymptotic distribution of the ith component of $\hat{\boldsymbol{\theta}}_{\mathrm{ML}}$ is

$$\hat{\theta}_i \overset{\mathrm{a}}{\sim} \mathrm{N}\big(\theta_i, I^{ii}\big),$$

where I^{ii} denotes the ith diagonal element of the inverse observed Fisher information:

$$I^{ii} = \big[\boldsymbol{I}(\hat{\boldsymbol{\theta}}_{\mathrm{ML}})^{-1}\big]_{ii}.$$

In addition to Result 5.1, this property provides an alternative justification of our definition of Wald confidence intervals given in Sect. 5.2.

There is also a multidimensional extension of the Cramér–Rao inequality from Sect. 4.2.1, which ensures that $\hat{\theta}_i$ is asymptotically efficient, i.e. has an asymptotically smallest variance among all asymptotically unbiased estimators.

5.4.3 The Multivariate Delta Method

Application of the multivariate delta method (see Appendix A.4.5) to $\hat{\boldsymbol{\theta}}_{\mathrm{ML}} \overset{\mathrm{a}}{\sim} \mathrm{N}_p(\boldsymbol{\theta}, \boldsymbol{I}(\hat{\boldsymbol{\theta}}_{\mathrm{ML}})^{-1})$ for some continuously differentiable function $\boldsymbol{g} : \mathbb{R}^p \to \mathbb{R}^q$ with $q \leq p$ gives:

$$\boldsymbol{g}(\hat{\boldsymbol{\theta}}_{\mathrm{ML}}) \overset{\mathrm{a}}{\sim} \mathrm{N}_q\big(\boldsymbol{g}(\boldsymbol{\theta}), \boldsymbol{D}(\hat{\boldsymbol{\theta}}_{\mathrm{ML}})\boldsymbol{I}(\hat{\boldsymbol{\theta}}_{\mathrm{ML}})^{-1}\boldsymbol{D}(\hat{\boldsymbol{\theta}}_{\mathrm{ML}})^\top\big).$$

Here $\boldsymbol{D}(\boldsymbol{\theta})$ denotes the $q \times p$ Jacobian of $\boldsymbol{g}(\boldsymbol{\theta})$. The square root of the diagonal elements of the $q \times q$ matrix $\boldsymbol{D}(\hat{\boldsymbol{\theta}}_{\mathrm{ML}})\boldsymbol{I}(\hat{\boldsymbol{\theta}}_{\mathrm{ML}})^{-1}\boldsymbol{D}(\hat{\boldsymbol{\theta}}_{\mathrm{ML}})^\top$ are the standard errors of the q components of $\boldsymbol{g}(\hat{\boldsymbol{\theta}}_{\mathrm{ML}})$.

Example 5.13 (Analysis of survival times) We revisit the gamma model from Example 5.9 for the PBC survival times shown in Table 1.4, see also Example 2.3. The unknown parameter vector is $\boldsymbol{\theta} = (\mu, \phi)^\top$, we are interested in the transformation

$$\boldsymbol{g}(\boldsymbol{\theta}) = (\alpha, \beta)^\top = \left(\frac{\mu}{\phi}, \frac{1}{\phi}\right)^\top.$$

Now

$$\boldsymbol{D}(\boldsymbol{\theta}) = \begin{pmatrix} \frac{1}{\phi} & -\frac{\mu}{\phi^2} \\ 0 & -\frac{1}{\phi^2} \end{pmatrix},$$

and $\boldsymbol{D}(\hat{\boldsymbol{\theta}}_{\text{ML}})\boldsymbol{I}(\hat{\boldsymbol{\theta}}_{\text{ML}})^{-1}\boldsymbol{D}(\hat{\boldsymbol{\theta}}_{\text{ML}})^{\top}$ is a consistent estimate of the asymptotic covariance matrix of $\boldsymbol{g}(\hat{\boldsymbol{\theta}}_{\text{ML}})$. The square roots of the diagonal elements are the corresponding standard errors $\text{se}(\hat{\alpha}_{\text{ML}}) = 0.1593$ and $\text{se}(\hat{\beta}_{\text{ML}}) = 8.91 \cdot 10^{-5}$. Thus, the 95 % Wald confidence interval for α has limits

$$\hat{\alpha}_{\text{ML}} \pm 1.96 \cdot \text{se}(\hat{\alpha}_{\text{ML}}) = 0.97 \pm 0.31 = 0.66 \text{ and } 1.28.$$

Note that the value $\alpha = 1$, which corresponds to exponentially distributed survival times, is covered by the 95 % confidence interval. ■

5.4.4 The Likelihood Ratio Statistic

The likelihood ratio statistic can be defined in the multivariate case with parameter vector $\boldsymbol{\theta}$ just as for a scalar parameter θ:

$$W = 2\log\left\{\frac{L(\hat{\boldsymbol{\theta}}_{\text{ML}})}{L(\boldsymbol{\theta})}\right\} = -2\tilde{l}(\boldsymbol{\theta}).$$

Combining Eqs. (5.13) and (5.4) shows that the number of degrees of freedom of the asymptotic χ^2 distribution is equal to the dimension p of $\boldsymbol{\theta}$:

$$W \approx (\hat{\boldsymbol{\theta}}_{\text{ML}} - \boldsymbol{\theta})^{\top}\boldsymbol{I}(\hat{\boldsymbol{\theta}}_{\text{ML}})(\hat{\boldsymbol{\theta}}_{\text{ML}} - \boldsymbol{\theta}) \overset{\text{a}}{\sim} \chi^2(p).$$

In analogy to (4.17) and (4.18), respectively, we can now construct $\gamma \cdot 100$ % likelihood confidence regions for a multidimensional parameter $\boldsymbol{\theta}$:

$$\begin{aligned}&\left\{\boldsymbol{\theta} : \tilde{l}(\boldsymbol{\theta}) \geq -\frac{1}{2}\chi^2_{\gamma}(p)\right\}\\ &= \left\{\boldsymbol{\theta} : \tilde{L}(\boldsymbol{\theta}) \geq \exp\left[-\frac{1}{2}\chi^2_{\gamma}(p)\right]\right\}.\end{aligned} \tag{5.15}$$

Of practical use is typically only the case $p = 2$, where interestingly

$$\exp\left[-\frac{1}{2}\chi^2_{\gamma}(2)\right] = 1 - \gamma.$$

Therefore, all values of a two-dimensional parameter vector $\boldsymbol{\theta} = (\theta_1, \theta_2)^{\top}$ with relative likelihood larger than $1 - \gamma$ will form a $\gamma \cdot 100$ % likelihood confidence region for $\boldsymbol{\theta}$.

Example 5.14 (Analysis of survival times) Figure 5.7 shows a contour plot of the relative likelihood for the parameter vector $\boldsymbol{\theta} = (\alpha, \mu)^{\top}$ in a Weibull model for survival times, compare Example 2.3. The 95 % confidence region is defined as all values of $\boldsymbol{\theta}$ with relative likelihood larger than 0.05 and is shaded in the figure. ■

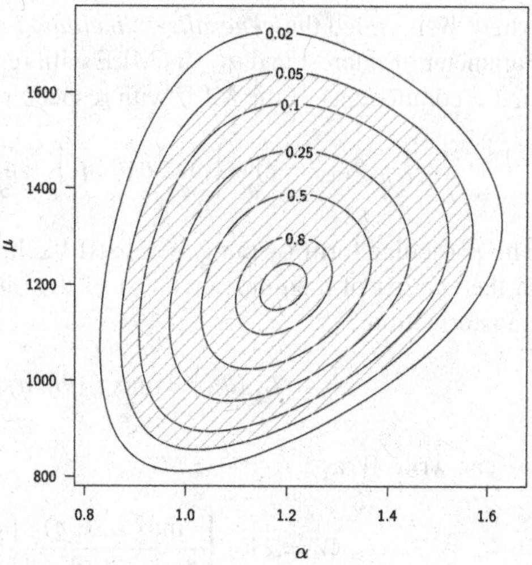

Fig. 5.7 Relative likelihood with 95 % confidence region (*shaded*) for the parameter vector $\boldsymbol{\theta} = (\alpha, \mu)^\top$ in the Weibull model from Example 2.3

Table 5.1 Thresholds $\exp\{-\frac{1}{2}\chi^2_\gamma(p)\}$ for the calibration of the relative likelihood of a parameter of dimension p at confidence level γ

γ	$p=1$	$p=2$	$p=3$	$p=4$
90 %	0.259	0.100	0.044	0.020
95 %	0.147	0.050	0.020	0.009
99 %	0.036	0.010	0.003	0.001

The general result (5.15) illustrates that a frequentist calibration of the relative likelihood depends on the dimension of the unknown parameter vector. Table 5.1 lists the corresponding thresholds $\exp\{-\frac{1}{2}\chi^2_\gamma(p)\}$ for various values of the confidence level γ and the dimension p of the parameter vector.

5.5 The Generalised Likelihood Ratio Statistic

Let $\boldsymbol{\theta}$ denote the q-dimensional parameter vector of interest, and $\boldsymbol{\eta}$ the r-dimensional nuisance parameter vector. The total parameter space has thus dimension $p = q + r$.

The joint likelihood $L(\boldsymbol{\theta}, \boldsymbol{\eta})$ can be used to derive the profile likelihood of $\boldsymbol{\theta}$:

$$L_p(\boldsymbol{\theta}) = \max_{\boldsymbol{\eta}} L(\boldsymbol{\theta}, \boldsymbol{\eta}) = L\{\boldsymbol{\theta}, \hat{\boldsymbol{\eta}}_{\mathrm{ML}}(\boldsymbol{\theta})\}.$$

We will sketch in the following that the profile likelihood $L_p(\boldsymbol{\theta})$ can be treated as a normal likelihood. In particular, we will show in Result 5.4 that

$$W = -2\tilde{l}_p(\boldsymbol{\theta}) = -2\log\{\tilde{L}_p(\boldsymbol{\theta})\} = 2\log\left\{\frac{L_p(\hat{\boldsymbol{\theta}}_{\mathrm{ML}})}{L_p(\boldsymbol{\theta})}\right\} \overset{\mathrm{a}}{\sim} \chi^2(q),$$

where W is called the *generalised likelihood ratio statistic*. Here $\boldsymbol{\theta}$ denotes the true parameter of interest and $\hat{\boldsymbol{\theta}}_{\text{ML}}$ its MLE with respect to $L_p(\boldsymbol{\theta})$. This is identical to the first q components of the MLE with respect to the joint likelihood $L(\boldsymbol{\theta}, \boldsymbol{\eta})$, i.e.

$$L_p(\hat{\boldsymbol{\theta}}_{\text{ML}}) = \max_{\boldsymbol{\theta}} \left\{ \max_{\boldsymbol{\eta}} L(\boldsymbol{\theta}, \boldsymbol{\eta}) \right\} = \max_{\boldsymbol{\theta}, \boldsymbol{\eta}} L(\boldsymbol{\theta}, \boldsymbol{\eta}) = L(\hat{\boldsymbol{\theta}}_{\text{ML}}, \hat{\boldsymbol{\eta}}_{\text{ML}}).$$

The particular form of the generalised likelihood ratio statistic W can be viewed in the context of a significance test of the null hypothesis $H_0 : \boldsymbol{\theta} = \boldsymbol{\theta}_0$. Using the transformation

$$L_p(\boldsymbol{\theta}_0) = \max_{\boldsymbol{\eta}} L(\boldsymbol{\theta}_0, \boldsymbol{\eta}) = \max_{H_0} L(\boldsymbol{\theta}, \boldsymbol{\eta}),$$

we can write W as

$$W = 2 \log \left\{ \frac{\max L(\boldsymbol{\theta}, \boldsymbol{\eta})}{\max_{H_0} L(\boldsymbol{\theta}, \boldsymbol{\eta})} \right\} = 2 \log \left\{ \frac{L_p(\hat{\boldsymbol{\theta}}_{\text{ML}})}{L_p(\boldsymbol{\theta}_0)} \right\}.$$

A large value of W implies that H_0 has a relatively small likelihood. Calibration of W is done using the asymptotic $\chi^2(q)$ distribution, which will be derived subsequently.

In some cases we may even have an exact distribution of W for finite sample size, as the following example shows.

Example 5.15 (Two-sample t test) Let $X_{1:n_1}$ denote a random sample from an $\mathrm{N}(\mu_1, \sigma^2)$ distribution and $X_{(n_1+1):(n_1+n_2)}$ a random sample from an $\mathrm{N}(\mu_2, \sigma^2)$ distribution with μ_1, μ_2 and σ^2 all unknown. Assume that the two random samples are independent from each other. Consider the null hypothesis $H_0 : \mu = \mu_1 = \mu_2$. Using the reparametrisation $\mu_2 = \mu_1 + c$, the null hypothesis can be rewritten as $H_0 : c = 0$, which shows that H_0 fixes one parameter of the unrestricted model at a particular value ($c = 0$). Therefore, we are in the framework described above.

The likelihood is

$$\begin{aligned} L(\mu_1, \mu_2, \sigma^2) \\ = (2\pi)^{-\frac{n}{2}} (\sigma^2)^{-\frac{n}{2}} \exp\left\{ -\frac{1}{2} \frac{\sum_{i=1}^{n_1} (x_i - \mu_1)^2 + \sum_{i=n_1+1}^{n} (x_i - \mu_2)^2}{\sigma^2} \right\}. \end{aligned}$$

Under H_0, the MLE of σ^2 is

$$\hat{\sigma}_0^2 = \frac{1}{n} \sum_{i=1}^{n} (x_i - \hat{\mu})^2,$$

where $n = n_1 + n_2$, and

$$\hat{\mu} = \frac{1}{n} \sum_{i=1}^{n} x_i = \bar{x}$$

is the MLE of μ. We therefore have

$$\max_{H_0} L\big(\mu_1, \mu_2, \sigma^2\big) = L\big(\hat{\mu}, \hat{\mu}, \hat{\sigma}_0^2\big)$$

$$= (2\pi)^{-\frac{n}{2}} \big(\hat{\sigma}_0^2\big)^{-\frac{n}{2}} \exp\left\{ -\frac{1}{2} \frac{\sum_{i=1}^n (x_i - \bar{x})^2}{\frac{1}{n}\sum_{i=1}^n (x_i - \bar{x})^2} \right\}$$

$$= (2\pi)^{-\frac{n}{2}} \big(\hat{\sigma}_0^2\big)^{-\frac{n}{2}} \exp\left(-\frac{n}{2} \right).$$

Without the H_0 restriction, we have the MLEs

$$\hat{\mu}_1 = \frac{1}{n_1} \sum_{i=1}^{n_1} x_i,$$

$$\hat{\mu}_2 = \frac{1}{n_2} \sum_{i=n_1+1}^{n} x_i \quad \text{and}$$

$$\hat{\sigma}^2 = \frac{1}{n} \left\{ \sum_{i=1}^{n_1} (x_i - \hat{\mu}_1)^2 + \sum_{i=n_1+1}^{n} (x_i - \hat{\mu}_2)^2 \right\},$$

and we obtain

$$\max L\big(\mu_1, \mu_2, \sigma^2\big) = L\big(\hat{\mu}_1, \hat{\mu}_2, \hat{\sigma}^2\big)$$

$$= (2\pi)^{-\frac{n}{2}} \big(\hat{\sigma}^2\big)^{-n/2} \cdot \exp\left(-\frac{n}{2} \right).$$

The generalised likelihood ratio statistic is therefore

$$W = 2\log\left\{ \frac{\max L(\mu_1, \mu_2, \sigma^2)}{\max_{H_0} L(\mu_1, \mu_2, \sigma^2)} \right\} = 2\log\left\{ \left(\frac{\hat{\sigma}^2}{\hat{\sigma}_0^2} \right)^{-\frac{n}{2}} \right\}$$

$$= n \log\left(\frac{\hat{\sigma}_0^2}{\hat{\sigma}^2} \right).$$

Now

$$n\hat{\sigma}_0^2 = \sum_{i=1}^{n} (x_i - \hat{\mu})^2$$

$$= \sum_{i=1}^{n_1} (x_i - \hat{\mu}_1 + \hat{\mu}_1 - \hat{\mu})^2 + \sum_{i=n_1+1}^{n} (x_i - \hat{\mu}_2 + \hat{\mu}_2 - \hat{\mu})^2$$

$$= \sum_{i=1}^{n_1} (x_i - \hat{\mu}_1)^2 + n_1(\hat{\mu}_1 - \hat{\mu})^2 + \sum_{i=n_1+1}^{n} (x_i - \hat{\mu}_2)^2 + n_2(\hat{\mu}_2 - \hat{\mu})^2$$

$$
\begin{aligned}
&= n\hat{\sigma}^2 + n_1(\hat{\mu}_1 - \hat{\mu})^2 + n_2(\hat{\mu}_2 - \hat{\mu})^2 \\
&= n\hat{\sigma}^2 + \frac{n_1 n_2}{n}(\hat{\mu}_1 - \hat{\mu}_2)^2 \\
&= n\hat{\sigma}^2 + \frac{1}{\frac{1}{n_1} + \frac{1}{n_2}}(\hat{\mu}_1 - \hat{\mu}_2)^2,
\end{aligned}
$$

where the penultimate line follows from $\hat{\mu} = \frac{1}{n}(n_1\hat{\mu}_1 + n_2\hat{\mu}_2)$ and the combination of quadratic forms in Appendix B.1.5. So we can rewrite W as

$$
\begin{aligned}
W &= n \log\left\{1 + \frac{1}{\frac{1}{n_1} + \frac{1}{n_2}} \frac{(\hat{\mu}_1 - \hat{\mu}_2)^2}{n\hat{\sigma}^2}\right\} \\
&= n \log\left(1 + \frac{1}{n-2} t^2\right),
\end{aligned}
$$

where

$$
t = \frac{\hat{\mu}_1 - \hat{\mu}_2}{\sqrt{(\frac{1}{n_1} + \frac{1}{n_2})\frac{n}{n-2}\hat{\sigma}^2}}
$$

is the test statistic of the *two-sample t test*. Note that W is a monotone function of t^2 and hence of $|t|$. Under H_0, the exact distribution of t is a standard t distribution with $n_1 + n_2 - 2$ degrees of freedom, i.e. $t \sim \mathrm{t}(n_1 + n_2 - 2)$. This also induces an exact distribution for W under H_0.

Asymptotically ($n \to \infty$), the t distribution converges to the standard normal distribution, so t^2 has asymptotically a $\chi^2(1)$ distribution. Now, for large n, we have

$$
W = n \log\left(1 + \frac{1}{n-2} t^2\right) \approx \log\left\{\left(1 + \frac{t^2}{n}\right)^n\right\} \approx \log\{\exp(t^2)\} = t^2,
$$

which illustrates that W and t^2 have the same asymptotic distribution. We would also have obtained the asymptotic $\chi^2(1)$ distribution of W from a more general result, which will be described below. ■

Example 5.16 (Blood alcohol concentration) Consider the alcohol concentration data from Sect. 1.1.7. We would like to test whether there is a real difference in the transformation factor between the genders. We obtain a two-sample t test statistic $t = -3.59$ with P-value 0.00042 on 183 degrees of freedom. The generalised likelihood ratio statistic is $W = 12.61$ with P-value 0.00038. So the P-values are very similar, and the observed data set is very unlikely under the null hypothesis of equal mean transformation factors across the genders. ■

As noted before, the asymptotic distribution of the generalised likelihood ratio statistic is a χ^2 distribution.

Result 5.4 (Distribution of the generalised likelihood ratio statistic) *Under regularity conditions and assuming that* $H_0: \boldsymbol{\theta} = \boldsymbol{\theta}_0$ *holds, we have that, as* $n \to \infty$,

$$W = 2\log\left\{\frac{\max L(\boldsymbol{\theta}, \boldsymbol{\eta})}{\max_{H_0} L(\boldsymbol{\theta}, \boldsymbol{\eta})}\right\} \xrightarrow{D} \chi^2(q).$$

Proof To prove Result 5.4, let $(\hat{\boldsymbol{\theta}}_{\text{ML}}, \hat{\boldsymbol{\eta}}_{\text{ML}})^\top$ denote the unrestricted MLE and $\hat{\boldsymbol{\eta}}_0 = \hat{\boldsymbol{\eta}}_{\text{ML}}(\boldsymbol{\theta}_0)$ the restricted MLE of $\boldsymbol{\eta}$ under the null hypothesis $H_0: \boldsymbol{\theta} = \boldsymbol{\theta}_0$. Under H_0, the true parameter is denoted as $(\boldsymbol{\theta}_0, \boldsymbol{\eta})^\top$. We know that, under H_0,

$$\begin{pmatrix} \hat{\boldsymbol{\theta}}_{\text{ML}} \\ \hat{\boldsymbol{\eta}}_{\text{ML}} \end{pmatrix} \overset{\text{a}}{\sim} \text{N}_p\left(\begin{pmatrix} \boldsymbol{\theta}_0 \\ \boldsymbol{\eta} \end{pmatrix}, \boldsymbol{I}(\hat{\boldsymbol{\theta}}_{\text{ML}}, \hat{\boldsymbol{\eta}}_{\text{ML}})^{-1}\right), \tag{5.16}$$

where we partition the inverse observed Fisher information matrix as in Result 5.1:

$$\boldsymbol{I}(\hat{\boldsymbol{\theta}}_{\text{ML}}, \hat{\boldsymbol{\eta}}_{\text{ML}})^{-1} = \begin{pmatrix} \boldsymbol{I}^{11} & \boldsymbol{I}^{12} \\ \boldsymbol{I}^{21} & \boldsymbol{I}^{22} \end{pmatrix}.$$

We now apply a second-order Taylor approximation from Appendix B.2.3 to $\tilde{l}_p(\boldsymbol{\theta})$ around $\boldsymbol{\theta} = \hat{\boldsymbol{\theta}}_{\text{ML}}$. From Result 5.1 we know that the Hessian of $\tilde{l}_p(\boldsymbol{\theta})$ at the MLE is $-(\boldsymbol{I}^{11})^{-1}$, which gives us the quadratic approximation

$$\tilde{l}_p(\boldsymbol{\theta}) \approx -\frac{1}{2}(\boldsymbol{\theta} - \hat{\boldsymbol{\theta}}_{\text{ML}})^\top (\boldsymbol{I}^{11})^{-1} (\boldsymbol{\theta} - \hat{\boldsymbol{\theta}}_{\text{ML}})$$

for the relative profile log-likelihood. The generalised likelihood ratio statistic $W = -2\tilde{l}_p(\boldsymbol{\theta}_0)$ can therefore be approximated as

$$W \approx (\hat{\boldsymbol{\theta}}_{\text{ML}} - \boldsymbol{\theta}_0)^\top (\boldsymbol{I}^{11})^{-1} (\hat{\boldsymbol{\theta}}_{\text{ML}} - \boldsymbol{\theta}_0).$$

But we have from (5.16) that the marginal distribution of the MLE for $\boldsymbol{\theta}$ is $\hat{\boldsymbol{\theta}}_{\text{ML}} \overset{\text{a}}{\sim} \text{N}_q(\boldsymbol{\theta}_0, \boldsymbol{I}^{11})$. Hence, we finally obtain

$$W \overset{\text{a}}{\sim} \chi^2(q). \qquad \square$$

Example 5.17 (Goodness-of-fit) We will now use the generalised likelihood ratio statistic to derive a well-known *goodness-of-fit* statistic. Consider the MN blood group data from Sect. 1.1.4 and suppose that we want to investigate if there is evidence that the underlying population is not in Hardy–Weinberg equilibrium. The restricted model now assumes that the population is in Hardy–Weinberg equilibrium. From Example 2.7 we know that $\hat{\upsilon}_{\text{ML}} \approx 0.570$ with log-likelihood value

$$l(\hat{\upsilon}_{\text{ML}}) = x_1 \log(\hat{\upsilon}_{\text{ML}}^2) + x_2 \log\{2\hat{\upsilon}_{\text{ML}}(1 - \hat{\upsilon}_{\text{ML}})\} + x_3 \log\{(1 - \hat{\upsilon}_{\text{ML}})^2\} = -754.17.$$

Without restriction we have a trinomial model with MLE $\hat{\boldsymbol{\pi}}_{\mathrm{ML}} = \boldsymbol{x}/n$, where $n = x_1 + x_2 + x_3$, and log-likelihood value

$$l(\hat{\boldsymbol{\pi}}_{\mathrm{ML}}) = \sum_{i=1}^{3} x_i \log(\hat{\pi}_i) = -753.19.$$

The generalised likelihood ratio statistic is therefore $2\{l(\hat{\boldsymbol{\pi}}_{\mathrm{ML}}) - l(\hat{\upsilon}_{\mathrm{ML}})\} = 1.959$, from which we can easily compute the corresponding P-value 0.16, using the distribution function of the $\chi^2(1)$ distribution. This P-value is fairly large, so we conclude that there is no evidence against Hardy–Weinberg equilibrium. ■

In this example we have derived a special case of *Wilk's G^2 statistic*. The observed frequencies $x_1, \dots, x_k$ are compared with the fitted ("expected") frequencies e_i in a restricted model with, say, r parameters. The G^2 statistic is then

$$G^2 = 2 \sum_{i=1}^{k} x_i \log\left(\frac{x_i}{e_i}\right),$$

using the convention that $0 \log(0) = 0$. This is also known as the *deviance* and is an alternative to the *χ^2-statistic*

$$\chi^2 = \sum_{i=1}^{k} \frac{(x_i - e_i)^2}{e_i}.$$

Both have under H_0 an asymptotic $\chi^2(k - 1 - r)$ distribution. In the above example the χ^2 statistic gives a quite similar value with $\chi^2 = 1.956$ (P-value 0.16).

Example 5.18 (Screening for colon cancer) We now apply the χ^2 and G^2 statistic to the data on colon cancer screening from Sect. 1.1.5 to investigate the plausibility of the underlying binomial and beta-binomial models. The computations are straightforward, so we briefly discuss only the derivation of the degrees of freedom. The data are given in $k = 6$ categories, so the saturated multinomial model has five degrees of freedom. The truncated binomial model has one parameter, and the truncated beta-binomial model has two. The corresponding goodness-of-fit tests therefore have $5 - 1 = 4$ and $5 - 2 = 3$ degrees of freedom, respectively.

There is strong evidence against the truncated binomial model from Example 2.12 ($\chi^2 = 332.8$, $G^2 = 185.1$ at four degrees of freedom, with both P-values <0.0001). This confirms our initial concerns at the end of Example 2.12. The more flexible truncated beta-binomial model from Example 5.10 gives a much better fit with $\chi^2 = 2.12$ and $G^2 = 2.19$ at three degrees of freedom (P-values 0.55 and 0.53, respectively). ■

5.6 Conditional Likelihood

Conditional likelihood is an alternative approach to make inference for a parameter of interest $\boldsymbol{\theta}$ in the presence of a nuisance parameter $\boldsymbol{\eta}$. Briefly, the idea is to find a suitable one-to-one transformation (y_1, y_2), say, of the data x such that the conditional distribution of $y_1 \mid y_2$ does not depend on the nuisance parameter $\boldsymbol{\eta}$. If, in addition, the distribution of y_2 carries little or no information about $\boldsymbol{\theta}$, then we can treat y_2 as fixed and base inference only on the conditional likelihood $L_c(\boldsymbol{\theta})$ (or the conditional log-likelihood $l_c(\boldsymbol{\theta})$) implied by the distribution of $y_1 \mid y_2$.

Example 5.19 (Poisson model) Suppose we want to compare disease counts x and y in two areas with known expected numbers of cases e_x and e_y. The common statistical model is to assume that the corresponding random variables X and Y are independent Poisson with means $e_x\lambda_x$ and $e_y\lambda_y$, respectively. The parameter of interest is then the rate ratio $\theta = \lambda_x/\lambda_y$, and we can consider λ_y as nuisance parameter if we write $e_x\lambda_x$ as $e_x\theta\lambda_y$. Intuitively, the sum $Z = X + Y$ carries only little information about the rate ratio, so we consider the conditional distribution of $X \mid Z = z$, which is known to be a binomial $\operatorname{Bin}(z, \pi)$ distribution (cf. Appendix A.5.1) with "number of trials" equal to $z = x + y$ and "success probability" equal to

$$\pi = \frac{e_x\theta\lambda_y}{e_x\theta\lambda_y + e_y\lambda_y} = \frac{e_x\theta}{e_x\theta + e_y},$$

which does not depend on the nuisance parameter λ_y. The corresponding odds $\pi/(1-\pi)$ are $(e_x\theta)/e_y$, so the conditional log-likelihood for θ has the same form as the odds in a binomial model, which we derived in Example 2.6:

$$l_c(\theta) = x\log\left(\frac{e_x\theta}{e_y}\right) - (x+y)\log\left(1 + \frac{e_x\theta}{e_y}\right). \tag{5.17}$$

We can easily derive the MLE of θ by equating the empirical odds x/y with $\pi/(1-\pi) = (e_x\theta)/e_y$:

$$\hat{\theta}_{\mathrm{ML}} = \frac{x/e_x}{y/e_y}.$$

Conditional likelihood confidence intervals can be obtained using the conditional log-likelihood (5.17).

Now we consider profile likelihood as an alternative approach to eliminate the nuisance parameter λ_y. First, we need to derive the MLE of λ_y for fixed θ. The joint log-likelihood of θ and λ_y given the data x and y is

$$\begin{aligned} l(\theta, \lambda_y) &= x\log(\theta\lambda_y) - e_x\theta\lambda_y + y\log(\lambda_y) - e_y\lambda_y \\ &= x\log(\theta) + (x+y)\log(\lambda_y) - (e_x\theta + e_y)\lambda_y, \end{aligned}$$

from which we easily obtain

$$\hat{\lambda}_y(\theta) = \frac{x+y}{e_x\theta + e_y}.$$

This leads to the profile log-likelihood

$$\begin{aligned} l_p(\theta) &= l\{\theta, \hat{\lambda}_y(\theta)\} \\ &= x\log(\theta) + (x+y)\log\left(\frac{x+y}{e_x\theta+e_y}\right) - (x+y) \\ &= x\log(\theta) - (x+y)\log(e_x\theta+e_y) + (x+y)\log(x+y) - (x+y). \end{aligned}$$

The last two terms do not depend on θ, and so can be ignored. We are also at liberty to add the constant

$$x\{\log(e_x) + \log(e_y)\},$$

which gives, after some rearrangement,

$$l_p(\theta) = x\log\left(\frac{e_x\theta}{e_y}\right) - (x+y)\log\left(1 + \frac{e_x\theta}{e_y}\right).$$

The conditional log-likelihood $l_c(\theta)$ is therefore identical to the profile log-likelihood $l_p(\theta)$ in this case. ■

Conditional likelihood can also be used for the comparison of two binomial samples.

Example 5.20 (Comparison of proportions) Suppose we want to compare two binomial samples $X \sim \text{Bin}(m, \pi_x)$ and $Y \sim \text{Bin}(n, \pi_y)$ using the log odds ratio

$$\psi = \log\left\{\frac{\pi_x/(1-\pi_x)}{\pi_y/(1-\pi_y)}\right\}$$

as in Example 5.8. As in the previous Example 5.19, conditioning on the value $Z = z$ of the sum $Z = X + Y$ gives a likelihood that only depends on ψ, cf. Example 3.17, where we have shown that $X \mid Z = z$ follows a hypergeometric distribution under the assumption $\psi = 0$. In the general case the distribution of $X \mid Z = z$ turns out to be noncentral hypergeometric NCHypGeom$(z, m+n, m, \exp(\psi))$ with probability mass function

$$f(x \mid x+y=z) = \frac{\binom{m}{x}\binom{n}{z-x}\exp(\psi x)}{\sum_{s=0}^{z}\binom{m}{s}\binom{n}{z-s}\exp(\psi s)}.$$

This conditional likelihood is not identical but close to the profile likelihood derived in Example 5.8. Figure 5.8 compares the resulting relative log-likelihoods for the "Tervila" data analysed in Example 5.8. We see that the differences between the two functions are very small relative to the function values plotted in Fig. 5.4b. ■

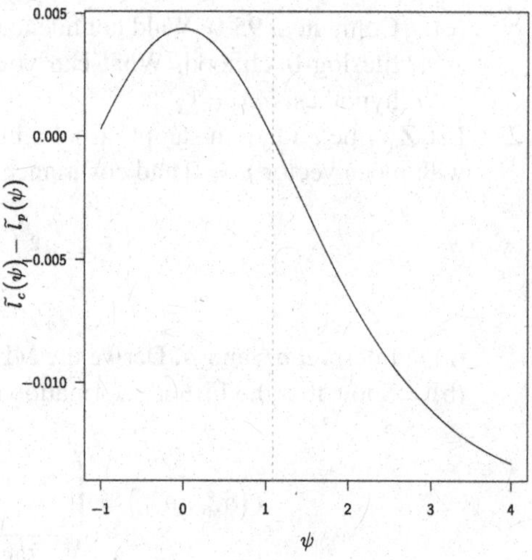

Fig. 5.8 Difference of the relative profile and conditional log-likelihood for the "Tervila" study data. The MLE location is marked with a *dashed vertical line*

5.7 Exercises

1. In a cohort study on the incidence of ischaemic heart disease (IHD), 337 male probands were enrolled. Each man was categorised as non-exposed (group 1, daily energy consumption ≥2750 kcal) or exposed (group 2, daily energy consumption <2750 kcal) to summarise his average level of physical activity. For each group, the number of person years ($Y_1 = 2768.9$ and $Y_2 = 1857.5$), and the number of IHD cases ($D_1 = 17$ and $D_2 = 28$) was registered thereafter.

   We assume that $D_i \mid \lambda_i \stackrel{\text{ind}}{\sim} \text{Po}(\lambda_i Y_i)$, $i = 1, 2$, where $\lambda_i > 0$ is the group-specific incidence rate.

   (a) For each group, derive the MLE $\hat{\lambda}_i$ and a corresponding 95 % Wald confidence interval for $\log(\lambda_i)$ with subsequent back-transformation to the λ_i-scale.

   (b) In order to analyse whether $\lambda_1 = \lambda_2$, we reparametrise the model with $\lambda = \lambda_1$ and $\theta = \lambda_2/\lambda_1$. Show that the joint log-likelihood kernel of λ and θ has the following form:

   $$l(\lambda, \theta) = D\log(\lambda) + D_2\log(\theta) - \lambda Y_1 - \theta\lambda Y_2,$$

   where $D = D_1 + D_2$.

   (c) Compute the MLE $(\hat{\lambda}, \hat{\theta})$, the observed Fisher information matrix $I(\hat{\lambda}, \hat{\theta})$ and derive expressions for both profile log-likelihood functions $l_p(\lambda) = l\{\lambda, \hat{\theta}(\lambda)\}$ and $l_p(\theta) = l\{\hat{\lambda}(\theta), \theta\}$.

   (d) Plot both functions $l_p(\lambda)$ and $l_p(\theta)$ and also create a contour plot of the relative log-likelihood $\tilde{l}(\lambda, \theta)$ using the R-function `contour`. Add the points $\{\lambda, \hat{\theta}(\lambda)\}$ and $\{\hat{\lambda}(\theta), \theta\}$ to the contour plot, analogously to Fig. 5.3a.

(e) Compute a 95 % Wald confidence interval for $\log(\theta)$ based on the profile log-likelihood. What can you say about the P-value for the null hypothesis $\lambda_1 = \lambda_2$?

2. Let $\boldsymbol{Z}_{1:n}$ be a random sample from a bivariate normal distribution $\mathrm{N}_2(\boldsymbol{\mu}, \boldsymbol{\Sigma})$ with mean vector $\boldsymbol{\mu} = 0$ and covariance matrix

$$\boldsymbol{\Sigma} = \sigma^2 \begin{pmatrix} 1 & \rho \\ \rho & 1 \end{pmatrix}.$$

(a) Interpret σ^2 and ρ. Derive the MLE $(\hat{\sigma}^2_{\mathrm{ML}}, \hat{\rho}_{\mathrm{ML}})$.

(b) Show that the Fisher information matrix is

$$\boldsymbol{I}\left(\hat{\sigma}^2_{\mathrm{ML}}, \hat{\rho}_{\mathrm{ML}}\right) = \begin{pmatrix} \frac{n}{\hat{\sigma}^4_{\mathrm{ML}}} & -\frac{n\hat{\rho}_{\mathrm{ML}}}{\hat{\sigma}^2_{\mathrm{ML}}(1-\hat{\rho}^2_{\mathrm{ML}})} \\ -\frac{n\hat{\rho}_{\mathrm{ML}}}{\hat{\sigma}^2_{\mathrm{ML}}(1-\hat{\rho}^2_{\mathrm{ML}})} & \frac{n(1+\hat{\rho}^2_{\mathrm{ML}})}{(1-\hat{\rho}^2_{\mathrm{ML}})^2} \end{pmatrix}.$$

(c) Show that

$$\mathrm{se}(\hat{\rho}_{\mathrm{ML}}) = \frac{1-\hat{\rho}^2_{\mathrm{ML}}}{\sqrt{n}}.$$

3. Calculate again the Fisher information of the profile log-likelihood in Result 5.2, but this time without using Result 5.1. Use instead the fact that $\hat{\alpha}_{\mathrm{ML}}(\delta)$ is a point where the partial derivative of $l(\alpha, \delta)$ with respect to α equals zero.

4. Let $X \sim \mathrm{Bin}(m, \pi_x)$ and $Y \sim \mathrm{Bin}(n, \pi_y)$ be independent binomial random variables. In order to analyse the null hypothesis $H_0 : \pi_x = \pi_y$, one often considers the *relative risk* $\theta = \pi_x/\pi_y$ or the *log relative risk* $\psi = \log(\theta)$.

(a) Compute the MLE $\hat{\psi}_{\mathrm{ML}}$ and its standard error for the log relative risk estimation. Proceed as in Example 5.8.

(b) Compute a 95 % confidence interval for the relative risk θ given the data in Table 3.1.

(c) Also compute the profile likelihood and the corresponding 95 % profile likelihood confidence interval for θ.

5. Suppose that ML estimates of the *sensitivity* π_x and *specificity* π_y of a diagnostic test for a specific disease are obtained from independent binomial samples $X \sim \mathrm{Bin}(m, \pi_x)$ and $Y \sim \mathrm{Bin}(n, \pi_y)$, respectively.

(a) Use Result 5.2 to compute the standard error of the logarithm of the *positive* and *negative likelihood ratio*, defined as $\mathrm{LR}^+ = \pi_x/(1-\pi_y)$ and $\mathrm{LR}^- = (1-\pi_x)/\pi_y$. Suppose $m = n = 100$, $x = 95$ and $y = 90$. Compute a point estimate and the limits of a 95 % confidence interval for both LR^+ and LR^-, using the standard error from above.

(b) The *positive predictive value* PPV is the probability of disease, given a positive test result. The equation

$$\frac{\mathrm{PPV}}{1-\mathrm{PPV}} = \mathrm{LR}^+ \cdot \omega$$

relates PPV to LR^+ and to the pre-test odds of disease ω. Likewise, the following equation holds for the *negative predictive value* NPV, the probability to be disease-free, given a negative test result:

$$\frac{1-\mathrm{NPV}}{\mathrm{NPV}} = \mathrm{LR}^- \cdot \omega.$$

Suppose $\omega = 1/1000$. Use the 95 % confidence interval for LR^+ and LR^-, obtained in 5(a), to compute the limits of a 95 % confidence interval for both PPV and NPV.

6. In the placebo-controlled clinical trial of diuretics during pregnancy to prevent preeclampsia by Fallis et al. (cf. Table 1.1), 6 out of 38 treated women and 18 out of 40 untreated women got preeclampsia.
   (a) Formulate a statistical model assuming independent binomial distributions in the two groups. Translate the null hypothesis "there is no difference in preeclampsia risk between the two groups" into a statement on the model parameters.
   (b) Let θ denote the risk difference between treated and untreated women. Derive the MLE $\hat{\theta}_{\mathrm{ML}}$ and a 95 % Wald confidence interval for θ. Also, give the MLE for the number needed to treat (NNT), which is defined as $1/\theta$.
   (c) Write down the joint log-likelihood kernel $l(\pi_1, \theta)$ of the risk parameter π_1 in the treatment group and θ. In order to derive the profile log-likelihood of θ, consider θ as fixed and write down the score function for π_1,

$$S_{\pi_1}(\pi_1, \theta) = \frac{\partial}{\partial \pi_1} l(\pi_1, \theta).$$

   Which values are allowed for π_1 when θ is fixed?
   (d) Write an R-function that solves $S_{\pi_1}(\pi_1, \theta) = 0$ (use `uniroot`) and thus gives an estimate $\hat{\pi}_1(\theta)$. Hence, write an R-function for the profile log-likelihood $l_p(\theta) = l\{\hat{\pi}_1(\theta), \theta\}$.
   (e) Compute a 95 % profile likelihood confidence interval for θ using numerical tools. Compare it with the Wald interval. What can you say about the P-value for the null hypothesis from Exercise 6(a)?
7. The AB0 blood group system was described by Karl Landsteiner in 1901, who was awarded the Nobel Prize for this discovery in 1930. It is the most important blood type system in human blood transfusion and comprises four different groups: A, B, AB and 0.

   Blood groups are inherited from both parents. Blood groups A and B are dominant over 0 and codominant to each other. Therefore a phenotype blood group A may have the genotype AA or A0; for phenotype B, the genotype is BB or B0, for phenotype AB, there is only the genotype AB, and for phenotype 0, there is only the genotype 00.

Table 5.2 Probability of offspring's blood group given allele frequencies in parental generation and sample realisations

Blood group	Probability	Observation
A = {AA, A0}	$\pi_1 = p^2 + 2pr$	$x_1 = 182$
B = {BB, B0}	$\pi_2 = q^2 + 2qr$	$x_2 = 60$
AB = {AB}	$\pi_3 = 2pq$	$x_3 = 17$
0 = {00}	$\pi_4 = r^2$	$x_4 = 176$

Let p, q and r be the proportions of alleles A, B, and 0 in a population, so $p+q+r=1$ and $p, q, r > 0$. Then the probabilities of the four blood groups for the offspring generation are given in Table 5.2. Moreover, the realisations in a sample of size $n = 435$ are reported.

(a) Explain how the probabilities in Table 5.2 arise. What assumption is tacitly made?

(b) Write down the log-likelihood kernel of $\boldsymbol{\theta} = (p, q)^\top$. To this end, assume that $\boldsymbol{x} = (x_1, x_2, x_3, x_4)^\top$ is a realisation from a multinomial distribution with parameters $n = 435$ and $\boldsymbol{\pi} = (\pi_1, \pi_2, \pi_3, \pi_4)^\top$.

(c) Compute the MLEs of p and q numerically, using the R function `optim`. Use the option `hessian = TRUE` in `optim` and process the corresponding output to receive the standard errors of $\hat{p}_{\mathrm{ML}}$ and $\hat{q}_{\mathrm{ML}}$.

(d) Finally, compute $\hat{r}_{\mathrm{ML}}$ and $\mathrm{se}(\hat{r}_{\mathrm{ML}})$. Make use of Sect. 5.4.3.

(e) Create a contour plot of the relative log-likelihood and mark the 95 % likelihood confidence region for $\boldsymbol{\theta}$. Use the R-functions `contourLines` and `polygon` for sketching the confidence region.

(f) Use the χ^2 and the G^2 test statistic to analyse the plausibility of the modelling assumptions.

8. Let $T \sim \mathrm{t}(n-1)$ be a standard t random variable with $n-1$ degrees of freedom.

(a) Derive the density function of the random variable

$$W = n \log\left(1 + \frac{T^2}{n-1}\right),$$

see Example 5.15, and compare it graphically with the density function of the $\chi^2(1)$ distribution for different values of n.

(b) Show that as $n \to \infty$, W follows indeed a $\chi^2(1)$ distribution.

9. Consider the χ^2 statistic given k categories with n observations. Let

$$D_n = \sum_{i=1}^{k} \frac{(n_i - np_{i0})^2}{np_{i0}} \quad \text{and} \quad W_n = 2\sum_{i=1}^{k} n_i \log\left(\frac{n_i}{np_{i0}}\right).$$

Show that $W_n - D_n \xrightarrow{P} 0$ as $n \to \infty$.

10. In a psychological experiment the forgetfulness of probands is tested with the recognition of syllable triples. The proband has ten seconds to memorise the triple, afterwards it is covered. After a waiting time of t seconds, it is

checked whether the proband remembers the triple. For each waiting time t, the experiment is repeated n times.

Let $y = (y_1, \dots, y_m)$ be the relative frequencies of correctly remembered syllable triples for the waiting times of $t = 1, \dots, m$ seconds. The *power model* now assumes that

$$\pi(t; \boldsymbol{\theta}) = \theta_1 t^{-\theta_2}, \quad 0 \leq \theta_1 \leq 1,\ \theta_2 > 0,$$

is the probability to correctly remember a syllable triple after the waiting time $t \geq 1$.

(a) Derive an expression for the log-likelihood $l(\boldsymbol{\theta})$ where $\boldsymbol{\theta} = (\theta_1, \theta_2)$.

(b) Create a contour plot of the log-likelihood in the parameter range $[0.8, 1] \times [0.3, 0.6]$ with $n = 100$ and

$$y = (0.94, 0.77, 0.40, 0.26, 0.24, 0.16), \qquad t = (1, 3, 6, 9, 12, 18).$$

11. Often the *exponential model* is used instead of the power model (described in Exercise 10), assuming:

$$\pi(t; \boldsymbol{\theta}) = \min\{1, \theta_1 \exp(-\theta_2 t)\}, \quad t > 0,\ \theta_1 > 0 \text{ and } \theta_2 > 0.$$

(a) Create a contour plot of the log-likelihood in the parameter range $[0.8, 1.4] \times [0, 0.4]$ for the same data as in Exercise 10.

(b) Use the R-function `optim` to numerically compute the MLE $\hat{\boldsymbol{\theta}}_{\mathrm{ML}}$. Add the MLE to the contour plot from 11(a).

(c) For $0 \leq t \leq 20$, create a plot of $\pi(t; \hat{\boldsymbol{\theta}}_{\mathrm{ML}})$ and add the observations y.

12. Let $X_{1:n}$ be a random sample from a log-normal $\mathrm{LN}(\mu, \sigma^2)$ distribution, cf. Table A.2.

(a) Derive the MLE of μ and σ^2. Use the connection between the densities of the normal distribution and the log normal distribution. Also compute the corresponding standard errors.

(b) Derive the profile log-likelihood functions of μ and σ^2 and plot them for the following data:

$$x = (225, 171, 198, 189, 189, 135, 162, 136, 117, 162).$$

Compare the profile log-likelihood functions with their quadratic approximations.

13. We assume an exponential model for the survival times in the randomised placebo-controlled trial of Azathioprine for primary biliary cirrhosis (PBC) from Sect. 1.1.8. The survival times (in days) of the $n = 90$ patients in the placebo group are denoted by x_i with censoring indicators γ_i $(i = 1, \dots, n)$, while the survival times of the $m = 94$ patients in the treatment group are denoted by y_i and have censoring indicators δ_i $(i = 1, \dots, m)$. The (partly unobserved) uncensored survival times follow exponential models with rates η and $\theta\eta$ in the placebo and treatment group, respectively $(\eta, \theta > 0)$.

(a) Interpret η and θ. Show that their joint log-likelihood is

$$l(\eta, \theta) = (n\bar{\gamma} + m\bar{\delta})\log(\eta) + m\bar{\delta}\log(\theta) - \eta(n\bar{x} + \theta m\bar{y}),$$

where $\bar{\gamma}$, $\bar{\delta}$, $\bar{x}$, $\bar{y}$ are the averages of the γ_i, δ_i, x_i and y_i, respectively.

(b) Calculate the MLE $(\hat{\eta}_{\text{ML}}, \hat{\theta}_{\text{ML}})$ and the observed Fisher information matrix $\boldsymbol{I}(\hat{\eta}_{\text{ML}}, \hat{\theta}_{\text{ML}})$.

(c) Show that

$$\text{se}(\hat{\theta}_{\text{ML}}) = \hat{\theta}_{\text{ML}} \cdot \sqrt{\frac{n\bar{\gamma} + m\bar{\delta}}{n\bar{\gamma} \cdot m\bar{\delta}}}$$

and derive a general formula for a $\gamma \cdot 100\,\%$ Wald confidence interval for θ. Use Appendix B.1.1 to compute the required entry of the inverse observed Fisher information.

(d) Consider the transformation $\psi = \log(\theta)$. Derive a $\gamma \cdot 100\,\%$ Wald confidence interval for ψ using the delta method.

(e) Derive the profile log-likelihood for θ. Implement an R-function that calculates a $\gamma \cdot 100\,\%$ profile likelihood confidence interval for θ.

(f) Calculate 95 % confidence intervals for θ based on Exercises 13(c), 13(d) and 13(e). Also compute for each of the three confidence intervals 13(c), 13(d) and 13(e) the corresponding P-value for the null hypothesis that the exponential distribution for the PBC survival times in the treatment group is not different from the placebo group.

14. Let $X_{1:n}$ be a random sample from the $\text{N}(\mu, \sigma^2)$ distribution.

(a) First assume that σ^2 is known. Derive the likelihood ratio statistic for testing specific values of μ.

(b) Show that, in this special case, the likelihood ratio statistic is an exact pivot and exactly has a $\chi^2(1)$ distribution.

(c) Show that, in this special case, the corresponding likelihood ratio confidence interval equals the Wald confidence interval.

(d) Now assume that μ is known. Derive the likelihood ratio statistic for testing specific values of σ^2.

(e) Compare the likelihood ratio statistic and its distribution with the exact pivot mentioned in Example 4.21. Derive a general formula for a confidence interval based on the exact pivot, analogously to Example 3.8.

(f) Consider the transformation factors from Table 1.3 and assume that the "mean" is the known μ and the "standard deviation" is $\hat{\sigma}_{\text{ML}}$. Compute both a 95 % likelihood ratio confidence interval and a 95 % confidence interval based on the exact pivot for σ^2. Illustrate the likelihood ratio confidence interval by plotting the relative log-likelihood and the cut-value, similar to Fig. 4.8. In order to compute the likelihood ratio confidence interval, use the R-function `uniroot` (cf. Appendix C.1.2).

15. Consider K independent groups of normally distributed observations with group-specific means and variances, i.e. let $X^{(k)}_{1:n_k}$ be a random sample from $\mathrm{N}(\mu_k, \sigma_k^2)$ for group $k = 1, \dots, K$. We want to test the null hypothesis that the variances are identical, i.e. $H_0 : \sigma_k^2 = \sigma^2$.
    (a) Write down the log-likelihood kernel for the parameter vector $\boldsymbol{\theta} = (\mu_1, \dots, \mu_K, \sigma_1^2, \dots, \sigma_K^2)^\top$. Derive the MLE $\hat{\boldsymbol{\theta}}_{\mathrm{ML}}$ by solving the score equations $S_{\mu_k}(\boldsymbol{\theta}) = 0$ and then $S_{\sigma_k^2}(\boldsymbol{\theta}) = 0$, for $k = 1, \dots, K$.
    (b) Compute the MLE $\hat{\boldsymbol{\theta}}_0$ under the restriction $\sigma_k^2 = \sigma^2$ of the null hypothesis.
    (c) Show that the generalised likelihood ratio statistic for testing $H_0 : \sigma_k^2 = \sigma^2$ is

$$W = \sum_{k=1}^{K} n_k \log\left(\hat{\sigma}_0^2 / \hat{\sigma}_k^2\right),$$

    where $\hat{\sigma}_0^2$ and $\hat{\sigma}_k^2$ are the ML variance estimates for the kth group with and without the H_0 assumption, respectively. What is the approximate distribution of W under H_0?
    (d) Consider the special case with $K = 2$ groups having equal sizes $n_1 = n_2 = n$. Show that W is large when the ratio

$$R = \frac{\sum_{i=1}^{n}(x_i^{(1)} - \bar{x}_1)^2}{\sum_{i=1}^{n}(x_i^{(2)} - \bar{x}_2)^2}$$

    is large or small. Show that W is minimal if $R = 1$. Which value has W for $R = 1$?
    (e) Bartlett's modified likelihood ratio test statistic (Bartlett 1937) is $B = T/C$, where in

$$T = \sum_{k=1}^{K} (n_k - 1) \log\left(s_0^2 / s_k^2\right)$$

    compared to W, the numbers of observations n_k have been replaced by the degrees of freedom $n_k - 1$, and the sample variances $s_k^2 = (n_k - 1)^{-1} \sum_{i=1}^{n_k}(x_i^{(k)} - \bar{x}_k)^2$ define the pooled sample variance $s_0^2 = \{\sum_{k=1}^{K}(n_k - 1)\}^{-1} \sum_{k=1}^{K}(n_k - 1)s_k^2$. The correction factor

$$C = 1 + \frac{1}{3(K-1)} \left\{ \left(\sum_{k=1}^{K} \frac{1}{n_k - 1} \right) - \frac{1}{\sum_{k=1}^{K}(n_k - 1)} \right\}$$

    is used because T/C converges more rapidly to the asymptotic $\chi^2(K-1)$ distribution than T alone.

Write two R-functions that take the vectors of the group sizes $(n_1, \ldots, n_K)$ and the sample variances $(s_1^2, \ldots, s_K^2)$ and return the values of the statistics W and B, respectively.

(f) In the H_0 setting with $K = 3$, $n_k = 5$, $\mu_k = k$, $\sigma^2 = 1/4$, simulate 10 000 data sets and compute the statistics W and B in each case. Compare the empirical distributions with the approximate $\chi^2(K-1)$ distribution. Is B closer to $\chi^2(K-1)$ than W in this case?

(g) Consider the alcohol concentration data from Sect. 1.1.7. Quantify the evidence against equal variances of the transformation factor between the genders using P-values based on the test statistics W and B.

16. In a 1:1 *matched case-control study*, one control (i.e. a disease-free individual) is matched to each case (i.e. a diseased individual) based on certain individual characteristics, e.g. age or gender. Exposure history to a potential risk factor is then determined for each individual in the study. If exposure E is binary (e.g. smoking history? yes/no), then it is common to display the data as frequencies of case-control pairs, depending on exposure history:

		History of control	
		Exposed	Unexposed
History of case	Exposed	a	b
	Unexposed	c	d

For example, a is the number of case-control pairs with positive exposure history of both the case and the control.

Let ω_1 and ω_0 denote the odds for a case and a control, respectively, to be exposed, such that

$$\Pr(E \mid \text{case}) = \frac{\omega_1}{1+\omega_1} \quad \text{and} \quad \Pr(E \mid \text{control}) = \frac{\omega_0}{1+\omega_0}.$$

To derive conditional likelihood estimates of the odds ratio $\psi = \omega_1/\omega_0$, we argue conditional on the number N_E of exposed individuals in a case-control pair. If $N_E = 2$, then both the case and the control are exposed, so that the corresponding a case-control pairs do not contribute to the conditional likelihood. This is also the case for the d case-control pairs where both the case and the control are unexposed ($N_E = 0$). In the following we therefore only consider the case $N_E = 1$, in which case either the case or the control is exposed, but not both.

(a) Conditional on $N_E = 1$, show that the probability that the case rather than the control is exposed is $\omega_1/(\omega_0 + \omega_1)$. Show that the corresponding conditional odds are equal to the odds ratio ψ.

(b) Write down the binomial log-likelihood in terms of ψ and show that the MLE of the odds ratio ψ is $\hat{\psi}_{\text{ML}} = b/c$ with standard error $\text{se}\{\log(\hat{\psi}_{\text{ML}})\} = \sqrt{1/b + 1/c}$. Derive the Wald test statistic for H_0: $\log(\psi) = 0$.

(c) Derive the standard error $\mathrm{se}(\hat{\psi}_{\mathrm{ML}})$ of $\hat{\psi}_{\mathrm{ML}}$ and derive the Wald test statistic for H_0: $\psi = 1$. Compare your result with the Wald test statistic for H_0: $\log(\psi) = 0$.

(d) Finally, compute the score test statistic for H_0: $\psi = 1$ based on the expected Fisher information of the conditional likelihood.

17. Let $Y_i \stackrel{\text{ind}}{\sim} \mathrm{Bin}(1, \pi_i)$, $i = 1, \dots, n$, be the binary response variables in a *logistic regression model*, where the probabilities $\pi_i = F(\boldsymbol{x}_i^\top \boldsymbol{\beta})$ are parametrised via the inverse logit function

$$F(x) = \frac{\exp(x)}{1 + \exp(x)}$$

by the regression coefficient vector $\boldsymbol{\beta} = (\beta_1, \dots, \beta_p)^\top$. The vector $\boldsymbol{x}_i = (x_{i1}, \dots, x_{ip})^\top$ contains the values of the p covariates for the ith observation.

(a) Show that F is indeed the inverse of the logit function $\mathrm{logit}(x) = \log\{x/(1-x)\}$, and that $\frac{d}{dx}F(x) = F(x)\{1 - F(x)\}$.

(b) Use the results on multivariate derivatives outlined in Appendix B.2.2 to show that the log-likelihood, score vector and Fisher information matrix of $\boldsymbol{\beta}$, given the realisation $\boldsymbol{y} = (y_1, \dots, y_n)^\top$, are

$$l(\boldsymbol{\beta}) = \sum_{i=1}^{n} y_i \log(\pi_i) + (1 - y_i) \log(1 - \pi_i),$$

$$\boldsymbol{S}(\boldsymbol{\beta}) = \sum_{i=1}^{n} (y_i - \pi_i)\boldsymbol{x}_i = \boldsymbol{X}^\top(\boldsymbol{y} - \boldsymbol{\pi}) \quad \text{and}$$

$$\boldsymbol{I}(\boldsymbol{\beta}) = \sum_{i=1}^{n} \pi_i(1 - \pi_i)\boldsymbol{x}_i\boldsymbol{x}_i^\top = \boldsymbol{X}^\top \boldsymbol{W} \boldsymbol{X},$$

respectively, where $\boldsymbol{X} = (\boldsymbol{x}_1, \dots, \boldsymbol{x}_n)^\top$ is the design matrix, $\boldsymbol{\pi} = (\pi_1, \dots, \pi_n)^\top$, $\boldsymbol{W} = \mathrm{diag}\{\pi_i(1 - \pi_i)\}_{i=1}^n$.

(c) Show that the statistic $\boldsymbol{T}(y) = \sum_{i=1}^n y_i \boldsymbol{x}_i$ is minimal sufficient for $\boldsymbol{\beta}$.

(d) Implement an R-function that maximises the log-likelihood using the Newton–Raphson algorithm (see Appendix C.1.3) by iterating

$$\boldsymbol{\beta}^{(t+1)} = \boldsymbol{\beta}^{(t)} + \boldsymbol{I}(\boldsymbol{\beta}^{(t)})^{-1}\boldsymbol{S}(\boldsymbol{\beta}^{(t)}), \quad t = 1, 2, \dots$$

until the new estimate $\boldsymbol{\beta}^{(t+1)}$ and the old one $\boldsymbol{\beta}^{(t)}$ are almost identical and $\hat{\boldsymbol{\beta}}_{\mathrm{ML}} = \boldsymbol{\beta}^{(t+1)}$. Start with $\boldsymbol{\beta}^{(1)} = 0$.

(e) Consider the data set `amlxray` on the connection between X-ray usage and acute myeloid leukaemia in childhood, which is available in the R-package `faraway`. Here $y_i = 1$ if the disease was diagnosed for the ith child and $y_i = 0$ otherwise (`disease`). We include an intercept in the regression model, i.e. we set $x_1 = 1$. We want to analyse the association of the diabetes status with the covariates x_2 (age in years), x_3 (1 if the child is male and 0 otherwise, Sex), x_4 (1 if the mother ever have an X-ray and 0 otherwise, Mray) and x_5 (1 if the father ever have an X-ray and 0 otherwise, Fray).

Interpret $\beta_2, \dots, \beta_5$ by means of odds ratios. Compute the MLE $\hat{\boldsymbol{\beta}}_{\text{ML}} = (\hat{\beta}_1, \dots, \hat{\beta}_5)^\top$ and standard errors $\text{se}(\hat{\beta}_i)$ for all coefficient estimates $\hat{\beta}_i$ and construct 95 % Wald confidence intervals for β_i ($i = 1, \dots, 5$). Interpret the results and compare them with those from the R-function `glm` (using the `binomial` family).

(f) Implement an R-function that returns the profile log-likelihood of one of the p parameters. Use it to construct 95 % profile likelihood confidence intervals for them. Compare with the Wald confidence intervals from above and with the results from the R-function `confint` applied to the `glm` model object.

(g) We want to test if the inclusion of the covariates x_2, x_3, x_4 and x_5 improves the fit of the model to the data. To this end, we consider the null hypothesis $H_0: \beta_2 = \dots = \beta_5 = 0$.

How can this be expressed in the form $H_0: \boldsymbol{C}\boldsymbol{\beta} = \boldsymbol{\delta}$, where $\boldsymbol{C}$ is a $q \times p$ contrast matrix (of rank $q \leq p$), and $\boldsymbol{\delta}$ is a vector of length q? Use a result from Appendix A.2.4 to show that under H_0,

$$(\boldsymbol{C}\hat{\boldsymbol{\beta}}_{\text{ML}} - \boldsymbol{\delta})^\top \{\boldsymbol{C}\boldsymbol{I}(\hat{\boldsymbol{\beta}}_{\text{ML}})^{-1}\boldsymbol{C}^\top\}^{-1}(\boldsymbol{C}\hat{\boldsymbol{\beta}}_{\text{ML}} - \boldsymbol{\delta}) \overset{\text{a}}{\sim} \chi^2(q). \tag{5.18}$$

(h) Compute two P-values quantifying the evidence against H_0, one based on the squared Wald statistic (5.18) and the other based on the generalised likelihood ratio statistic.

(i) Since the data is actually from a matched case-control study, where pairs of one case and one control have been matched (by age, race and county of residence; the variable `ID` denotes the matched pairs), it is more appropriate to apply conditional logistic regression. Compute the corresponding MLEs and 95 % confidence intervals with the R-function `clogit` from the package `survival` and compare the results.

18. In clinical dose-finding studies, the relationship between the dose $d \geq 0$ of the medication and the average response $\mu(d)$ in a population is to be inferred. Considering a continuously measured response y, then a simple model for the individual measurements assumes $y_{ij} \overset{\text{ind}}{\sim} \text{N}(\mu(d_{ij}; \boldsymbol{\theta}), \sigma^2)$, $i = 1, \dots, K$, $j = 1, \dots, n_i$. Here n_i is the number of patients in the ith dose group with dose d_i (placebo group has $d = 0$). The *Emax model* has the functional form

$$\mu(d; \boldsymbol{\theta}) = \theta_1 + \theta_2 \frac{d}{d + \theta_3}.$$

(a) Plot the function $\mu(d; \boldsymbol{\theta})$ for different choices of the parameters $\theta_1, \theta_2, \theta_3 > 0$. Give reasons for the interpretation of θ_1 as the mean placebo response, θ_2 as the maximum treatment effect, and θ_3 as the dose giving 50 % of the maximum treatment effect.

(b) Compute the expected Fisher information for the parameter vector $\boldsymbol{\theta}$. Using this result, implement an R function that calculates the approximate covariance matrix of the MLE $\hat{\boldsymbol{\theta}}_{\text{ML}}$ for a given set of doses $d_1, \dots, d_K$, a total sample size $N = \sum_{i=1}^{K} n_i$, allocation weights $w_i = n_i/N$ and given error variance σ^2.

(c) Assume that $\theta_1 = 0$, $\theta_2 = 1$, $\theta_3 = 0.5$ and $\sigma^2 = 1$. Calculate the approximate covariance matrix, first, for $K = 5$ doses 0, 1, 2, 3, 4 and, second, for doses 0, 0.5, 1, 2, 4, both times with balanced allocations $w_i = 1/5$ and total sample size $N = 100$. Compare the approximate standard deviations of the MLEs of the parameters between the two designs, also compare the determinants of the two calculated matrices.

(d) Using the second design, determine the required total sample size N so that the standard deviation for estimation of θ_2 is 0.35 (so that the half-length of a 95 % confidence interval is about 0.7).

5.8 References

Multiparameter likelihood inference is described comprehensively in Pawitan (2001) and Davison (2003). The proof of Result 5.1 is adapted from Patefield (1977). An efficient algorithm for the computation of profile likelihood intervals is described in Venzon and Moolgavkar (1988).

Bayesian Inference

6

Contents

Frequentist inference treats the data X as random. Point and interval estimates of the parameter θ are viewed as functions of the data X, in order to obtain frequentist properties of the estimates. The parameter θ is unknown but treated as fixed, not as random.

Things are just the other way round in the Bayesian approach to statistical inference, named after Thomas Bayes (1702–1761). The unknown parameter θ is now a random variable with appropriate *prior distribution* $f(\theta)$. The *posterior distribution* $f(\theta \mid x)$, computed with Bayes' theorem, summarises the information about θ after observing the data $X = x$. Note that Bayesian inference conditions on the observation $X = x$, in contrast to frequentist inference.

L. Held, D. Sabanés Bové, *Likelihood and Bayesian Inference*,
Statistics for Biology and Health, https://doi.org/10.1007/978-3-662-60792-3_6,

For the time being, we assume in the following that θ is a scalar. For simplicity, we will always speak of a (prior or posterior) density function, even when it is in fact a probability mass function of a discrete parameter.

6.1 Bayes' Theorem

We start this chapter with a brief review of Bayes' theorem for simple events, as outlined in Appendix A.1.2.

For any two events A and B with $0 < \Pr(A) < 1$ and $\Pr(B) > 0$, we have

$$\Pr(A \mid B) = \frac{\Pr(B \mid A)\Pr(A)}{\Pr(B)}. \tag{6.1}$$

The use of Bayes' theorem in diagnostic testing is an established part of clinical reasoning, as illustrated in the following example.

Example 6.1 (Diagnostic test) Suppose a simple diagnostic test for a specific disease, which produces either a positive or a negative test result, is known to have the 90 % *sensitivity*. This means that if the person being tested has the disease ($D+$), the probability of a positive test result ($T+$) is 90 %: $\Pr(T+ \mid D+) = 0.9$. Now assume that the test also has the 90 % *specificity* and write $D-$ if the person being tested is free of the disease. Similarly, let $T-$ denote a negative test result. The 90 % specificity now translates to $\Pr(T- \mid D-) = 0.9$.

We can use Bayes' formula to compute the conditional probability of disease given a positive test result:

$$\Pr(D+ \mid T+) = \frac{\Pr(T+ \mid D+)\Pr(D+)}{\Pr(T+)}. \tag{6.2}$$

All we need to do is to multiply the *prevalence* $\Pr(D+)$ with the sensitivity $\Pr(T+ \mid D+)$ and divide through $\Pr(T+)$. For the time being, let us assume that the prevalence $\Pr(D+)$ is 1 % for the disease considered. The denominator $\Pr(T+)$ in (6.2), the unconditional probability of a positive test result, is unknown but can be computed via the *law of total probability*, cf. Eq. (A.3) in Appendix A.1.1:

$$\Pr(T+) = \Pr(T+ \mid D+)\Pr(D+) + \Pr(T+ \mid D-)\Pr(D-). \tag{6.3}$$

Thus, we can calculate $\Pr(T+)$ if we know the sensitivity $\Pr(T+ \mid D+)$, the prevalence $\Pr(D+)$, and $\Pr(T+ \mid D-) = 1 - \Pr(T- \mid D-)$, i.e. 1 minus the specificity. In the above example,

$$\Pr(T+) = 0.9 \cdot 0.01 + 0.1 \cdot 0.99 = 0.108,$$

and hence

$$\Pr(D+ \mid T+) = \frac{0.9 \cdot 0.01}{0.108} \approx 0.083,$$

i.e. the *positive predictive value* is 8.3 %. So if the test was positive, then the disease risk increases from 1.0 % to 8.3 %.

It is up to the reader to write down an equivalent formula to (6.2) to compute the *negative predictive value* $\Pr(D- \mid T-)$, which turns out to be approximately 99.89 %. So if the test was negative, then the disease risk decreases from 1.0 % to $\Pr(D+ \mid T-) = 1 - \Pr(D- \mid T-) = 1 - 0.9989 = 0.0011$, i.e. 0.11 %. The disease risk changes in the expected direction depending on the test result. ■

Example 6.1 exemplifies the process of *Bayesian updating*: We update the prior risk $\Pr(D+)$ in the light of a positive test result $T+$ to obtain the posterior risk $\Pr(D+ \mid T+)$, the conditional probability of disease given a positive test result, also known as the positive predictive value.

However, Eq. (6.2), with the denominator $\Pr(T+)$ replaced by (6.3), is somewhat complex and not particularly intuitive. A simpler version of Bayes' theorem can be obtained if we switch from probabilities to odds:

$$\frac{\Pr(D+ \mid T+)}{\Pr(D- \mid T+)} = \frac{\Pr(T+ \mid D+)}{\Pr(T+ \mid D-)} \cdot \frac{\Pr(D+)}{\Pr(D-)}. \tag{6.4}$$

Here $\Pr(D+)/\Pr(D-)$ are the *prior odds*, $\Pr(D+ \mid T+)/\Pr(D- \mid T+)$ are the *posterior odds*, and $\Pr(T+ \mid D+)/\Pr(T+ \mid D-)$ is the so-called *likelihood ratio* for a positive test result. Bayesian updating is thus just one simple mathematical operation: Multiply the prior odds with the likelihood ratio to obtain the posterior odds. This also explains the formula for the positive predictive value given in Exercise 5 of Chap. 5 on page 156.

Example 6.2 (Diagnostic test) In Example 6.1 the prior odds are 1/99 and the likelihood ratio (for a positive test result) is $0.9/0.1 = 9$. The posterior odds (given a positive test result) are therefore $9 \cdot 1/99 = 1/11 \approx 0.09$, which of course corresponds to the posterior probability of 8.3 % calculated above. So the prior odds of 1 to 99 change to posterior odds of 1 to 11 in the light of a positive test result. If the test result was negative, then the prior odds need to be multiplied with the likelihood ratio for a negative test result, which is $\Pr(T- \mid D+)/\Pr(T- \mid D-) = 0.1/0.9 = 1/9$. This leads to posterior odds of $1/9 \cdot 1/99 = 1/891$. We leave it up to the reader to check that this corresponds to a negative predictive value of approximately 99.89 %. ■

Formula (6.2) is specified for a positive test result $T+$ and a diseased person $D+$ but is equally valid if we replace a positive test result $T+$ by a negative one, i.e. $T-$, or a diseased person $D+$ by a non-diseased one, i.e. $D-$, or both. Thus, a more general description of Bayes' theorem is given by

$$f(D = d \mid T = t) = \frac{f(T = t \mid D = d) \cdot f(D = d)}{f(T = t)}, \tag{6.5}$$

where D and T are binary random variables taking values d and t, respectively. In the diagnostic setting outlined above, d and t can be either + ("positive") or − ("negative"). Note that we have switched notation from $\Pr(\cdot)$ to $f(\cdot)$ to emphasise that (6.5) relates to general probability mass functions of the random variables D and T, and not only to probabilities of the events $D+$ and $T+$, say.

A more compact version of (6.5) is

$$f(d \mid t) = \frac{f(t \mid d) \cdot f(d)}{f(t)}, \tag{6.6}$$

cf. Eq. (A.8). Note that this equation also holds if the random variables D or T have more than two possible values. The formula is also correct if it involves continuous random variables, in which case $f(\cdot)$ denotes a density function, see Eq. (A.10). This simple formula forms the basis for the whole rest of this chapter.

6.2 Posterior Distribution

The posterior distribution is the most important quantity in Bayesian inference. It contains all the information available about the unknown parameter θ after having observed the data $X = x$. Certain characteristics of the posterior distribution can be used to derive Bayesian point and interval estimates.

Definition 6.1 (Posterior distribution) Let $X = x$ denote the observed realisation of a (possibly multivariate) random variable X with density function $f(x \mid \theta)$. Specifying a prior distribution with density function $f(\theta)$ allows us to compute the density function $f(\theta \mid x)$ of the *posterior distribution* using Bayes' theorem

$$f(\theta \mid x) = \frac{f(x \mid \theta) f(\theta)}{\int f(x \mid \theta) f(\theta)\, d\theta}, \tag{6.7}$$

cf. Eq. (A.10). For discrete parameters θ, the integral in the denominator has to be replaced with a sum. ◆

The term $f(x \mid \theta)$ in (6.7) is simply the likelihood function $L(\theta)$ previously denoted by $f(x; \theta)$. Since θ is now random, we explicitly condition on a specific value θ and write $L(\theta) = f(x \mid \theta)$. The denominator in (6.7) can be written as

$$\int f(x \mid \theta) f(\theta)\, d\theta = \int f(x, \theta)\, d\theta = f(x),$$

which emphasises that it does not depend on θ. Note that Eq. (6.7) is then equivalent to Eq. (6.6). The quantity $f(x)$ is known as the *marginal likelihood* and is important for Bayesian model selection, cf. Sect. 7.2.

The density of the posterior distribution is therefore proportional to the product of the likelihood and the density of the prior distribution, with proportionality constant $1/f(x)$. This is usually denoted as

$$f(\theta \mid x) \propto f(x \mid \theta) f(\theta) \quad \text{or} \quad f(\theta \mid x) \propto L(\theta) f(\theta), \tag{6.8}$$

where "$\propto$" stands for "is proportional to" and implies that $1/\int L(\theta) f(\theta)\, d\theta$ is the normalising constant to ensure that $\int f(\theta \mid x)\, d\theta = 1$, such that $f(\theta \mid x)$ is a valid density function.

Posterior distribution
The density function of the posterior distribution can be obtained through multiplication of the likelihood function and the density function of the prior distribution with subsequent normalisation.

Statistical inference about θ is based solely on the posterior distribution. Suitable point estimates are location parameters, such as the mean, median or mode, of the posterior distribution. We will formally define those now for a scalar parameter θ:

Definition 6.2 (Bayesian point estimates) The *posterior mean* $\mathsf{E}(\theta \mid x)$ is the expectation of the posterior distribution:

$$\mathsf{E}(\theta \mid x) = \int \theta f(\theta \mid x)\, d\theta.$$

The *posterior mode* $\text{Mod}(\theta \mid x)$ is the mode of the posterior distribution:

$$\text{Mod}(\theta \mid x) = \arg\max_{\theta} f(\theta \mid x).$$

The *posterior median* $\text{Med}(\theta \mid x)$ is the median of the posterior distribution, i.e. any number a that satisfies

$$\int_{-\infty}^{a} f(\theta \mid x)\, d\theta = 0.5 \quad \text{and} \quad \int_{a}^{\infty} f(\theta \mid x)\, d\theta = 0.5. \tag{6.9}$$

◆

Implicitly we often assume that the posterior mean is finite, in which case it is also unique. However, the posterior mode and the posterior median are not necessarily unique. Indeed, a posterior distribution can have several modes and is then called *multimodal*. As an example where the median is not unique, consider a continuous real-valued parameter, where the (posterior) density is zero in a central interval. If 50 % of the probability mass are distributed to the left and right of this centre, then any value a in the central interval fulfils (6.9), so is a posterior median.

Bayesian interval estimates are also derived from the posterior distribution. To distinguish them from confidence intervals, which have a different interpretation, they are called *credible intervals*. Here is the definition for a scalar parameter θ:

Definition 6.3 (Credible interval) For fixed $\gamma \in (0, 1)$, a $\gamma \cdot 100\,\%$ *credible interval* is defined through two real numbers t_l and t_u, that fulfil

$$\int_{t_l}^{t_u} f(\theta \mid x)\, d\theta = \gamma. \tag{6.10}$$

The quantity γ is called the *credibility level* of the credible interval $[t_l, t_u]$. ♦

The definition implies that the random variable $\theta \mid x$ is contained in a $\gamma \cdot 100\,\%$ credible interval with probability γ. The easiest way to construct credible intervals is to choose t_l as the $(1-\gamma)/2$ quantile and t_u as the $(1+\gamma)/2$ quantile of the posterior distribution. To compute such *equal-tailed* credible intervals, one needs to be able to compute quantiles of the posterior distribution, see Appendix A.5 for frequently used R-functions.

This simple and intuitive interpretation of credible intervals is to be compared with the rather complex Definition 3.6 of frequentist confidence intervals. Now we can say that an unknown parameter lies in a certain credible interval with probability γ. This interpretation is not correct for confidence intervals.

Example 6.3 (Inference for a proportion) Inference for a proportion is based on a random sample of size n and determines the number $X = x$ of individuals in this sample with a certain event of interest. It is often reasonable to assume that $X \sim \mathrm{Bin}(n, \pi)$, where $\pi \in (0, 1)$ is the unknown probability of the event. It is tempting to select a beta distribution as a prior distribution for π because the support of a beta distribution equals the parameter space $(0, 1)$. So let a priori $\pi \sim \mathrm{Be}(\alpha, \beta)$ with suitably chosen parameters $\alpha, \beta > 0$. Then

$$f(x \mid \pi) = \binom{n}{x} \pi^x (1-\pi)^{n-x}, \qquad x = 0, 1, \dots, n,$$

$$f(\pi) = \frac{1}{\mathrm{B}(\alpha, \beta)} \pi^{\alpha-1} (1-\pi)^{\beta-1}, \quad 0 < \pi < 1,$$

so the posterior distribution is

$$\begin{aligned} f(\pi \mid x) &\propto f(x \mid \pi) \cdot f(\pi) \\ &\propto \pi^x (1-\pi)^{n-x} \cdot \pi^{\alpha-1} (1-\pi)^{\beta-1} \\ &= \pi^{\alpha+x-1} (1-\pi)^{\beta+n-x-1}. \end{aligned}$$

This can be easily identified as yet another beta distribution with parameters $\alpha + x$ and $\beta + n - x$:

$$\pi \mid x \sim \mathrm{Be}(\alpha + x, \beta + n - x). \tag{6.11}$$

Compared with the prior distribution $\mathrm{Be}(\alpha, \beta)$ of π, the number of successes x is added to the first prior parameter, while the number of failures $n - x$ is added to the second prior parameter. The prior and posterior density functions are displayed in Fig. 6.1 for a simple example.

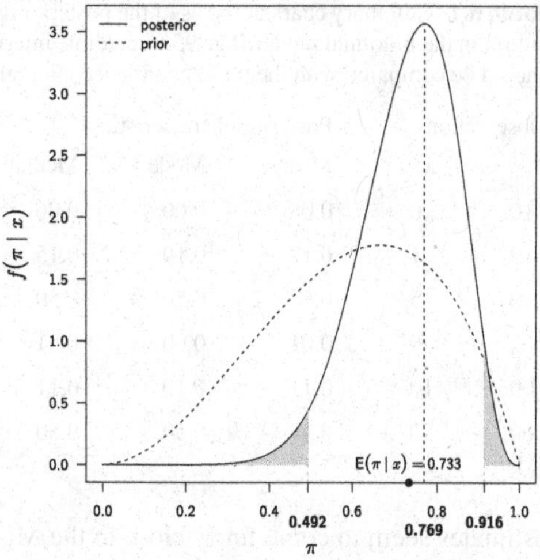

Fig. 6.1 Posterior Be(11, 4) density of π (*solid line*) for a Be(3, 2) prior distribution (*dashed line*) and observation $x = 8$ in a binomial experiment with $n = 10$ trials. The posterior mean is 0.733, and the posterior mode is 0.769. The equal-tailed 95 % credible interval [0.492, 0.916] is also shown

It is convenient to think of the Be(α, β) prior distribution as that which would have arisen if we had started with an "improper" Be(0, 0) prior and then observed α successes in $\alpha + \beta$ trials. Thus, $n_0 = \alpha + \beta$ can be viewed as a *prior sample size*. This interpretation of the prior parameters is useful in order to assess the weight attached to the prior distribution, as we will see in the following.

There are simple explicit formulas for the mean and the mode of the Be(α, β) distribution, see Appendix A.5.2 for details. For example, the mean is simply $\alpha/(\alpha + \beta)$. Therefore, the mean of $\pi \mid x \sim \text{Be}(\alpha + x, \beta + n - x)$ is

$$\frac{\alpha + x}{\alpha + \beta + n}.$$

Re-writing this posterior mean as

$$\frac{\alpha + x}{\alpha + \beta + n} = \frac{\alpha + \beta}{\alpha + \beta + n} \cdot \frac{\alpha}{\alpha + \beta} + \frac{n}{\alpha + \beta + n} \cdot \frac{x}{n}$$

shows that it is a weighted average of the prior mean $\alpha/(\alpha + \beta)$ and the MLE $\bar{x} = x/n$ with weights proportional to the prior sample size $n_0 = \alpha + \beta$ and the data sample size n, respectively. The *relative prior sample size* $n_0/(n_0 + n)$ quantifies the weight of the prior mean in the posterior mean. Note that the relative prior sample size decreases with increasing data sample size n.

The case $\alpha = \beta = 1$ is of particular interest, as it corresponds to a uniform prior distribution on the interval (0, 1), an apparently natural choice in the absence of any prior information. This is in fact exactly the prior used by Bayes in his famous essay (Bayes 1763). The prior sample size n_0 is 2, one success and one failure. The posterior mean is now $(x + 1)/(n + 2)$, and the posterior mode equals the MLE $\bar{x}$.

The posterior median and other posterior quantiles can be easily calculated numerically using the quantile function of the beta distribution (`qbeta` in `R`). Some examples are shown in Table 6.1 assuming a uniform prior distribution. The Bayes

Table 6.1 Summary characteristics of the posterior distribution of π under a uniform prior distribution in the binomial model. The 95 % credible interval based on the 2.5 % and 97.5 % quantiles should be compared with the 95 % confidence intervals from Table 4.2

Observation		Posterior characteristics				
n	x	Mean	Mode	Median	2.5 % quantile	97.5 % quantile
10	0	0.08	0.00	0.06	0.00	0.28
10	1	0.17	0.10	0.15	0.02	0.41
10	5	0.50	0.50	0.50	0.23	0.77
100	0	0.01	0.00	0.01	0.00	0.04
100	10	0.11	0.10	0.11	0.06	0.17
100	50	0.50	0.50	0.50	0.40	0.60

estimates seem to come fairly close to the MLEs x/n for increasing sample size n. This empirical result will be discussed in more detail in Sect. 6.6.2. ■

Example 6.4 (Diagnostic test) We now revisit the diagnostic testing problem discussed in Example 6.1 under the more realistic scenario that the disease prevalence $\pi = \Pr(D+)$ is not known but only estimated from a *prevalence study*. We will describe how the Bayesian approach can be used to assess the uncertainty of the positive and negative predictive values in this case.

For example, suppose that there was $x = 1$ diseased individual in a prevalence study with $n = 100$ participants. The Be(0.5, 5) prior distribution for π expresses our initial prior beliefs that the prevalence is below 33 % with approximately 95 % probability and below 5 % with approximately 50 % probability. The posterior distribution of π is then $\text{Be}(\tilde{\alpha}, \tilde{\beta})$ with updated parameters $\tilde{\alpha} = 0.5 + x = 1.5$ and $\tilde{\beta} = 5 + n - x = 104$. The posterior mean, median and mode are 0.014, 0.011 and 0.005, respectively. The difference between the point estimates indicates that the posterior distribution is quite skewed. The equal-tailed 95 % credible interval is $[0.001, 0.044]$.

In the posterior odds (6.4), the prevalence π enters:

$$\omega = \frac{\Pr(D+ \mid T+)}{\Pr(D- \mid T+)} = \frac{\Pr(T+ \mid D+)}{1 - \Pr(T- \mid D-)} \cdot \frac{\pi}{1-\pi}.$$

Therefore, the posterior odds ω are easily obtained from π and we can transform ω into the corresponding probability $\Pr(D+ \mid T+)$ as follows:

$$\theta = \Pr(D+ \mid T+) = \frac{\omega}{1+\omega} = \left(1 + \omega^{-1}\right)^{-1}. \tag{6.12}$$

Suppose we want to replace the fixed prevalence $\pi = 1/100$ with the posterior distribution $\pi \sim \text{Be}(\tilde{\alpha}, \tilde{\beta})$ to appreciate the uncertainty involved in the estimation of π. Note that (6.12) is a monotone function of π. Therefore, any quantile of the posterior distribution of π can be transformed to the θ-scale. Based on the corresponding quantiles of π stated above, we easily obtain the posterior median 0.09 and the 95 % equal-tailed credible interval [0.01, 0.29] for θ.

It is also possible to analytically compute the implied distribution of $\theta = \Pr(D+ \mid T+)$. In Exercise 2 it is shown that the density of θ is

$$f(\theta) = c \cdot \theta^{-2} \cdot f_F\big(c(1/\theta - 1); 2\tilde{\beta}, 2\tilde{\alpha}\big), \tag{6.13}$$

where

$$c = \frac{\tilde{\alpha} \Pr(T+ \mid D+)}{\tilde{\beta}\{1 - \Pr(T- \mid D-)\}},$$

and $f_F(x; 2\tilde{\beta}, 2\tilde{\alpha})$ is the density of the F distribution with parameters $2\tilde{\beta}$ and $2\tilde{\alpha}$, cf. Appendix A.5.2. Note that c does not depend on θ. One can analogously proceed for the negative predictive value $\Pr(D- \mid T-)$, see Exercise 2. The posterior mean and the posterior mode of θ can now be computed using numerical integration and optimisation, respectively. The posterior mean turns out to be 0.109 while the posterior mode is 0.049. The positive predictive value 8.3 % obtained in Example 6.1 with fixed prevalence $\pi = 0.01$ lies in between.

For the negative predictive value, we obtain the posterior mean 0.9984 and the posterior mode 0.9995. This is to be compared with the negative predictive value 99.89 % for fixed prevalence π, cf. again Example 6.1. We further obtain the posterior median 0.9987 and the 95 % equal-tailed credible interval [0.9949, 0.9999].

It is also simple to generate a random sample from $\Pr(D+ \mid T+)$ and $\Pr(D- \mid T-)$ using samples from the beta distribution of π. The following R-code illustrates this.

```
## prior parameters
a <- 0.5
b <- 5
## data
x <- 1
n <- 100
## posterior parameters
apost <- a+x
bpost <- b+n-x
## sample size
nsample <- 10^5
## prevalence values sampled from Be(1.5, 99.5) distribution
prev <- rbeta(nsample, apost, bpost)
## set values for sensitivity and specificity
sens <- 0.9
spec <- 0.9
## compute resulting positive and negative predictive value samples
ppv <- sens * prev / (sens * prev + (1 - spec) * (1 - prev))
npv <- spec * (1 - prev) / ((1 - sens) * prev + spec * (1 - prev))
```

Histograms of these samples are shown in Fig. 6.2. The histograms are in perfect agreement with the true density functions. Such samples form the basis of *Monte Carlo techniques* used for Bayesian inference, as discussed in more detail in Chap. 8. ■

In Example 6.3 the posterior mode of the binomial proportion equals the MLE when a uniform prior is used. This result holds in general:

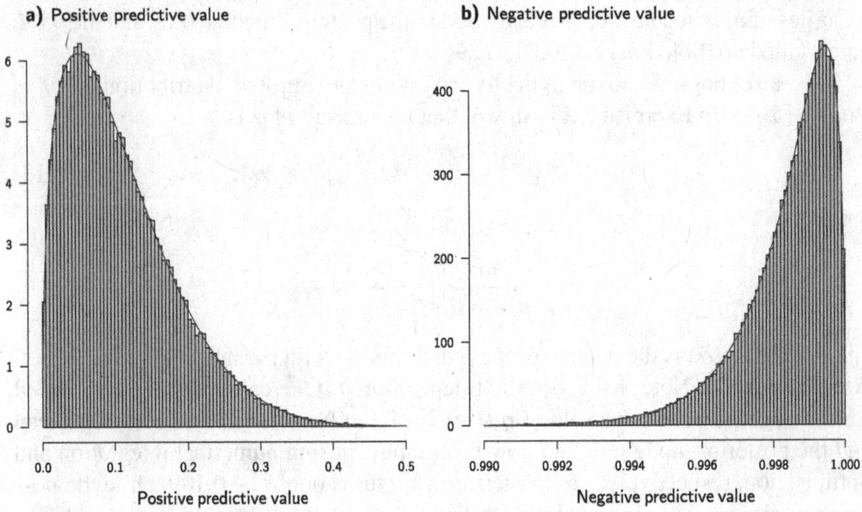

Fig. 6.2 Densities and samples histograms of the positive and negative predictive value if the prevalence follows the Be(1.5, 104) distribution, and the sensitivity and specificity are both 90 %

Result 6.1 *Under a uniform prior, the posterior mode equals the MLE.*

Proof This can easily be seen from the fact that if the prior on θ is uniform, the density $f(\theta)$ does not depend on θ. From Eq. (6.8) it follows that

$$f(\theta \mid x) \propto L(\theta).$$

Hence, the mode of the posterior distribution must equal the value that maximises the likelihood function, which is the MLE. □

There are infinitely many $\gamma \cdot 100$ % credible intervals for fixed γ, at least if the parameter θ is continuous. In the previous example with $\gamma = 0.95$ we cut off 2.5 % probability mass of each tail of the posterior distribution to obtain an equal-tailed credible interval. Alternatively, we could, for example, cut off 5 % probability mass on the left side of the posterior distribution and fix the right limit of the credible interval at the upper bound of the parameter space (which was 1 in Example 6.4). Under some regularity conditions, the following additional requirement ensures the uniqueness of credible intervals.

Definition 6.4 (Highest posterior density interval) Let $\gamma \in (0, 1)$ be a fixed credibility level. A $\gamma \cdot 100$ % credible interval $I = [t_l, t_u]$ is called a *highest posterior density interval* (HPD interval) if

$$f(\theta \mid x) \geq f(\tilde{\theta} \mid x)$$

for all $\theta \in I$ and all $\tilde{\theta} \notin I$. ◆

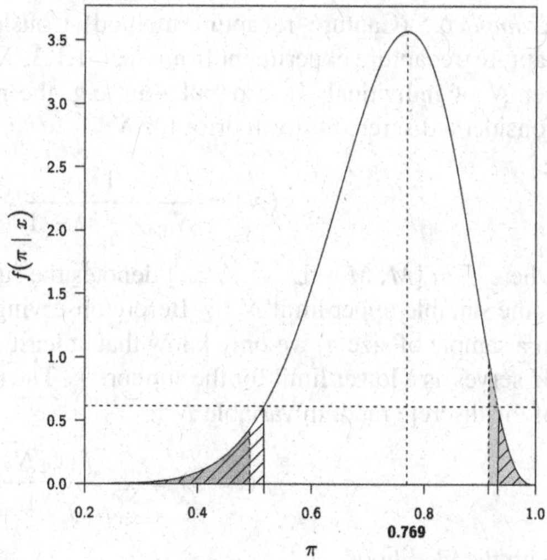

Fig. 6.3 Comparison of an equal-tailed and a 95 % HPD credible interval for the Be(11, 4) posterior distribution of a probability π (cf. Fig. 6.1). Both *grey areas* cover 2.5 % probability mass and embed the equal-tailed credible interval [0.492, 0.916]. The *shaded areas* define the HPD interval [0.517, 0.932], which contains all values of π with $f(\pi \mid x) > 0.614$. Note that the range of the HPD interval is 0.416, slightly smaller than the range of the equal-tailed interval, which is 0.424

An HPD interval contains all those parameter values that have higher posterior density than all parameter values not contained in the interval. Under some additional regularity conditions, the posterior density ordinates at the limits of an HPD interval are equal, i.e.

$$f(t_l \mid x) = f(t_u \mid x). \tag{6.14}$$

There are counterexamples where this property of HPD intervals is not fulfilled. For example, suppose that the posterior distribution is exponential. The lower limit t_l of an HPD interval at any level will then be zero because the density of the exponential distribution is monotonically decreasing. However, as we will see in Sect. 6.6.2, the posterior distribution of a continuous parameter is asymptotically normal, i.e. unimodal, if the sample size increases. Equation (6.14) will therefore typically hold for larger sample sizes.

Numerical computation of HPD intervals typically requires iterative algorithms. Figure 6.3 compares an HPD interval with an equal-tailed credible interval.

For discrete parameter spaces Θ, the definition of credible intervals has to be modified since it might not be possible to find any interval with exact credibility level γ as required by Eq. (6.10). A $\gamma \cdot 100$ % credible interval $I = [t_l, t_u]$ for θ is then defined through

$$\sum_{\theta \in I \cap \Theta} f(\theta \mid x) \geq \gamma. \tag{6.15}$$

Similarly, the posterior median $\text{Med}(\theta \mid x)$ will typically not be unique if the parameter is discrete. One can add an additional arbitrary requirement to make it unique, for example define $\text{Med}(\theta \mid x)$ as the smallest possible value with at least 50 % posterior probability mass below it.

Example 6.5 (Capture–recapture method) Consider now Bayesian inference in the capture–recapture experiment from Sect. 1.1.3. The unknown parameter is the number N of individuals in a population, e.g. the number of fish in a lake. We first consider a discrete uniform prior for N:

$$f(N) = \frac{1}{N_{\max} - M + 1} \quad \text{for } N \in \mathcal{T},$$

where $\mathcal{T} = \{M, M+1, \dots, N_{\max}\}$ denotes the support of the prior distribution with some suitable upper limit $N_{\max}$. Before observing the number of marked fish $X = x$ in a sample of size n, we only know that at least M (marked) fish are in the lake, so M serves as a lower limit for the support $\mathcal{T}$. The posterior probability mass function of the discrete random variable N is

$$f(N \mid x) = \frac{f(x \mid N) f(N)}{\sum_{N \in \mathcal{T}} f(x \mid N) f(N)}$$

with the likelihood

$$f(x \mid N) = \frac{\binom{M}{x}\binom{N-M}{n-x}}{\binom{N}{n}} \quad \text{for } N \in \{\max(n, M+n-x), \max(n, M+n-x)+1, \dots\}$$

already derived in Example 2.2. The support $\mathcal{P}$ of the posterior distribution is the intersection of the support $\mathcal{T}$ of the prior distribution and the parameter values allowed in the likelihood. As $\max(n, M+n-x) \geq M+n-x \geq M$, it is

$$\mathcal{P} = \{\max(n, M+n-x), \dots, N_{\max}\}.$$

Since the prior probability function $f(N)$ does not depend on N, the posterior probability function is proportional to the likelihood:

$$f(N \mid x) = \frac{f(x \mid N)}{\sum_{N \in \mathcal{P}} f(x \mid N)}.$$

Figure 6.4a gives the posterior distribution for $M = 26$, $n = 63$ and $x = 5$. Three different point estimates and a 95 % HPD interval are shown. Note that the posterior mode equals the MLE $\hat{N}_{\text{ML}} = 327$ (cf. Example 2.2).

To avoid the specification of the fairly arbitrary upper limit $N_{\max}$, a useful alternative to the uniform prior is a geometric distribution $N \sim \text{Geom}(\pi)$, truncated to $N \geq M$, i.e.

$$f(N) \propto \pi(1-\pi)^{N-1} \quad \text{for } N = M, M+1, \dots,$$

cf. Fig. 6.4b. Under this prior, the point estimates and the limits of the 95 % credible interval are slightly shifted to the left. The posterior distribution is slightly more concentrated with a shorter range of the 95 % HPD interval. ■

As we have seen in the previous example, different prior distributions can be chosen for a given data model. The choice of the prior distribution is the topic of the next section.

a) Uniform prior on the interval [26, 1 500]. Possible point estimates are $\text{Mod}(N \mid x) = 327$, $\text{Med}(N \mid x) = 426$ and $\text{E}(N \mid x) = 488.7$. The 95% HPD credible interval is [165, 997].

b) Truncated geometric prior distribution with $\pi = 0.001$. Possible point estimates are $\text{Mod}(N \mid x) = 313$, $\text{Med}(N \mid x) = 392$ and $\text{E}(N \mid x) = 446.5$. The 95% HPD credible interval is [161, 869].

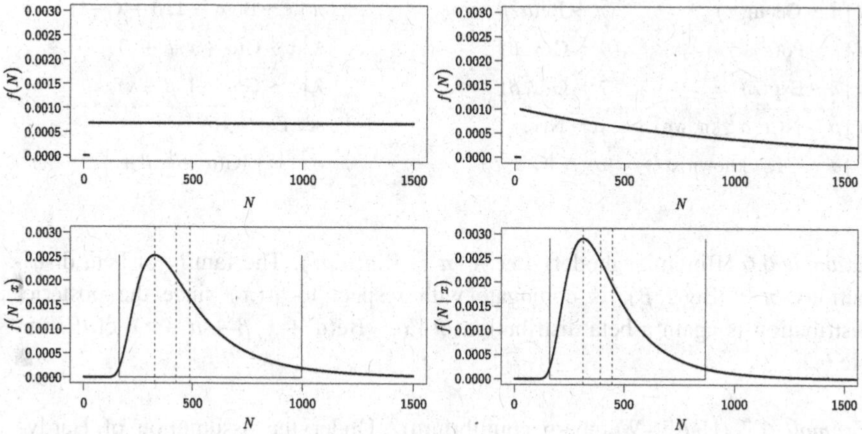

Fig. 6.4 Capture–recapture experiment: $M = 26$ fish have been marked, and $x = 5$ of them have been caught in a sample of $n = 63$. The chosen prior probability function is shown in the *upper panels*. The posterior probability functions can be seen in the *lower panels* and are to be compared with the likelihood function in Fig. 2.2

6.3 Choice of the Prior Distribution

Bayesian inference allows the probabilistic specification of prior beliefs through a prior distribution. It is often useful and justified to restrict the range of possible prior distributions to a specific family with one or two parameters, say. The choice of this family can be based on the type of likelihood function encountered. We will now discuss such a choice.

6.3.1 Conjugate Prior Distributions

A pragmatic approach to choosing a prior distribution is to select a member of a specific family of distributions such that the posterior distribution belongs to the same family. This is called a *conjugate prior distribution.*

Definition 6.5 (Conjugate prior distribution) Let $L(\theta) = f(x \mid \theta)$ denote a likelihood function based on the observation $X = x$. A class $\mathcal{G}$ of distributions is called *conjugate with respect to* $L(\theta)$ if the posterior distribution $f(\theta \mid x)$ is in $\mathcal{G}$ for all x whenever the prior distribution $f(\theta)$ is in $\mathcal{G}$. ◆

The family $\mathcal{G} = \{\text{all distributions}\}$ is trivially conjugate with respect to any likelihood function. In practice one tries to find smaller sets $\mathcal{G}$ that are specific to the likelihood $L_x(\theta)$.

Table 6.2 Summary of conjugate prior distributions for different likelihood functions

Likelihood	Conjugate prior distribution	Posterior distribution
$X \mid \pi \sim \text{Bin}(n, \pi)$	$\pi \sim \text{Be}(\alpha, \beta)$	$\pi \mid x \sim \text{Be}(\alpha + x, \beta + n - x)$
$X \mid \pi \sim \text{Geom}(\pi)$	$\pi \sim \text{Be}(\alpha, \beta)$	$\pi \mid x \sim \text{Be}(\alpha + 1, \beta + x - 1)$
$X \mid \lambda \sim \text{Po}(e \cdot \lambda)$	$\lambda \sim \text{G}(\alpha, \beta)$	$\lambda \mid x \sim \text{G}(\alpha + x, \beta + e)$
$X \mid \lambda \sim \text{Exp}(\lambda)$	$\lambda \sim \text{G}(\alpha, \beta)$	$\lambda \mid x \sim \text{G}(\alpha + 1, \beta + x)$
$X \mid \mu \sim \text{N}(\mu, \sigma^2 \text{ known})$	$\mu \sim \text{N}(\nu, \tau^2)$	see Eq. (6.16)
$X \mid \sigma^2 \sim \text{N}(\mu \text{ known}, \sigma^2)$	$\sigma^2 \sim \text{IG}(\alpha, \beta)$	$\sigma^2 \mid x \sim \text{IG}(\alpha + \frac{1}{2}, \beta + \frac{1}{2}(x - \mu)^2)$

Example 6.6 (Binomial model) Let $X \mid \pi \sim \text{Bin}(n, \pi)$. The family of beta distributions, $\pi \sim \text{Be}(\alpha, \beta)$, is conjugate with respect to $L(\pi)$ since the posterior distribution is again a beta distribution: $\pi \mid x \sim \text{Be}(\alpha + x, \beta + n - x)$, cf. Example 6.3. ■

Example 6.7 (Hardy–Weinberg equilibrium) Under the assumption of Hardy–Weinberg equilibrium, the likelihood has the form

$$L(\upsilon) = \upsilon^{2x_1 + x_2}(1 - \upsilon)^{x_2 + 2x_3}$$

with allele frequency $\upsilon \in (0, 1)$ (cf. Example 2.7). It is easy to see that a beta prior distribution $\text{Be}(\alpha, \beta)$ for υ results in a beta posterior distribution,

$$\upsilon \mid \boldsymbol{x} \sim \text{Be}(\alpha + 2x_1 + x_2, \beta + x_2 + 2x_3),$$

so the beta distribution is a conjugate class for this likelihood. ■

Before we study further conjugate prior distributions, we note that it is sufficient to study conjugacy for one member X_i of a random sample $X_{1:n}$. Indeed, if the prior is conjugate, the posterior after observing the first observation, say, is by definition of the same type and serves as the new prior distribution for the next observation. The new posterior distribution, now incorporating also the second observation, is again within the conjugate class, and so on. Only the parameters of the distribution will change in such a sequential processing of the data.

Table 6.2 gives further examples of conjugate prior distributions with the corresponding likelihood function. We now look at one entry of this table in more detail.

Example 6.8 (Normal model) Let X denote a sample from a normal $\text{N}(\mu, \sigma^2)$ distribution with unknown mean μ and known variance σ^2. The corresponding likelihood function is

$$L(\mu) \propto \exp\left\{-\frac{1}{2\sigma^2}(x - \mu)^2\right\}.$$

Combined with a normal prior distribution with mean ν and variance τ^2 for the unknown mean μ, i.e.

$$f(\mu) \propto \exp\left\{-\frac{1}{2\tau^2}(\mu-\nu)^2\right\},$$

the posterior density of μ is given by

$$\begin{aligned} f(\mu \mid x) &\propto L(\mu) \cdot f(\mu) \\ &\propto \exp\left[-\frac{1}{2}\left\{\frac{1}{\sigma^2}(x-\mu)^2 + \frac{1}{\tau^2}(\mu-\nu)^2\right\}\right] \\ &\propto \exp\left[-\frac{1}{2}\left(\frac{1}{\sigma^2}+\frac{1}{\tau^2}\right)\left\{\mu - \left(\frac{1}{\sigma^2}+\frac{1}{\tau^2}\right)^{-1} \cdot \left(\frac{x}{\sigma^2}+\frac{\nu}{\tau^2}\right)\right\}^2\right], \end{aligned}$$

see (B.5) for justification of the last rearrangement. So the posterior distribution is also normal:

$$\mu \mid x \sim \mathrm{N}\left(\left(\frac{1}{\sigma^2}+\frac{1}{\tau^2}\right)^{-1} \cdot \left(\frac{x}{\sigma^2}+\frac{\nu}{\tau^2}\right), \left(\frac{1}{\sigma^2}+\frac{1}{\tau^2}\right)^{-1}\right). \tag{6.16}$$

As in the binomial model, the posterior mean is a weighted average of the prior mean ν and the data x with weights proportional to $1/\tau^2$ and $1/\sigma^2$, respectively.

Equation (6.16) simplifies if one uses *precisions*, i.e. inverse variances, rather than the variances themselves. Indeed, let $\kappa = 1/\sigma^2$ and $\delta = 1/\tau^2$; then

$$\mu \mid x \sim \mathrm{N}\left(\frac{\kappa x + \delta \nu}{\kappa + \delta}, (\kappa+\delta)^{-1}\right).$$

Therefore, a Bayesian analysis of normal observations often uses precision parameters rather than variance parameters.

The above result can be easily extended to a random sample $X_{1:n}$ from an $\mathrm{N}(\mu, \sigma^2)$ distribution with known variance σ^2. For simplicity, we work with precisions rather than variances and use the fact that $\bar{x}$ is sufficient for μ, so the likelihood function of μ is

$$L(\mu) \propto \exp\left\{-\frac{n}{2\sigma^2}(\mu-\bar{x})^2\right\},$$

cf. Result 2.4. Combined with the prior $\mu \sim \mathrm{N}(\nu, \tau^2)$, we easily obtain

$$\mu \mid x_{1:n} \sim \mathrm{N}\left(\frac{n\kappa\bar{x} + \delta\nu}{n\kappa + \delta}, (n\kappa+\delta)^{-1}\right).$$

The corresponding formula with variances rather than precisions reads

$$\mu \mid x_{1:n} \sim \mathrm{N}\left(\left(\frac{n}{\sigma^2}+\frac{1}{\tau^2}\right)^{-1} \cdot \left(\frac{n\bar{x}}{\sigma^2}+\frac{\nu}{\tau^2}\right), \left(\frac{n}{\sigma^2}+\frac{1}{\tau^2}\right)^{-1}\right).$$

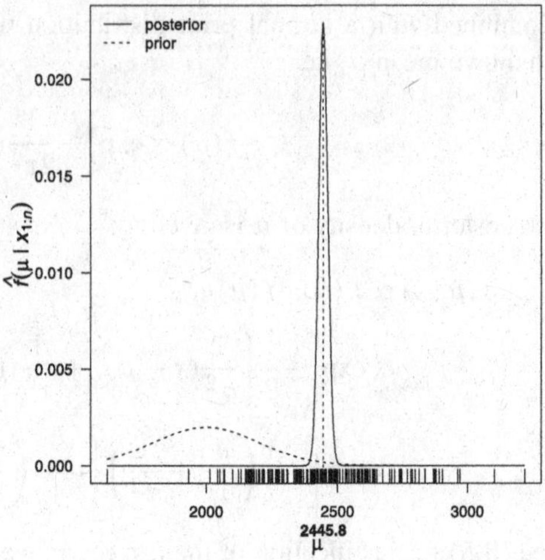

Fig. 6.5 Posterior density for mean transformation factor μ (*solid line*) with posterior expectation $\mathsf{E}(\mu \mid x_{1:n}) = 2445.8$ (*vertical dashed line*). The observations $x_{1:n}$ are marked by small vertical lines at the x-axis. The density of the prior $\mu \sim \mathrm{N}(2000, 200^2)$ is also shown (*dashed line*)

Note that the posterior mean is a weighted average of the prior mean ν and the MLE $\bar{x}$, with weights proportional to $\delta = 1/\tau^2$ and $n\kappa = n/\sigma^2$, respectively. A larger sample size n thus leads to a higher weight of the MLE and to a decreasing posterior variance $(n\kappa + \delta)^{-1}$. This behaviour of the posterior distribution is intuitively reasonable. Furthermore, we can interpret $n_0 = \delta/\kappa$ as a prior sample size to obtain the relative prior sample size $n_0/(n_0 + n)$, cf. Example 6.3. ■

Example 6.9 (Blood alcohol concentration) We illustrate Example 6.8 by applying it to the alcohol concentration data from Sect. 1.1.7. If we consider the estimated standard deviation $\sigma = 237.8$ of the transformation factors as known and specify a normal prior with mean 2000 and standard deviation 200 for the mean transformation factor μ, then we obtain the prior and posterior densities in Fig. 6.5. ■

Example 6.10 (Comparison of proportions) We know from Example 5.8 that the MLE $\hat{\psi}_{\mathrm{ML}} = \log\{(a \cdot d)/(b \cdot c)\}$ of the log odds ratio is approximately normal with mean equal to the true log odds ratio ψ and variance $\sigma^2 = a^{-1} + b^{-1} + c^{-1} + d^{-1}$.

For illustration, consider the data from the Tervila study from Table 1.1, summarised in Table 3.1. Here we obtain $\hat{\psi}_{\mathrm{ML}} = 1.09$ and $\sigma^2 = 0.69$. The MLE of the odds ratio is $\hat{\theta}_{\mathrm{ML}} = \exp(\hat{\psi}_{\mathrm{ML}}) = 2.97$ with 95 % Wald confidence interval [0.59, 15.07].

A Bayesian analysis selects a suitable prior that represents realistic prior beliefs about the quantity of interest. A *semi-Bayes* analysis now uses the MLE $\hat{\psi}_{\mathrm{ML}}$ as the data, instead of the underlying two binomial samples. This has the advantage that the new likelihood depends directly on the parameter of interest ψ, with no additional nuisance parameter. A suitable prior distribution is now placed directly on ψ, rather than working with a multivariate prior for the success probabilities π_1 and π_2 of the two underlying binomial experiments. The MLE $\hat{\psi}_{\mathrm{ML}}$ is thus regarded as the observed data, with likelihood depending on ψ, assuming that σ^2 is fixed at its estimate.

For example, we might use a normal prior for the log odds ratio ψ with mean zero and variance $\tau^2 = 1$. Note that the prior mean (or median) of zero for the log odds ratio corresponds to a prior median of one for the odds ratio, expressing prior indifference regarding positive or negative treatment effects. This particular prior implies that the odds ratio is a priori between 1/7 and 7 with approximately 95 % probability. These numbers arise from the fact that $\exp(z_{0.975}) \approx 7$ where $z_{0.975} \approx 1.96$ is the 97.5 % quantile of the standard normal distribution, the selected prior for the log odds ratio. Using Eq. (6.16), the posterior variance of the log odds ratio is $(1/\sigma^2 + 1/\tau^2)^{-1} = 0.41$ and the posterior mean is $(1/\sigma^2 + 1/\tau^2)^{-1} \cdot \hat{\psi}_{\mathrm{ML}}/\sigma^2 = 0.65$. Due to the normality of the posterior distribution, the posterior median and mode are identical to the posterior mean.

The posterior median estimate of the odds ratio is therefore $\mathrm{Med}(\theta \mid \hat{\psi}_{\mathrm{ML}}) = \exp(0.65) \approx 1.91$ with equal-tailed 95 % credible interval [0.55, 6.66]. Note that both the posterior median and the upper limit of the credible interval are considerably smaller than the MLE and the upper limit of the confidence interval, respectively, whereas the lower limit has barely changed. It is also straightforward to calculate the posterior mean and the posterior mode of the odds ratio $\theta = \exp(\psi)$ based on properties of the log-normal distribution, cf. Appendix A.5.2. One obtains $\mathsf{E}(\theta \mid \hat{\psi}_{\mathrm{ML}}) = 2.34$ and $\mathrm{Mod}(\theta \mid \hat{\psi}_{\mathrm{ML}}) = 1.27$. ■

6.3.2 Improper Prior Distributions

The prior distribution has an (intended) influence on the posterior distribution. If one wants to minimise the influence of the prior distribution, then it is common to specify a "vague" prior, for example one with very large variance. In the limit this may lead to an *improper* prior distribution, with a "density" function that does not integrate to unity. Due to the missing normalizing constant, such density functions are usually specified using the proportionality sign "$\propto$". If one uses improper priors, then it is necessary to check that at least the posterior distribution is proper. If this is the case, then improper priors can be used in a Bayesian analysis.

Example 6.11 (Comparison of proportions) It is easy to see that the likelihood analysis at the beginning of Example 6.10 has a Bayesian interpretation if one uses a normal prior for the log odds ratio ψ with zero mean and very large (infinite) variance. Indeed, if one lets the prior variance τ^2 in (6.16) go to infinity, then the posterior distribution is simply

$$\psi \mid \hat{\psi}_{\mathrm{ML}} \sim \mathrm{N}(\hat{\psi}_{\mathrm{ML}}, \sigma^2).$$

Note that the limit $\tau^2 \to \infty$ induces an improper *locally uniform* density function $f(\psi) \propto 1$ on the real line. Nevertheless, the posterior distribution is proper. Now all three Bayesian point estimates of the log odds ratio ψ are equal to the MLE $\hat{\psi}_{\mathrm{ML}}$. Furthermore, the equal-tailed credible intervals are numerically identical to Wald confidence intervals for any choice of the credibility/confidence level γ.

However, from a medical perspective it can be argued that this is not a realistic prior since it places the same prior weight on odds ratios between 0.5 and 2 as on odds ratios between 11 000 and 44 000, say. The reason is that these two intervals have the same width log(4) on the log scale and the prior density for the log odds ratio is constant across the whole real line. Such huge effect sizes are unrealistic and rarely encountered in clinical or epidemiological research. ■

We now add a formal definition of an improper prior distribution.

Definition 6.6 (Improper prior distribution) A prior distribution with density function $f(\theta) \geq 0$ is called *improper* if

$$\int_\Theta f(\theta)\, d\theta = \infty \quad \text{or} \quad \sum_{\theta \in \Theta} f(\theta) = \infty$$

for continuous or discrete parameters θ, respectively. ♦

Example 6.12 (Haldane's prior) The conjugate prior $\pi \sim \text{Be}(\alpha, \beta)$ in the binomial model $X \sim \text{Bin}(n, \pi)$ has the density

$$f(\pi) \propto \pi^{\alpha-1}(1-\pi)^{\beta-1}$$

and is proper for $\alpha > 0$ and $\beta > 0$. In the limiting case $\alpha = \beta = 0$, one obtains an improper prior distribution with density

$$f(\pi) \propto \pi^{-1}(1-\pi)^{-1},$$

known as *Haldane's prior*. We will denote this as a Be(0, 0) distribution, being aware that it is not really a member of the family of beta distributions. Due to Example 6.3, the posterior distribution is the $\text{Be}(x, n-x)$ distribution, which is proper if $x > 0$ and $n - x > 0$. We note that Haldane's prior is equivalent to an improper locally uniform prior $f(\phi) \propto 1$ on the whole real line for $\phi = \text{logit}(\pi)$, which can be shown using the change-of-variables formula (A.11) (see Example 6.13). ■

6.3.3 Jeffreys' Prior Distributions

In some situations it may be useful to choose a prior distribution that does not convey much information about the parameter because of weak or missing prior knowledge. A first naive choice is a (locally) uniform prior $f_\theta(\theta) \propto 1$, in which case the posterior is proportional to the likelihood function. Note that a locally uniform prior will be improper if the parameter space is not bounded.

However, there are problems associated with this approach. Suppose that $\phi = h(\theta)$ is a one-to-one differentiable transformation of θ, that has a (locally) uniform prior with density $f_\theta(\theta) \propto 1$. Using the change-of-variables formula (A.11), we obtain the corresponding prior for ϕ,

$$f_\phi(\phi) = f_\theta\{h^{-1}(\phi)\} \cdot \left|\frac{dh^{-1}(\phi)}{d\phi}\right|$$
$$\propto \left|\frac{dh^{-1}(\phi)}{d\phi}\right|.$$

Note that this term is not necessarily constant. Indeed, $f_\phi(\phi)$ will be independent of ϕ only if h is linear. If h is nonlinear, the prior density $f_\phi(\phi)$ will depend on ϕ and will thus not be (locally) uniform. However, if we had chosen a parametrisation with ϕ from the start, we would have chosen a (locally) uniform prior $f_\phi(\phi) \propto 1$. This lack of invariance under reparameterisation is an unappealing feature of the (locally) uniform prior distribution. Note that we implicitly assume that we can apply the change-of-variables formula to improper densities in the same way as to proper densities.

Example 6.13 (Binomial model) Let $X \sim \text{Bin}(n, \pi)$ with a uniform prior for π, i.e. $f_\pi(\pi) = 1$ for $\pi \in (0, 1)$. Consider now the logit transformation $\phi = h(\pi) = \log\{\pi/(1-\pi)\} \in \mathbb{R}$. Applying the change-of-variables formula, we obtain

$$f_\phi(\phi) = \frac{\exp(\phi)}{\{1+\exp(\phi)\}^2},$$

i.e. the log odds ϕ follow a priori a standard logistic distribution (cf. Appendix A.5.2).

On the other hand, if we select an improper locally uniform prior for ϕ, i.e. $f_\phi(\phi) \propto 1$, then

$$f_\pi(\pi) \propto \pi^{-1}(1-\pi)^{-1},$$

again using the change-of-variables formula with $\pi = h^{-1}(\phi) = \exp(\phi)/\{1 + \exp(\phi)\}$. This density can be identified as Haldane's prior (see Example 6.12), which is of course also improper. ■

It turns out that a particular choice of prior distribution is invariant under reparametrisation. This is Jeffreys' prior (after Sir Harold Jeffreys, 1891–1989):

Definition 6.7 (Jeffreys' prior) Let X be a random variable with likelihood function $f(x \mid \theta)$ where θ is an unknown scalar parameter. *Jeffreys' prior* is defined as

$$f(\theta) \propto \sqrt{J(\theta)}, \tag{6.17}$$

where $J(\theta)$ is the expected Fisher information of θ. Equation (6.17) is also known as *Jeffreys' rule*. ♦

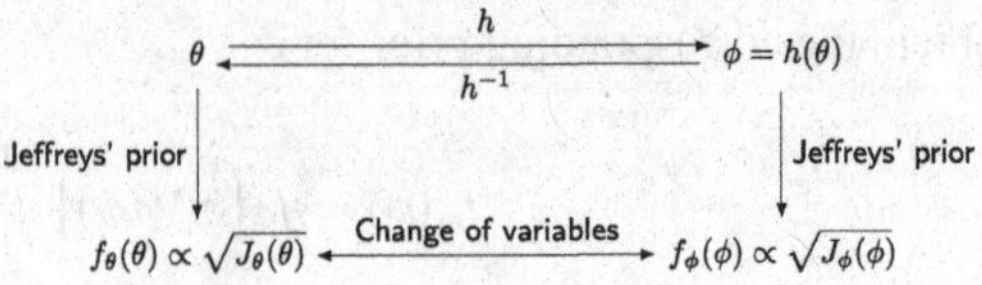

Fig. 6.6 Diagram illustrating the invariance of Jeffreys' prior

Jeffreys' prior is proportional to the square root of the expected Fisher information, which may give an improper prior distribution. At first sight, it is surprising that this choice is supposed to be invariant under reparametrisation, but we will see in the following result that this is indeed the case.

Result 6.2 (Invariance of Jeffreys' prior) *Jeffreys' prior is invariant under a one-to-one reparametrisation of* θ*: If*

$$f_\theta(\theta) \propto \sqrt{J_\theta(\theta)},$$

then the density function of $\phi = h(\theta)$ *is*

$$f_\phi(\phi) \propto \sqrt{J_\phi(\phi)},$$

where $J_\phi(\phi)$ *denotes the expected Fisher information of* ϕ.

Proof From $f_\theta(\theta) \propto \sqrt{J_\theta(\theta)}$ it follows by the change-of-variables formula (A.11) and Result 4.3 that

$$\begin{aligned} f_\phi(\phi) &= f_\theta(\theta) \cdot \left| \frac{dh^{-1}(\phi)}{d\phi} \right| \\ &\propto \sqrt{J_\theta(\theta)} \cdot \left| \frac{dh^{-1}(\phi)}{d\phi} \right| = \sqrt{J_\theta(\theta) \cdot \left\{ \frac{dh^{-1}(\phi)}{d\phi} \right\}^2} \\ &= \sqrt{J_\phi(\phi)}, \end{aligned}$$

which is the claim of Result 6.2. □

If we had chosen the parametrisation $\phi = h(\theta)$ from the start, application of Jeffreys' rule would have given the same prior distribution: $f_\phi(\phi) \propto \sqrt{J_\phi(\phi)}$. Figure 6.6 illustrates this invariance property of Jeffreys' prior: There are two ways to move from θ to $f_\phi(\phi)$, but both give the same result $f_\phi(\phi) \propto \sqrt{J_\phi(\phi)}$.

Invariance under one-to-one transformations seems to be a minimal requirement for a default prior, but does it really represent a sufficiently "non-informative" prior? We will now outline such a justification of Jeffreys' prior based on a different argument.

In Sect. 4.1 we have discussed the expected Fisher information $J_\theta(\theta)$ as a (frequentist) measure of the (average) information in the data with respect to the unknown parameter θ. If the Fisher information does not depend on the true parameter θ, then the (average) information provided by the data is the same whatever the particular value of θ is. In this case it seems reasonable to select a (possibly improper) locally uniform distribution for θ as a default prior.

However, if $J_\theta(\theta)$ depends on θ, then it seems natural to first apply the variance-stabilising transformation $\phi = h(\theta)$ (cf. Sect. 4.3) to remove this dependence. Then $J_\phi(\phi)$ does not depend on ϕ, and we therefore select a locally uniform prior for ϕ, i.e. $f_\phi(\phi) \propto 1$. Through a change of variables and with Eq. (4.14) it follows that

$$f_\theta(\theta) \propto f_\phi(\phi) \cdot \left|\sqrt{J_\theta(\theta)}\right| \propto \sqrt{J_\theta(\theta)},$$

i.e. the argument leads directly to Jeffreys' prior. To put it the other way round, the derivative of a variance-stabilising transformation $h(\theta)$ equals Jeffreys' prior for the original parameter θ. For example, the variance-stabilising transformation for the mean λ of Poisson distributed data is $h(\lambda) = \sqrt{\lambda}$ (cf. Example 4.14), so Jeffreys' prior is given by

$$f_\lambda(\lambda) \propto \frac{d\sqrt{\lambda}}{d\lambda} = \frac{1}{2}\lambda^{-1/2} \propto \lambda^{-1/2}.$$

This somewhat surprising result gives an interesting connection between likelihood and Bayesian inference.

Jeffreys' rule gives commonly accepted default priors for scalar parameters. Quite often one obtains improper priors, which can be identified as limiting cases of proper conjugate priors. Here are two examples.

Example 6.14 (Normal model) Let $X_{1:n}$ denote a random sample from a $\mathrm{N}(\mu, \sigma^2)$ distribution with unknown mean μ and known variance σ^2. From Example 4.3 we know that $J(\mu) = n/\sigma^2$ does not depend on μ, so Jeffreys' prior for μ is locally uniform on the whole real line $\mathbb{R}$, i.e. $f(\mu) \propto 1$, and is an improper prior distribution. This is the limiting prior distribution if we assume the conjugate prior $\mu \sim \mathrm{N}(\nu, \tau^2)$ and let $\tau^2 \to \infty$. The resulting posterior distribution is (cf. Example 6.11)

$$\mu \mid x_{1:n} \sim \mathrm{N}\left(\bar{x}, \frac{\sigma^2}{n}\right).$$

The posterior mean, median and mode are therefore all equal to the MLE $\bar{x}$.

Suppose now that the mean μ is known but the variance σ^2 is unknown. Then $J(\sigma^2) = n/(2\sigma^4)$, compare again Example 4.3, so Jeffreys' prior is

$$f(\sigma^2) \propto 1/\sigma^2,$$

an improper distribution. This is the limiting prior distribution if we initially assume that the conjugate prior $\sigma^2 \sim \mathrm{IG}(\alpha, \beta)$ and let $\alpha, \beta \to 0$. Smaller values of σ^2 have higher prior weight than larger values. Combined with the likelihood

$$L(\sigma^2) = (\sigma^2)^{-n/2} \exp\left\{-\frac{1}{2\sigma^2}\sum_{i=1}^{n}(x_i - \mu)^2\right\},$$

we obtain the posterior density

$$f(\sigma^2 \,|\, x_{1:n}) \propto (\sigma^2)^{-(1+n/2)} \exp\left\{-\frac{1}{2\sigma^2}\sum_{i=1}^n (x_i-\mu)^2\right\},$$

which can be identified as an inverse gamma distribution (cf. Appendix A.5.2) with parameters $n/2$ and $\sum_{i=1}^n (x_i-\mu)^2/2$:

$$\sigma^2 \,|\, x_{1:n} \sim \mathrm{IG}\left(\frac{n}{2}, \frac{\sum_{i=1}^n (x_i-\mu)^2}{2}\right).$$

Note that this is a regular distribution if $n \geq 1$ (and, strictly speaking, if $x_i \neq \mu$ for at least one observation, but this is the case with probability 1 already for $n = 1$). The mean of the inverse gamma distribution $\mathrm{IG}(\alpha, \beta)$ is $\beta/(\alpha-1)$ if $\alpha > 1$, and the mode is $\beta/(\alpha+1)$. Therefore, for $n > 2$, the posterior mean is

$$\mathsf{E}(\sigma^2 \,|\, x_{1:n}) = \frac{\sum_{i=1}^n (x_i-\mu)^2}{n-2},$$

and the posterior mode is

$$\mathrm{Mod}(\sigma^2 \,|\, x_{1:n}) = \frac{\sum_{i=1}^n (x_i-\mu)^2}{n+2}.$$

Note that the MLE $\hat{\sigma}^2_{\mathrm{ML}} = \sum_{i=1}^n (x_i-\mu)^2/n$ (compare Example 2.9) lies between these two estimates and will have a very similar numerical value, provided that the sample size n is not very small. ■

Table 6.3 lists Jeffreys' prior distributions for further likelihood functions. Only in the binomial case Jeffreys' prior, the $\mathrm{Be}(0.5, 0.5)$ distribution, turns out to be proper. All other priors in Table 6.3 are improper distributions. They can be viewed as limiting cases of the corresponding conjugate proper prior distributions, and this is how they are listed. Bayesian point estimates using Jeffreys' prior are often very close or even identical to the MLE. Table 6.4 illustrates this for the posterior mean.

It is also interesting to compare Bayesian credible intervals using Jeffreys' prior with the corresponding frequentist confidence intervals. Quite surprisingly, it turns out that the frequentist properties of such Bayesian procedures may be as good or even better than those of their truly frequentist counterparts, as the following example illustrates.

Example 6.15 (Inference for a proportion) For a binomial observation $X \sim \mathrm{Bin}(n, \pi)$, Table 6.5 compares HPD and equal-tailed credible intervals using Jeffreys' prior $\pi \sim \mathrm{Be}(1/2, 1/2)$ with the corresponding Wald and likelihood confidence intervals. The different techniques lead to nearly identical intervals if the sample size n is relatively large. Note that the HPD interval tends to be the shortest,

Table 6.3 Jeffreys' prior for several likelihood functions. All the prior distributions are improper, except for the binomial likelihood

Likelihood	Jeffreys' prior	Density of Jeffreys' prior
$\text{Bin}(n, \pi)$	$\pi \sim \text{Be}(\frac{1}{2}, \frac{1}{2})$	$f(\pi) \propto \{\pi(1-\pi)\}^{-\frac{1}{2}}$
$\text{Geom}(\pi)$	$\pi \sim \text{Be}(0, \frac{1}{2})$	$f(\pi) \propto \pi^{-1}(1-\pi)^{-\frac{1}{2}}$
$\text{Po}(\lambda)$	$\lambda \sim \text{G}(\frac{1}{2}, 0)$	$f(\lambda) \propto \lambda^{-\frac{1}{2}}$
$\text{Exp}(\lambda)$	$\lambda \sim \text{G}(0, 0)$	$f(\lambda) \propto \lambda^{-1}$
$\text{N}(\mu, \sigma^2 \text{ known})$	$\mu \sim \text{N}(0, \infty)$	$f(\mu) \propto 1$
$\text{N}(\mu \text{ known}, \sigma^2)$	$\sigma^2 \sim \text{IG}(0, 0)$	$f(\sigma^2) \propto \sigma^{-2}$

Table 6.4 Comparison of MLEs and the posterior mean using Jeffreys' prior

Likelihood	$\hat{\theta}_{\text{ML}}$	Posterior mean using Jeffreys' prior
$\text{Bin}(n, \pi)$	$\bar{x}$	$\frac{n}{n+1}(\bar{x} + \frac{1}{2n})$
$\text{Geom}(\pi)$	$1/\bar{x}$	$1/(\bar{x} + \frac{1}{2n})$
$\text{Po}(\lambda)$	$\bar{x}$	$\bar{x} + \frac{1}{2n}$
$\text{Exp}(\lambda)$	$1/\bar{x}$	$1/\bar{x}$
$\text{N}(\mu, \sigma^2 \text{ known})$	$\bar{x}$	$\bar{x}$
$\text{N}(\mu \text{ known}, \sigma^2)$	$\frac{1}{n}\sum_{i=1}^n (x_i - \mu)^2$	$\frac{1}{n-2}\sum_{i=1}^n (x_i - \mu)^2$

Table 6.5 A comparison of different credible and confidence intervals for $X \sim \text{Bin}(n, \pi)$ and $\pi \sim \text{Be}(1/2, 1/2)$. If $n = 100$ and $x = 50$, the four approaches yield nearly identical intervals

Observation		Interval at 95 % level of type			
n	x	Equal-tailed	HPD	Wald	Likelihood
10	0	0.000 to 0.217	0.000 to 0.171	0.000 to 0.000	0.000 to 0.175
10	1	0.011 to 0.381	0.000 to 0.331	−0.086 to 0.286	0.006 to 0.372
10	5	0.224 to 0.776	0.224 to 0.776	0.190 to 0.810	0.218 to 0.782
100	0	0.000 to 0.025	0.000 to 0.019	0.000 to 0.000	0.000 to 0.019
100	10	0.053 to 0.170	0.048 to 0.164	0.041 to 0.159	0.051 to 0.169
100	50	0.403 to 0.597	0.403 to 0.597	0.402 to 0.598	0.403 to 0.597

compare Table 6.6. But how good are the frequentist properties of Bayesian credible intervals based on Jeffreys' prior? To address this question, we have computed the actual coverage probabilities for $n = 50$, just as we did in Example 4.22 for confidence intervals. Figure 6.7 displays the actual coverage of the HPD and the equal-tailed credible interval. They behave similarly to the likelihood confidence interval in Fig. 4.9e with slightly lower coverage of the HPD interval as a result of the smaller interval width. It is to be noted, though, that the coverage is actually comparable if not even better than that from the likelihood-based confidence intervals discussed in Example 4.22. ■

Table 6.6 Comparison of the widths of the different confidence and credible intervals from Table 6.5

Observation		Width of 95 % interval of type			
n	x	Equal-tailed	HPD	Wald	Likelihood
10	0	0.217	0.171	0.000	0.175
10	1	0.370	0.330	0.372	0.366
10	5	0.553	0.553	0.620	0.565
100	0	0.025	0.019	0.000	0.019
100	10	0.118	0.116	0.118	0.117
100	50	0.194	0.194	0.196	0.194

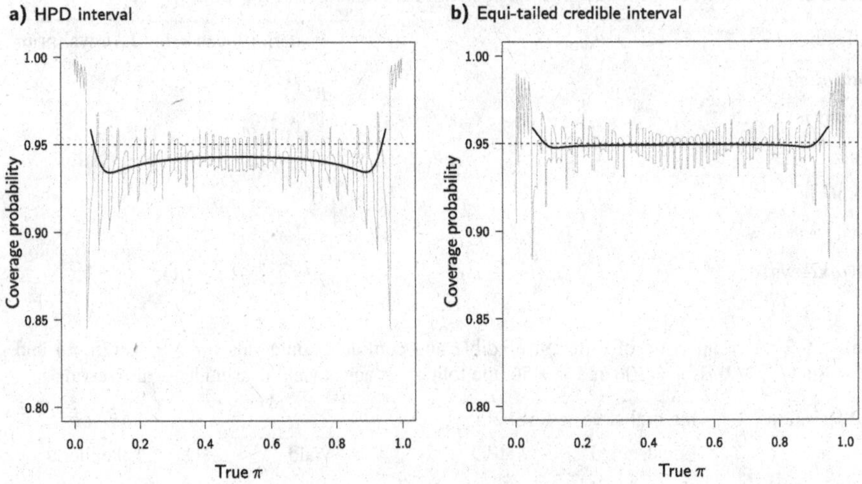

Fig. 6.7 Actual (*grey*) and locally smoothed (*black*) coverage probabilities of 95 % credible intervals based on Jeffreys' prior for $n = 50$. For $X = 0$ or $X = n$, the limits of the HPD interval do not have equal posterior density ordinates. In these two cases the HPD interval is $[0, b_{0.95}(0.5, 50.5)]$ or $[b_{0.05}(50.5, 0.5), 1]$, respectively, where $b_\alpha(\alpha, \beta)$ denotes the α quantile of the $\mathrm{Be}(\alpha, \beta)$ distribution

We know from Result 6.2 that one advantage of Jeffreys' prior is that it is invariant under one-to-one transformations of the original parameter θ. For example, one might be interested in the standard deviation σ or the precision $\kappa = 1/\sigma^2$ of the normal distribution rather than the variance σ^2.

Example 6.16 (Normal model) Jeffreys' prior for the variance σ^2 of a normal distribution $\mathrm{N}(\mu, \sigma^2)$ is

$$f(\sigma^2) \propto 1/\sigma^2$$

when μ is fixed. A change of variables gives us Jeffreys' prior for the standard deviation $\sigma = h(\sigma^2) = \sqrt{\sigma^2}$:

$$f(\sigma) \propto \left| \frac{dh^{-1}(\sigma)}{d\sigma} \right| \cdot 1/\sigma^2 = 2\sigma \cdot 1/\sigma^2 \propto 1/\sigma.$$

For the precision $\kappa = h(\sigma^2) = 1/\sigma^2$, one obtains

$$f(\kappa) \propto \left| \frac{dh^{-1}(\kappa)}{d\kappa} \right| \cdot 1/\sigma^2 = \kappa^{-2} \cdot \kappa = 1/\kappa.$$

We note that Jeffreys' priors for the variance σ^2, the standard deviation σ and the precision κ of a normal distribution are all proportional to the respective reciprocal parameter. Another change of variables shows that the corresponding priors for $\log(\sigma^2)$, $\log(\sigma)$ and $\log(\kappa)$ are hence all locally uniform on the real line $\mathbb{R}$. ■

The relationship between the variance-stabilising transformation and Jeffreys' prior can also be helpful to derive Jeffreys' prior in the presence of nuisance parameters. To do this, we adopt a semi-Bayes approach and directly start with the variance-stabilising transformation of the parameter of interest. The following example illustrates this for the correlation parameter of the bivariate normal distribution.

Example 6.17 (Jeffreys' prior for a correlation parameter) In Example 4.16 we have shown that a variance-stabilising transformation of the correlation parameter $\rho \in (-1, 1)$ of a normal distribution is

$$h(\rho) = \tanh^{-1}(\rho) = 0.5 \log\left(\frac{1+\rho}{1-\rho}\right).$$

The derivative is

$$\frac{dh(\rho)}{d\rho} = \frac{1}{1-\rho^2},$$

so Jeffreys' prior for ρ is

$$f(\rho) \propto \frac{1}{1-\rho^2},$$

again an improper distribution. This prior distribution gives more weight to extreme values than to values close to zero. For example, $f(\pm 0.9)/f(0) = 5.3$, $f(\pm 0.99)/f(0) = 50.3$ and $f(\pm 0.999)/f(0)$ is even 500.3, whereas this ratio would always equal the unity for the uniform prior distribution. ■

6.4 Properties of Bayesian Point and Interval Estimates

To estimate an unknown parameter θ, there are at least three possible Bayesian point estimates available, the posterior mean, mode and median. Which one should we choose in a specific application? To answer this question, we take a decision-theoretic view and first introduce the notion of a *loss function*.

6.4.1 Loss Function and Bayes Estimates

To simplify the notation, we denote in this section a point estimate of θ with a rather than with $\hat{\theta}$.

Definition 6.8 (Loss function) A *loss function* $l(a, \theta) \in \mathbb{R}$ quantifies the loss encountered when estimating the true parameter θ by a. ♦

If $a = \theta$, the associated loss is typically set to zero: $l(\theta, \theta) = 0$. A commonly used loss function is the *quadratic loss function* $l(a, \theta) = (a - \theta)^2$. Alternatively, one might use the *linear loss function* $l(a, \theta) = |a - \theta|$ or the *zero–one loss function*

$$l_\varepsilon(a, \theta) = \begin{cases} 0, & |a - \theta| \le \varepsilon, \\ 1, & |a - \theta| > \varepsilon, \end{cases}$$

where we have to suitably choose the additional parameter $\varepsilon > 0$.

We now choose the point estimate a, such that it minimises the a posteriori expected loss with respect to $f(\theta \mid x)$. Such a point estimate is called a *Bayes estimate*.

Definition 6.9 (Bayes estimate) A *Bayes estimate* of θ with respect to a loss function $l(a, \theta)$ minimises the expected loss with respect to the posterior distribution $f(\theta \mid x)$, i.e. it minimises

$$\mathsf{E}\{l(a, \theta) \mid x\} = \int_\Theta l(a, \theta) f(\theta \mid x)\, d\theta.$$ ♦

It turns out that the commonly used Bayesian point estimates can be viewed as Bayes estimates with respect to one of the loss functions described above.

Result 6.3 *The posterior mean is the Bayes estimate with respect to quadratic loss. The posterior median is the Bayes estimate with respect to linear loss. The posterior mode is the Bayes estimate with respect to zero–one loss, as* $\varepsilon \to 0$.

Proof We first derive the posterior mean $\mathsf{E}(\theta \mid x)$ as the Bayes estimate with respect to quadratic loss. The expected quadratic loss is

$$\mathsf{E}\{l(a, \theta) \mid x\} = \int l(a, \theta) f(\theta \mid x)\, d\theta = \int (a - \theta)^2 f(\theta \mid x)\, d\theta.$$

Setting the derivative with respect to a to zero leads to

$$2 \cdot \int (a-\theta) f(\theta \mid x)\, d\theta = 0 \iff a - \int \theta f(\theta \mid x)\, d\theta = 0.$$

It immediately follows that $a = \int \theta f(\theta \mid x)\, d\theta = \mathsf{E}(\theta \mid x)$.

Consider now the expected linear loss

$$\begin{aligned}
\mathsf{E}\{l(a,\theta) \mid x\} &= \int l(a,\theta) f(\theta \mid x)\, d\theta = \int |a-\theta| f(\theta \mid x)\, d\theta \\
&= \int_{\theta \le a} (a-\theta) f(\theta \mid x)\, d\theta + \int_{\theta > a} (\theta - a) f(\theta \mid x)\, d\theta.
\end{aligned}$$

The derivative with respect to a can be calculated using Leibniz's integral rule (cf. Appendix B.2.4):

$$\begin{aligned}
&\frac{\partial}{\partial a} \mathsf{E}\{l(a,\theta) \mid x\} \\
&= \frac{\partial}{\partial a} \int_{-\infty}^{a} (a-\theta) f(\theta \mid x)\, d\theta + \frac{\partial}{\partial a} \int_{a}^{\infty} (\theta - a) f(\theta \mid x)\, d\theta \\
&= \int_{-\infty}^{a} f(\theta \mid x)\, d\theta - \big(a - (-\infty)\big) f(-\infty \mid x) \cdot 0 + (a-a) f(a \mid x) \cdot 1 \\
&\quad - \int_{a}^{\infty} f(\theta \mid x)\, d\theta - (a-a) f(a \mid x) \cdot 1 + (\infty - a) f(\infty \mid x) \cdot 0 \\
&= \int_{-\infty}^{a} f(\theta \mid x)\, d\theta - \int_{a}^{\infty} f(\theta \mid x)\, d\theta.
\end{aligned}$$

Setting this equal to zero yields the posterior median $a = \mathrm{Med}(\theta \mid x)$ as the solution for the estimate.

Finally, the expected zero–one loss is

$$\begin{aligned}
\mathsf{E}\{l(a,\theta) \mid x\} &= \int l_\varepsilon(a,\theta) f(\theta \mid x)\, d\theta \\
&= \int_{-\infty}^{a-\varepsilon} f(\theta \mid x)\, d\theta + \int_{a+\varepsilon}^{+\infty} f(\theta \mid x)\, d\theta \\
&= 1 - \int_{a-\varepsilon}^{a+\varepsilon} f(\theta \mid x)\, d\theta.
\end{aligned}$$

This will be minimised if the integral $\int_{a-\varepsilon}^{a+\varepsilon} f(\theta \mid x)\, d\theta$ is maximised. For small ε the integral is approximately $2\varepsilon f(a \mid x)$, which is maximised through the posterior mode $a = \mathrm{Mod}(\theta \mid x)$. □

The question arises if credible intervals can also be optimal with respect to certain loss functions. For simplicity, we assume again that the unknown parameter $\theta \in \Theta$ is a scalar with associated posterior density $f(\theta \mid x)$. First, we introduce the notion of a *credible region*, a straightforward generalisation of a credible interval. Similarly, a *highest posterior density region* (HPD region) can be defined.

Definition 6.10 (Credible region) A subset $C \subseteq \Theta$ with

$$\int_C f(\theta \mid x)\, d\theta = \gamma$$

is called $\gamma \cdot 100\ \%$ *credible region* for θ with respect to $f(\theta \mid x)$. If C is a real interval, then C is also called a *credible interval*. ♦

There is no unique $\gamma \cdot 100\ \%$ credible region for fixed γ. It is natural to force uniqueness through the specification of a certain loss function $l(C, \theta)$. The following loss function penalises (for fixed γ) the size $|C|$ of the credible region, i.e. smaller regions are preferred. In addition, the loss function gives another penalty if the true parameter θ does not lie within the credible region. We restrict our attention to continuous posterior distributions.

Result 6.4 *Let $f(\theta \mid x)$ denote the posterior density function of θ and let, for fixed $\gamma \in (0, 1)$,*

$$\mathcal{A} = \{C \mid \Pr(\theta \in C \mid x) = \gamma\}$$

denote the set of all $\gamma \cdot 100\ \%$ credible regions for θ. Consider now the loss function

$$l(C, \theta) = |C| - \mathsf{I}_C(\theta) \quad \textit{for } C \in \mathcal{A}, \theta \in \Theta.$$

C is optimal with respect to $l(C, \theta)$ if and only if, for all $\theta_1 \in C$ and $\theta_2 \notin C$,

$$f(\theta_1 \mid x) \geq f(\theta_2 \mid x), \tag{6.18}$$

i.e. if C is an HPD region.

Proof To prove Result 6.4, consider some set $C \in \mathcal{A}$ with expected loss

$$\int_\Theta l(C, \theta) f(\theta \mid x)\, d\theta = |C| - \int_\Theta \mathsf{I}_C(\theta) f(\theta \mid x)\, d\theta = |C| - \Pr(\theta \in C \mid x) = |C| - \gamma.$$

For fixed γ, the $\gamma \cdot 100\ \%$ credible region C with smallest size $|C|$ will therefore minimise the expected loss. Now let $C \in \mathcal{A}$ with property (6.18), and let $D \in \mathcal{A}$ be some other element in $\mathcal{A}$. We need to show that $|C| \leq |D|$. Let $A \dot{\cup} B$ denote the disjoint union of A and B, i.e. $A \dot{\cup} B = A \cup B$ and $A \cap B = \emptyset$. Then:

$$C = C \cap \Theta = C \cap (D \dot{\cup} D^c) = (C \cap D) \dot{\cup} (C \cap D^c)$$
$$\text{and analogously} \quad D = (C \cap D) \dot{\cup} (C^c \cap D).$$

It therefore remains to show that $|C \cap D^c| \leq |C^c \cap D|$. From (6.18) it follows that

$$\sup_{C^c \cap D} f(\theta \mid x) \leq \inf_{C \cap D^c} f(\theta \mid x), \tag{6.19}$$

and from $C, D \in \mathcal{A}$ it follows that

$$\Pr(\theta \in C \mid x) = \Pr(\theta \in D \mid x) = \gamma.$$

Using the above decompositions of C and D, we also have

$$\begin{aligned}\Pr(\theta \in C \cap D \mid x) + \Pr(\theta \in C \cap D^c \mid x) &= \Pr(\theta \in C \mid x)\\ &= \Pr(\theta \in D \mid x)\\ &= \Pr(\theta \in C \cap D \mid x) + \Pr(\theta \in C^c \cap D \mid x)\end{aligned}$$

and hence

$$\Pr(\theta \in C \cap D^c \mid x) = \Pr(\theta \in C^c \cap D \mid x). \tag{6.20}$$

In total we obtain

$$\begin{aligned}\inf_{C \cap D^c} f(\theta \mid x) \cdot |C \cap D^c| &\leq \int_{C \cap D^c} f(\theta \mid x)\, d\theta\\ &= \int_{C^c \cap D} f(\theta \mid x)\, d\theta \quad \text{with (6.20)}\\ &\leq \sup_{C^c \cap D} f(\theta \mid x) \cdot |C^c \cap D|\\ &\leq \inf_{C \cap D^c} f(\theta \mid x) \cdot |C^c \cap D| \quad \text{with (6.19).}\end{aligned}$$

Therefore, $|C \cap D^c| \leq |C^c \cap D|$ and hence $|C| \leq |D|$. The proof in the other direction is similar. □

6.4.2 Compatible and Invariant Bayes Estimates

In practice point and interval estimates are often given jointly. The question arises if all Bayesian point and interval estimates are compatible. A minimal requirement appears to be that the point estimate is always within the credible interval. This is fulfilled only by some combinations. The posterior mode, for example, will always be within any HPD interval, and the posterior median is always within any equal-tailed credible interval. In contrast, in extreme cases the posterior mean may not necessarily be within the HPD nor the equal-tailed credible interval.

It is also interesting to study the behaviour of the different point and interval estimates under a one-to-one transformation $\phi = h(\theta)$ of the parameter θ. It turns out that the posterior mode and the posterior mean are in general *not* invariant, for example, $\mathsf{E}\{h(\theta) \mid x\} = h\{\mathsf{E}(\theta \mid x)\}$ does not hold in general. In fact, $\mathsf{E}\{h(\theta) \mid x\} <$

$h\{\mathsf{E}(\theta \mid x)\}$ if h is strictly concave; if h is strictly convex, then the inequality sign is in the other direction; cf. Appendix A.3.7. However, all characteristics based on quantiles of the posterior distribution, such as the posterior median and equal-tailed credible intervals, are invariant under (continuous) one-to-one transformations.

Example 6.18 (Inference for a proportion) The posterior median of the posterior distribution $\pi \mid x \sim \mathrm{Be}(11, 4)$ derived in Example 6.3 and shown in Fig. 6.1 is $\mathrm{Med}(\pi \mid x) = 0.744$. The posterior median of the odds $\omega = \pi/(1-\pi)$ is therefore $0.744/(1-0.744) = 2.905$. ■

6.5 Bayesian Inference in Multiparameter Models

Bayesian inference in multiparameter models with parameter vector $\boldsymbol{\theta}$ proceeds as for scalar parameters θ: Multiplication of a multivariate prior density $f(\boldsymbol{\theta})$ with the likelihood and subsequent normalisation gives us the posterior density of $\boldsymbol{\theta}$. Before we discuss how to eliminate nuisance parameters, we comment on the choice of the prior distribution for parameter vectors $\boldsymbol{\theta}$.

6.5.1 Conjugate Prior Distributions

Conjugate prior distributions are available also for multivariate likelihood functions. Here are a few examples.

Example 6.19 (Normal model) Let $L(\boldsymbol{\mu})$ denote the likelihood function of an observation $\boldsymbol{x}$ from a multivariate normal distribution $\mathrm{N}_p(\boldsymbol{\mu}, \boldsymbol{\Sigma})$. The unknown mean vector $\boldsymbol{\mu}$ is a vector of dimension p, while the covariance matrix $\boldsymbol{\Sigma}$ is assumed to be known. The $\mathrm{N}_p(\boldsymbol{\nu}, \mathcal{T})$ distribution is conjugate to $L(\boldsymbol{\mu})$ since the posterior distribution of $\boldsymbol{\mu}$ is again p-variate normal:

$$\boldsymbol{\mu} \mid \boldsymbol{x} \sim \mathrm{N}_p\big((\boldsymbol{\Sigma}^{-1} + \mathcal{T}^{-1})^{-1} \cdot (\boldsymbol{\Sigma}^{-1}\boldsymbol{x} + \mathcal{T}^{-1}\boldsymbol{\nu}), (\boldsymbol{\Sigma}^{-1} + \mathcal{T}^{-1})^{-1}\big).$$

This result can easily be derived using Eq. (B.4). It can also be generalised to a random sample $\boldsymbol{X}_{1:n}$ from an $\mathrm{N}_p(\boldsymbol{\mu}, \boldsymbol{\Sigma})$ distribution:

$$\boldsymbol{\mu} \mid \boldsymbol{x}_{1:n} \sim \mathrm{N}_p\big((n\boldsymbol{\Sigma}^{-1} + \mathcal{T}^{-1})^{-1} \cdot (n\boldsymbol{\Sigma}^{-1}\bar{\boldsymbol{x}} + \mathcal{T}^{-1}\boldsymbol{\nu}), (n\boldsymbol{\Sigma}^{-1} + \mathcal{T}^{-1})^{-1}\big),$$

where $\bar{\boldsymbol{x}} = \frac{1}{n}\sum_{i=1}^{n} \boldsymbol{x}_i$ denotes the mean of the realisations $\boldsymbol{x}_{1:n}$. This generalises the conjugate analysis for the univariate normal distribution, cf. Example 6.8. ■

Example 6.20 (Multinomial model) Consider a multinomially distributed random variable $\boldsymbol{X} \sim \mathrm{M}_p(n, \boldsymbol{\theta} = \boldsymbol{\pi})$ with unknown probability vector $\boldsymbol{\pi} = (\pi_1, \dots, \pi_p)^\top$. The *Dirichlet distribution* $\boldsymbol{\pi} \sim \mathrm{D}_p(\alpha_1, \dots, \alpha_p)$ (cf. Appendix A.5.3) with density

$$f(\boldsymbol{\pi}) \propto \prod_{i=1}^{p} \pi_i^{\alpha_i - 1}, \tag{6.21}$$

where $\pi_i > 0$ ($i = 1, \dots, p$) and $\sum_{i=1}^{p} \pi_i = 1$, is conjugate to the multinomial likelihood. Indeed, the likelihood function of a multinomial observation $\boldsymbol{x}$ is $L(\boldsymbol{\pi}) = \prod_{i=1}^{p} \pi_i^{x_i}$, so the posterior distribution of $\boldsymbol{\pi}$ can be easily derived using (6.21):

$$\boldsymbol{\pi} \mid \boldsymbol{x} \sim \mathrm{D}_p(\alpha_1 + x_1, \dots, \alpha_p + x_p), \tag{6.22}$$

a Dirichlet distribution with parameters $\alpha_i + x_i$, $i = 1, \dots, p$. Note that for $p = 2$, we obtain the special case of a binomial likelihood with a beta prior. The posterior distribution (6.11) of a binomial success probability is hence a special case of (6.22). The prior sample size of the Dirichlet prior (6.21) is now represented by $n_0 = \sum_i^p \alpha_i$, cf. the discussion in Example 6.3. ■

Example 6.21 (Normal model) Of particular interest is Bayesian inference for a random sample $X_{1:n}$ of a normal distribution $\mathrm{N}(\mu, 1/\kappa)$ if both parameters μ and κ are unknown. To keep notation simple we parametrise the normal distribution here in terms of the precision κ rather than the variance $\sigma^2 = 1/\kappa$. Our goal is to specify a conjugate bivariate prior distribution $f(\boldsymbol{\theta})$ on $\mathbb{R} \times \mathbb{R}^+$ for the parameter vector $\boldsymbol{\theta} = (\mu, \kappa)^\top$.

The *normal-gamma distribution* $\mathrm{NG}(\nu, \lambda, \alpha, \beta)$ is defined through the factorization

$$f(\boldsymbol{\theta}) = f(\kappa) \cdot f(\mu \mid \kappa), \tag{6.23}$$

where $\kappa \sim \mathrm{G}(\alpha, \beta)$ and $\mu \mid \kappa \sim \mathrm{N}(\nu, (\lambda \cdot \kappa)^{-1})$. The explicit functional form of the density is hence

$$\begin{aligned} f(\boldsymbol{\theta}) &\propto \kappa^{\alpha-1} \exp(-\beta\kappa) \cdot (\lambda\kappa)^{1/2} \exp\left\{-\frac{\lambda\kappa}{2}(\mu - \nu)^2\right\} \\ &= \kappa^{\alpha-1/2} \lambda^{1/2} \exp(-\beta\kappa) \exp\left\{-\frac{\lambda\kappa}{2}(\mu - \nu)^2\right\}. \end{aligned} \tag{6.24}$$

The density of a normal-gamma distribution is shown in Fig. 6.8.

It is straightforward to show that the posterior distribution is again normal-gamma,

$$\begin{aligned} \boldsymbol{\theta} \mid x_{1:n} \sim \mathrm{NG}\Bigg(&(\lambda + n)^{-1}(\lambda\nu + n\bar{x}), \lambda + n, \alpha + \frac{n}{2}, \\ &\beta + \frac{n\hat{\sigma}^2_{\mathrm{ML}} + (\lambda + n)^{-1} n\lambda(\nu - \bar{x})^2}{2} \Bigg), \end{aligned} \tag{6.25}$$

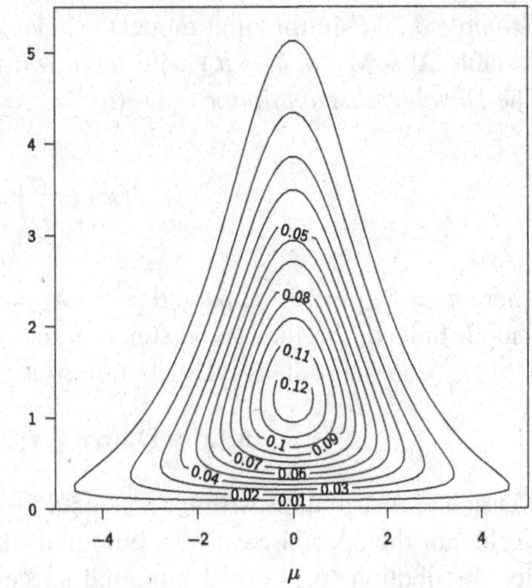

Fig. 6.8 Density function $f(\boldsymbol{\theta})$ of a normal-gamma distribution $\mathrm{NG}(\nu, \lambda, \alpha, \beta)$ for $\boldsymbol{\theta} = (\mu, \kappa)^\top$ with parameters $\nu = 0, \lambda = 0.5$ and $\alpha = 2, \beta = 1.2$. The mode of this distribution has coordinates $\mu = 0$ and $\kappa = (2 - 0.5)/1.2 = 1.25$

so the normal-gamma distribution is conjugate to the normal likelihood when both mean and variance are unknown. In (6.25), $\hat{\sigma}^2_{\mathrm{ML}} = n^{-1} \sum_{i=1}^n (x_i - \bar{x})^2$ is the usual MLE of the variance σ^2.

It is interesting that all parameters ν, λ, α and β of the normal-gamma distribution are updated additively based on the information in the data. In particular, the first parameter of the posterior distribution (6.25), the posterior mean of μ, is a weighted average of the prior mean ν and the sample average $\bar{x}$ with weights proportional to the prior precision λ and the sample size n, respectively. ■

6.5.2 Jeffreys' and Reference Prior Distributions

Jeffreys' prior for a parameter vector $\boldsymbol{\theta}$ is usually defined as

$$f(\boldsymbol{\theta}) \propto \sqrt{\left|J(\boldsymbol{\theta})\right|},$$

where $|J(\boldsymbol{\theta})|$ denotes the determinant of the expected Fisher information matrix. This definition is a generalisation of Jeffreys' rule (6.17) for scalar parameters.

Example 6.22 (Multinomial model) The expected Fisher information $J(\boldsymbol{\pi})$ of a multinomial observation $\boldsymbol{X} \sim \mathrm{M}_p(n, \boldsymbol{\pi})$ with unknown probability vector $\boldsymbol{\pi} = (\pi_1, \ldots, \pi_p)^\top$ has already been derived in Example 5.12. Its determinant is

$$\left|J(\boldsymbol{\theta})\right| = \frac{n}{\prod_{i=1}^p \pi_i},$$

so Jeffreys' prior is

$$f(\boldsymbol{\pi}) = \prod_{i=1}^{p} \pi_i^{-\frac{1}{2}}.$$

This is the kernel of a Dirichlet distribution $\boldsymbol{\pi} \sim \mathrm{D}_p(1/2, \dots, 1/2)$ with all p parameters equal to $1/2$. As in the binomial case, this distribution is proper. The posterior is, using Eq. (6.22),

$$\boldsymbol{\pi} \mid \boldsymbol{x} \sim \mathrm{D}_p(1/2 + x_1, \dots, 1/2 + x_p).$$ ■

However, application of Jeffreys' rule to vector-valued parameters $\boldsymbol{\theta}$ is controversial, although it fulfils the important invariance property. In the next example we discuss one particular problem of the multivariate Jeffreys' prior.

Example 6.23 (Normal model) Let $X_{1:n}$ denote a random sample of a $\mathrm{N}(\mu, \sigma^2)$ distribution with unknown parameter vector $\boldsymbol{\theta} = (\mu, \sigma^2)^\top$. We know from Example 5.11 that

$$J(\boldsymbol{\theta}) = \begin{pmatrix} \frac{n}{\sigma^2} & 0 \\ 0 & \frac{n}{2\sigma^4} \end{pmatrix},$$

so Jeffreys' prior is

$$f(\boldsymbol{\theta}) \propto \sqrt{|J(\boldsymbol{\theta})|} = \sqrt{\frac{n^2}{2\sigma^6}} \propto \sigma^{-3}.$$

So μ and σ^2 are a priori independent with a locally uniform prior density for μ and prior density proportional to $(\sigma^2)^{-3/2}$ for σ^2.

This result is in conflict with Jeffreys' prior if the parameter μ is known. Indeed, it is always possible to factorise the prior $f(\mu, \sigma^2) \propto (\sigma^2)^{-3/2}$ in the form $f(\mu) f(\sigma^2 \mid \mu)$. Due to prior independence of μ and σ^2, the conditional prior $f(\sigma^2 \mid \mu)$ must therefore be equal to the marginal prior, which is proportional to $(\sigma^2)^{-3/2}$. But the conditional prior for $\sigma^2 \mid \mu$ should also equal Jeffreys' prior for known μ, which is, however, proportional to $(\sigma^2)^{-1}$. ■

A better approach is the so-called *reference prior distribution*. Loosely speaking, the reference prior minimises the influence of the prior distribution on the posterior distribution. Information of a probability distribution is measured using the *entropy*, compare Appendix A.3.8.

The reference prior is identical to Jeffreys' prior for scalar parameters, however this analogy does no longer hold if the parameter is a vector. Then we need to separate the parameter vector $\boldsymbol{\theta}$ into a parameter of interest and a nuisance parameter. For example, a rigorous derivation of the reference prior for the correlation coefficient, informally described in Example 6.17, is based on a multivariate normal distribution with five unknown parameters, the correlation and the remaining four nuisance parameters, which are the means and variances of the two marginal normal distributions. The following example discusses the reference prior for the univariate normal distribution when both parameters are unknown.

Example 6.24 (Normal model) An alternative to Jeffreys' prior from Example 6.23 is the reference prior

$$f(\boldsymbol{\theta}) \propto \sigma^{-2}$$

for $\boldsymbol{\theta} = (\mu, \sigma^2)^\top$. Formally, this can be obtained through multiplication of Jeffreys' prior for μ (with known σ^2) with Jeffreys' prior for σ^2 (with μ known). In this special case the reference prior remains the same, whether we treat μ or σ^2 as parameter of interest and the other one as nuisance parameter.

Using the precision $\kappa = \sigma^{-2}$ rather than the variance, the reference prior is

$$f(\boldsymbol{\theta}) \propto \kappa^{-1}.$$

Formally, this is the normal-gamma distribution $\mathrm{NG}(0, 0, -1/2, 0)$, compare Eq. (6.24). Of course, this is an improper distribution, but this representation is useful to derive the posterior with (6.25):

$$\boldsymbol{\theta} \mid x_{1:n} \sim \mathrm{NG}\left(\bar{x}, n, \frac{1}{2}(n-1), \frac{1}{2}\sum_{i=1}^{n}(x_i - \bar{x})^2\right), \tag{6.26}$$

which is a proper distribution if $n \geq 2$. ■

6.5.3 Elimination of Nuisance Parameters

Suppose we are only interested in the first component $\boldsymbol{\theta}$ of the parameter vector $(\boldsymbol{\theta}, \boldsymbol{\eta})^\top$. Elimination of the nuisance parameter $\boldsymbol{\eta}$ is straightforward: all we need to do is to integrate the joint posterior density $f(\boldsymbol{\theta}, \boldsymbol{\eta} \mid x)$ with respect to $\boldsymbol{\eta}$. This gives us the marginal posterior density of the parameter of interest.

More specifically, let

$$f(\boldsymbol{\theta}, \boldsymbol{\eta} \mid x) = \frac{f(x \mid \boldsymbol{\theta}, \boldsymbol{\eta}) \cdot f(\boldsymbol{\theta}, \boldsymbol{\eta})}{f(x)}$$

denote the joint posterior density of $\boldsymbol{\theta}$ and $\boldsymbol{\eta}$. The marginal posterior of $\boldsymbol{\theta}$ is then

$$f(\boldsymbol{\theta} \mid x) = \int f(\boldsymbol{\theta}, \boldsymbol{\eta} \mid x)\, d\boldsymbol{\eta}. \tag{6.27}$$

If $\boldsymbol{\theta}$ is actually a scalar θ, then $f(\theta \mid x)$ can be used to calculate the commonly used point and interval estimates.

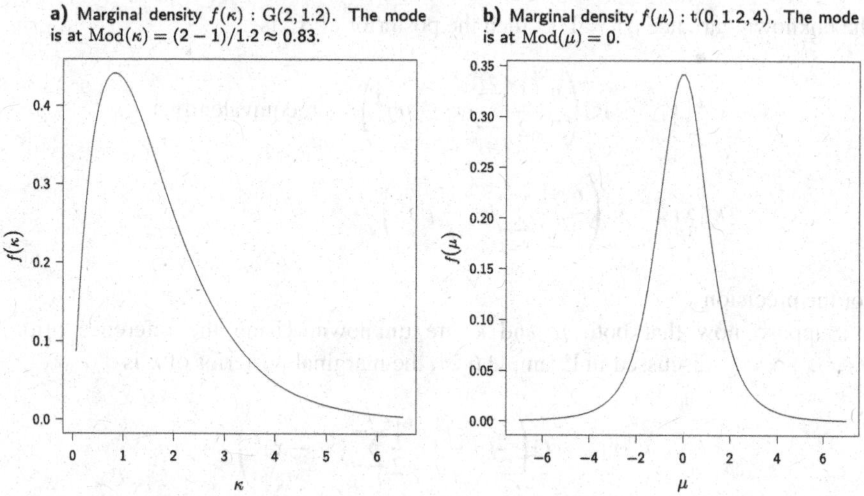

Fig. 6.9 Marginal distributions of a joint normal-gamma distribution NG($\nu, \lambda, \alpha, \beta$) with parameters $\nu = 0$, $\lambda = 0.5$ and $\alpha = 2$, $\beta = 1.2$

Conceptually, the elimination of nuisance parameters is very simple in the Bayesian inference approach. However, calculation of the integral (6.27) may be difficult and may require numerical techniques. But in some circumstances, in particular if conjugate priors are used, analytic calculation may be possible. Here is an example.

Example 6.25 (Normal-gamma distribution) Let $\boldsymbol{\theta} = (\mu, \kappa)^\top \sim \text{NG}(\nu, \lambda, \alpha, \beta)$ follow a normal-gamma distribution. The normal-gamma distribution is defined through the factorisation (6.23) as a product of a marginal gamma distribution of κ and a conditional normal distribution of μ given κ, so the marginal distribution of κ is obviously $\kappa \sim \text{G}(\alpha, \beta)$.

Analytic integration is required to compute the marginal density of μ, which turns out to belong to a t distribution. Indeed, using the parametrisation from Appendix A.5.2 we have that

$$\mu \sim \text{t}\left(\nu, \frac{\beta}{\alpha \cdot \lambda}, 2\alpha\right). \tag{6.28}$$

For illustration, Fig. 6.9 displays the marginal densities of μ and κ for the joint normal-gamma density shown in Fig. 6.8. ■

Example 6.26 (Normal model) Let $X_{1:n}$ denote a random sample from an $\text{N}(\mu, \kappa^{-1})$ distribution with known precision $\kappa = 1/\sigma^2$. We know from Example 6.14 that Jeffreys' prior for μ leads to the posterior $\mu \mid x_{1:n} \sim \text{N}(\bar{x}, \sigma^2/n)$. However, if the mean μ is known and Jeffreys' prior $f(\sigma^2) \propto 1/\sigma^2$ is used for the unknown variance $\sigma^2 = 1/\kappa$, then the posterior of σ^2 is

$$\sigma^2 \mid x_{1:n} \sim \text{IG}\left(\frac{n}{2}, \frac{1}{2}\sum_{i=1}^{n}(x_i - \mu)^2\right) \quad \text{or, equivalently,}$$

$$\kappa \mid x_{1:n} \sim \text{G}\left(\frac{n}{2}, \frac{1}{2}\sum_{i=1}^{n}(x_i - \mu)^2\right)$$

for the precision κ.

Suppose now that both μ and κ are unknown. Using the reference prior $f(\mu, \kappa) \propto \kappa^{-1}$ discussed in Example 6.24, the marginal posterior of κ is

$$\kappa \mid x_{1:n} \sim \text{G}\left(\frac{1}{2}(n-1), \frac{1}{2}\sum_{i=1}^{n}(x_i - \bar{x})^2\right)$$

as a direct consequence from (6.26). The first parameter is half of $n-1$ rather than n, so in complete analogy to the frequentist analysis, one degree of freedom is lost if μ is treated as unknown. The marginal posterior of $\sigma^2 = 1/\kappa$ is

$$\sigma^2 \mid x_{1:n} \sim \text{IG}\left(\frac{1}{2}(n-1), \frac{1}{2}\sum_{i=1}^{n}(x_i - \bar{x})^2\right) \tag{6.29}$$

with mean $\mathsf{E}(\sigma^2 \mid x_{1:n}) = \sum_{i=1}^{n}(x_i - \bar{x})^2/(n-3)$ and mode $\text{Mod}(\sigma^2 \mid x_{1:n}) = \sum_{i=1}^{n}(x_i - \bar{x})^2/(n+1)$. The unbiased estimate $S^2 = \sum_{i=1}^{n}(x_i - \bar{x})^2/(n-1)$ lies between the posterior mean and the posterior mode, and this is also true for the MLE $\hat{\sigma}^2_{\text{ML}} = \sum_{i=1}^{n}(x_i - \bar{x})^2/n$. See Fig. 6.10b for an illustration.

The marginal posterior distribution of μ is a t distribution:

$$\mu \mid x_{1:n} \sim \text{t}\left(\bar{x}, \frac{\sum_{i=1}^{n}(x_i - \bar{x})^2}{n(n-1)}, n-1\right). \tag{6.30}$$

This can be shown straightforwardly by applying Example 6.25 to (6.26). Due to the symmetry of the t distribution around $\bar{x}$, the posterior mean, mode and median are all identical to $\bar{x}$ (if $n > 2$), and any $\gamma \cdot 100\,\%$ HPD interval will also be equal-tailed.

It is interesting that the frequentist approach discussed in Example 3.8 gives numerically exactly the same result: The distribution of the pivot

$$\sqrt{n}\frac{\bar{X} - \mu}{\sqrt{\frac{1}{n-1}\sum_{i=1}^{n}(X_i - \bar{X})^2}} = \frac{\bar{X} - \mu}{\sqrt{\frac{1}{(n-1)n}\sum_{i=1}^{n}(X_i - \bar{X})^2}} \sim \text{t}(n-1) = \text{t}(0, 1, n-1)$$

is due to

$$Y \sim \text{t}(\mu, \sigma^2, \alpha) \quad \Rightarrow \quad (Y - \mu)/\sigma \sim \text{t}(0, 1, \alpha)$$

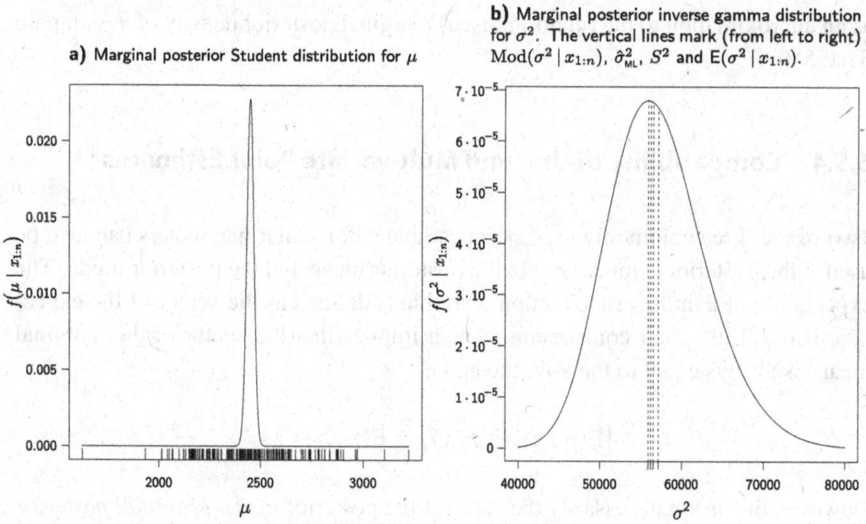

Fig. 6.10 Marginal posterior distributions for the mean and variance of the transformation factors in the alcohol concentration data set

equivalent to the posterior distribution of the centred and standardised mean μ:

$$\frac{\mu - \bar{x}}{\sqrt{\frac{1}{(n-1)n}\sum_{i=1}^{n}(x_i - \bar{x})^2}} \,\bigg|\, x_{1:n} \sim \text{t}(0, 1, n-1).$$

Since the standard t distribution is symmetric around the origin, this shows that the limits of the equal-tailed $\gamma \cdot 100\,\%$ credible interval for μ will be equal to the limits of the corresponding frequentist $\gamma \cdot 100\,\%$ confidence interval. ■

Example 6.27 (Blood alcohol concentration) In Example 6.9 we have considered the variance $\sigma^2 = 1/\kappa$ of the transformation factors as known. Now we want to estimate both the mean μ and the variance σ^2. In Fig. 6.10 the marginal posterior densities resulting from the reference prior $f(\mu, \sigma^2) \propto (\sigma^2)^{-1}$ are shown. They can be obtained by plugging in the sufficient statistics $n = 185$, $\bar{x} = 2449.18$ and $s^2 = 237.77^2$ into (6.30) and (6.29). ■

There is an interesting connection between Bayesian and frequentist techniques for the elimination of nuisance parameters. The Bayesian approach requires integration of the joint posterior density with respect to the nuisance parameter η. However, integration can be avoided by using the Laplace approximation (see Appendix C.2.2) for fixed $\boldsymbol{\theta}$. If one ignores the term in the Laplace approximation that depends on the curvature of the posterior density and approximates the posterior density by the likelihood function, then maximisation of the likelihood function with respect to η for fixed $\boldsymbol{\theta}$ is required. This approach leads to the profile likelihood as an approximation to the (unnormalised) marginal posterior density of $\boldsymbol{\theta}$, compare Sect. 5.3.

6.5.4 Compatibility of Uni- and Multivariate Point Estimates

Two of the three commonly used point estimates for scalar parameters can also be used if the posterior is multivariate: the posterior mean and the posterior mode. The expectation of a multivariate random variable is defined as the vector of the expectations of all its scalar components, which implies that the vector of the marginal means is always equal to the joint mean, i.e.

$$\big\{\mathsf{E}(\mu \mid x), \mathsf{E}(\rho \mid x)\big\} = \mathsf{E}(\mu, \rho \mid x).$$

However, this is not necessarily the case for the posterior mode. *Marginal posterior modes* may not equal the components of the *joint mode*, i.e.

$$\big\{\text{Mod}(\mu \mid x), \text{Mod}(\rho \mid x)\big\} \neq \text{Mod}(\mu, \rho \mid x),$$

as the following example illustrates.

Example 6.28 (Normal-gamma distribution) Let $\boldsymbol{\theta} = (\mu, \kappa)^\top \sim \text{NG}(\nu, \lambda, \alpha, \beta)$. For $\alpha > 1/2$, the mode of the normal-gamma distribution is $\text{Mod}(\boldsymbol{\theta}) = (\nu, (\alpha - 1/2)/\beta)^\top$, cf. Appendix A.5.3. The marginal distribution of μ is the t distribution (6.28), which also has mode $\text{Mod}(\mu) = \nu$. However, the marginal distribution of κ is the gamma distribution $\kappa \sim \text{G}(\alpha, \beta)$ with mode $\text{Mod}(\kappa) = (\alpha - 1)/\beta$ (if $\alpha > 1$), which is not equal to the second component of the mode of the joint distribution, $(\alpha - 1/2)/\beta$. Figures 6.8 and 6.9 illustrate this particular feature. ■

6.6 Some Results from Bayesian Asymptotics

In this section we will discuss asymptotic properties of posterior distributions and characteristics thereof. One important question is whether Bayesian point estimates are consistent, i.e. whether they converge to the true value if the sample size goes to infinity. We will see that this is indeed the case under certain regularity conditions.

Another important asymptotic property concerns the whole posterior distribution. We will sketch that any posterior distribution is asymptotically normal if the Fisher regularity conditions described in Definition 4.1 hold and the prior is not degenerate. This result parallels the asymptotic normal distribution of the MLE discussed in Sect. 4.2.3. In particular, it allows for a Bayesian interpretation of the MLE and its standard error.

6.6.1 Discrete Asymptotics

For discrete parameters, we cannot expect an asymptotically normal posterior distribution. However, the following result shows that the posterior mass gets more and more concentrated around the true parameter value for increasing sample size.

Result 6.5 (Posterior consistency) *Let $\theta \in \Theta = \{\theta_0, \theta_1, \theta_2, \dots\}$ denote an unknown discrete parameter with countable parameter space Θ. Let $p_i = f(\theta_i) = \Pr(\theta = \theta_i)$ denote the corresponding prior probability mass function for $i = 0, 1, 2, \dots$. Let $X_{1:n}$ denote a random sample from a distribution with density function $f(x \mid \theta)$ and suppose that θ_0 is the true parameter value. Then*

$$\lim_{n\to\infty} f(\theta_0 \mid x_{1:n}) = 1 \quad \textit{and} \quad \lim_{n\to\infty} f(\theta_i \mid x_{1:n}) = 0 \quad \textit{for all } i \neq 0.$$

Proof In order to show posterior consistency, we first consider n as fixed. The posterior probability of θ_i is

$$\begin{aligned} f(\theta_i \mid x_{1:n}) &= \frac{f(x_{1:n} \mid \theta_i) f(\theta_i)}{f(x_{1:n})} = \frac{p_i f(x_{1:n} \mid \theta_i)/f(x_{1:n} \mid \theta_0)}{\sum_j p_j f(x_{1:n} \mid \theta_j)/f(x_{1:n} \mid \theta_0)} \\ &= \frac{p_i \prod_{k=1}^n f(x_k \mid \theta_i)/f(x_k \mid \theta_0)}{\sum_j p_j \prod_{k=1}^n f(x_k \mid \theta_j)/f(x_k \mid \theta_0)} = \frac{\exp\{\log(p_i) + S_i^{(n)}\}}{\sum_j \exp\{\log(p_j) + S_j^{(n)}\}}, \end{aligned}$$

where $S_j^{(n)} = \sum_{k=1}^n \log\{f(x_k \mid \theta_j)/f(x_k \mid \theta_0)\}$ for $j = 1, \dots, n$.

Now let

$$\mathsf{D}(\theta_0 \parallel \theta_i) = \int f(x \mid \theta_0) \log\left\{\frac{f(x \mid \theta_0)}{f(x \mid \theta_i)}\right\} dx = \mathsf{E}_{X \mid \theta_0}\left[\log\left\{\frac{f(X \mid \theta_0)}{f(X \mid \theta_i)}\right\}\right]$$

denote the Kullback–Leibler discrepancy between $f(x \mid \theta_0)$ and $f(x \mid \theta_i)$ (cf. Appendix A.3.8). The information inequality implies that $\mathsf{D}(\theta_0 \parallel \theta_0) = 0$ and $\mathsf{D}(\theta_0 \parallel \theta_i) > 0$ for all $i \neq 0$. With the law of large numbers (Appendix A.4.3), we have

$$\lim_{n\to\infty} \frac{1}{n} S_j^{(n)} = \mathsf{E}_{X \mid \theta_0}\left[\log\left\{\frac{f(X \mid \theta_j)}{f(X \mid \theta_0)}\right\}\right] = -\mathsf{D}(\theta_0 \parallel \theta_j) \begin{cases} = 0 & \text{for } j = 0, \\ < 0 & \text{for } j \neq 0, \end{cases}$$

and therefore

$$\lim_{n\to\infty} S_j^{(n)} = \begin{cases} 0 & \text{for } j = 0, \\ -\infty & \text{for } j \neq 0, \end{cases}$$

so that

$$\lim_{n\to\infty} f(\theta_i \mid x_{1:n}) = \begin{cases} 1 & \text{for } i = 0, \\ 0 & \text{for } i \neq 0. \end{cases}$$

The posterior probability of the true value θ_0 hence converges to 1 as $n \to \infty$. □

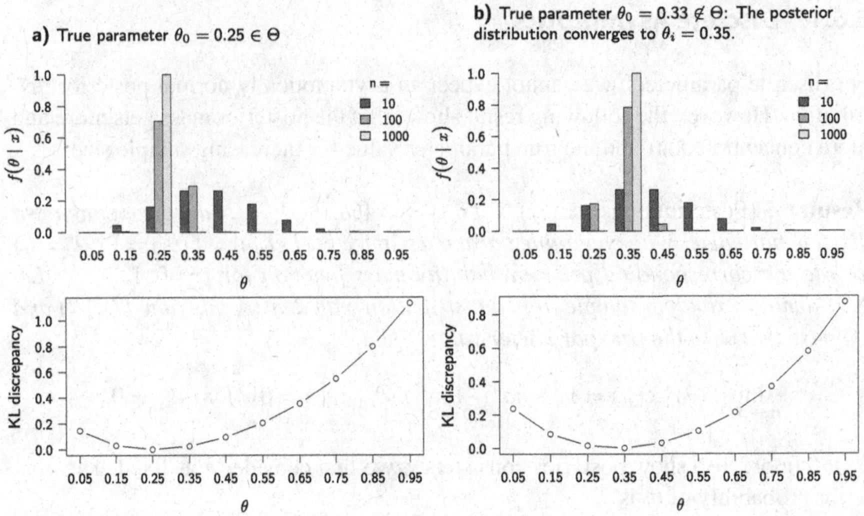

Fig. 6.11 Inference for the proportion θ after simulation from a binomial distribution $\text{Bin}(n, \theta)$ with $n = 10, 100, 1000$ and using a discrete uniform distribution on $\Theta = \{0.05, 0.15, \ldots, 0.95\}$: The posterior distribution (*top*) converges to the value $\theta \in \Theta$ with smallest Kullback–Leibler discrepancy (*bottom*) to the true model

Figure 6.11a illustrates this convergence property in a specific setting.

An interesting generalisation of the above proof concerns the case $\theta_0 \notin \Theta$: The posterior distribution will then converge to the value $\theta_i \in \Theta$ with smallest Kullback–Leibler discrepancy to the true model. This phenomenon is illustrated in Fig. 6.11b.

6.6.2 Continuous Asymptotics

In this section we will sketch that an unknown continuous parameter is—under suitable regularity conditions—asymptotically normally distributed. We consider the general case of an unknown parameter vector $\boldsymbol{\theta}$.

Let $X_{1:n}$ denote a random sample from a distribution with probability mass or density function $f(x \mid \boldsymbol{\theta})$. The posterior density is then

$$f(\boldsymbol{\theta} \mid x_{1:n}) \propto f(\boldsymbol{\theta}) f(x_{1:n} \mid \boldsymbol{\theta}) = \exp\bigl[\underbrace{\log\{f(\boldsymbol{\theta})\}}_{(1)} + \underbrace{\log\{f(x_{1:n} \mid \boldsymbol{\theta})\}}_{(2)}\bigr],$$

where $f(x_{1:n} \mid \boldsymbol{\theta}) = \prod_{i=1}^{n} f(x_i \mid \boldsymbol{\theta})$. A quadratic approximation using a Taylor expansion of the terms (1) and (2) around their maxima $\boldsymbol{m}_0$ (the prior mode) and the MLE $\hat{\boldsymbol{\theta}}_n = \hat{\boldsymbol{\theta}}_{\text{ML}}(x_{1:n})$, respectively, gives

$$\log\{f(\boldsymbol{\theta})\} \approx \log\{f(\boldsymbol{m}_0)\} - \frac{1}{2}(\boldsymbol{\theta} - \boldsymbol{m}_0)^\top \boldsymbol{I}_0(\boldsymbol{\theta} - \boldsymbol{m}_0) \quad \text{and}$$

$$\log\{f(x_{1:n} \mid \boldsymbol{\theta})\} \approx \log\{f(x_{1:n} \mid \hat{\boldsymbol{\theta}}_n)\} - \frac{1}{2}(\boldsymbol{\theta} - \hat{\boldsymbol{\theta}}_n)^\top \boldsymbol{I}(\hat{\boldsymbol{\theta}}_n)(\boldsymbol{\theta} - \hat{\boldsymbol{\theta}}_n).$$

Here $\boldsymbol{I}_0$ denotes the negative curvature of $\log\{f(\boldsymbol{\theta})\}$ at the mode $\boldsymbol{m}_0$, and $\boldsymbol{I}(\hat{\boldsymbol{\theta}}_n) = \boldsymbol{I}(\hat{\boldsymbol{\theta}}_n; x_{1:n})$ is the observed Fisher information matrix. Under regularity conditions, it follows that the posterior density is asymptotically proportional to

$$\exp\left[-\frac{1}{2}\{(\boldsymbol{\theta} - \boldsymbol{m}_0)^\top \boldsymbol{I}_0(\boldsymbol{\theta} - \boldsymbol{m}_0) + (\boldsymbol{\theta} - \hat{\boldsymbol{\theta}}_n)^\top \boldsymbol{I}(\hat{\boldsymbol{\theta}}_n)(\boldsymbol{\theta} - \hat{\boldsymbol{\theta}}_n)\}\right]$$
$$\propto \exp\left\{-\frac{1}{2}(\boldsymbol{\theta} - \boldsymbol{m}_n)^\top \boldsymbol{I}_n(\boldsymbol{\theta} - \boldsymbol{m}_n)\right\},$$

where

$$\boldsymbol{I}_n = \boldsymbol{I}_0 + \boldsymbol{I}(\hat{\boldsymbol{\theta}}_n) \quad \text{and}$$
$$\boldsymbol{m}_n = \boldsymbol{I}_n^{-1}\{\boldsymbol{I}_0\boldsymbol{m}_0 + \boldsymbol{I}(\hat{\boldsymbol{\theta}}_n)\hat{\boldsymbol{\theta}}_n\},$$

compare Appendix B.1.5. If n is large, the posterior distribution of $\boldsymbol{\theta}$ is therefore approximately normal with mean $\boldsymbol{m}_n$ and covariance matrix $\boldsymbol{I}_n^{-1}$:

$$\boldsymbol{\theta} \mid x_{1:n} \overset{a}{\sim} \mathrm{N}(\boldsymbol{m}_n, \boldsymbol{I}_n^{-1}).$$

Further approximations are possible:

1. For large n, the prior precision $\boldsymbol{I}_0$ will be negligible compared to the observed Fisher information $\boldsymbol{I}(\hat{\boldsymbol{\theta}}_n)$. We then have

$$\boldsymbol{\theta} \mid x_{1:n} \overset{a}{\sim} \mathrm{N}(\hat{\boldsymbol{\theta}}_n, \boldsymbol{I}(\hat{\boldsymbol{\theta}}_n)^{-1}).$$

Asymptotic normality of the posterior distribution

Any posterior distribution is (under suitable regularity conditions) asymptotically normal with mean equal to the MLE and covariance equal to the inverse observed Fisher information matrix:

$$\boldsymbol{\theta} \mid x_{1:n} \overset{a}{\sim} \mathrm{N}(\hat{\boldsymbol{\theta}}_n, \boldsymbol{I}(\hat{\boldsymbol{\theta}}_n)^{-1}).$$

This result gives a Bayesian interpretation of the MLE as asymptotic posterior mode (or mean). In addition, for large sample size, the limits of a Wald confidence interval for any component of θ will become numerically identical to the limits of an HPD (or equal-tailed) credible interval if confidence and credibility levels are identical.

There are at least three similar statements regarding asymptotic normality of the posterior distribution:

2. In complete analogy to likelihood asymptotics, the observed Fisher information matrix $\boldsymbol{I}(\hat{\boldsymbol{\theta}}_n)$ can be replaced by the expected Fisher information matrix $\boldsymbol{J}(\hat{\boldsymbol{\theta}}_n)$, so
$$\boldsymbol{\theta} \mid x_{1:n} \overset{\text{a}}{\sim} \mathrm{N}\big(\hat{\boldsymbol{\theta}}_n, \boldsymbol{J}(\hat{\boldsymbol{\theta}}_n)^{-1}\big).$$
3. If the posterior mode $\mathrm{Mod}(\boldsymbol{\theta} \mid x_{1:n})$ and the negative curvature $\boldsymbol{C}_n$ of the log posterior density at the mode are available, e.g. by numerical techniques, then also
$$\boldsymbol{\theta} \mid x_{1:n} \overset{\text{a}}{\sim} \mathrm{N}\big(\mathrm{Mod}(\boldsymbol{\theta} \mid x_{1:n}), \boldsymbol{C}_n^{-1}\big).$$
4. Often the posterior mean $\mathsf{E}(\boldsymbol{\theta} \mid x_{1:n})$ and the posterior covariance $\mathrm{Cov}(\boldsymbol{\theta} \mid x_{1:n})$ can be computed analytically or can at least be approximated with Monte Carlo techniques. The following approximation may then be useful:
$$\boldsymbol{\theta} \mid x_{1:n} \overset{\text{a}}{\sim} \mathrm{N}\big(\mathsf{E}(\boldsymbol{\theta} \mid x_{1:n}), \mathrm{Cov}(\boldsymbol{\theta} \mid x_{1:n})\big).$$

Example 6.29 (Binomial model) Consider the binomial model $X \mid \pi \sim \mathrm{Bin}(n, \pi)$. We know that the likelihood corresponds to a random sample of size n from a Bernoulli distribution with parameter π.

Given the observation $X = x$, the MLE is $\hat{\pi}_{\mathrm{ML}} = x/n$, and we know from Examples 2.10 and 4.1 that here
$$I(\hat{\pi}_{\mathrm{ML}}) = J(\hat{\pi}_{\mathrm{ML}}) = \frac{n}{\hat{\pi}_{\mathrm{ML}}(1 - \hat{\pi}_{\mathrm{ML}})}.$$
Thus, the above approximations 1 and 2 are identical here. Under a conjugate beta prior, $\pi \sim \mathrm{Be}(\alpha, \beta)$, the posterior equals
$$\pi \mid x \sim \mathrm{Be}(\alpha + x, \beta + n - x),$$
with known mean, mode and variance. So we can compute approximation 4. We can as well compute the negative curvature at the mode:
$$-\frac{d^2}{d\pi^2} \log\left\{\frac{1}{\mathrm{B}(\alpha + x, \beta + n - x)} \pi^{\alpha+x-1}(1 - \pi)^{\beta+n-x-1}\right\}\bigg|_{\pi = \frac{\alpha+x-1}{\alpha+\beta+n-2}}$$
$$= \frac{(\alpha + \beta + n - 2)^3}{(\alpha + x - 1)(\beta + n - x - 1)}.$$
Hence, we can also investigate the approximation 3 described above. All three approximations are compared in Fig. 6.12. ■

6.7 Empirical Bayes Methods

Empirical Bayes methods are a combination of the Bayesian approach with likelihood techniques. The general idea is to estimate parameters of the prior distribution from multiple experiments, rather than fixing them based on prior knowledge.

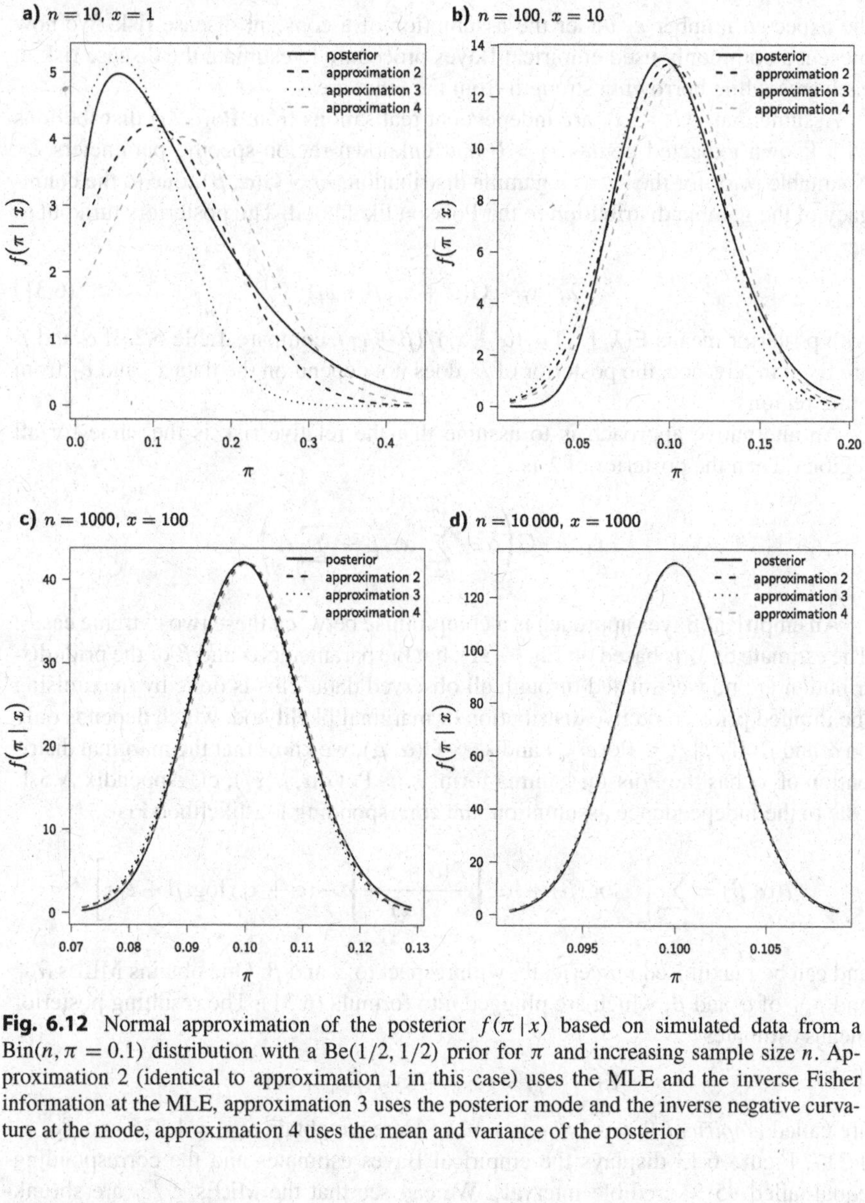

Fig. 6.12 Normal approximation of the posterior $f(\pi \mid x)$ based on simulated data from a $\text{Bin}(n, \pi = 0.1)$ distribution with a $\text{Be}(1/2, 1/2)$ prior for π and increasing sample size n. Approximation 2 (identical to approximation 1 in this case) uses the MLE and the inverse Fisher information at the MLE, approximation 3 uses the posterior mode and the inverse negative curvature at the mode, approximation 4 uses the mean and variance of the posterior

Strictly speaking, this is not a fully Bayesian approach, but it can be shown that empirical Bayes estimates have attractive theoretical properties. Empirical Bayes techniques are often used in various applications.

Example 6.30 (Scottish lip cancer) Consider Example 1.1.6, where we have analysed the incidence of lip cancer in $n = 56$ regions of Scotland. For each region $i = 1, \dots, n$, the observed number of lip cancer cases x_i are available as well as the expected number e_i under the assumption of a constant disease risk. We now present a commonly used empirical Bayes procedure to estimate the disease risk in each area while borrowing strength from the other areas.

Assume that $x_1, \dots, x_n$ are independent realisations from $\text{Po}(e_i \lambda_i)$ distributions with known expected counts $e_i > 0$ and unknown region-specific parameters λ_i. A suitable prior for the λ_is is a gamma distribution, $\lambda_i \sim \text{G}(\alpha, \beta)$, due to the conjugacy of the gamma distribution to the Poisson likelihood. The posteriors turn out to be

$$\lambda_i \mid x_i \sim \text{G}(\alpha + x_i, \beta + e_i) \tag{6.31}$$

with posterior means $\mathsf{E}(\lambda_i \mid x_i) = (\alpha + x_i)/(\beta + e_i)$, compare Table 6.2. If α and β are fixed in advance, the posterior of λ_i does not depend on the data x_j and e_j from other regions $j \neq i$.

An alternative approach is to assume that the relative risk is the same for all regions. Then the posterior of λ is

$$\lambda \mid x_{1:n} \sim \text{G}\left(\alpha + \sum_{i=1}^{n} x_i, \beta + \sum_{i=1}^{n} e_i\right).$$

An empirical Bayes approach is a compromise between these two extreme cases: The estimate of λ_i is based on Eq. (6.31), but the parameters α and β of the prior distribution are now estimated through all observed data. This is done by maximising the implied prior predictive distribution or marginal likelihood, which depends only on α and β: If $x_i \mid \lambda_i \sim \text{Po}(e_i \lambda_i)$ and $\lambda_i \sim \text{G}(\alpha, \beta)$, we know that the marginal distribution of x_i has the Poisson-gamma form: $x_i \sim \text{PoG}(\alpha, \beta, e_i)$, cf. Appendix A.5.1. Due to the independence assumption, the corresponding log-likelihood is

$$l(\alpha, \beta) = \sum_{i=1}^{n} \left[\alpha \log(\beta) + \log\left\{ \frac{\Gamma(\alpha + x_i)}{\Gamma(\alpha)} \right\} - (\alpha + x_i) \log(\beta + e_i) \right]$$

and can be maximised numerically with respect to α and β. One obtains MLEs $\hat{\alpha}_{\text{ML}}$ and $\hat{\beta}_{\text{ML}}$ of α and β, which are plugged into formula (6.31). The resulting posterior means estimates

$$\mathsf{E}(\lambda_i \mid x_i) = (\hat{\alpha}_{\text{ML}} + x_i)/(\hat{\beta}_{\text{ML}} + e_i)$$

are called *empirical Bayes estimates* of λ_i. Here we obtain $\hat{\alpha}_{\text{ML}} = 1.876$ and $\hat{\beta}_{\text{ML}} = 1.317$. Figure 6.13 displays the empirical Bayes estimates and the corresponding equal-tailed 95 % credible intervals. We can see that the MLEs x_i/e_i are shrunk towards the prior mean, i.e. the empirical Bayes estimates lie between these two extremes. This phenomenon is called *shrinkage*.

It is illustrative to consider the partial derivative of the log-likelihood $l(\alpha, \beta)$ with respect to β and to set it to zero. This score equation will hold for the MLE $\hat{\beta}_{\text{ML}}$, i.e.

$$\frac{1}{n} \sum_{i=1}^{n} \frac{\hat{\alpha}_{\text{ML}} + x_i}{\hat{\beta}_{\text{ML}} + e_i} = \frac{\hat{\alpha}_{\text{ML}}}{\hat{\beta}_{\text{ML}}}.$$

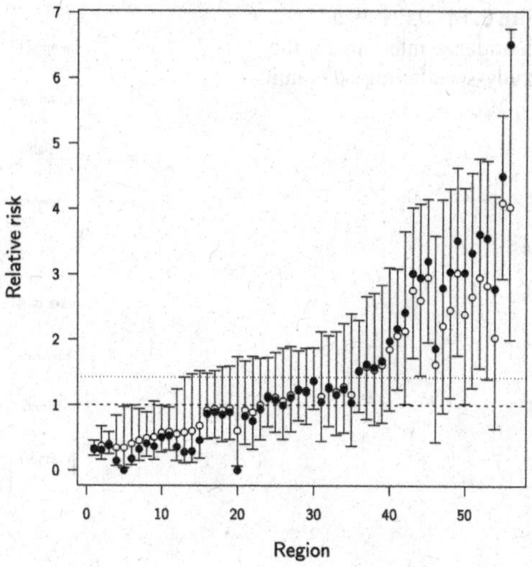

Fig. 6.13 95 % equal-tailed credible intervals for Scottish lip cancer incidence rates λ_i ($i = 1, \ldots, 56$), calculated with an empirical Bayes approach. The *dotted line* marks the MLE $\hat{\alpha}_{\text{ML}}/\hat{\beta}_{\text{ML}} = 1.4240$ of the prior mean. *Open circles* denote the posterior mean estimates of λ_i, while *filled circles* denote the MLEs x_i/e_i. The regions are ordered as in Fig. 4.3

So the average of the empirical Bayes estimates equals the MLE $\hat{\alpha}_{\text{ML}}/\hat{\beta}_{\text{ML}}$ of the prior mean α/β. For the Scotland lip cancer data, this prior mean equals 1.424. ■

Example 6.31 (Comparison of proportions) We would like to combine the results from the nine different clinical studies on the use of diuretics in pregnancy to prevent preeclampsia, which are shown in Table 1.1, in a *meta-analysis*. Our goal is to estimate the log odds ratio ψ based on the available data from all studies.

First, we assume that the unknown disease risks for treated and control women do not vary across the different studies. In total there are 291 diseased and 3468 healthy women, which were treated with diuretics, while there were 345 diseased and 2838 healthy women in the control group. Using the results from Example 5.8 on these total counts, we obtain the MLE $\hat{\psi}_{\text{ML}} = \log\{(291 \cdot 2838)/(3468 \cdot 345)\} = -0.37$ of the log odds ratio with 95 % Wald confidence interval from -0.53 to -0.21.

Alternatively, we can proceed similarly to the semi-Bayes approach in Example 6.10. Let a_i, b_i, c_i, d_i denote the corresponding cell entries in the ith study; then the MLE of ψ_i is $\hat{\psi}_i = \log\{(a_i \cdot d_i)/(b_i \cdot c_i)\}$. The difference to the analysis above is that we do not pool the whole data. Instead, we compute separate MLEs for the studies but assume that the underlying true treatment effect does not vary across studies. This already slightly relaxes the unrealistic assumption from above, where the data was treated as obtained from a single large study. The approximate likelihood of the observed log odds ratio $\hat{\psi}_i$ is thus

$$\hat{\psi}_i \mid \psi \sim \text{N}(\psi, \sigma_i^2),$$

where $\sigma_i^2 = a_i^{-1} + b_i^{-1} + c_i^{-1} + d_i^{-1}$. All studies are based on relatively large sample sizes, so the above normal approximation is likely to be fairly accurate. Figure 6.14 shows the study-specific log odds ratio estimates $\hat{\psi}_i$ with corresponding Wald confidence intervals. Under a locally uniform reference prior for ψ, its posterior is normal

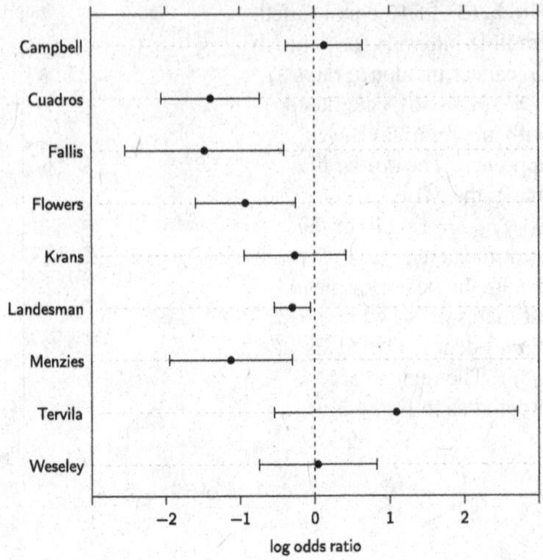

Fig. 6.14 95 % Wald confidence intervals for the study-specific log odds ratios ψ_i

with mean

$$\hat{\psi} = \frac{\sum_{i=1}^{9} w_i \hat{\psi}_i}{\sum_{i=1}^{9} w_i} \tag{6.32}$$

and variance $1/\sum_{i=1}^{9} w_i$, where $w_i = 1/\sigma_i^2$ is the precision of the ith study-specific estimate. This is the so-called *fixed effect model* to meta-analysis, which gives an overall treatment effect estimate as the weighted average of the study-specific treatment effects with weights proportional to the inverse squared standard errors. Based on this approach, we obtain the overall estimate $\hat{\psi} = -0.40$ with 95 % credible interval for ψ from -0.57 to -0.22, so very similar values as before.

We now allow the underlying true treatment effects to vary from study to study, i.e.

$$\begin{aligned} \hat{\psi}_i \mid \psi_i &\sim \mathrm{N}(\psi_i, \sigma_i^2), \\ \psi_i &\sim \mathrm{N}(\nu, \tau^2), \end{aligned}$$

and ν, the average treatment effect across all studies, is now of primary interest. The study-specific effects ψ_i are allowed to vary randomly around ν, therefore such a model is called a *random effects model*. If ν and τ^2 are known, we can easily derive the posterior distribution

$$\psi_i \mid \hat{\psi}_i \sim \mathrm{N}(\tilde{\nu}_i, \tilde{\sigma}_i^2) \tag{6.33}$$

of each study effect ψ_i, compare Example 6.8. Here $\tilde{\sigma}_i^2 = 1/(1/\sigma_i^2 + 1/\tau^2)$ and $\tilde{\nu}_i = \tilde{\sigma}_i^2(\hat{\psi}_i/\sigma_i^2 + \nu/\tau^2)$. So the posterior mean is a weighted average of the study-specific log odds ratio estimate $\hat{\psi}_i$ and the prior mean ν with weights proportional to $1/\sigma_i^2$ and $1/\tau^2$, respectively.

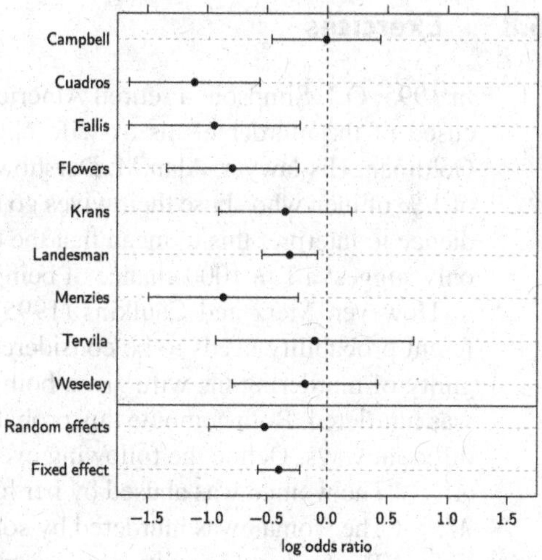

Fig. 6.15 95 % credible intervals for the log odds ratios ψ_i in an empirical Bayes random effects model. Also shown is a 95 % profile likelihood confidence interval for the mean effect ν ("random effects") as well as a 95 % Wald confidence interval for ψ under the assumption of equal study effects ("fixed effect")

Of course, application of this formula requires knowledge of ν and τ^2. These can be estimated in an empirical Bayes fashion by (numerical) maximisation of the marginal likelihood. For fixed τ^2, the marginal distribution of the ith log odds ratio is known to be $\hat{\psi}_i \sim \mathrm{N}(\nu, \sigma_i^2 + \tau^2)$. Then the estimate $\hat{\nu}_{\mathrm{ML}}(\tau^2)$ of the underlying treatment effect is as in the fixed effects model (6.32) a weighted mean of the individual study effects $\hat{\psi}_i$, but now with weights proportional to $w_i = \sigma_i^2 + \tau^2$ rather than $w_i = \sigma_i^2$. In other words, in moving from a fixed effect to a random effects analysis, the weights given to each study become more evenly distributed, so that in a random effects analysis small studies receive relatively large weight. Estimation of the variance τ^2 between true study effects can be done by numerically maximising the profile log-likelihood

$$l_p(\tau^2) = -\frac{1}{2}\sum_{i=1}^{n}\left[\log(\sigma_i^2+\tau^2) + \frac{\{\hat{\psi}_i - \hat{\nu}_{\mathrm{ML}}(\tau^2)\}^2}{\sigma_i^2+\tau^2}\right].$$

Empirical Bayes estimates of the individual study effects ψ_i are finally obtained by plugging the MLEs $\hat{\nu}_{\mathrm{ML}}$ and $\hat{\tau}^2_{\mathrm{ML}}$ into (6.33) in place of the fixed values ν and τ^2.

For the preeclampsia data, we obtain $\hat{\nu}_{\mathrm{ML}} = -0.52$ and $\hat{\tau}^2_{\mathrm{ML}} = 0.24$. Note that the MLE $\hat{\nu}_{\mathrm{ML}} = -0.52$ in the model with random effects is smaller than under a fixed effect model ($\hat{\psi}_{\mathrm{ML}} = -0.37$). Figure 6.15 displays 95 % empirical Bayes credible intervals for the individual study effects. Five of them lie below zero, so for these studies, we can identify a positive treatment effect. Note, however, that the intervals tend to be too small, as they do not take into account the uncertainty in the estimation of ν and τ^2. Also displayed is a 95 % confidence interval based on the profile likelihood of the mean study effect ν. Note that this is substantially wider than the corresponding one for the fixed effect ψ under a homogeneity assumption. ■

6.8 Exercises

1. In 1995, O.J. Simpson, a retired American football player and actor, was accused of the murder of his ex-wife Nicole Simpson and her friend Ronald Goldman. His lawyer, Alan M. Dershowitz stated on T.V. that only one-tenth of 1 % of men who abuse their wives go on to murder them. He wanted his audience to interpret this to mean that the evidence of abuse by Simpson would only suggest a 1 in 1000 chance of being guilty of murdering her.

   However, Merz and Caulkins (1995) and Good (1995) argue that a different probability needs to be considered: the probability that the husband is guilty of murdering his wife given both that he abused his wife *and* his wife was murdered. Both compute this probability using Bayes theorem but in two different ways. Define the following events:

   A: "The woman was abused by her husband."
   M: "The woman was murdered by somebody."
   G: "The husband is guilty of murdering his wife."

   (a) Merz and Caulkins (1995) write the desired probability in terms of the corresponding odds as

   $$\frac{\Pr(G \mid A, M)}{\Pr(G^c \mid A, M)} = \frac{\Pr(A \mid G, M)}{\Pr(A \mid G^c, M)} \cdot \frac{\Pr(G \mid M)}{\Pr(G^c \mid M)}. \tag{6.34}$$

   They use the fact that, of the 4936 women who were murdered in 1992, about 1430 were killed by their husband. In a newspaper article, Dershowitz stated that "It is, of course, true that, among the small number of men who do kill their present or former mates, a considerable number did first assault them." Merz and Caulkins (1995) interpret "a considerable number" to be $1/2$. Finally, they assume that the probability of a wife being abused by her husband, given that she was murdered by somebody else, is the same as the probability of a randomly chosen woman being abused, namely 0.05.

   Calculate the odds (6.34) based on this information. What is the corresponding probability of O.J. Simpson being guilty, given that he has abused his wife and she has been murdered?

   (b) Good (1995) uses the alternative representation

   $$\frac{\Pr(G \mid A, M)}{\Pr(G^c \mid A, M)} = \frac{\Pr(M \mid G, A)}{\Pr(M \mid G^c, A)} \cdot \frac{\Pr(G \mid A)}{\Pr(G^c \mid A)}. \tag{6.35}$$

   He first needs to estimate $\Pr(G \mid A)$ and starts with Dershowitz's estimate of $1/1000$ that the abuser will murder his wife. He assumes that the probability that this will happen in the year in question is at least $1/10$. Thus, $\Pr(G \mid A)$ is at least $1/10\,000$. If the husband did not murder his wife, then the fact that he abused her becomes irrelevant to the probability of M, so $\Pr(M \mid G^c, A) = \Pr(M \mid G^c) \approx \Pr(M)$. Since there are about 25 000 murders a year in the U.S. population of 250 000 000, Good (1995) estimates $\Pr(M \mid G^c, A)$ to be $1/10\,000$.

Calculate the odds (6.35) based on this information. What is the corresponding probability of O.J. Simpson being guilty, given that he has abused his wife and she has been murdered?

(c) Good (1996) revised this calculation, noting that approximately only a quarter of murdered victims are female, so $\Pr(M \mid G^c, A)$ reduces to $1/20\,000$. He also corrected $\Pr(G \mid A)$ to $1/2000$, when he realised that Dershowitz's estimate was an annual and not a lifetime risk. Calculate the probability of O.J. Simpson being guilty based on this updated information.

2. Consider Example 6.4. Here we will derive the implied distribution of $\theta = \Pr(D+ \mid T+)$ if the prevalence is $\pi \sim \mathrm{Be}(\tilde{\alpha}, \tilde{\beta})$.

(a) Deduce with the help of Appendix A.5.2 that

$$\gamma = \frac{\tilde{\alpha}}{\tilde{\beta}} \cdot \frac{1-\pi}{\pi}$$

follows an F distribution with parameters $2\tilde{\beta}$ and $2\tilde{\alpha}$, denoted by $\mathrm{F}(2\tilde{\beta}, 2\tilde{\alpha})$.

(b) Show that as a function of γ, the transformation (6.12) reduces to

$$\theta = g(\gamma) = (1+\gamma/c)^{-1},$$

where

$$c = \frac{\tilde{\alpha}\Pr(T+ \mid D+)}{\tilde{\beta}\{1-\Pr(T- \mid D-)\}}.$$

(c) Show that

$$\frac{d}{d\gamma} g(\gamma) = -\frac{1}{c(1+\gamma/c)^2}$$

and that $g(\gamma)$ is a strictly monotonically decreasing function of γ.

(d) Use the change-of-variables formula (A.11) to derive the density of θ in (6.13).

(e) Analogously proceed with the negative predictive value $\tau = \Pr(D- \mid T-)$ to show that the density of τ is

$$f(\tau) = d \cdot \tau^{-2} \cdot f_{\mathrm{F}}\big(d(1/\tau - 1); 2\tilde{\alpha}, 2\tilde{\beta}\big),$$

where

$$d = \frac{\tilde{\beta}\Pr(T- \mid D-)}{\tilde{\alpha}\{1-\Pr(T+ \mid D+)\}},$$

and $f_{\mathrm{F}}(x; 2\tilde{\alpha}, 2\tilde{\beta})$ is the density of the F distribution with parameters $2\tilde{\alpha}$ and $2\tilde{\beta}$.

3. Suppose that the heights of male students are normally distributed with mean 180 and unknown variance σ^2. We believe that σ^2 is in the range $[22, 41]$ with approximately 95 % probability. Thus, we assign an inverse-gamma distribution $\text{IG}(38, 1110)$ as prior distribution for σ^2.
   (a) Verify with R that the parameters of the inverse-gamma distribution lead to a prior probability of approximately 95 % that $\sigma^2 \in [22, 41]$.
   (b) Derive and plot the posterior density of σ^2 corresponding to the following data:

   $$183, 173, 181, 170, 176, 180, 187, 176, 171, 190, 184, 173, 176, 179, 181, 186.$$

   (c) Compute the posterior density of the standard deviation σ.
4. Assume that n throat swabs have been tested for influenza. We denote by X the number of throat swabs that yield a positive result and assume that X is binomially distributed with parameters n and unknown probability π, so that $X \mid \pi \sim \text{Bin}(n, \pi)$.
   (a) Determine the expected Fisher information and obtain Jeffreys' prior.
   (b) Reparametrise the binomial model using the log odds $\eta = \log\{\pi/(1 - \pi)\}$, leading to

   $$f(x \mid \eta) = \binom{n}{x} \exp(\eta x)\{1 + \exp(\eta)\}^{-n}.$$

   Obtain Jeffreys' prior distribution directly for this likelihood and not with the change-of-variables formula.
   (c) Take the prior distribution of 4(a) and apply the change-of-variables formula to obtain the induced prior for η. Because of the invariance under reparametrisation this prior density should be the same as in part 4(b).
5. Suppose that the survival times $X_{1:n}$ form a random sample from an exponential distribution with parameter λ.
   (a) Derive Jeffreys' prior for λ and show that it is improper.
   (b) Suppose that the survival times are only partially observed until the rth death such that $n - r$ observations are actually censored. Write down the corresponding likelihood function and derive the posterior distribution under Jeffreys' prior.
   (c) Show that the posterior is improper if all observations are censored.
6. After observing a patient, his/her LDL cholesterol level θ is estimated by a. Due to the increased health risk of high cholesterol levels, the consequences of underestimating a patient's cholesterol level are considered more serious than those of overestimation. That is to say, $|a - \theta|$ should be penalised more when $a \leq \theta$ than when $a > \theta$. Consider the following loss function parameterised in terms of $c, d > 0$:

   $$l(a, \theta) = \begin{cases} -c(a - \theta) & \text{if } a - \theta \leq 0, \\ d(a - \theta) & \text{if } a - \theta > 0. \end{cases}$$

(a) Plot $l(a, \theta)$ as a function of $a - \theta$ for $c = 3$ and $d = 1$.
(b) Compute the Bayes estimate with respect to the loss function $l(a, \theta)$.

7. Our goal is to estimate the allele frequency at one bi-allelic marker, which has either allele A or B. DNA sequences for this location are provided for n individuals. We denote the observed number of allele A by X and the underlying (unknown) allele frequency with π. A formal model specification is then a binomial distribution $X \mid \pi \sim \text{Bin}(n, \pi)$, and we assume a beta prior distribution $\pi \sim \text{Be}(\alpha, \beta)$ where $\alpha, \beta > 0$.
   (a) Derive the posterior distribution of π and determine the posterior mean and mode.
   (b) For some genetic markers, the assumption of a beta prior may be restrictive and a bimodal prior density, e.g., might be more appropriate. For example, we can easily generate a bimodal shape by considering a mixture of two beta distributions:

$$f(\pi) = w f_{\text{Be}}(\pi; \alpha_1, \beta_1) + (1 - w) f_{\text{Be}}(\pi; \alpha_2, \beta_2)$$

   with mixing weight $w \in (0, 1)$.
   (i) Derive the posterior distribution of π.
   (ii) The posterior distribution is a mixture of two familiar distributions. Identify these distributions and the corresponding posterior weights.
   (iii) Determine the posterior mean of π.
   (iv) Write an R-function that numerically computes the limits of an equal-tailed credible interval.
   (v) Let $n = 10$ and $x = 3$. Assume an even mixture ($w = 0.5$) of two beta distributions, Be(10, 20) and Be(20, 10). Plot the prior and posterior distributions in one figure.

8. The negative binomial distribution is used to represent the number of trials, x, needed to get r successes, with probability π of success in any one trial. Let X be negative binomial, $X \mid \pi \sim \text{NBin}(r, \pi)$, so that

$$f(x \mid \pi) = \binom{x-1}{r-1} \pi^r (1 - \pi)^{x-r}$$

with $0 < \pi < 1$, $r \in \mathbb{N}$ and support $\mathcal{T} = \{r, r+1, \dots\}$. As a prior distribution, assume $\pi \sim \text{Be}(\alpha, \beta)$,

$$f(\pi) = \text{B}(\alpha, \beta)^{-1} \pi^{\alpha-1} (1 - \pi)^{\beta-1},$$

with $\alpha, \beta > 0$.

(a) Derive the posterior density $f(\pi \mid x)$. Which distribution is this and what are its parameters?
(b) Define conjugacy and explain why, or why not, the beta prior is conjugate with respect to the negative binomial likelihood.
(c) Show that the expected Fisher information is proportional to $\pi^{-2}(1-\pi)^{-1}$ and derive therefrom Jeffreys' prior and the resulting posterior distribution.

9. Let $X_{1:n}$ denote a random sample from a uniform distribution on the interval $[0, \theta]$ with unknown upper limit θ. Suppose we select a *Pareto distribution* $\operatorname{Par}(\alpha, \beta)$ with parameters $\alpha > 0$ and $\beta > 0$ as a prior distribution for θ, cf. Table A.2 in Sect. A.5.2.
(a) Show that $T(X_{1:n}) = \max\{X_1, \ldots, X_n\}$ is sufficient for θ.
(b) Derive the posterior distribution of θ and identify the distribution type.
(c) Determine posterior mode $\operatorname{Mod}(\theta \mid x_{1:n})$, posterior mean $\mathsf{E}(\theta \mid x_{1:n})$, and the general form of the 95 % HPD interval for θ.

10. We continue Exercise 1 in Chap. 5, so we assume that the number of IHD cases is $D_i \mid \lambda_i \overset{\text{ind}}{\sim} \operatorname{Po}(\lambda_i Y_i), i = 1, 2$, where $\lambda_i > 0$ is the group-specific incidence rate. We use independent Jeffreys' priors for the rates λ_1 and λ_2.
(a) Derive the posterior distribution of λ_1 and λ_2. Plot these in R for comparison.
(b) Derive the posterior distribution of the relative risk $\theta = \lambda_2/\lambda_1$ as follows:
  (i) Derive the posterior distributions of $\tau_1 = \lambda_1 Y_1$ and $\tau_2 = \lambda_2 Y_2$.
  (ii) An appropriate multivariate transformation of $\boldsymbol{\tau} = (\tau_1, \tau_2)^\top$ to work with is $\boldsymbol{g}(\boldsymbol{\tau}) = \boldsymbol{\eta} = (\eta_1, \eta_2)^\top$ with $\eta_1 = \tau_2/\tau_1$ and $\eta_2 = \tau_2 + \tau_1$ to obtain the joint density $f_{\boldsymbol{\eta}}(\boldsymbol{\eta}) = f_{\boldsymbol{\tau}}\{\boldsymbol{g}^{-1}(\boldsymbol{\eta})\}|(\boldsymbol{g}^{-1})'(\boldsymbol{\eta})|$, cf. Appendix A.2.3.
  (iii) Since $\eta_1 = \tau_2/\tau_1$ is the parameter of interest, integrate η_2 out of $f_{\boldsymbol{\eta}}(\boldsymbol{\eta})$ and show that the marginal density is

$$f(\eta_1) = \frac{\eta_1^{\alpha_1 - 1}(1+\eta_1)^{-\alpha_1-\alpha_2}}{\mathrm{B}(\alpha_1, \alpha_2)},$$

  which is a beta prime distribution with parameters α_1 and α_2.
  (iv) From this distribution of τ_2/τ_1 the posterior distribution of λ_2/λ_1 is then easily found.
(c) For the given data, compute a 95 % credible interval for θ and compare the results with those from Exercise 1 in Chap. 5.

11. Consider Exercise 10 in Chap. 3. Our goal is now to perform Bayesian inference with an improper discrete uniform prior for the unknown number N of beds:

$$f(N) \propto 1 \quad \text{for } N = 2, 3, \ldots .$$

(a) Why is the posterior mode equal to the MLE?

(b) Show that for $n > 1$, the posterior probability mass function is

$$f(N \mid x_n) = \frac{n-1}{x_n} \binom{x_n}{n} \binom{N}{n}^{-1} \quad \text{for } N \geq x_n.$$

(c) Show that the posterior expectation is

$$\mathsf{E}(N \mid x_n) = \frac{n-1}{n-2} \cdot (x_n - 1) \quad \text{for } n > 2.$$

(d) Compare the frequentist estimates from Exercise 10 in Chap. 3 with the posterior mode and mean for $n = 48$ and $x_n = 1812$. Numerically compute the associated 95 % HPD interval for N.

12. Assume that $X_1, \dots, X_n$ are independent samples from the binomial models $\text{Bin}(m, \pi_i)$ and assume that $\pi_i \overset{\text{iid}}{\sim} \text{Be}(\alpha, \beta)$. Compute empirical Bayes estimates $\hat{\pi}_i$ of π_i as follows:

(a) Show that the marginal distribution of X_i is beta-binomial, see Appendix A.5.1 for details. The first two moments of this distribution are

$$\mu_1 = \mathsf{E}(X_i) = m \frac{\alpha}{\alpha + \beta},$$

$$\mu_2 = \mathsf{E}\left(X_i^2\right) = m \frac{\alpha\{m(1+\alpha) + \beta\}}{(\alpha+\beta)(1+\alpha+\beta)}.$$

Solve for α and β using the sample moments $\widehat{\mu}_1 = n^{-1} \sum_{i=1}^n x_i$, $\widehat{\mu}_2 = n^{-1} \sum_{i=1}^n x_i^2$ to obtain estimates of α and β.

(b) Now derive the empirical Bayes estimates $\hat{\pi}_i$. Compare them with the corresponding MLEs.

6.9 References

Lee (2012) gives an accessible introduction to Bayesian inference. The introductory article to Bayesian inference by Edwards et al. (1963) is still worth reading. A comprehensive description can be found in Bernardo and Smith (2000) and O'Hagan and Forster (2004). Further classics are Jeffreys (1961), Box and Tiao (1973) and Robert (2001). See Mossman and Berger (2001) and Bayarri and Berger (2004) for background to Example 6.4, and Bernardo and Smith (2000, Sect. 5.4) for a rigorous treatment of reference priors. Empirical Bayes methods are described in Carlin and Louis (2008), but see also Davison (2003, Sect. 11.5). The empirical Bayes approach outlined in Example 6.30 is due to Clayton and Kaldor (1987).

Model Selection

7

Contents

Parametric statistical inference is based on the assumption of a model that describes the stochastic properties of the data. A fundamental question of statistics therefore is how one selects a model—given the data—from a set of candidate models? This chapter introduces likelihood and Bayesian methods to address this problem.

Frequently, it is the context that influences the choice of a particular type of model. For example, the assumption of a normal distribution is obviously inadequate for binary observations. However, there are many situations where the choice of a specific model for the available data is far from obvious. In the analysis of survival data, for example, any distribution with a positive support seems appropriate. In Example 5.9 we fitted the Weibull and the gamma model to survival times of patients treated with Azathioprine for primary biliary cirrhosis (PBC), see also Sect. 1.1.8. Both formulations are generalisations of the exponential model. But which model is best describing the data?

Recall from Sect. 5.5 that the generalised likelihood ratio (LR) statistic W can be used to compare two models, if the simpler model is a special case of the more complex model with some parameters being fixed. The dimension of the unknown parameter vector will then differ between the models, and the simpler model is called *nested* in the more general model.

L. Held, D. Sabanés Bové, *Likelihood and Bayesian Inference*,
Statistics for Biology and Health, https://doi.org/10.1007/978-3-662-60792-3_7,

Suppose M_1 is the simpler model and M_2 the more complex. We can now apply the generalised LR statistic

$$W = 2\log\left\{\frac{\max_{M_2} L(\boldsymbol{\theta})}{\max_{M_1} L(\boldsymbol{\theta})}\right\} = 2\left\{\max_{M_2} l(\boldsymbol{\theta}) - \max_{M_1} l(\boldsymbol{\theta})\right\}$$

to compare the two models. Here the log-likelihood $l(\boldsymbol{\theta})$ is understood as the log-likelihood in the more general model M_2. Model M_1 can be written as a restriction of model M_2, so the maximised log-likelihood in model M_1 is identical to $l(\boldsymbol{\theta})$ maximised under the restriction corresponding to M_1, which we denote as $\max_{M_1} l(\boldsymbol{\theta})$. Under the assumption that model M_1 is correct, W is asymptotically χ^2 distributed. The associated degrees of freedom are given by the difference of the number of unknown parameters of the two models considered. For example, if the more complex model M_2 has one parameter more than the simpler model M_1, a P-value can be computed based on the upper tail of the χ^2 distribution with one degree of freedom, evaluated at the observed value w of W: $\Pr(X \geq w)$ where $X \sim \chi^2(1)$. For example, if $w = 3.84$, then the P-value is 0.05.

Example 7.1 (Analysis of survival times) Coming back to the introductory example, we want to determine an adequate model for the survival data of the PBC patients, see Sect. 1.1.8. In particular, we need to take into account that the observations are right censored.

In the exponential model (M_1) of Example 2.8, it is possible to determine the MLE analytically. Substituting this estimator into the log-likelihood function gives $l(\hat{\lambda}_{\mathrm{ML}}) = -424.0243$.

We can use the code from Example 5.9 to determine the maximal log-likelihood in the Weibull model M_2.

```
start <- c(1000, 1)
resultWeibull <- optim(start, weibullLik, log=TRUE,
                       control=list(fnscale=-1), hessian=TRUE)
logLikWeibull <- resultWeibull$value
logLikWeibull
[1] -424.0043
```

So we obtain the maximised log-likelihood $l(\hat{\boldsymbol{\theta}}_{\mathrm{ML}}) = -424.0043$ at the MLE $\hat{\boldsymbol{\theta}}_{\mathrm{ML}} = (\hat{\mu}_{\mathrm{ML}}, \hat{\alpha}_{\mathrm{ML}})^\top$.

Since the Weibull model with $\alpha = 1$ corresponds to the exponential model, we can use the generalised LR statistic to compare the two models. We obtain the test statistic

$$W_{21} = 2\left\{\max_{M_2} l(\boldsymbol{\theta}) - \max_{M_1} l(\boldsymbol{\theta})\right\} = 2\{-424.0043 - (-424.0243)\} = 0.0400$$

with corresponding P-value 0.84 computed from the χ^2 distribution function with one degree of freedom. So we have no evidence against the simpler exponential model.

In Example 2.3 a gamma model for the uncensored data has been described. Analogously to the approach above, we can determine the maximal log-likelihood for the gamma model M_3 in the parametrisation using μ and ϕ:

```
## implement the (log) likelihood in the gamma model:
gammaLik <- function(muphi, log = TRUE){
    mu <- muphi[1]
    phi <- muphi[2]
    loglik <- with (pbcTreat,
                    sum(d * dgamma (time,
                                    shape = mu / phi,
                                    scale = phi,
                                    log = TRUE) +
                        (1 - d) * pgamma (time,
                                          shape = mu / phi,
                                          scale = phi,
                                          lower.tail = FALSE,
                                          log.p = TRUE)))
    if(log)
    {
        return(loglik)
    } else {
        return(exp(loglik))
    }
}
start <- c (1000, 1000)
resultGamma <- optim(start, gammaLik, control=list(fnscale=-1),
                     hessian=TRUE)
logLikGamma <- resultGamma$value
logLikGamma
[1] -424.0047
```

This is almost the same value as for the Weibull model. Also the gamma model is a generalisation of the exponential model, which is obtained when $\mu = \phi$. Hence, we can test this null hypothesis with the generalised LR statistic

$$W_{31} = 2\left\{\max_{M_3} l(\boldsymbol{\theta}) - \max_{M_1} l(\boldsymbol{\theta})\right\} = 2\left\{-424.0047 - (-424.0243)\right\} = 0.0393$$

and obtain the same P-value 0.84 as before, with the same conclusion that there is no evidence against the exponential model. ■

The LR statistic is a valuable method for model selection and is frequently used. However, for several reasons, it is not appropriate as a general method for model selection. First, only nested models can be compared, the comparison of non-nested models is not possible. For example, the LR statistic cannot be used to choose between the Weibull and the gamma model in Example 7.1 since none of the two models can be written as a special case of the other. Second, the procedure is based on a significance test, so it is asymmetric in nature: it can only provide evidence against, but not for, the simpler model. Finally, the LR test cannot be generalised to a comparison of more than two models, it can only be used for pairwise model comparisons.

The form of the LR statistic indicates that the maximal value of the likelihood function under the respective model $\max_{M_i} L(\boldsymbol{\theta})$ is the central quantity for model selection. However, a more complex model, in which the simpler model is nested, will always increase the likelihood at the corresponding MLE. It is therefore not a sensible strategy to choose the model with the largest maximal likelihood because one would always select the most complex model. Ultimately, the dimensions of the different models also have to be taken into account.

The penalisation of *model complexity* is in line with William of Ockham, whose famous quotation of 1320

> *"Pluralitas non ponenda est sine necessitate"*
> (plurality should not be posited without necessity)

was later interpreted statistically under the name *Ockham's razor*. John Ponce of Cork, for example, wrote in 1639:

> *"Variation must be taken as random until there is positive evidence to the contrary; and new parameters in laws, when they are suggested, must be tested one at a time unless there is specific reason to the contrary."*

We will see that all approaches to model selection that are described in this chapter, i.e. both likelihood methods (Sect. 7.1) and Bayesian methods (Sect. 7.2), do indeed penalise model complexity somehow. As a more fundamental question we have to discuss if there even is a "true" model or if all models are more or less bad descriptions of the available data. This aspect is reflected in Sect. 7.2.4, where not one single model is chosen, but where results from different models are combined.

7.1 Likelihood-Based Model Selection

The value of the likelihood function at the MLE $\hat{\boldsymbol{\theta}}_{\mathrm{ML}}$ describes the quality of the model fit. This value will in the following be combined with measures of model complexity resulting in different model selection criteria, which allow the comparison of non-nested models. The measure of model complexity will in some form depend on the dimension p of the parameter vector $\boldsymbol{\theta}$. The model with the best value of the criterion is then chosen as the best model. Note that it is now crucial to include all multiplicative constants in the likelihood since otherwise the comparison of two models based on a likelihood criterion is meaningless.

7.1.1 Akaike's Information Criterion

Akaike's information criterion (after Hirotugu Akaike, 1927–2009)

$$\mathrm{AIC} = -2l(\hat{\boldsymbol{\theta}}_{\mathrm{ML}}) + 2p \tag{7.1}$$

penalises the maximised log-likelihood with the number of parameters p. The criterion is negatively oriented, i.e. the model with minimal AIC is selected. Therefore, a difference of $2q$ is sufficient for a model with q additional parameters to be preferred. For example, for $q = 1$, a difference of 2 is sufficient, which corresponds to a P-value of 0.16 for the comparison of two nested models using the LR test. For $q = 2$, the corresponding P-value is 0.14. Table 7.1 shows that the corresponding P-values decrease with increasing q. This indicates that model selection based on AIC is not equivalent to or compatible with model selection based on the LR statistic.

Table 7.1 P-value thresholds based on the LR statistic with q degrees of freedom, which correspond to model selection based on AIC with penalty $2q$

q	1	2	3	4	5	6	7	8
P-value	0.16	0.14	0.11	0.092	0.075	0.062	0.051	0.042

Example 7.2 (Analysis of survival times) For comparing the three survival models with AIC, we take the maximised log-likelihood values from Example 7.1 and the number of parameters (2 for the gamma and Weibull models and 1 for the exponential model), and obtain the AIC values 850.0486, 852.0087 and 852.0094 for M_1, M_2 and M_3, respectively. The conclusion is the same as before: the exponential model M_1 is preferred because it has the smallest AIC. The Weibull model M_2 and the gamma model M_3 have nearly the same AIC values. ■

The original derivation of Akaike's information criterion is a generalisation of the maximum likelihood principle from parameter estimation to model selection. We will now sketch why this is so. In general, we must assume that none of the proposed models $f(x;\boldsymbol{\theta})$ will match the true data-generating model $g(x)$ exactly. Then the models are *misspecified*. The true density $g(x)$ is unknown, and we want to approximate it as well as possible with one of the parametric models, i.e. we want a model that is as close as possible to the underlying true mechanism. We restrict our attention to a random sample $X_{1:n}$ from $g(x)$.

We measure the closeness between the two densities $g(x)$ and $f(x;\boldsymbol{\theta})$ with the Kullback–Leibler discrepancy (see Appendix A.3.8):

$$\begin{aligned}
\mathsf{D}\big(g \parallel f(\cdot;\boldsymbol{\theta})\big) &= \int g(x)\log\left\{\frac{g(x)}{f(x;\boldsymbol{\theta})}\right\}dx \\
&= \int g(x)\log g(x)\,dx - \int g(x)\log f(x;\boldsymbol{\theta})\,dx \\
&= \mathsf{E}_g\big\{\log g(X)\big\} - \mathsf{E}_g\big\{\log f(X;\boldsymbol{\theta})\big\}.
\end{aligned} \tag{7.2}$$

Since g is fixed but unknown, minimisation of (7.2) reduces to maximising the mean log-likelihood $\mathsf{E}_g\{\log f(X;\boldsymbol{\theta})\}$, which also depends on the unknown distribution g. For a fixed model, let $\boldsymbol{\theta}_0$ denote the resulting optimal ("least false") parameter value. If the model is actually correct, then $g(x)$ and $f(x;\boldsymbol{\theta}_0)$ are identical. Consider now the MLE $\hat{\boldsymbol{\theta}}_{\text{ML}}$, that maximises the log-likelihood $l(\boldsymbol{\theta};x_{1:n}) = \sum_{i=1}^{n}\log f(x_i;\boldsymbol{\theta})$. Due to the law of large numbers (Appendix A.4.3), we have

$$\frac{1}{n}l(\boldsymbol{\theta};X_{1:n}) \xrightarrow{D} \mathsf{E}_g\big\{\log f(X;\boldsymbol{\theta})\big\},$$

and thus $\hat{\boldsymbol{\theta}}_{\text{ML}} \to \boldsymbol{\theta}_0$ as $n \to \infty$. This perspective justifies the use of the MLE even in settings when model misspecification is suspected. Moreover, the MLE can now be viewed as the parameter estimate which minimises approximately the Kullback–Leibler discrepancy to the truth: When the true distribution g is replaced with the empirical distribution $\hat{g}_n$ of the data, we have $\mathsf{E}_{\hat{g}_n}\{\log f(X;\boldsymbol{\theta})\} = \frac{1}{n}l(\boldsymbol{\theta};x_{1:n})$.

Now consider a collection of parametric models $f(x; \boldsymbol{\theta})$, from which we want to select the best approximating one. Again, we study the closeness of the models to the truth via the Kullback–Leibler discrepancy. From above we know that the MLE is a good estimate of the optimal parameter within one model, so we consider $\mathsf{D}(g \parallel f(\cdot; \hat{\boldsymbol{\theta}}_{\text{ML}}(X_{1:n})))$. We emphasise that this is a random variable depending on the data $X_{1:n}$ through the ML estimator $\hat{\boldsymbol{\theta}}_{\text{ML}}(X_{1:n})$. Since we would like to select a model that works well for new data Y, i.e. for predictive purposes, we would like to choose the model that minimises the expected Kullback–Leibler discrepancy. To this end, we only need to estimate the expectation of the mean log-likelihood

$$K = \mathsf{E}_g\{M(X_{1:n})\} = \mathsf{E}_g\big[\mathsf{E}_g\{\log f\big(Y; \hat{\boldsymbol{\theta}}_{\text{ML}}(X_{1:n})\big)\}\big] \tag{7.3}$$

because the first term in the Kullback–Leibler discrepancy (7.2) does not depend on the model. The inner expectation in (7.3) computes the mean log-likelihood M with respect to the data Y, while the outer expectation is taken with respect to the independent random sample $X_{1:n}$ from g. Replacing g with the empirical distribution $\hat{g}_n$ in (7.3), we obtain the estimate

$$\hat{K} = \mathsf{E}_{\hat{g}_n}\left\{\frac{1}{n} l\big(\hat{\boldsymbol{\theta}}_{\text{ML}}(X_{1:n}); x_{1:n}\big)\right\} = \frac{1}{n} l\big(\hat{\boldsymbol{\theta}}_{\text{ML}}(x_{1:n}); x_{1:n}\big).$$

It is intuitively clear that this estimate will be biased because we have used the same data $x_{1:n} = (x_1, \dots, x_n)$ twice, both for estimating the mean log-likelihood and for estimating its expectation. The AIC thus includes a bias correction for this estimate: We will now show that

$$\mathsf{E}(\hat{K} - K) \approx p/n, \tag{7.4}$$

where p is the number of parameters in the model, so that the bias-corrected estimate is

$$\hat{K} - p/n = \frac{1}{n}\{l(\hat{\boldsymbol{\theta}}_{\text{ML}}) - p\} = -\frac{1}{2n}\text{AIC}. \tag{7.5}$$

Hence, choosing the model that minimises the AIC is approximately the same as minimising the expected Kullback–Leibler discrepancy to the truth.

To prove Eq. (7.5), we use a Taylor expansion of $\log f(x; \hat{\boldsymbol{\theta}}_{\text{ML}})$ around $\boldsymbol{\theta}_0$. The difference between the estimate $\hat{K}(X_{1:n})$ and the mean log-likelihood $M(X_{1:n})$ can now be approximated by

$$\hat{K} - M \approx \frac{1}{n}\sum_{i=1}^{n} Z_i + (\hat{\boldsymbol{\theta}}_{\text{ML}} - \boldsymbol{\theta}_0)^\top \boldsymbol{G}_1(\hat{\boldsymbol{\theta}}_{\text{ML}} - \boldsymbol{\theta}_0), \tag{7.6}$$

where $\boldsymbol{G}_1 = \mathsf{E}_g\{\boldsymbol{I}_1(\boldsymbol{\theta}_0; X)\}$ is the expected unit information. Since the random variables $Z_i = \log f(X_i; \boldsymbol{\theta}_0) - \mathsf{E}_g\{\log f(X; \boldsymbol{\theta}_0)\}$ have zero mean, the expectation of (7.6) reduces to the expectation of the quadratic form $(\hat{\boldsymbol{\theta}}_{\text{ML}} - \boldsymbol{\theta}_0)^\top \boldsymbol{G}_1(\hat{\boldsymbol{\theta}}_{\text{ML}} - \boldsymbol{\theta}_0)$. Analogously to the case in Sect. 4.2.3 where the model is correctly specified, there is the following more general result for the distribution of the ML estimator in case of model misspecification:

$$\sqrt{n}(\hat{\boldsymbol{\theta}}_{\text{ML}} - \boldsymbol{\theta}_0) \xrightarrow{D} \boldsymbol{G}_1^{-1}\boldsymbol{U},$$

where $\boldsymbol{U} = \boldsymbol{S}(\boldsymbol{\theta}_0; X_{1:n})/\sqrt{n} \overset{\text{a}}{\sim} \text{N}_p(\boldsymbol{0}, \boldsymbol{H}_1)$ due to the central limit theorem (cf. Appendix A.4.4), because $\mathsf{E}_g\{\boldsymbol{S}_1(\boldsymbol{\theta}_0; X)\} = \boldsymbol{0}$ and $\boldsymbol{H}_1 = \text{Cov}_g\{\boldsymbol{S}_1(\boldsymbol{\theta}_0; X)\}$. Hence, we have

$$(\hat{\boldsymbol{\theta}}_{\text{ML}} - \boldsymbol{\theta}_0)^\top \boldsymbol{G}_1(\hat{\boldsymbol{\theta}}_{\text{ML}} - \boldsymbol{\theta}_0) \xrightarrow{D} \frac{1}{n}\boldsymbol{U}^\top \boldsymbol{G}_1^{-1}\boldsymbol{G}_1\boldsymbol{G}_1^{-1}\boldsymbol{U} = \frac{1}{n}\boldsymbol{U}^\top \boldsymbol{G}_1^{-1}\boldsymbol{U}. \tag{7.7}$$

Using the approximate normal distribution of $\boldsymbol{U}$ and a result from Appendix A.2.4, we obtain

$$\mathsf{E}_g\big(\boldsymbol{U}^\top \boldsymbol{G}_1^{-1}\boldsymbol{U}\big) \approx \text{tr}\big(\boldsymbol{G}_1^{-1}\boldsymbol{H}_1\big) = p^*, \tag{7.8}$$

where p^* can be interpreted as a generalised parameter dimension. Note that $\boldsymbol{G}_1 = \boldsymbol{H}_1 = \boldsymbol{J}_1(\boldsymbol{\theta}_0)$ is the expected unit Fisher information in the case where the model is correctly specified. Then we have exactly $p^* = \text{tr}(\boldsymbol{I}_p) = p$, the number of parameters in the model. This directly leads to (7.4). However, even in the case of misspecification, the AIC approximation $p^* \approx p$ is justifiable. More sophisticated estimates of (7.8) are only rarely used in practice and may have larger variance.

7.1.2 Cross Validation and AIC

Recall that the plain value $l(\hat{\boldsymbol{\theta}}_{\text{ML}})$ cannot be used as a model selection criterion since the value of the maximal log-likelihood automatically increases for a more complex model. The underlying problem is that the available data $x_{1:n}$ are used twice: on the one hand, to calculate the estimate $\hat{\boldsymbol{\theta}}_{\text{ML}}(x_{1:n})$ and, on the other hand, to calculate the model selection criterion, the log-likelihood $l(\boldsymbol{\theta}; x_{1:n})$. A better approach is to divide the data into a *training* and a *validation sample*. The training sample is used to estimate the parameters, and the validation sample is used to evaluate the model selection criterion. The splitting into training and validation part is then repeated such that each observation is contained once in a training sample, and the average value of the model criterion is calculated. Instead of an automatic increase, for more complex models, this average will decrease once the model is overfitted. This quite general method is known as *cross validation*.

AIC can approximately be interpreted as a cross validation criterion, as explained in the following. Suppose that in each cross validation iteration only one observation (x_i) is left out to create the validation sample and the remaining observations (x_{-i}) constitute the training sample. In general, this is called *leave-one-out* cross validation. Then it turns out that the resulting cross-validated average log-likelihood

$$\hat{K}_{\text{CV}} = \frac{1}{n}\sum_{i=1}^{n} l\big(\hat{\boldsymbol{\theta}}_{\text{ML}}(x_{-i}); x_i\big) \tag{7.9}$$

is an approximately unbiased estimate of the expected mean log-likelihood (7.3). This is intuitive because

$$\mathsf{E}_g(\hat{K}_{\text{CV}}) = \frac{1}{n}\sum_{i=1}^{n} \mathsf{E}_g\big[\mathsf{E}_g\big\{\log f\big(Y; \hat{\boldsymbol{\theta}}_{\text{ML}}(X_{1:(n-1)})\big)\big\}\big] \approx \frac{1}{n} nK = K$$

for large sample sizes n. That means that both the scaled AIC (7.5) and the cross-validated average log-likelihood (7.9) are approximately unbiased estimates of K and are hence equivalent model selection criteria for large sample sizes.

Example 7.3 (Analysis of survival times) We can compute the cross-validated average log-likelihood values (7.9) for the three survival models by a simple adaptation of the R-functions to compute the log-likelihood: Each function obtains an additional `subset` argument, which gives the indices of the data points that are used to compute the log-likelihood. Looping over all observations $i = 1, \dots, n$, for each observation x_i, the MLE is computed with the remaining observations x_{-i}, and the log-likelihood is evaluated at this estimate for the observation x_i. As an example, we show R-code for the gamma model:

```
## Copy function gammaLik to gammaLik2, add an argument "subset" and
## replace "pbcTreat" by "pbcTreat[subset, , drop=FALSE]" in the
## function.
cvLogliksGamma <- numeric(nrow(pbcTreat))
for(i in seq_len(nrow(pbcTreat)))
{
    ## compute the MLE from the training sample:
    subsetMle <- optim(c(1000, 1),
                       gammaLik2,
                       log=TRUE,
                       subset=
                       setdiff(x=seq_len(nrow(pbcTreat)),
                               y=i),
                       control=list(fnscale=-1),
                       hessian=TRUE)
    stopifnot(subsetMle$convergence==0)
    subsetMle <- subsetMle$par
    ## compute the log-likelihood for the validation observation:
    cvLogliksGamma[i] <- gammaLik2(subsetMle,
                                   log=TRUE,
                                   subset=i)
}
meanCvLoglikGamma <- mean(cvLogliksGamma)
```

Averaging these values for each model as in the last R-code line, we obtain −4.5219, −4.5330 and −4.5331 for the models M_1, M_2 and M_3, respectively. On the other hand, if we scale the AIC values from Example 7.2 by dividing with $-2n$, we obtain −4.5215, −4.5320 and −4.5320. These values are very close to the cross-validated average log-likelihood values. ■

We now sketch a proof of the equivalence of the cross-validated average log-likelihood and AIC. This can also be seen as an alternative derivation of the AIC. Let $\hat{\boldsymbol{\theta}}_i = \hat{\boldsymbol{\theta}}_{\text{ML}}(X_{-i})$ denote the ML estimator in the ith cross validation iteration, and write $l_i(\boldsymbol{\theta}) = \log f(x_i; \boldsymbol{\theta})$, $\boldsymbol{S}_i(\boldsymbol{\theta}) = \frac{\partial}{\partial \boldsymbol{\theta}} l_i(\boldsymbol{\theta})$ for the log-likelihood and score vector contribution of the ith observation, respectively. Using a first-order Taylor expansion of $l_i(\hat{\boldsymbol{\theta}}_i)$ around the ML estimator $\hat{\boldsymbol{\theta}}_{\mathrm{ML}} = \hat{\boldsymbol{\theta}}_{\mathrm{ML}}(X_{1:n})$ based on the complete data $X_{1:n}$, we obtain

$$\begin{aligned} n\hat{K}_{\mathrm{CV}} &= \sum_{i=1}^{n} l_i(\hat{\boldsymbol{\theta}}_i) \\ &= \sum_{i=1}^{n} l_i(\hat{\boldsymbol{\theta}}_{\mathrm{ML}}) + \frac{\partial}{\partial \boldsymbol{\theta}^\top} l_i(\boldsymbol{\theta}_i^*)(\hat{\boldsymbol{\theta}}_i - \hat{\boldsymbol{\theta}}_{\mathrm{ML}}) \\ &= l(\hat{\boldsymbol{\theta}}_{\mathrm{ML}}) + \sum_{i=1}^{n} \boldsymbol{S}_i(\boldsymbol{\theta}_i^*)^\top (\hat{\boldsymbol{\theta}}_i - \hat{\boldsymbol{\theta}}_{\mathrm{ML}}), \end{aligned} \tag{7.10}$$

where $\boldsymbol{\theta}_i^*$ lies somewhere on the line between $\hat{\boldsymbol{\theta}}_{\mathrm{ML}}$ and $\hat{\boldsymbol{\theta}}_i$ (see Appendix B.2.3). We also use a Taylor expansion around $\hat{\boldsymbol{\theta}}_{\mathrm{ML}}$ to rewrite the score vector as

$$\begin{aligned} \boldsymbol{S}(\hat{\boldsymbol{\theta}}_i) &= \frac{\partial}{\partial \boldsymbol{\theta}} l(\hat{\boldsymbol{\theta}}_i) \\ &= \frac{\partial}{\partial \boldsymbol{\theta}} l(\hat{\boldsymbol{\theta}}_{\mathrm{ML}}) + \frac{\partial^2}{\partial \boldsymbol{\theta} \partial \boldsymbol{\theta}^\top} l(\boldsymbol{\theta}_i^*)(\hat{\boldsymbol{\theta}}_i - \hat{\boldsymbol{\theta}}_{\mathrm{ML}}) \\ &\approx \boldsymbol{0} - n\boldsymbol{G}_1(\hat{\boldsymbol{\theta}}_i - \hat{\boldsymbol{\theta}}_{\mathrm{ML}}). \end{aligned} \tag{7.11}$$

Replacement of the observed information $-\frac{\partial^2}{\partial \boldsymbol{\theta} \partial \boldsymbol{\theta}^\top} l(\boldsymbol{\theta}_i)$ at $\boldsymbol{\theta}_i = \boldsymbol{\theta}_i^*$ by the expected information $n\boldsymbol{G}_1$ at $\boldsymbol{\theta}_0$ is justified because $\boldsymbol{\theta}_i^* \to \boldsymbol{\theta}_0$ as for $n \to \infty$ due to $\hat{\boldsymbol{\theta}}_i \to \hat{\boldsymbol{\theta}}_{\mathrm{ML}} \to \boldsymbol{\theta}_0$. Since $\hat{\boldsymbol{\theta}}_i$ maximises the reduced data log-likelihood $\sum_{j \neq i} l_j(\boldsymbol{\theta}) = l(\boldsymbol{\theta}) - l_i(\boldsymbol{\theta})$, we know that its derivative evaluated at $\hat{\boldsymbol{\theta}}_i$ is zero: $\boldsymbol{S}(\hat{\boldsymbol{\theta}}_i) - \boldsymbol{S}_i(\hat{\boldsymbol{\theta}}_i) = 0$. Hence, also $\boldsymbol{S}_i(\hat{\boldsymbol{\theta}}_i)$ is approximately (7.11), which yields

$$\hat{\boldsymbol{\theta}}_i - \hat{\boldsymbol{\theta}}_{\mathrm{ML}} \approx -\frac{1}{n} \boldsymbol{G}_1^{-1} \boldsymbol{S}_i(\hat{\boldsymbol{\theta}}_i).$$

If we plug this into (7.10), we obtain

$$\begin{aligned} n\hat{K}_{\mathrm{CV}} &\approx l(\hat{\boldsymbol{\theta}}_{\mathrm{ML}}) - \frac{1}{n} \sum_{i=1}^{n} \boldsymbol{S}_i(\boldsymbol{\theta}_i^*)^\top \boldsymbol{G}_1^{-1} \boldsymbol{S}_i(\hat{\boldsymbol{\theta}}_i) \\ &= l(\hat{\boldsymbol{\theta}}_{\mathrm{ML}}) - \frac{1}{n} \sum_{i=1}^{n} \mathrm{tr}\{\boldsymbol{G}_1^{-1} \boldsymbol{S}_i(\hat{\boldsymbol{\theta}}_i) \boldsymbol{S}_i(\boldsymbol{\theta}_i^*)^\top\} && (7.12) \\ &\approx l(\hat{\boldsymbol{\theta}}_{\mathrm{ML}}) - \mathrm{tr}(\boldsymbol{G}_1^{-1} \boldsymbol{H}_1), && (7.13) \\ &\approx l(\hat{\boldsymbol{\theta}}_{\mathrm{ML}}) - p = -\frac{1}{2}\mathrm{AIC}, && (7.14) \end{aligned}$$

where we used properties of the trace operation (see Appendix B.1.1) in (7.12). In (7.13) we replaced the observed score "covariance" matrix $\boldsymbol{S}_i(\hat{\boldsymbol{\theta}}_i)\boldsymbol{S}_i(\boldsymbol{\theta}_i^*)^\top$ with the expected one $\boldsymbol{H}_1$, arguing again with the convergence of $\boldsymbol{\theta}_i^*$ and $\hat{\boldsymbol{\theta}}_i$ to $\boldsymbol{\theta}_0$. This approximation is the same as in the AIC derivation, and we again replaced the trace by the dimension p in (7.14) to arrive at the scaled AIC.

7.1.3 Bayesian Information Criterion

As an alternative, the *Bayesian information criterion*

$$\mathrm{BIC} = -2l(\hat{\boldsymbol{\theta}}_{\mathrm{ML}}) + p\log(n)$$

is frequently used, where n denotes the size of the sample. Half of the negative BIC is also known as the *Schwarz criterion.* It has the same orientation as AIC, such that models with smaller BIC are preferred. It penalises model complexity in general (i.e. if $\log(n) \geq 2 \Leftrightarrow n \geq 8$) more distinctly than AIC. A derivation of BIC is outlined in Sect. 7.2.2.

Example 7.4 (Hardy–Weinberg equilibrium) In Example 5.17 we used the LR test to test the presence of the Hardy–Weinberg equilibrium for the MN blood group frequencies from Iceland. The value of the test statistic was $W = 1.96$ with one degree of freedom, which corresponds to a P-value of 0.16. We conclude that there is no evidence against the assumption of the Hardy–Weinberg equilibrium.

The corresponding maximal values of the log-likelihood have been -754.17 in the case of the Hardy–Weinberg equilibrium with one free parameter and -753.19 in the trinomial model with two free parameters. The corresponding AIC values hence are $2 \cdot 754.17 + 2 = 1510.34$ and $2 \cdot 753.19 + 4 = 1510.38$. AIC is therefore preferring the Hardy–Weinberg equilibrium, where the difference to the trinomial model is admittedly minimal. The use of BIC leads to a more clear-cut difference. With $n = 747$ and thus $\log(n) \approx 6.62$, we obtain BIC values of $2 \cdot 754.17 + 6.62 = 1514.96$ and $2 \cdot 753.19 + 2 \cdot 6.62 = 1519.61$ for the Hardy–Weinberg and the trinomial model, respectively. The Hardy–Weinberg model is clearly preferred to the trinomial model.

Now we evaluate if the Hardy–Weinberg equilibrium with $\upsilon = 1/2$ is present or if $\upsilon \neq 1/2$. The LR statistic is 29.05 (P-value < 0.0001), so there is strong evidence against $H_0 : \upsilon = 1/2$. This is in accordance with the standard error of $\hat{\upsilon}_{\mathrm{ML}}$, which allows us to obtain a 95 % Wald confidence interval for υ. This goes from 0.545 to 0.595, so it does not contain the value $\upsilon = 0.5$. Due to $p = 0$ parameters, AIC and BIC are identical and equal to twice the negative value of the log-likelihood: AIC = BIC = 1537.39, substantially larger than for the Hardy–Weinberg equilibrium and for the saturated trinomial model. Hence in this example, both AIC and BIC support the assumption of the Hardy–Weinberg equilibrium. ■

Example 7.5 (Analysis of survival times) We can also use BIC to compare the three survival models. Since we have $n = 94$ observations, we get $p\log(n) = 4.54p$ as the penalty term instead of $2p$ for the AIC, which we used in Example 7.2. We obtain the BIC values 852.5919, 857.0953 and 857.0960 for models M_1, M_2 and M_3, respectively. Now the difference between the more complex models M_2, M_3 and the simpler model M_1 is larger due to the larger penalty factor. However, the conclusion remains unchanged. ■

BIC is very similar to AIC in that it is composed of the model fit measured by the maximised log-likelihood $l(\hat{\boldsymbol{\theta}}_{\text{ML}})$ and a penalty term incorporating the model complexity measured by the number of parameters p. However, the penalty is higher in BIC, where p is multiplied by $\log(n)$ instead of the factor 2 in the AIC. The question is now how this affects the statistical properties of AIC and BIC.

As in the derivation of the AIC, let us assume that none of the models $M_1, \dots, M_K$ under study corresponds to the true model $g(x)$ that generated the data. That means that we cannot pick a model M_k such that $f(x; \boldsymbol{\theta}, M_k) = g(x)$ for some parameter value $\boldsymbol{\theta}$. Therefore, we would like to pick a model that is closest to the truth in terms of the Kullback–Leibler discrepancy. If we use AIC or BIC to guide this decision, the probability that we pick the model reaching the minimum Kullback–Leibler discrepancy goes to one with increasing sample size n.

However, there might be multiple models that are closest to the truth. In that case, we would like to pick the model that is most parsimonious, i.e. has the smallest number of parameters p among the closest models. If we use BIC for model selection, the probability that we pick the closest and most parsimonious model goes also to one with increasing sample size n. In contrast, the AIC does not guarantee this kind of model selection consistency. That means, AIC may select models that are too complex. This overfitting tendency of the AIC is due to the penalty term's independence of the sample size n. For BIC, the complexity penalty is increasing for growing n, which guards the criterion against overfitting.

7.2 Bayesian Model Selection

In comparison to likelihood-based model selection, Bayesian model selection incorporates two additional components. First, the prior distributions on the parameters in the different models are taken into account by averaging the ordinary likelihood with respect to the prior density to obtain the so-called marginal likelihood. The ratio of marginal likelihoods of two models is the Bayes factor defined in Sect. 7.2.1. BIC can be seen as an approximation of the marginal likelihood; the asymptotic arguments are outlined in Sect. 7.2.2. The other popular criterion, the AIC, also has a Bayesian counterpart, which is described in Sect. 7.2.3. Second, the models themselves are assigned prior probabilities. After updating these probabilities with the data, posterior model probabilities are obtained, which can also be used to average over multiple models instead of selecting one of them. Model averaging is discussed in Sect. 7.2.4.

7.2.1 Marginal Likelihood and Bayes Factor

From a Bayesian point of view, it is natural to assign prior probabilities $\Pr(M_1)$ and $\Pr(M_2)$ when one has to select one of two models M_1 and M_2, where naturally $\Pr(M_1)+\Pr(M_2)=1$ has to be satisfied. After observing the data x, the question is what the *posterior model probabilities* $\Pr(M_1 \mid x)$ and $\Pr(M_2 \mid x)$ are. Using Bayes' theorem (see Appendix A.1.2), they are easily calculated as

$$\Pr(M_i \mid x) = \frac{f(x \mid M_i)\Pr(M_i)}{\sum_{j=1}^{2} f(x \mid M_j)\Pr(M_j)}, \quad i = 1, 2.$$

The *posterior odds* $\Pr(M_1 \mid x)/\Pr(M_2 \mid x)$ can hence be written as the product of the so-called *Bayes factor* $\mathrm{BF}_{12} = f(x \mid M_1)/f(x \mid M_2)$ and the *prior odds* $\Pr(M_1)/\Pr(M_2)$:

$$\frac{\Pr(M_1 \mid x)}{\Pr(M_2 \mid x)} = \frac{f(x \mid M_1)}{f(x \mid M_2)} \cdot \frac{\Pr(M_1)}{\Pr(M_2)}.$$

The Bayes factor can therefore be interpreted as the ratio of the posterior odds of M_1 and the prior odds of M_1, i.e.

$$\begin{Bmatrix} \mathrm{BF}_{12} > 1 \\ \mathrm{BF}_{12} < 1 \end{Bmatrix} \text{ if the data } x \begin{Bmatrix} \text{increased} \\ \text{decreased} \end{Bmatrix} \text{ the probability of } M_1.$$

The Bayes factor is identical to the likelihood ratio if the models M_1 and M_2 are completely specified, i.e. do not contain unknown parameters. Otherwise, the *prior predictive distribution*

$$f(x \mid M_i) = \int f(x \mid \boldsymbol{\theta}_i, M_i) \cdot f(\boldsymbol{\theta}_i \mid M_i)\, d\boldsymbol{\theta}_i, \quad i = 1, 2, \qquad (7.15)$$

has to be evaluated at the observed data x, where $\boldsymbol{\theta}_i$ is the unknown parameter vector in model M_i. This value is called the *marginal likelihood* of the model M_i.

Marginal likelihood
The marginal likelihood $f(x \mid M)$ of a model M is the value of the prior predictive distribution at the observed data x.

For discrete data x, the marginal likelihood can therefore be interpreted as the probability of the data for a given model M_i. The marginal likelihood $f(x \mid M_i)$ can be derived by integration with respect to the prior distribution from the ordinary likelihood $f(x \mid \boldsymbol{\theta}_i, M_i)$. Note that the prior distribution $f(\boldsymbol{\theta}_i \mid M_i)$ cannot be improper since otherwise $f(x \mid M_i)$ would be indeterminate.

The term *Bayes factor* has been coined by Irving John Good (1916–2009). For numerical reasons, it is often the logarithm of the marginal likelihood or the Bayes factor that is being calculated. For the interpretation of Bayes factors it is common

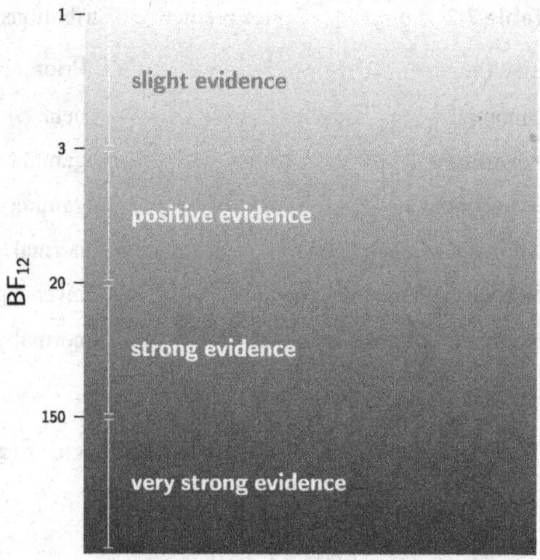

Fig. 7.1 Evidence for M_1 against M_2 for different Bayes factors BF_{12}

to use equidistant thresholds on the logarithmic scale, which form the basis of the categories shown in Fig. 7.1.

> **Bayes factor**
> The Bayes factor BF_{12} is the ratio of marginal likelihoods of two models M_1 and M_2.

The integration in (7.15) can be avoided in conjugate families. The reason for this is that $f(x) = \int f(x \mid \boldsymbol{\theta}) f(\boldsymbol{\theta})\, d\boldsymbol{\theta}$ (we leave out the additional conditioning on M_i for readability) appears as denominator in the posterior distribution

$$f(\boldsymbol{\theta} \mid x) = \frac{f(x \mid \boldsymbol{\theta}) f(\boldsymbol{\theta})}{f(x)},$$

and hence $f(x)$ is simply

$$f(x) = \frac{f(x \mid \boldsymbol{\theta}) f(\boldsymbol{\theta})}{f(\boldsymbol{\theta} \mid x)} \tag{7.16}$$

for all $\boldsymbol{\theta} \in \Theta$. Note that here the proportionality constants in $f(x \mid \boldsymbol{\theta})$, $f(\boldsymbol{\theta})$ and $f(\boldsymbol{\theta} \mid x)$ are important. However, these are known in conjugate cases, and Table 7.2 summarises the most common examples.

Example 7.6 (Hardy–Weinberg equilibrium) Under the assumption of a $\mathrm{Be}(\alpha, \beta)$ prior distribution for the parameter υ in the Hardy–Weinberg equilibrium, the posterior distribution is given as

$$\upsilon \mid \boldsymbol{x} \sim \mathrm{Be}(\alpha + 2x_1 + x_2, \beta + x_2 + 2x_3),$$

Table 7.2 Summary of prior predictive distributions in conjugate settings

Likelihood $f(x \mid \theta)$	Prior $f(\theta)$	Prior predictive $f(x)$
Binomial	beta	beta-binomial
Poisson	gamma	Poisson-gamma
Exponential	gamma	gamma-gamma
Normal (variance known)	normal	normal
Normal (variance unknown)	inverse gamma	*t* distribution
Normal (both unknown)	normal-gamma	*t* distribution

cf. Example 6.7. The marginal likelihood $f(\boldsymbol{x})$ is hence calculated easily through:

$$\begin{aligned} f(\boldsymbol{x}) &= \frac{f(\boldsymbol{x} \mid \upsilon) f(\upsilon)}{f(\upsilon \mid \boldsymbol{x})} \\ &= \frac{\frac{n!}{x_1! x_2! x_3!}(\upsilon^2)^{x_1}\{2\upsilon(1-\upsilon)\}^{x_2}\{(1-\upsilon)^2\}^{x_3} \cdot \frac{1}{\mathrm{B}(\alpha,\beta)}\upsilon^{\alpha-1}(1-\upsilon)^{\beta-1}}{\frac{1}{\mathrm{B}(\alpha+2x_1+x_2,\beta+x_2+2x_3)}\upsilon^{\alpha+2x_1+x_2-1}(1-\upsilon)^{\beta+x_2+2x_3-1}} \\ &= \frac{n!}{x_1! x_2! x_3!} \cdot \frac{2^{x_2}\,\mathrm{B}(\alpha+2x_1+x_2, \beta+x_2+2x_3)}{\mathrm{B}(\alpha, \beta)} \\ &= \frac{n!}{x_1! x_2! x_3!} \cdot \frac{2^{x_2}\Gamma(\alpha+\beta)}{\Gamma(\alpha)\Gamma(\beta)} \cdot \frac{\Gamma(\alpha+2x_1+x_2)\Gamma(\beta+x_2+2x_3)}{\Gamma(\alpha+\beta+2n)}. \end{aligned} \tag{7.17}$$

Under the assumption of a Dirichlet prior distribution, i.e. $\boldsymbol{\pi} \sim \mathrm{D}_3(\boldsymbol{\alpha})$ with $\boldsymbol{\alpha} = (\alpha_1, \alpha_2, \alpha_3)^\top$, in the general trinomial model $\boldsymbol{x} \sim \mathrm{M}_3(n, \boldsymbol{\pi})$ we obtain as a prior predictive distribution the so-called *multinomial Dirichlet distribution* with probability function

$$f(\boldsymbol{x}) = \frac{n!\Gamma(\sum_{j=1}^k \alpha_j)}{\prod_{j=1}^k \Gamma(\alpha_j)} \cdot \frac{\prod_{j=1}^k \Gamma(\alpha_j^*)}{\Gamma(\sum_{j=1}^k \alpha_j^*) \cdot \prod_{j=1}^k x_j!},$$

where $\alpha_j^* = \alpha_j + x_j$, cf. Appendix A.5.3.

Comparing the Hardy–Weinberg equilibrium model (M_1) with the general trinomial model (M_2) under $\mathrm{Be}(1, 1)$ and $\mathrm{D}_3((1, 1, 1)^\top)$ priors, respectively, we obtain a Bayes factor of $\mathrm{BF}_{12} = 4.3$, i.e. positive evidence for the simpler model M_1 of the Hardy–Weinberg equilibrium. Comparing M_1 with the model M_3 that assumes a factor $\upsilon = 1/2$ in the Hardy–Weinberg equilibrium, we obtain a Bayes factor of $\mathrm{BF}_{13} = 65\,336$ and hence very strong evidence for the more complex model M_1. In contrast to that, the use of $\mathrm{Be}(1/2, 1/2)$ and $\mathrm{D}_3((1/2, 1/2, 1/2)^\top)$ priors leads to the Bayes factors $\mathrm{BF}_{12} = 5.8$ and $\mathrm{BF}_{13} = 42\,017$, i.e. values of comparable size.

The analysis using the LR test in Example 7.4 gave a similar result: while there was no evidence for model M_2 against model M_1, there was strong evidence for model M_2 against M_3. ■

Example 7.7 (Blood alcohol concentration) In Example 6.9 we have conducted inference for the overall mean μ of the observed blood alcohol transformation factors, where the corresponding variance σ^2 was assumed to be known. Now we would like to compare both genders and calculate the posterior probability that the mean transformation factor for women (μ_1) is different than the one for men (μ_2). To this end, we compare two models, the first one having a single mean for the population and the second one having gender-specific means.

Initially, we suppose that all observations, regardless of the subgroup they belong to, are independent realisations of a normal distribution with unknown expected value μ. We denote this model of no differences between subgroups as M_1. For simplicity, we assume that the estimated standard deviation of the underlying normal distribution is known, i.e. $\kappa^{-1/2} = 237.8$; but conceptually the procedure is the same in the case of unknown variance, see Exercise 3. As a prior for μ, we choose the conjugate normal distribution with expected value $\nu = 2000$ and standard deviation $\delta^{-1/2} = 200$. This means that we expect a priori with 95-% probability an average transformation factor between 1600 and 2400. The marginal likelihood in this model can be explicitly calculated as

$$f(x_{1:n} \mid M_1) = \left(\frac{\kappa}{2\pi}\right)^{\frac{n}{2}} \left(\frac{\delta}{n\kappa + \delta}\right)^{\frac{1}{2}} \exp\left[-\frac{\kappa}{2}\left\{\sum_{i=1}^{n}(x_i - \bar{x})^2 + \frac{n\delta}{n\kappa + \delta}(\bar{x} - \nu)^2\right\}\right]. \tag{7.18}$$

For the available data, we obtain a log marginal likelihood value of $\log f(x \mid M_1) = -1279.14$.

We would like to compare the model M_1 with a model allowing different expected values in the different subgroups. In this model M_2, the data are partitioned into the two gender groups, and in both groups a conjugate prior $\mathrm{N}(\nu, \delta^{-1})$ for μ_1 and μ_2 is used. The marginal likelihood for each group is then calculated analogously to above, where of course only the data belonging to each subgroup are included in the calculation. If we suppose furthermore that μ_1 and μ_2 are a priori independent, then the marginal likelihood of the model M_2 is given as the product of the marginal likelihood values within both subgroups. For the available data, we obtain the value $\log f(x \mid M_2) = -1276.11$, i.e. a higher value than for model M_1. The Bayes factor of the model M_2 compared to the model M_1 is hence $\mathrm{BF}_{21} = \exp\{-1276.11 - (-1279.14)\} = 20.6$, strongly indicating that the model M_2 better describes the data. In other words, there is strong evidence for a difference in the average transformation factor between genders. Assigning both models the same prior probability $\Pr(M_1) = \Pr(M_2) = 0.5$, we obtain the posterior probabilities $\Pr(M_1 \mid x) = 0.0463$ and $\Pr(M_2 \mid x) = 0.9537$.

In this example, the posterior probabilities are ordered according to the number of unknown parameters in the model, i.e. higher flexibility in the model is rewarded with higher posterior probability. However, we emphasise that Bayesian model selection does not automatically favour the more complex model, see Example 7.8. This aspect will be discussed in more detail in Sect. 7.2.2. ■

Bayesian model selection strongly depends on the prior distribution for the model parameters. While for Bayesian estimation of parameters the influence of the prior distribution is asymptotically negligible for large samples, this is not true for Bayesian model selection. We already saw that the definition of the marginal likelihood does not allow improper priors. Of course, one could use proper but very vague (i.e. having large variation) priors that are still integrable. However, such an approach should be discouraged since it can be shown that for ever increasing prior variation, the posterior probability of the simplest model will always converge to 1, regardless of the information in the data. This phenomenon is known as *Lindley's paradox* and prohibits the use of vague prior distributions in Bayesian model selection.

Example 7.8 (Blood alcohol concentration) If we repeat the analysis from Example 7.7 using a prior variance of $\delta^{-1} = 10^{10}$, then the simpler model M_1 has the posterior probability 0.8360. Using $\delta^{-1} = 10^{100}$, the posterior probability of M_1 is effectively equal to 1. ■

7.2.2 Marginal Likelihood and BIC

AIC and BIC criteria are frequently used for model selection in classical approaches. Both combine the maximised log-likelihood with a term penalising the number of parameters in the model. The penalisation term receives more weight in BIC than in AIC. Model selection based on BIC is actually asymptotically equivalent to Bayesian model selection based on the marginal likelihood as described in the following.

For a model with parameter $\boldsymbol{\theta}$ of dimension p, we can write the marginal likelihood $f(x_{1:n})$ of the realisation $x_{1:n}$ from a random sample $X_{1:n}$ as

$$\begin{aligned} f(x_{1:n}) &= \int f(x_{1:n} \mid \boldsymbol{\theta}) f(\boldsymbol{\theta})\, d\boldsymbol{\theta} \\ &= \int \exp\{\log f(x_{1:n} \mid \boldsymbol{\theta}) + \log f(\boldsymbol{\theta})\}\, d\boldsymbol{\theta} \\ &= \int \exp\{-nk(\boldsymbol{\theta})\}\, d\boldsymbol{\theta}, \end{aligned}$$

where $k(\boldsymbol{\theta}) = -\{\log f(x_{1:n} \mid \boldsymbol{\theta}) + \log f(\boldsymbol{\theta})\}/n$ and $-nk(\boldsymbol{\theta})$ is the non-normalised log-posterior density. To this representation we can apply the Laplace approximation (see Appendix C.2.2). The minimum $\tilde{\boldsymbol{\theta}}$ of $k(\boldsymbol{\theta})$ is a maximum of $-nk(\boldsymbol{\theta})$ and hence equal to the posterior mode. Denoting the Hessian of k at the point $\tilde{\boldsymbol{\theta}}$ by $\boldsymbol{K}$ (see Appendix B.2.2), we obtain the following approximation for the log-marginal likelihood:

$$\begin{aligned} \log f(x_{1:n}) &\approx \log\left[\left(\frac{2\pi}{n}\right)^{\frac{p}{2}} |\boldsymbol{K}|^{-\frac{1}{2}} \exp\{-nk(\tilde{\boldsymbol{\theta}})\}\right] \\ &= \frac{p}{2}\log(2\pi) - \frac{p}{2}\log(n) - \frac{1}{2}\log|\boldsymbol{K}| + \log f(x_{1:n} \mid \tilde{\boldsymbol{\theta}}) + \log f(\tilde{\boldsymbol{\theta}}). \end{aligned}$$

The terms $p/2 \cdot \log(2\pi)$ and $\log f(\tilde{\boldsymbol{\theta}})$ can be neglected if the sample size n is large. Further, one can show that the determinant $|\boldsymbol{K}|$ of the $p \times p$ Hessian is bounded from above by a constant and is hence also negligible. The contribution of the prior to the posterior is also small for large n, so we can replace $\tilde{\boldsymbol{\theta}}$ with the MLE $\hat{\boldsymbol{\theta}}_{\text{ML}}$ (see Sect. 6.6.2). Combining these approximations, we obtain the Bayesian information criterion of Sect. 7.1.3

$$-2 \log f(x_{1:n}) \approx \text{BIC} = -2 \log f(x_{1:n} \mid \hat{\boldsymbol{\theta}}_{\text{ML}}) + p \log(n).$$

The error of this approximation is of order $O(1)$ under certain regularity conditions (see Appendix B.2.6), so is constant for increasing sample size. However, the values of the log marginal likelihood and BIC are increasing with larger sample size, so the relative error decreases with increasing sample size n:

$$\frac{2 \log f(x_{1:n}) - \text{BIC}}{2 \log f(x_{1:n})} \to 0.$$

Hence, in the same way as the prior becomes less important compared to the likelihood in its contribution to the posterior, so does the BIC approximation improve for increasing sample sizes. Therefore, selecting the model with smallest BIC is indeed asymptotically equivalent to selecting the *maximum a posteriori* (MAP) model with the largest posterior model probability.

We note that $\exp(-\text{BIC}/2)$ can be interpreted as an approximate marginal likelihood, from which posterior probabilities can be calculated. Moreover, half the difference of the BIC values of two different models is an approximation of the log Bayes factor. The Bayes factor, as a ratio of two marginal likelihoods, incorporates the parameter priors, which even for large sample sizes have a non-negligible influence on the marginal likelihood. However, it is important to recall that in the calculation of BIC these priors do not enter. Posterior probabilities derived from BIC values will therefore in general only be rough approximations of those derived from full Bayesian approaches. In the following example the agreement is rather good, though.

Example 7.9 (Blood alcohol concentration) We intend to evaluate how an approximate Bayesian model selection procedure based on BIC performs compared to an exact Bayesian approach. We obtain BIC values of 2553.6 and 2546.7 for the models M_1 and M_2 resulting in approximate log-marginal likelihood values of $f(x \mid M_1) \approx -1276.8$ and $f(x \mid M_2) \approx -1273.4$. Qualitatively, these values are similar to the values obtained in Example 7.7. Transforming the BIC values to posterior model probabilities while assuming equal prior model probabilities, we obtain $\Pr(M_1 \mid x) \approx 0.0307$ and $\Pr(M_2 \mid x) \approx 0.9693$. These values are again quite close to the results of Example 7.7.

As a comparison, the AIC values of the two models M_1 and M_2 are respectively 2550.4 and 2540.3, and hence again model M_2 is preferred. ■

We noted above that the accuracy of the BIC approximation to the log marginal likelihood is of order $O(1)$ in general. For a specific choice of a parameter prior, the accuracy is higher: Using the so-called *unit information prior*, we have $-2\log f(x) - \text{BIC} = O(n^{-1/2})$, that the approximation error gets smaller for increasing sample size. The unit information prior

$$\boldsymbol{\theta} \sim \mathrm{N}_p\big(\boldsymbol{\theta}_0, \boldsymbol{J}_1(\boldsymbol{\theta}_0)^{-1}\big),$$

contains as much information as one unit of the data, quantified by the expected unit Fisher information $\boldsymbol{J}_1(\boldsymbol{\theta}_0)$ at the prior mean $\boldsymbol{\theta}_0$. This approach can be extended to the case where the models to be compared have a common nuisance parameter.

Example 7.10 (Normal model) Assume the normal model $\mathrm{N}(\mu, \sigma^2)$ with unknown mean μ and known variance σ^2 for the random sample $X_{1:n}$, as we did in Example 6.8. From Example 2.9 we know that the expected unit Fisher information is $J_1(\mu) = 1/\sigma^2$. Hence, the unit information prior is

$$\mu \sim \mathrm{N}\big(\mu_0, \sigma^2\big)$$

and thus has the same variance as the likelihood, to which it is also conjugate. The log marginal likelihood can easily be derived from Eq. (7.18):

$$\log f(x_{1:n}) = \frac{n}{2}\log\left(\frac{\kappa}{2\pi}\right) - \frac{1}{2}\log(n+1) - \frac{\kappa}{2}\left\{\sum_{i=1}^{n}(x_i - \bar{x})^2 + \frac{n}{n+1}(\bar{x} - \mu_0)^2\right\},$$

where $\kappa = 1/\sigma^2$ denotes the precision.

We obtain BIC by plugging the MLE $\hat{\mu}_{\mathrm{ML}} = \bar{x}$ into the normal log-likelihood and accounting for the $p = 1$ parameter:

$$\text{BIC} = -2l(\hat{\mu}_{\mathrm{ML}}) + \log(n) = n\log\left(\frac{2\pi}{\kappa}\right) + \kappa\sum_{i=1}^{n}(x_i - \bar{x})^2 + \log(n).$$

The difference of the scaled log-marginal likelihood and BIC is hence

$$\begin{aligned}
-2\log f(x_{1:n}) - \text{BIC} &= \log(n+1) + \kappa\frac{n}{n+1}(\bar{x} - \mu_0)^2 - \log(n) \\
&= \log\left(\frac{n+1}{n}\right) + \kappa\frac{n}{n+1}(\bar{x} - \mu_0)^2.
\end{aligned}$$

Now, as $n \to \infty$, we have $(n+1)/n \to 1$, and thus the log term goes to zero. Moreover, $\kappa n/(n+1)$ goes to κ. Furthermore, we need the assumption that $\hat{\mu}_{\mathrm{ML}}$ converges to μ_0 at an appropriate rate, which holds if μ_0 is the true mean parameter (this assumption can also be relaxed for the alternative model). Then $(\bar{x} - \mu_0)^2$ will also be small for large enough sample size n. ■

7.2.3 Deviance Information Criterion

An alternative to BIC is the *deviance information criterion*

$$\mathrm{DIC} = -2l(\bar{\boldsymbol{\theta}}) + 2p_D,$$

where $\bar{\boldsymbol{\theta}} = \mathsf{E}(\boldsymbol{\theta} \mid y)$ is the posterior mean of the parameter vector, and p_D is an estimate of the effective number of parameters in the model. Note that this resembles very much the AIC definition in (7.1). p_D is defined as the posterior expected deviance

$$p_D = \mathsf{E}\{D(\boldsymbol{\theta}, \bar{\boldsymbol{\theta}}) \mid y\} = \int D(\boldsymbol{\theta}, \bar{\boldsymbol{\theta}}) f(\boldsymbol{\theta} \mid y)\, d\boldsymbol{\theta},$$

where

$$D(\boldsymbol{\theta}, \bar{\boldsymbol{\theta}}) = 2\{l(\bar{\boldsymbol{\theta}}) - l(\boldsymbol{\theta})\}$$

is the deviance of $\boldsymbol{\theta}$ versus the point estimate $\bar{\boldsymbol{\theta}}$. While analytic computation of DIC is rarely possible, it can easily be approximated if parameter samples, say $\boldsymbol{\theta}^{(1)}, \dots, \boldsymbol{\theta}^{(B)}$, from the posterior are available. Then $\bar{\boldsymbol{\theta}} \approx B^{-1} \sum_{b=1}^{B} \boldsymbol{\theta}^{(b)}$ is approximated by the average of the samples, and $p_D \approx B^{-1} \sum_{b=1}^{B} D(\boldsymbol{\theta}^{(b)}, \bar{\boldsymbol{\theta}})$ is approximated by the average of the sampled deviances. See Sect. 8.3 for more details.

The proximity of DIC to AIC can be established as follows. Using a second-order Taylor expansion of $D(\boldsymbol{\theta}, \bar{\boldsymbol{\theta}})$ in $\boldsymbol{\theta}$ around $\bar{\boldsymbol{\theta}}$ as e.g. in (5.4) in Sect. 5.1, we obtain the approximation

$$D(\boldsymbol{\theta}, \bar{\boldsymbol{\theta}}) \approx (\boldsymbol{\theta} - \bar{\boldsymbol{\theta}})^\top \boldsymbol{I}(\bar{\boldsymbol{\theta}})(\boldsymbol{\theta} - \bar{\boldsymbol{\theta}}).$$

Using the asymptotic normality of the posterior from Sect. 6.6.2, we have $\boldsymbol{\theta} \mid y \overset{a}{\sim} \mathrm{N}_p(\bar{\boldsymbol{\theta}}, \boldsymbol{I}(\bar{\boldsymbol{\theta}})^{-1})$ for large sample sizes. Therefore, we have

$$(\boldsymbol{\theta} - \bar{\boldsymbol{\theta}})^\top \boldsymbol{I}(\bar{\boldsymbol{\theta}})(\boldsymbol{\theta} - \bar{\boldsymbol{\theta}}) \xrightarrow{D} \boldsymbol{U}^\top \boldsymbol{G}_1^{-1} \boldsymbol{U}.$$

This is analogous to (7.7) in the AIC derivation. From the AIC derivation we know that the expected value of the right-hand side is $p^* = \mathrm{tr}(\boldsymbol{G}_1^{-1} \boldsymbol{H}_1)$. Hence, $p_D \approx p^*$ and DIC $\approx$ AIC for large sample sizes under the regularity conditions in Sect. 6.6.

Example 7.11 (Hardy–Weinberg equilibrium) We now want to use DIC to decide between the Hardy–Weinberg model and the trinomial model for the blood group frequencies $x_1 = 233$, $x_2 = 385$ and $x_3 = 129$ from Sect. 1.1.4. In Example 6.7 we have seen that the beta prior is conjugate to the Hardy–Weinberg model with log-likelihood kernel

$$l(\upsilon) = (2x_1 + x_2)\log(\upsilon) + (x_2 + 2x_3)\log(1 - \upsilon)$$

derived in Example 2.7. We choose a uniform prior for υ, i.e. $\upsilon \sim \mathrm{Be}(1, 1)$, which results in the posterior

$$\upsilon \mid \boldsymbol{x} \sim \mathrm{Be}(1 + 2x_1 + x_2, 1 + x_2 + 2x_3).$$

In Example 6.20 we have seen that the Dirichlet prior is conjugate to the trinomial likelihood

$$l(\boldsymbol{\pi}) = x_1 \log(\pi_1) + x_2 \log(\pi_2) + x_3 \log(\pi_3),$$

cf. Example 5.2. Again we choose a uniform prior on the probability simplex, i.e. $\boldsymbol{\pi} \sim \mathrm{D}_3(1, 1, 1)$. This results in the posterior

$$\boldsymbol{\pi} \mid \boldsymbol{x} \sim \mathrm{D}_3(1 + x_1, 1 + x_2, 1 + x_3).$$

In Example 8.8 we will compute the DIC values for these two models using samples from the posterior distributions. Note that the posterior expectations are analytically available because we are working with conjugate priors. Moreover, the same log-likelihood constants must be used in both models: If we directly used the log-likelihood kernel from above for the Hardy–Weinberg model, then we would miss the constant $x_2 \log(2)$, which is incorporated in the trinomial log-likelihood. Then the DIC values would no longer be on the same scale.

The resulting values, 1510.33 for the Hardy–Weinberg model and 1510.37 for the trinomial model, are very close to the AIC values determined in Example 7.4. The reason is that the uniform prior we have used has only little information compared to the information from the data. Therefore, the posterior expectations of the parameters are almost identical to the MLEs. Also, the estimated numbers of parameters ($p_D = 0.99$ and 1.99 for the two models) are almost identical to the true numbers of parameters (1 and 2) for both models. ■

7.2.4 Model Averaging

Suppose we have computed the posterior probabilities

$$\Pr(M_k \mid x) = \frac{\Pr(x \mid M_k) \Pr(M_k)}{\sum_{j=1}^{K} \Pr(x \mid M_j) \Pr(M_j)}$$

of all models $M_1, \dots, M_K$. The question arises what to do with these probabilities. The answer involves again a discussion of the notion of a true model.

The simplest perspective is that the model collection contains the true model generating the data. So we are sure that we have included the true model in our collection, but we do not know which one of the models it is. Intuitively, one can then choose the model with the highest posterior probability, the MAP-model. In that case, one only uses the probabilities $\Pr(M_k \mid x)$ for ranking the models and picks the model that is on top of the list. This approach can be justified using a decision-theoretic view. Recalling Sect. 6.4.1, which introduced loss functions and resulting optimal Bayes estimates, consider the zero–one loss function $l(a, \theta)$, where a is the model choice, and θ is the true model. If we choose the true model, i.e. $a = \theta$, then we have zero loss, and in all other cases we have a positive loss of value one. Then the Bayes estimate, which minimises the posterior expected loss, is the MAP-model.

Note that the "parameter" θ, which is in fact the model here, is discrete, so that we do not have to work with an $\varepsilon > 0$ for defining the zero–one loss function.

The implicit use of the zero–one loss function for deciding on the optimal model choice may be inappropriate if the models are close in their description of the data generating process. More application-specific loss functions could be constructed, which might lead to another model being chosen as the Bayes estimate. We do not proceed further here with this discussion, but describe an alternative approach for using the posterior model probabilities. Especially in situations where the posterior model probabilities are quite similar in size, so that no clear "winner" model arises, choosing the MAP-model and discarding the other models appears as an inappropriate approach. Do we really need to select a single model at all? What is the ultimate goal of the statistical analysis? Often an unknown quantity, say $\lambda = h(\boldsymbol{\theta})$, is the object of interest, and uncertainty statements about λ are the ultimate goal of the statistical analysis. In that case, we can simply calculate the marginal posterior distribution of λ, which is a discrete mixture of K model-specific posterior distributions with weights given by the posterior model probabilities:

$$f(\lambda \mid x) = \sum_{k=1}^{K} f(\lambda \mid x, M_k) \cdot \mathsf{Pr}(M_k \mid x).$$

This is a *Bayesian model average*. It fully takes into account the model uncertainty in the estimation of λ.

We can easily compute the model-averaged posterior expectation and variance of λ, using the law of iterated expectations and the law of total variance from Appendix A.3.4, respectively. First, we have

$$\mathsf{E}(\lambda \mid x) = \mathsf{E}\{\mathsf{E}(\lambda \mid x, M)\} = \sum_{k=1}^{K} \mathsf{E}(\lambda \mid x, M_k) \mathsf{Pr}(M_k \mid x),$$

where the outer expectation is with respect to the distribution of M given x, i.e. the posterior model distribution. Second, we have

$$\begin{aligned}
\operatorname{Var}(\lambda \mid x) &= \mathsf{E}\{\operatorname{Var}(\lambda \mid x, M)\} + \operatorname{Var}\{\mathsf{E}(\lambda \mid x, M)\} \\
&= \mathsf{E}\{\operatorname{Var}(\lambda \mid x, M) + \mathsf{E}(\lambda \mid x, M)^2\} - [\mathsf{E}\{\mathsf{E}(\lambda \mid x, M)\}]^2 \\
&= \sum_{k=1}^{K} \{\operatorname{Var}(\lambda \mid x, M_k) + \mathsf{E}(\lambda \mid x, M_k)^2\} \mathsf{Pr}(M_k \mid x) - \mathsf{E}(\lambda \mid x)^2.
\end{aligned}$$

So we can easily compute the model-averaged central moments for the quantity λ from the model-specific expectations $\mathsf{E}(\lambda \mid x, M_k)$ and variances $\operatorname{Var}(\lambda \mid x, M_k)$, using the weights $\mathsf{Pr}(M_k \mid x)$ from the model average.

Example 7.12 (Blood alcohol concentration) In Example 7.7 the model M_2 has been determined as a MAP-model. The primary interest in this example, though, is the average transformation factor μ_i in the two gender groups $i = 1, 2$. For exam-

Table 7.3 Estimated mean transformation factors in the alcohol concentration data. The models are M_1: no difference between genders and M_2: difference between genders

Gender	Transformation factor within model		
	M_1	M_2	Model-averaged
Female	2445.8	2305.5	2311.9
Male	2445.8	2473.1	2471.9

ple, in the model M_2 the posterior mean transformation factor is $\hat{\mu}_1 = 2305.5$ for females. The estimates of the average transformation factors for both models are given in Table 7.3. Additionally, the model-averaged estimates are given as well. They are quite similar to the estimates obtained under the MAP-model M_2 because of its large posterior probability. The influence of the model M_1 is negligible because it has a small posterior probability. ■

The perspective we considered so far, which assumes that the true model is contained in our chosen model space $M_1, \dots, M_k$, is known as "M-closed". It is a rather unrealistic perspective because the real world is almost always much more complex than our simple statistical models. While often the most important and prevailing features can be captured quite well with these models and we can proceed as if the truth was included in the model space, sometimes this may not be adequate. For example, we might have a more complicated model, say M_t, in which we believe in, but need to restrict ourselves for some reason to simpler models $M_1, \dots, M_K$. This is the "M-completed" perspective, under which we might evaluate all simple models in the light of our belief model M_t. The computations are more intricate because the expected losses of the possible model choices must be evaluated with respect to the model M_t. Usually, this does not allow for closed-form solutions for the Bayes estimates. The third case is the "M-open" perspective, which does not assume any belief model at all. This cautious view leads to cross-validation as the only way to evaluate the models.

While Bayesian model averaging arises naturally from integrating out the model from the posterior distribution, there are also proposals for frequentist model averaging. The weights for these frequentist model averages are usually derived from information criteria as for example AIC or BIC. For BIC, the rationale is due to the fact that BIC is an approximation of twice the log marginal likelihood of a model, see Sect. 7.2.2. Hence, we have

$$f(x \mid M_k) \approx \exp(-\text{BIC}_k/2),$$

where BIC_k denotes the BIC for the model M_k. Assuming a flat prior on the model space, i.e. $\Pr(M_k) = 1/K$, the weight w_k for the model M_k is its resulting approximate posterior model probability

$$w_k = \frac{\exp(-\text{BIC}_k/2)}{\sum_{j=1}^{K} \exp(-\text{BIC}_j/2)}.$$

There are proposals to replace BIC with AIC for calculating the weights w_k. Other information criteria can be used as well in principle. However, the analogy to Bayesian model averaging is lost, but frequentist properties of the resulting estimators can still be studied.

7.3 Exercises

1. Derive Eq. (7.18).
2. Let $Y_i \stackrel{\text{ind}}{\sim} \mathrm{N}(\mu_i, \sigma^2)$, $i = 1, \dots, n$, be the response variables in a *normal regression model*, where the variance σ^2 is assumed known, and the conditional means are $\mu_i = \boldsymbol{x}_i^\top \boldsymbol{\beta}$. The design vectors $\boldsymbol{x}_i$ and the coefficient vector $\boldsymbol{\beta}$ have dimension p and are defined as for the logistic regression model (Exercise 17 in Chap. 5).
   (a) Derive AIC for this normal regression model.
   (b) *Mallow's C_p statistic*
   $$C_p = \frac{\mathrm{SS}}{\hat{\sigma}^2_{\mathrm{ML}}} + 2p - n$$
   is often used to assess the fit of a regression model. Here $\mathrm{SS} = \sum_{i=1}^n (y_i - \hat{\mu}_i)^2$ is the *residual sum of squares*, and $\hat{\sigma}^2_{\mathrm{ML}}$ is the MLE of the variance σ^2. How does AIC relate to C_p?
   (c) Now assume that σ^2 is unknown as well. Show that AIC is given by
   $$\mathrm{AIC} = n \log(\hat{\sigma}^2_{\mathrm{ML}}) + 2p + n + 2.$$
3. Repeat the analysis of Example 7.7 with unknown variance κ^{-1} using the conjugate normal-gamma distribution (see Example 6.21) as a prior distribution for κ and μ.
   (a) First, calculate the marginal likelihood of the model by using the rearrangement of Bayes' theorem in (7.16).
   (b) Next, calculate explicitly the posterior probabilities of the two (a priori equally probable) models M_1 and M_2 using an $\mathrm{NG}(2000, 5, 1, 50000)$ distribution as a prior for κ and μ.
   (c) Evaluate the behaviour of the posterior probabilities depending on varying parameters of the prior normal-gamma distribution.
4. Let $X_{1:n}$ be a random sample from a normal distribution with expected value μ and known variance κ^{-1}, for which we want to compare two models. In the first model (M_1) the parameter μ is fixed to $\mu = \mu_0$. In the second model (M_2) we suppose that the parameter μ is unknown with prior distribution $\mu \sim \mathrm{N}(\nu, \delta^{-1})$, where ν and δ are fixed.
   (a) Determine analytically the Bayes factor BF_{12} of model M_1 compared to model M_2.
   (b) As an example, calculate the Bayes factor for the centred alcohol concentration data using $\mu_0 = 0$, $\nu = 0$ and $\delta = 1/100$.

(c) Show that the Bayes factor tends to ∞ as $\delta \to 0$ irrespective of the data and the sample size n.

5. In order to compare the models

$$M_0 : X \sim \mathrm{N}(0, \sigma^2) \quad \text{and}$$
$$M_1 : X \sim \mathrm{N}(\mu, \sigma^2)$$

with known σ^2, we calculate the Bayes factor BF_{01}.

(a) Show that

$$\mathrm{BF}_{01} \geq \exp\left\{-\frac{1}{2}z^2\right\}$$

for arbitrary prior distribution on μ, where $z = x/\sigma$ is standard normal under the model M_0. The expression $\exp(-1/2z^2)$ is called the *minimum Bayes factor* (Goodman 1999).

(b) Calculate for selected values of z the two-sided P-value $2\{1 - \Phi(|z|)\}$, the minimum Bayes factor and the corresponding posterior probability of M_0, assuming equal prior probabilities $\Pr(M_0) = \Pr(M_1) = 1/2$. Compare the results.

6. Consider the models

$$M_0 : p \sim \mathrm{U}(0, 1) \quad \text{and}$$
$$M_1 : p \sim \mathrm{Be}(\theta, 1),$$

where $0 < \theta < 1$. This scenario aims to reflect the distribution of a two-sided P-value p under the null hypothesis (M_0) and some alternative hypothesis (M_1), where smaller P-values are more likely (Sellke et al. 2001). This is captured by the decreasing density of the $\mathrm{Be}(\theta, 1)$ for $0 < \theta < 1$. Note that the data are now represented by the P-value.

(a) Show that the Bayes factor for M_0 versus M_1 is

$$\mathrm{BF}(p) = \left\{\int_0^1 \theta p^{\theta-1} f(\theta)d\theta\right\}^{-1}$$

for some prior density $f(\theta)$ for θ.

(b) Show that the minimum Bayes factor mBF over all prior densities $f(\theta)$ has the form

$$\mathrm{mBF}(p) = \begin{cases} -ep\log p & \text{for } p < e^{-1}, \\ 1 & \text{otherwise,} \end{cases}$$

where $e = \exp(1)$ is Euler's number.

(c) Compute and interpret the minimum Bayes factor for selected values of p (e.g. $p = 0.05$, $p = 0.01$, $p = 0.001$).

7. Box (1980) suggested a method to investigate the compatibility of a prior with the observed data. The approach is based on computation of a P-value obtained from the prior predictive distribution $f(x)$ and the actually observed datum x_o. Small p-values indicate a *prior-data conflict* and can be used for *prior criticism.*

   Box's p-value is defined as the probability of obtaining a result with prior predictive ordinate $f(X)$ equal to or lower than at the actual observation x_o:

$$\Pr\{f(X) \le f(x_o)\},$$

   where X is distributed according to the prior predictive distribution $f(x)$, so $f(X)$ is a random variable. Suppose that both likelihood and prior are normal, i.e. $X \mid \mu \sim \mathrm{N}(\mu, \sigma^2)$ and $\mu \sim \mathrm{N}(\nu, \tau^2)$. Show that Box's p-value is the upper tail probability of a $\chi^2(1)$ distribution evaluated at

$$\frac{(x_o - \nu)^2}{\sigma^2 + \tau^2}.$$

7.4 References

Davison (2003, Sect. 4.7) is a good overview over different aspects of model selection. A more detailed exposition is given in Claeskens and Hjort (2008) and Burnham and Anderson (2002). The original references for AIC and BIC are Akaike (1974) and Schwarz (1978), respectively, and the unit information prior interpretation of BIC is described in Kass and Wasserman (1995). Stone (1977) described the relationship of cross validation and AIC. A nice overview article on Bayesian model selection is Kass and Raftery (1995). Early contributions on minimum Bayes factors are Edwards et al. (1963) and Berger and Sellke (1987). Different perspectives on the notion of a "true" model are discussed in Bernardo and Smith (2000, Sect. 6.1.2). DIC has been proposed by Spiegelhalter et al. (2002), and frequentist model averaging with AIC weights by Buckland et al. (1997).

Numerical Methods for Bayesian Inference

8

Contents

In Chap. 6 we have combined likelihood functions with conjugate priors such that calculation of the posterior distribution was analytically possible. Similarly, Sect. 7.2 described Bayesian model choice based on closed formulas for the marginal likelihood due to the conjugacy of prior and likelihood. A potential problem in the application of Bayesian inference to more complex (non-conjugate) models is the integration necessary to compute the normalising constant of the posterior distribution in Bayes' theorem, i.e. the marginal likelihood. The calculation of certain characteristics of the posterior distribution such as the posterior mean or mode may require additional numerical techniques.

In this chapter we will discuss numerical techniques to perform such integrations. We will first describe standard (deterministic) methods, in particular the Laplace approximation. We will then move on to Monte Carlo and Markov chain Monte Carlo methods, which enable us to avoid explicit integration by simulating from the posterior distribution. Finally, we describe methods for numerical calculation of the marginal likelihood, which is a central quantity in Bayesian model selection.

L. Held, D. Sabanés Bové, *Likelihood and Bayesian Inference*,
Statistics for Biology and Health, https://doi.org/10.1007/978-3-662-60792-3_8,

8.1 Standard Numerical Techniques

Numerical integration techniques can be used if the dimension of the parameter vector is small or modest. In addition, optimisation and root finding methods are useful. Appendix C gives a summary of such techniques.

We will first sketch a general recipe to analyse a posterior distribution with numerical techniques. The following pseudo code illustrates a general strategy to calculate posterior quantities of interest. Although it uses the R language, it can easily be adapted for any other programming language.

Suppose that a specific model implies a likelihood (or log-likelihood) function $L(\theta)$ (or $l(\theta)$) and that the prior density $f(\theta)$ is also known. These functions should be implemented on the log scale:

```
log.likelihood <- function(theta, data){...}
log.prior <- function(theta){...}
```

It is often convenient to first derive the posterior mode $\text{Mod}(\theta \mid x)$ based on maximisation of the unnormalised posterior density $L(\theta) \cdot f(\theta)$ or, equivalently but numerically preferable, the unnormalised log-posterior density $l(\theta) + \log f(\theta)$. It is advisable (especially for application of the integration routines below) to allow vectors of parameter values as input for this function. We can easily achieve this with the `Vectorize` function in R:

```
log.unnorm.posterior <- function(theta, data)
    log.likelihood(theta, data) + log.prior(theta)
log.unnorm.posterior <- Vectorize(log.unnorm.posterior, "theta")
result.opt <-  optimize(log.unnorm.posterior, maximum=TRUE, data=...,
                        lower=..., upper=...)
post.mode <- result.opt$maximum
ordinate <- result.opt$objective
```

The ordinate at the posterior mode can be used to scale the unnormalised log-posterior such that it attains zero at the posterior mode and is negative everywhere else. This typically stabilises the numerical integration necessary to obtain the normalised posterior:

```
unnorm.posterior <- function(theta, data)
    exp(log.unnorm.posterior(theta, data) - ordinate)
norm.const <- integrate(unnorm.posterior, data=...,
                        lower=..., upper=...)$value
norm.posterior <- function(theta, data)
    unnorm.posterior(theta, data) / norm.const
```

Note that the limits `lower` and `upper` must span the whole support of the posterior. The posterior mean can of course also be obtained via numerical integration:

```
post.mean <- integrate(function(theta) theta * norm.posterior(theta,
                                                              data=...)
,
                       lower=..., upper=...)$value
```

The posterior distribution function and its inverse, the quantile function, can be implemented in the following scheme:

```
post.cdf <- function(x, data)
    integrate(norm.posterior, data=data,
              lower=..., upper=x)$value
post.quantile <- function(q, data)
    uniroot(function(x) post.cdf(x, data) - q,
            lower=..., upper=...)$root
```

The quantile function can then be used to compute the posterior median and equal-tailed credible intervals, e.g.:

```
post.median <- post.quantile(0.5, data=...)
post.ci <- function(gamma, data)
    c(post.quantile((1 - gamma) / 2, data),
      post.quantile((1 + gamma) / 2, data))
post.95ci <- post.ci(0.95, data=...)
```

For the calculation of HPD intervals, see Example 8.9 for an illustration.

The following two examples illustrate the application in the context of the colon cancer screening example, see Sect. 1.1.5.

Example 8.1 (Screening for colon cancer) It is not obvious if there is a conjugate prior distribution for the truncated binomial likelihood from Example 2.12. The log-likelihood was given in (2.7), and the data are $N = 6$, $n = 196$, $Z_1 = 37$, $Z_2 = 22$, $Z_3 = 25$, $Z_4 = 29$, $Z_5 = 34$ and $Z_6 = 49$. An R implementation of the log-likelihood function is taken from Example 2.12:

```
## Truncated binomial log-likelihood function
## pi: the parameter, the probability of a positive test result
## data: vector with counts Z_1, ..., Z_N
log.likelihood <- function(pi, data)
{
    n <- sum(data)
    k <- length(data)
    vec <- seq_len(k)
    result <-
        sum(data * (vec * log(pi) + (k - vec) * log(1 - pi))) -
            n * log(1 - (1 - pi)^k)
    return(result)
}
log.likelihood <- Vectorize(log.likelihood, "pi")
```

If we choose the beta prior Be(0.5, 0.5) for the sensitivity π and combine it with the likelihood, we first obtain the unnormalised log posterior:

```
log.prior <- function(pi)
    dbeta(pi, 0.5, 0.5, log=TRUE)
log.unnorm.posterior <- function(pi, data)
    log.likelihood(pi, data) + log.prior(pi)
```

We can calculate the normalising constant in Bayes' theorem numerically, which leads to the posterior density shown in Fig. 8.1.

```
## the data:
counts
[1] 37 22 25 29 34 49
## get posterior mode and its density ordinate:
result.opt <-  optimize(log.unnorm.posterior, maximum=TRUE, data=counts
 ,
                        lower=0, upper=1)
post.mode <- result.opt$maximum
post.mode
[1] 0.6241869
```

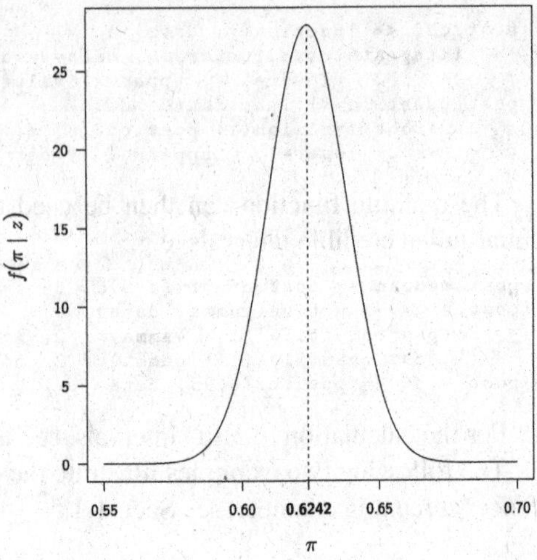

Fig. 8.1 Posterior density of π using a $\pi \sim \text{Be}(0.5, 0.5)$ prior and a truncated binomial likelihood for the colon cancer screening data

```
ordinate <- result.opt$objective
## use that to compute the normalised posterior density:
unnorm.posterior <- function(pi, data)
    exp(log.unnorm.posterior(pi, data) - ordinate)
norm.const <- integrate(unnorm.posterior, data=counts,
                        lower=0, upper=1)$value
norm.posterior <- function(pi, data)
    unnorm.posterior(pi, data) / norm.const
```

Following further our general recipe from above, we can calculate the posterior mean and median of π:

```
## posterior mean calculation as in the pseudo code:
post.mean <- integrate(function(pi) pi * norm.posterior(pi,
                                                        data=counts),
                       lower=0, upper=1)$value
post.mean
[1] 0.6239458
## likewise for the cdf and quantile functions, and hence the median:
post.cdf <- function(x, data)
    integrate(norm.posterior, data=data,
              lower=0, upper=x)$value
## numerical problems occur if we go exactly to the boundaries here,
## therefore go away some small epsilon:
eps <- 1e-10
post.quantile <- function(q, data)
    uniroot(function(x) post.cdf(x, data) - q,
            lower=0 + eps, upper=1 - eps)$root
post.median <- post.quantile(0.5, data=counts)
post.median
[1] 0.6240066
```

We find that the posterior distribution is nearly symmetric around the posterior mean 0.6239 and median 0.6240, respectively. The posterior mode 0.6242 is very close to the MLE derived in Example 2.12. To obtain the equal-tailed 95 % credible interval, we code:

```
post.ci <- function(gamma, data)
    c(post.quantile((1 - gamma) / 2, data),
      post.quantile((1 + gamma) / 2, data))
post.95ci <- post.ci(0.95, data=counts)
```

Fig. 8.2 Joint posterior density $f(\mu, \rho \mid x)$ assuming a $\mu, \rho \stackrel{\text{ind}}{\sim} \text{Be}(1, 1)$ prior

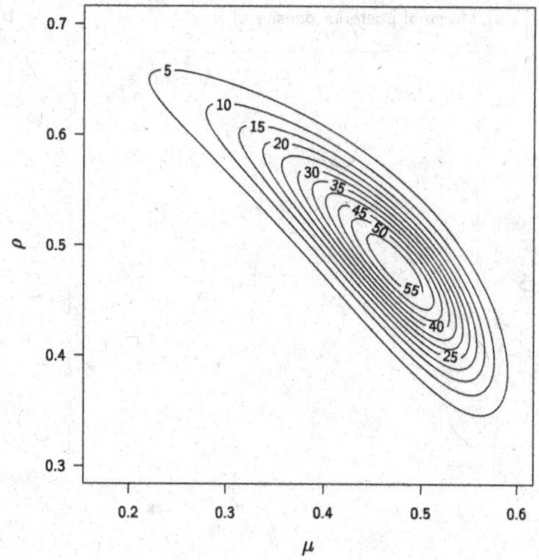


```
post.95ci
[1] 0.5956939 0.6517494
```

If the interest is in point estimates of the corresponding false negative fraction ξ, then application of the change-of-variables formula (cf. Appendix A.2.3) gives the posterior density of ξ, from which the posterior mean, say, can be calculated. The posterior median is easier to derive since it is invariant to one-to-one transformations (see Sect. 6.4.2); therefore,

$$\text{Med}(\xi \mid x) = \left\{1 - \text{Med}(\pi \mid x)\right\}^N = (1 - 0.6240)^6 = 0.0028.$$

The corresponding posterior median estimate of the number Z_0 of undetected cancer cases in the original screening study is therefore 0.56, so also very close to the MLE of Z_0, cf. Example 2.12. ■

We will now consider an example with two unknown parameters. Here numerical integration is necessary to derive the marginal posterior distributions. Since the R code follows the same scheme as in the example above, we will only show some parts of it.

Example 8.2 (Screening for colon cancer) Assuming two independent Be(1, 1) priors, i.e. uniform prior distributions for the parameters μ and ρ of the beta-binomial likelihood in Example 5.10, the joint posterior density is shown in Fig. 8.2. Computation is of course based on Bayes' theorem

$$f(\boldsymbol{\theta} \mid x) = \frac{f(x \mid \boldsymbol{\theta}) f(\boldsymbol{\theta})}{f(x)}$$

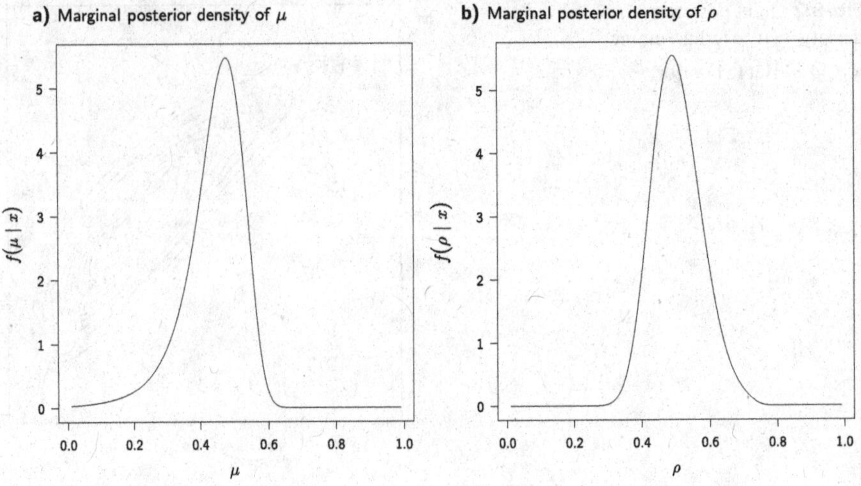

Fig. 8.3 Marginal posterior density of μ and ρ

with $\boldsymbol{\theta} = (\mu, \rho)^\top$. The denominator $f(x)$ can be calculated using two-dimensional numerical integration over $\Theta = [0, 1] \times [0, 1]$ (for example, using the package `cubature` in R, cf. Appendix C.2.1):

$$f(x) = \int_\Theta f(x, \boldsymbol{\theta})\, d\boldsymbol{\theta} = \int_\Theta f(x \,|\, \boldsymbol{\theta}) f(\boldsymbol{\theta})\, d\boldsymbol{\theta}. \tag{8.1}$$

The marginal posterior densities

$$f(\mu \,|\, x) = \int_0^1 f(\mu, \rho \,|\, x)\, d\rho \quad \text{and} \quad f(\rho \,|\, x) = \int_0^1 f(\mu, \rho \,|\, x)\, d\mu$$

can be calculated with a second numerical integration and are shown in Fig. 8.3. For example, the marginal posterior density of ρ can be calculated at grid points (which are stored in the vector `rgrid`) with the following code:

```
posterior.r <- numeric(length(rgrid))
for(j in seq_along(rgrid))
{
    posterior.r[j] <- integrate(posterior.norm.mu, myr=rgrid[j],
                                norm=norm, counts=data,
                                lower=0, upper=1,
                                rel.tol=1e-6)[["value"]]
}
```

Here `posterior.norm.mu` contains the joint posterior density function of μ and ρ, where the first argument (over which the function is integrated) is a vector of μ values, and the second argument (called `myr`) is a single value of ρ.

Bayesian inference for the false negative fraction $\xi = \mathrm{B}(\alpha, \beta + N)/\mathrm{B}(\alpha, \beta)$ is now possible with the multivariate change-of-variables formula (A.12), using the relationships

$$\alpha = \mu \cdot \frac{1-\rho}{\rho} \quad \text{and} \quad \beta = (1-\mu) \cdot \frac{1-\rho}{\rho},$$

which follow from $\mu = \alpha/(\alpha + \beta)$ and $\rho = (\alpha + \beta + 1)^{-1}$. However, this will require the partial derivatives of the beta function $\mathrm{B}(x, y)$, which may be quite involved. An alternative Monte Carlo approach is described in Example 8.14. ■

8.2 Laplace Approximation

We will now outline how to apply the Laplace approximation to calculate characteristics of the posterior distribution of an unknown scalar parameter θ. Application of the Laplace approximation to this task involves optimisation, rather than integration, which is typically much easier. Analytical results are available to study the approximation error in detail. The Laplace approximation can be applied also for multiparameter problems, which will be demonstrated in Example 8.4.

Let $f(\theta \mid x_{1:n})$ denote a posterior distribution obtained from a realisation $x_{1:n}$ from a random sample $X_{1:n}$ from some probability mass or density function $f(x \mid \theta)$. If n is large enough then, under certain regularity conditions, the posterior will be unimodal, cf. Sect. 6.6.2.

Suppose, we are interested in the posterior expectation

$$\mathsf{E}\{g(\theta) \mid x_{1:n}\} = \int g(\theta) f(\theta \mid x_{1:n})\, d\theta, \tag{8.2}$$

of a certain positive function $g(\theta)$ of the parameter θ. For example, if θ is positive and $g(\theta) = \theta$, we obtain the posterior mean $\mathsf{E}(\theta \mid x_{1:n})$. To calculate (8.2), we write

$$f(\theta \mid x_{1:n}) = \frac{f(x_{1:n} \mid \theta) f(\theta)}{\int f(x_{1:n} \mid \theta) f(\theta)\, d\theta},$$

to obtain

$$\mathsf{E}\{g(\theta) \mid x_{1:n}\} = \frac{\int g(\theta) f(x_{1:n} \mid \theta) f(\theta)\, d\theta}{\int f(x_{1:n} \mid \theta) f(\theta)\, d\theta}, \tag{8.3}$$

i.e. the ratio of two integrals. Now (8.3) can be written as

$$\mathsf{E}\{g(\theta) \mid x_{1:n}\} = \frac{\int \exp\{-nk_g(\theta)\}\, d\theta}{\int \exp\{-nk(\theta)\}\, d\theta}, \tag{8.4}$$

where

$$\begin{aligned} -nk(\theta) &= \log\{f(x_{1:n} \mid \theta)\} + \log\{f(\theta)\} \quad \text{and} \\ -nk_g(\theta) &= \log\{g(\theta)\} + \log\{f(x_{1:n} \mid \theta)\} + \log\{f(\theta)\}. \end{aligned} \tag{8.5}$$

Let $\hat{\theta}$ and $\hat{\theta}_g$ denote the locations of the minima of $k(\theta)$ and $k_g(\theta)$, respectively, i.e. the values where the terms $-nk(\theta)$ and $-nk_g(\theta)$ are maximal. Further, let

$$\hat{\kappa} = \frac{d^2 k(\hat{\theta})}{d\theta^2} \quad \text{and} \quad \hat{\kappa}_g = \frac{d^2 k_g(\hat{\theta}_g)}{d\theta^2}$$

denote the curvatures of $k(\theta)$ and $k_g(\theta)$, respectively, at the corresponding minima. Separate application of the Laplace approximation (see Appendix C.2.2) to both the numerator and denominator gives

$$\mathsf{E}\{g(\theta) \mid x_{1:n}\} \approx \sqrt{\frac{\hat{\kappa}}{\hat{\kappa}_g}} \exp\bigl[-n\{k_g(\hat{\theta}_g) - k(\hat{\theta})\}\bigr]. \tag{8.6}$$

Although the approximation error of the Laplace approximation for the two integrals in (8.4) is of order $O(n^{-1})$, the leading terms in the two errors cancel in (8.6). As a result, the error of the Laplace approximation of the posterior mean is only of order $O(n^{-2})$. Similar results can be obtained for the posterior variance.

We note that the Laplace approximation (8.6) is not invariant to changes in the parametrisation chosen for the likelihood and prior density. For example, suppose we reparametrise θ one-to-one to $\phi = h(\theta)$, i.e. $\theta = h^{-1}(\phi)$. Applying the substitution rule for integration to both numerator and denominator of (8.3), we obtain

$$\mathsf{E}\{g(\theta) \mid x_{1:n}\} = \frac{\int g\{h^{-1}(\phi)\} f\{x_{1:n} \mid h^{-1}(\phi)\} f(\phi)\, d\phi}{\int f\{x_{1:n} \mid h^{-1}(\phi)\} f(\phi)\, d\phi},$$

where $f(\phi)$ is the appropriate, Jacobian adjusted, prior density of ϕ, cf. Appendix A.2.3. If $h(\theta)$ is well selected, the Laplace approximation may become more accurate. A possible candidate for $h(\theta)$ is the variance-stabilising transformation, if available. See Example 8.3 for further illustration.

We note that a similar formula as (8.6) can be obtained for an unknown parameter vector $\boldsymbol{\theta}$ by replacing $\hat{\kappa}$ and $\hat{\kappa}_g$ with the determinants of the corresponding Hessian matrices (cf. Appendix B.2.2). As before, g must be a positive real-valued function.

Example 8.3 (Binomial model) Consider the posterior $\pi \mid x \sim \mathrm{Be}(x + 0.5, n - x + 0.5)$ resulting from combining a $\mathrm{Be}(0.5, 0.5)$ prior for the success probability π with a Bernoulli random sample of size n, which is equivalent to one binomial sample $X \sim \mathrm{Bin}(n, \pi)$. The posterior expectation is analytically known as $\mathsf{E}(\pi \mid x) = (x + 0.5)/(n + 1)$, which makes it possible to quantify the approximation error of the Laplace approximation.

To derive the Laplace approximation of $\mathsf{E}(\pi \mid x)$, we note that

$$f(x \mid \pi) = \binom{n}{x} \pi^x (1 - \pi)^{n-x}, \tag{8.7}$$

$$f(\pi) = \mathrm{B}(0.5, 0.5)^{-1} \{\pi(1 - \pi)\}^{-\frac{1}{2}} \tag{8.8}$$

and $g(\pi) = \pi$, so the log integrands in (8.4) are

$$
\begin{aligned}
-nk(\pi) &= \log\{f(\pi)\} + \log\{f(x \mid \pi)\} \\
&= -0.5\log(\pi) - 0.5\log(1-\pi) + x\log(\pi) + (n-x)\log(1-\pi) + \text{const} \\
&= (x-0.5)\log(\pi) + (n-x-0.5)\log(1-\pi) + \text{const} \quad \text{and} \\
-nk_g(\pi) &= \log(\pi) - nk(\pi) \\
&= (x+0.5)\log(\pi) + (n-x-0.5)\log(1-\pi) + \text{const},
\end{aligned}
$$

with derivatives

$$
\begin{aligned}
-n\frac{dk(\pi)}{d\pi} &= \frac{x-0.5}{\pi} - \frac{n-x-0.5}{(1-\pi)} = \frac{\pi(1-n)+x-0.5}{\pi(1-\pi)} \quad \text{and} \\
-n\frac{dk_g(\pi)}{d\pi} &= -n\frac{dk(\pi)}{d\pi} + \frac{1-\pi}{\pi(1-\pi)} = \frac{-n\pi+x+0.5}{\pi(1-\pi)},
\end{aligned}
$$

respectively. The roots of these derivatives are $\hat{\pi} = (x-0.5)/(n-1)$ and $\hat{\pi}_g = (x+0.5)/n$, respectively, and those values therefore minimise $k(\pi)$ and $k_g(\pi)$. The curvature of $k(\pi)$ at $\hat{\pi}$ is

$$
\begin{aligned}
\hat{\kappa} &= \frac{d^2k(\hat{\pi})}{d\pi^2} \\
&= -\frac{1}{n}\cdot\frac{(1-n)\hat{\pi}(1-\hat{\pi})}{\{\hat{\pi}(1-\hat{\pi})\}^2} = \frac{n-1}{n}\cdot\{\hat{\pi}(1-\hat{\pi})\}^{-1} \\
&= \frac{(n-1)^3}{n(x-0.5)(n-x-0.5)},
\end{aligned}
$$

and similarly we obtain $\hat{\kappa}_g = n^2/\{(x+0.5)(n-x-0.5)\}$. Collecting all these results, we obtain the Laplace approximation $\hat{\mathsf{E}}_1(\pi \mid x)$ of the posterior mean $\mathsf{E}(\pi \mid x)$:

$$
\begin{aligned}
&\hat{\mathsf{E}}_1(\pi \mid x) \\
&= \sqrt{\frac{(n-1)^3(x+0.5)(n-x-0.5)}{n^3(x-0.5)(n-x-0.5)}}\cdot\hat{\pi}_g\left(\frac{\hat{\pi}_g}{\hat{\pi}}\right)^{x-0.5}\left(\frac{1-\hat{\pi}_g}{1-\hat{\pi}}\right)^{n-x-0.5} \\
&= \sqrt{\frac{(n-1)^3(x+0.5)}{n^3(x-0.5)}}\cdot\frac{x+0.5}{n}\left\{\frac{(n-1)(x+0.5)}{n(x-0.5)}\right\}^{x-0.5}\left(\frac{n-1}{n}\right)^{n-x-0.5} \\
&= \frac{(x+0.5)^{x+1}(n-1)^{n+0.5}}{(x-0.5)^x n^{n+3/2}}.
\end{aligned}
$$

Table 8.1 Comparison of two Laplace approximations $\hat{\mathsf{E}}_1(\pi \mid x)$ and $\hat{\mathsf{E}}_2(\pi \mid x)$ with the true posterior mean $\mathsf{E}(\pi \mid x)$ in a binomial experiment with Jeffreys' prior. The corresponding relative error of the approximation is printed in brackets

Observation		Posterior mean		
x/n	n	$\mathsf{E}(\pi \mid x)$	$\hat{\mathsf{E}}_1(\pi \mid x)$	$\hat{\mathsf{E}}_2(\pi \mid x)$
0.6	5	0.5833	0.5630 (−0.0349)	0.5797 (−0.0062)
0.6	20	0.5952	0.5940 (−0.0021)	0.5950 (−0.0005)
0.6	100	0.5990	0.5990 (−0.0001)	0.5990 (−0.0000)
0.8	5	0.7500	0.7208 (−0.0389)	0.7464 (−0.0048)
0.8	20	0.7857	0.7838 (−0.0024)	0.7854 (−0.0004)
0.8	100	0.7970	0.7970 (−0.0001)	0.7970 (−0.0000)
1	5	0.9167	0.8793 (−0.0408)	0.9129 (−0.0041)
1	20	0.9762	0.9737 (−0.0025)	0.9759 (−0.0003)
1	100	0.9950	0.9949 (−0.0001)	0.9950 (−0.0000)

Alternatively, we can consider the Laplace approximation of the posterior mean using the variance-stabilising transformation $\phi = \arcsin(\sqrt{\pi})$ (cf. Example 4.15), which turns out to be (Exercise 2)

$$\hat{\mathsf{E}}_2(\pi \mid x) = \frac{(x+1)^{x+1} n^{n+0.5}}{x^x (n+1)^{n+3/2}}. \tag{8.9}$$

Table 8.1 compares the true posterior mean with both Laplace approximations for different values of n and x. Also shown is the relative error

$$\frac{\hat{\mathsf{E}}(\pi \mid x) - \mathsf{E}(\pi \mid x)}{\mathsf{E}(\pi \mid x)}$$

of each approximation. The reparametrised Laplace approximation is more accurate than the Laplace approximation under the original parametrisation. We can see that the accuracy of both approximations improves considerably with increasing sample size n for fixed proportions x/n. In the limit, the error of the Laplace approximation will disappear due to the asymptotic normality of the posterior distribution, under which the Laplace approximation is exact (compare Sect. 6.6.2). ■

In the previous example we could obtain analytical results for mode and curvature of the integrands of both integrals in (8.4). If this is not possible, then numerical maximisation is required. The same techniques as for maximising likelihood functions can be applied, so in R we will use the functions `optimize` or `optim`. The following example illustrates this approach.

Example 8.4 (Screening for colon cancer) In order to approximate the posterior expectation of μ and ρ in Example 8.2, we first calculate the mode and curvature of the unnormalised log-posterior density function `log.unnorm.posterior`, which represents $-nk(\boldsymbol{\theta})$ in (8.4):

```
optimObj <- optim(c(0.5, 0.5), log.unnorm.posterior, counts = data,
                  control = list(fnscale = -1), hessian = TRUE)
(mode <- optimObj$par)
[1] 0.4749234 0.4816687
curvature <- optimObj$hessian
(logDetCurvature <- as.numeric(determinant(curvature)$modulus))
[1] 11.96775
```

The variable `mode` now contains the mode $\hat{\boldsymbol{\theta}} = (0.475, 0.482)^\top$, and the variable `curvature` contains the corresponding Hessian matrix $\hat{\boldsymbol{K}}$ with log determinant $\log\{|\hat{\boldsymbol{K}}|\} = 11.968$, saved under the name `logDetCurvature`. Note that we save the original log value of the determinant to avoid loss of numerical precision below. The R-function `determinant` also returns the sign of the determinant in the list element `sign`, which we know to be positive in this case.

First, we are interested in the posterior expectation of μ, so we set $g(\boldsymbol{\theta}) = \mu$. After suitable definition of $-nk_g(\boldsymbol{\theta})$ we proceed as above:

```
log.mu.times.unnorm.posterior <- function (theta, counts)
    log (theta[1]) + log.unnorm.posterior (theta, counts)
muOptimObj <- optim(c(0.5, 0.5), log.mu.times.unnorm.posterior,
                    counts = data, control = list (fnscale = -1),
                    hessian = TRUE)
(muMode <- muOptimObj$par)
[1] 0.4841198 0.4739346
muCurvature <- muOptimObj$hessian
(muLogDetCurvature <- as.numeric(determinant(muCurvature)$modulus))
[1] 12.09874
```

The mode of this function is $\hat{\boldsymbol{\theta}}_g = (0.484, 0.474)^\top$ with log determinant of the Hessian matrix equal to $\log\{|\hat{\boldsymbol{K}}_g|\} = 12.099$. We now have all quantities necessary to calculate the multiparameter version of (8.6). We implement it on the log-scale to avoid loss of numerical precision.

$$\log\bigl[\mathsf{E}\{g(\boldsymbol{\theta})\,|\,x\}\bigr] \approx 1/2\log\{|\hat{\boldsymbol{K}}|\} - 1/2\log\{|\hat{\boldsymbol{K}}_g|\} - n\{k_g(\hat{\boldsymbol{\theta}}_g) - k(\hat{\boldsymbol{\theta}})\}.$$

```
logPosteriorExpectationMu <-
    1/2 * logDetCurvature - 1/2 * muLogDetCurvature +
    muOptimObj$value - optimObj$value
(posteriorExpectationMu <- exp(logPosteriorExpectationMu))
[1] 0.4491689
```

The posterior mean $\mathsf{E}(\mu\,|\,x) \approx 0.449$ is hence smaller than the posterior mode $\text{Mod}(\mu\,|\,x) = 0.475$, which is also reflected in the skewness of the marginal posterior $f(\mu\,|\,x)$ shown in Fig. 8.3a.

To compute the posterior mean of the second parameter ρ, we can proceed analogously and obtain the approximation $\mathsf{E}(\rho\,|\,x) \approx 0.505$, slightly larger than the corresponding mode and again in accordance with the skewness of the marginal posterior density shown in Fig. 8.3b. ∎

8.3 Monte Carlo Methods

In order to calculate characteristics of the posterior distribution, we often need to integrate certain functions. We have discussed numerical techniques and the Laplace approximation as potentially useful approaches to integration when analytic computation is not possible. The third approach is using the so-called *Monte Carlo methods*.

8.3.1 Monte Carlo Integration

Assume first that it is easy to generate independent samples $\theta^{(1)}, \dots, \theta^{(M)}$ from the posterior distribution $f(\theta \mid x)$ of interest. A *Monte Carlo estimate* of the posterior mean

$$\mathsf{E}(\theta \mid x) = \int \theta f(\theta \mid x)\, d\theta \tag{8.10}$$

is then given by

$$\hat{\mathsf{E}}(\theta \mid x) = \frac{1}{M} \sum_{m=1}^{M} \theta^{(m)}.$$

The law of large numbers (cf. Appendix A.4.3) ensures that this estimate is simulation-consistent, i.e. the estimate converges to the true posterior mean as $M \to \infty$.

This approach is called *Monte Carlo integration* and avoids the analytical integration in (8.10). More general, for any suitable function g,

$$\hat{\mathsf{E}}\{g(\theta) \mid x\} = \frac{1}{M} \sum_{m=1}^{M} g\bigl(\theta^{(m)}\bigr) \tag{8.11}$$

is a simulation-consistent estimate of $\mathsf{E}\{g(\theta) \mid x\}$, again easily shown with the law of large numbers. For example, with $g(\theta) = \theta^2$, we obtain an estimate of $\mathsf{E}(\theta^2 \mid x)$, which can be used to estimate the posterior variance: $\widehat{\mathsf{Var}}(\theta \mid x) = \hat{\mathsf{E}}(\theta^2 \mid x) - \hat{\mathsf{E}}(\theta \mid x)^2$. Another example is the indicator function $g(\theta) = \mathsf{I}_{(-\infty, \theta_0]}(\theta)$ to estimate $\Pr(\theta \leq \theta_0 \mid x)$.

The estimate $\hat{\mathsf{E}}\{g(\theta) \mid x\}$ is unbiased and has the variance

$$\mathsf{Var}\bigl[\hat{\mathsf{E}}\{g(\theta) \mid x\}\bigr] = \frac{1}{M} \int \bigl[g(\theta) - \mathsf{E}\{g(\theta) \mid x\}\bigr]^2 f(\theta \mid x)\, d\theta.$$

The associated *Monte Carlo standard error* is

$$\operatorname{se}\bigl[\hat{\mathsf{E}}\{g(\theta) \mid x\}\bigr] = \frac{1}{\sqrt{M}} \sqrt{\sum_{m=1}^{M} \bigl[g\bigl(\theta^{(m)}\bigr) - \hat{\mathsf{E}}\{g(\theta) \mid x\}\bigr]^2 / (M-1)},$$

which is of order $O(M^{-1/2})$.

We note that independent samples are not necessary for simulation consistency, which may still hold if the samples $\theta^{(1)}, \dots, \theta^{(M)}$ are dependent. However, the accuracy of a Monte Carlo estimate of $\mathsf{E}\{g(\theta) \mid x\}$ will be reduced if the transformed samples $g(\theta^{(1)}), \dots, g(\theta^{(M)})$ are positively correlated.

Example 8.5 (Binomial model) Suppose that we have obtained a $\text{Be}(4.5, 1.5)$ posterior distribution for the success probability π in a binomial model. Assume that we want to estimate the posterior expectation and the posterior probability $\Pr(\pi < 0.5 \mid x)$ by Monte Carlo sampling. Both quantities can be written in the general form $\mathsf{E}\{g(\pi) \mid x\}$ with $g(\pi) = \pi$ and $g(\pi) = \mathsf{I}_{[0,0.5)}(\pi)$, respectively. Here we generate $M = 10\,000$ independent samples, which typically gives sufficient accuracy for most quantities of interest.

```
M <- 10000
theta <- rbeta(M, 4.5, 1.5)
(Etheta <- mean(theta))
[1] 0.748382
(se.Etheta <- sqrt(var(theta)/M))
[1] 0.001649922
(Ptheta <- mean(theta < 0.5))
[1] 0.0875
(se.Ptheta <- sqrt(var(theta < 0.5)/M))
[1] 0.002825805
```

We obtain the Monte Carlo estimates $\hat{\mathsf{E}}(\pi \mid x) = 0.7484$ and $\hat{\Pr}(\pi < 0.5 \mid x) = 0.0875$ with Monte Carlo standard errors $\text{se}\{\hat{\mathsf{E}}(\pi \mid x)\} = 0.0016$ and $\text{se}\{\hat{\Pr}(\pi < 0.5 \mid x)\} = 0.0028$, respectively. The true values $\mathsf{E}(\pi \mid x) = 4.5/(1.5 + 4.5) = 0.75$ and $\Pr(\pi < 0.5 \mid x) \approx 0.087713$ (calculated using the R call `pbeta(0.5,4.5,1.5)`) are both less than two Monte Carlo standard errors away from their respective estimates. ■

Example 8.6 (Diagnostic test) We reconsider the problem to calculate the positive predictive value of a diagnostic test under the scenario that the disease prevalence $\pi = \Pr(D+)$ is estimated from a prevalence study. We now extend the approach from Example 6.4 and assume that also the sensitivity and specificity of the diagnostic test, previously both fixed at 90 %, are actually derived from a diagnostic study. Suppose that this study reported that 36 of 40 people who had the disease also had positive tests, but only 4 of 40 people who did not have the disease had positive tests. Again using Jeffreys' prior, we obtain independent $\text{Be}(36.5, 4.5)$ posteriors both for sensitivity and specificity. A Monte Carlo approach allows us to integrate this additional uncertainty in the estimation of the positive predictive value $\theta = \Pr(D+ \mid T+)$. The following R-code illustrates this:

```
M <- 10000
## prev: samples from Beta(1.5, 99.5) distribution
prev <- rbeta(n, 1.5, 104)
## first use fixed values for sensitivity and specificity
sens <- 0.9
spec <- 0.9
## and calculate positive predictive value (PPV)
ppv <- sens * prev / (sens * prev + (1-spec)*(1-prev))
## now assume distributions for sensitivity and specificity
sens <- rbeta(n, 36.5, 4.5)
```

Fig. 8.4 Boxplots of Monte Carlo samples from the positive predictive value under (**a**) fixed sensitivity and specificity and (**b**) incorporating the uncertainty attached to these estimates. In both cases the uncertainty of the prevalence estimates is taken into account. The boxplot (**a**) corresponds to Fig. 6.2

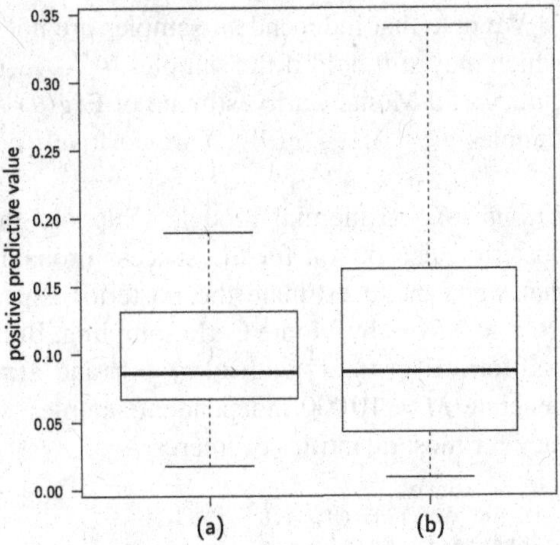


```
spec <- rbeta(n, 36.5, 4.5)
## and calculate the resulting samples for PPV
ppv2 <- sens * prev / (sens * prev + (1-spec)*(1-prev))
```

The samples are visualised as boxplots in Fig. 8.4. The boxplots illustrate that the point estimate of the positive predictive value rarely changes after incorporating the uncertainty with respect to sensitivity and specificity. However, the variance of the samples increases, so the uncertainty about this point estimate increases. In particular, larger positive predictive values become more likely. ■

Example 8.7 (Blood alcohol concentration) In Example 7.7 we have already tried to answer the question whether there exists a difference in mean transformation factors between men and women. There we computed the Bayes factor between a model with only one population mean and a model with gender-specific means. Here we take a different route: We are going to compute the posterior distribution of the difference of the gender-specific means using Monte Carlo methodology.

If we specify independent non-informative prior distributions $f(\mu_i, \sigma_i^2) \propto \sigma_i^{-2}$ $(i = 1, 2)$ in the two groups, we obtain from Example 6.26 the marginal posterior distributions of μ_i by plugging the sufficient statistics of the respective data subsets into (6.30) as

$$\mu_1 \mid x_1 \sim \mathrm{t}\big(2318.5, (38.32)^2, 32\big) \quad \text{and} \tag{8.12}$$

$$\mu_2 \mid x_2 \sim \mathrm{t}\big(2477.5, (18.86)^2, 151\big). \tag{8.13}$$

The posterior distribution of the difference $\theta = \mu_1 - \mu_2$ can now easily be estimated by simulation of M random numbers, each from (8.12), and (8.13), and subsequent computation of the differences $\theta^{(m)} = \mu_1^{(m)} - \mu_2^{(m)}$. For the simulation, we use the R-function `rst`, cf. Appendix A.5.2. In Fig. 8.5 we plot the histogram of simulated differences $\theta^{(m)}$, $m = 1, \dots, M = 10\,000$. The posterior probability $\Pr(\theta > 0 \mid x)$ that the mean transformation factor for women is smaller than for men is estimated as 0.0003, i.e. only 3 of the $M = 10\,000$ samples from $f(\theta \mid x)$ are positive.

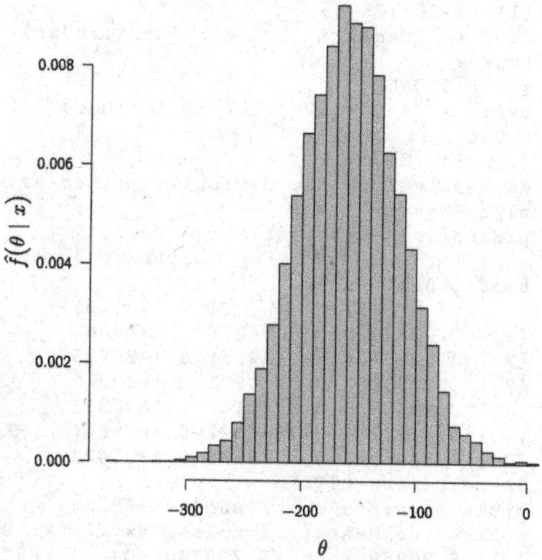

Fig. 8.5 Histogram of Monte Carlo samples $\theta^{(m)}$, $m = 1, \ldots, M$, from the posterior distribution $f(\theta \mid x)$

A rigorous frequentist analysis of this model with unequal variances in the two groups is impossible due to the famous *Behrens–Fisher problem*. However, if we assume equal variances in the two groups, we can apply the two-sample t test as we did in Example 5.16. There is an analogous Bayesian approach for this easier problem, which is an extension of the one-sample problem described in Example 6.24 to two samples. In this case the posterior distribution can be computed analytically. ■

Example 8.8 (Hardy–Weinberg equilibrium) We will now show how to compute the DIC values in Example 7.11 using Monte Carlo integration:

```
## data:
x <- c(233, 385, 129)
n <- sum(x)
## log-likelihoods:
triLoglik <- function(pi)
{
    if(is.vector(pi))
    {
        pi <- t(pi)
    }

    apply(pi, 1,
          FUN=function(onePi) sum(x * log(onePi)))
}
hwLoglik <- function(upsilon)
{
    pi <- cbind(upsilon * upsilon,
                2 * upsilon * (1 - upsilon),
                (1 - upsilon) * (1 - upsilon))
    triLoglik(pi)
}
## sample from the Hardy-Weinberg posterior:
aPost <- 1 + 2 * x[1] + x[2]
bPost <- 1 + x[2] + 2 * x[3]
upsilonSamples <- rbeta(n=10000, aPost, bPost)
## calculate DIC:
upsilonBar <- aPost / (aPost + bPost)
upsilonBar - mean(upsilonSamples) ## check: OK
```

```
[1] 0.0001126755
hwDf <- mean(2 * (hwLoglik(upsilonBar) - hwLoglik(upsilonSamples)))
hwDf
[1] 0.990882
hwDic <- - 2 * hwLoglik(upsilonBar) + 2 * hwDf
hwDic
[1] 1510.329
## sample from the Dirichlet posterior:
alphaPost <- 1 + x
piSamples <- rdirichlet(n=10000,
                        alphaPost)
head(piSamples)
          [,1]      [,2]      [,3]
[1,] 0.2889879 0.5354279 0.1755842
[2,] 0.2926704 0.5642561 0.1430735
[3,] 0.3133803 0.5250923 0.1615275
[4,] 0.2967536 0.5212148 0.1820316
[5,] 0.3427231 0.4894122 0.1678647
[6,] 0.3196069 0.5078508 0.1725423
## calculate DIC:
piBar <- alphaPost / sum(alphaPost)
piBar - colMeans(piSamples) ## check: OK
[1]  8.369696e-05 -2.725740e-04  1.888771e-04
triDf <- mean(2 * (triLoglik(piBar) - triLoglik(piSamples)))
triDf
[1] 1.988951
triDic <- - 2 * triLoglik(piBar) + 2 * triDf
triDic
[1] 1510.369
```

Note that we checked whether the Monte Carlo estimates of the posterior expectations are close to the analytically computed values: the differences were very small for both models, which indicates that the sample size ($B = 10\,000$) is large enough to obtain good accuracy for the DIC estimates. ■

It is surprising, how easy it is to obtain Monte Carlo estimates of posterior characteristics. Estimates of the posterior median and other posterior quantiles, as well as of the posterior variance etc. can be calculated analogously using the samples $\theta^{(1)}, \dots, \theta^{(M)}$. Even HPD intervals can be estimated consistently, at least for continuous parameters. To do so, we use the property of HPD intervals to have minimal length among all credible intervals at a certain fixed level. We hence calculate all possible credible intervals (based on the ordered samples) with fixed coverage of the whole sample and choose the one with minimal length as a Monte Carlo estimate of the true HPD interval. For example, suppose that $M = 100$ and we want to estimate the 95 % HPD interval. We first order the sample and obtain the ordered sample $\theta^{[1]} \leq \cdots \leq \theta^{[100]}$. Possible empirical credible intervals of 95 % empirical coverage are $[\theta^{[1]}, \theta^{[95]}]$, $[\theta^{[2]}, \theta^{[96]}]$, $[\theta^{[3]}, \theta^{[97]}]$, $[\theta^{[4]}, \theta^{[98]}]$, $[\theta^{[5]}, \theta^{[99]}]$, $[\theta^{[6]}, \theta^{[100]}]$, and we pick the one with smallest length as a Monte Carlo estimate of the 95 % HPD interval.

We note that one can roughly estimate the posterior mode by the centre of an empirical HPD interval at a very small level, say 1 %. Although this is a quite crude approach, the alternative based on kernel density estimation has also drawbacks because it depends on the bandwidth of the kernel.

Example 8.9 (Binomial model) In Example 8.5 we estimated the posterior expectations of (transformations of) the proportion π in the binomial model using Monte Carlo methodology. Now we would like to estimate the 95 % HPD interval for π by Monte Carlo and compare it to the 95 % equal-tailed credible interval:

```
## sort parameter samples
thetaorder <- theta[order(theta)]
## determine number and sizes of all possible credible intervals
M <- 10000
level <- 0.95
n.cis <- round(M * (1-level)) + 1
size <- numeric(n.cis)
for(i in seq_len(n.cis)){
    lower <- thetaorder[i]
    upper <- thetaorder[M - n.cis + i]
    size[i] <- upper - lower
}
## get the one with smallest size: the HPD interval
size.min <- which.min(size)
HPD.lower <- thetaorder[size.min]
HPD.upper <- thetaorder[M - n.cis + size.min]
## also compute the equal-tailed interval
ET.lower <- thetaorder[(M * (1-level)) / 2]
ET.upper <- thetaorder[M - (M * (1-level)) / 2]
## compare the results:
c(HPD.lower, HPD.upper)
[1] 0.435557 0.998196
> c(ET.lower, ET.upper)
[1] 0.3647148 0.9763416
```

So the 95 % HPD interval is [0.4356, 0.9982], which is shifted to the right compared with the 95 % equal-tailed credible interval [0.3647, 0.9763].

Alternatively, root finding methods (cf. Appendix C.1.2) can be used to compute the HPD interval numerically. To this end, the following R-code first defines a function outerdens, which computes the probability of all π that fulfil $f_{\text{Be}}(\pi; \alpha, \beta) < h$, i.e. the probability of all π that would not be contained in the corresponding HPD interval $\{\pi : f_{\text{Be}}(\pi; \alpha, \beta) > h\}$ (cf. Fig. 6.3). Then in the second function betaHpd, the value h_{opt} is determined that gives an outer probability of 5 %, so that the credibility level of the inner π values is 95 %. The boundaries of the HPD interval are then obtained as the values π fulfilling $f_{\text{Be}}(\pi; \alpha, \beta) = h_{\text{opt}}$.

```
outerdens <- function(h, alpha, beta)
{
    ## compute the mode of this beta distribution
    mode <- max(min((alpha-1)/(alpha + beta - 2),
                    1), 0)

    ## to compute the intersection points of the height h
    ## and the density function, only go up to epsilon
    ## to the mode:
    eps <- 1e-15

    ## compute the lower intersection point
    lower <-
        if(mode <= 0)
        {
            0             # when the density is monotonically decreasing
```

```
        } else {
            uniroot(function(x){dbeta(x, alpha, beta) - h},
                    interval = c(0, mode - eps))$root
        }

    ## compute the upper intersection point
    upper <-
        if(mode >= 1)
        {
            1           # when the density is monotonically increasing
        } else {
            uniroot(function(x){dbeta(x, alpha, beta) - h},
                    interval = c(mode + eps, 1))$root
        }

    ## compute the probability outside of the interval (lower, upper):
    prob <- pbeta(lower, alpha, beta) +
        pbeta(upper, alpha, beta, lower.tail = FALSE)

    ## return everything
    return(c(prob=prob,
             lower=lower,
             upper=upper))
}
betaHpd <- function(alpha,
                    beta,
                    level=0.95)
{
    ## compute the mode of this beta distribution,
    ## but go epsilon away from 0 or 1
    eps <- 1e-15
    mode <- max(min((alpha-1)/(alpha + beta - 2),
                    1-eps), 0+eps)

    ## determine h_opt:
    result <- uniroot(function(h){outerdens(h, alpha, beta)["prob"] -
                                      (1 - level)},
                      ## search in the interval (eps, f(mode) - eps)
                      interval =
                      c(eps,
                        dbeta(mode, alpha, beta) - eps))
    height <- result$root

    ## this gives the HPD interval
    hpd <- outerdens(height, alpha, beta)[c("lower", "upper")]
    hpd
}
alpha <- 4.5
beta <- 1.5
hpd <- betaHpd(alpha, beta)
hpd
    lower     upper
0.4359927 0.9982651
```

So the result is [0.4360, 0.9983], which is very close to the Monte Carlo estimate obtained above. The equal-tailed credible interval is [0.3714, 0.9775], which can be calculated with the R-function `qbeta`. This is slightly shifted to the right compared with the Monte Carlo result above.

Note that both credible intervals are larger than the corresponding ones in Fig. 6.3, where the uncertainty in the sensitivity and specificity estimates was not incorporated. ■

8.3.2 Importance Sampling

Importance Sampling is an extension of ordinary Monte Carlo integration. Up to now we have assumed that independent samples from the posterior $f(\theta \mid x)$ are available to estimate

$$\mathsf{E}\{g(\theta) \mid x\} = \int g(\theta) f(\theta \mid x)\, d\theta. \tag{8.14}$$

However, perhaps only samples from some other (and for the moment arbitrary) density $f^*(\theta)$ can be produced, but not from $f(\theta \mid x)$. We can re-write Eq. (8.14) to obtain

$$\mathsf{E}\{g(\theta) \mid x\} = \int g(\theta) \frac{f(\theta \mid x)}{f^*(\theta)} f^*(\theta)\, d\theta,$$

so based on the sample $\theta^{(1)}, \dots, \theta^{(M)}$ from $f^*(\theta)$, a suitable Monte Carlo estimate of (8.14) is

$$\hat{\mathsf{E}}\{g(\theta) \mid x\} = \frac{1}{\sum_{m=1}^{M} w_m} \sum_{m=1}^{M} w_m g(\theta^{(m)}), \tag{8.15}$$

where $w_m = f(\theta^{(m)} \mid x)/f^*(\theta^{(m)})$. So the importance sampling estimate of (8.14) is a weighted average of $g(\theta^{(m)})$, $m = 1, \dots, M$, with *importance weights* w_m. For $f^*(\theta) = f(\theta \mid x)$, we obtain the ordinary Monte Carlo estimate (8.11).

In expectation, the weights w_m are equal to one since

$$\int \frac{f(\theta \mid x)}{f^*(\theta)} f^*(\theta)\, d\theta = \int f(\theta \mid x)\, d\theta = 1,$$

and therefore $\sum_{m=1}^{M} w_m \approx M$. An alternative importance sampling estimate, sometimes reported in the literature, is therefore (8.15) with $1/\sum_{m=1}^{M} w_m$ replaced by $1/M$. The difference between these two estimates is usually small.

It can be shown that the importance sampling estimate (8.15) of $\mathsf{E}\{g(\theta) \mid x\}$ is unbiased and has the *Monte Carlo standard error*

$$\text{se}\left[\hat{\mathsf{E}}\{g(\theta) \mid x\}\right] = \frac{1}{\sum_{m=1}^{M} w_m} \sqrt{\sum_{m=1}^{M} w_m^2 \left[g(\theta^{(m)}) - \hat{\mathsf{E}}\{g(\theta) \mid x\}\right]^2}.$$

Example 8.10 (Binomial model) Let us reconsider Monte Carlo estimation of the posterior mean $\mathsf{E}(\pi \mid x)$ and the posterior probability $\Pr(\pi < 0.5 \mid x)$ from Example 8.5. We now want to use importance sampling based on independent samples from a standard uniform distribution. Computation of (8.15) is done with the following R-code:

```
u <- runif(M)
w <- dbeta(u, 4.5, 1.5)
(sum(w))
```

```
[1] 10151.34
(Etheta.u <- sum(u * w) / sum(w))
[1] 0.7494834
(se.Etheta.u <- sqrt(sum((u - Etheta.u)^2 * w^2)) / sum(w)
[1] 0.001780926
(Ptheta.u <- sum((u < 0.5) * w) / sum(w))
[1] 0.08824196
(se.Ptheta.u <- sqrt(sum(((u < 0.5) - Ptheta.u)^2 * w^2)) / sum(w)
[1] 0.00211711
```

First, note that the sum of the weights is 10151.3, so as expected quite close to the number of samples $M = 10\,000$. The importance sampling estimates appear to be also of quite good accuracy with standard errors similar to ordinary Monte Carlo. Note that the standard error for the mean estimate is slightly larger than with ordinary Monte Carlo, while the standard error for the $\Pr(\pi < 0.5 \mid x)$ estimate is slightly smaller. ■

In the previous example, we noted a decrease in the standard error for estimating the probability $\Pr(\pi < 0.5 \mid x)$ when importance sampling from the uniform distribution is used. This illustrates the possibility to improve the Monte Carlo precision by choosing a suitable importance sampling distribution $f^*(\theta)$. If the density $f^*(\theta)$ is approximately proportional to $|g(\theta)| f(\theta \mid x)$, then the importance sampling estimate (8.15) can be shown to have minimal variance. Note that, when $g(\theta) > 0$, the proportionality constant for the normalisation of this optimal $f^*(\theta)$ is $1/\int g(\theta) f(\theta \mid x)\, d\theta$, the reciprocal of the quantity we want to estimate. Unless we can simulate from $g(\theta) f(\theta \mid x)$ without knowledge of the normalising constant, this result seems therefore of limited practical value. For more details, we point the interested reader to the relevant literature listed at the end of this chapter.

8.3.3 Rejection Sampling

Importance sampling offers a way to estimate posterior expectations of certain functions of the parameter θ but does not provide samples from the posterior distribution of interest. Such samples are, however, necessary for certain posterior characteristics such as posterior quantiles or HPD intervals. We want to discuss a general approach to generate samples from some target distribution with density $f_X(x)$, say, without actually sampling from $f_X(x)$. The algorithm is called *rejection sampling*. Our target $f_X(x)$ is usually the posterior distribution of some scalar parameter, but the technique can be applied more widely. The goal is to effectively simulate X from $f_X(x)$ using two independent random numbers $U \sim \mathrm{U}(0, 1)$ and Z with density $f_Z(z)$. The *proposal distribution* $f_Z(z)$ can be chosen arbitrarily under the assumption that there exists an $a \geq 1$ with

$$f_X(z) \leq a \cdot f_Z(z) \quad \text{for all } z \in \mathbb{R}. \tag{8.16}$$

The rejection sampling algorithm to simulate X from $f_X(x)$ hence proceeds as follows:

1. Generate independent random variables Z from $f_Z(z)$ and $U \sim \mathrm{U}(0, 1)$.

2. If $U \leq f_X(Z)/\{a \cdot f_Z(Z)\}$, set $X = Z$ (*acceptance step*).
3. Otherwise, go back to 1 (*rejection step*).

The quantity $\alpha(z) = f_X(z)/\{a \cdot f_Z(z)\}$ is called the *acceptance probability* since the realisation $Z = z$ from $f_Z(z)$ is accepted as a random number from $f_X(x)$ with this probability.

To understand the principle of rejection sampling, consider first

$$\begin{aligned} &\Pr\{Z \leq x \mid a \cdot U \cdot f_Z(Z) \leq f_X(Z)\} \\ &= \frac{\Pr\{Z \leq x, a \cdot U \cdot f_Z(Z) \leq f_X(Z)\}}{\Pr\{a \cdot U \cdot f_Z(Z) \leq f_X(Z)\}} \\ &= \frac{\int_{-\infty}^{x} \Pr\{a \cdot U \cdot f_Z(Z) \leq f_X(Z) \mid Z = z\} f_Z(z)\, dz}{\int_{-\infty}^{+\infty} \Pr\{a \cdot U \cdot f_Z(Z) \leq f_X(Z) \mid Z = z\} f_Z(z)\, dz}. \end{aligned} \tag{8.17}$$

Now

$$\Pr\{a \cdot U \cdot f_Z(Z) \leq f_X(Z) \mid Z = z\} = \Pr\left\{U \leq \frac{f_X(z)}{a \cdot f_Z(z)}\right\} = \frac{f_X(z)}{a \cdot f_Z(z)}$$

because $f_X(z)/\{a \cdot f_Z(z)\} \leq 1$ and $U \sim \mathrm{U}(0, 1)$ by assumption. Equation (8.17) is therefore equal to

$$\frac{\int_{-\infty}^{x} \frac{f_X(z)}{a \cdot f_Z(z)} f_Z(z)\, dz}{\int_{-\infty}^{+\infty} \frac{f_X(z)}{a \cdot f_Z(z)} f_Z(z)\, dz} = \frac{\int_{-\infty}^{x} f_X(z)\, dz}{\int_{-\infty}^{+\infty} f_X(z)\, dz} = \int_{-\infty}^{x} f_X(z)\, dz = F_X(x),$$

i.e. conditional on the event $E = \{a \cdot U \cdot f_Z(Z) \leq f_X(Z)\}$, the random variable Z has a distribution function F_X with associated density f_X. Every pair (U, Z) fulfils the condition E with probability a^{-1}:

$$\Pr\{a \cdot U \cdot f_Z(Z) \leq f_X(Z)\} = \int_{-\infty}^{\infty} \frac{f_X(z)}{a \cdot f_Z(z)} f_Z(z)\, dz = \int_{-\infty}^{\infty} \frac{f_X(z)}{a}\, dz = a^{-1}.$$

The single trials are independent, so the number of trials up to the first success is geometrically distributed with parameter $1/a$. The expected number of trials up to the first success is therefore a; if a is large, the algorithm is hence not very efficient. The constant a should thus be chosen as small as possible while fulfilling condition (8.16), so typically $a = \max_{z \in \mathbb{R}} f_X(z)/f_Z(z)$, as in the following example.

Example 8.11 (Binomial model) To simulate from the posterior distribution $\theta \mid x \sim \mathrm{Be}(4.5, 1.5)$, as we already did in Example 8.10, we now use rejection sampling based on a uniform proposal distribution. First, we need to determine the constant a. Due to $f_Z(\theta) = 1$, the ordinate of the target density $f(\theta \mid x)$ at its mode $\mathrm{Mod}(\theta \mid x) = (4.5 - 1)/(4.5 + 1.5 - 2) = 7/8$ can be used:

```
## posterior parameters and mode:
alpha <- 4.5
beta <- 1.5
```

```
mode <- (alpha - 1) / (alpha + beta - 2)
a <- dbeta(mode, alpha, beta)
## number of samples to be produced:
M <- 10000
## vector where the samples will be stored:
theta <- numeric(M)
## also save the number of trials for each sample:
trials <- numeric(M)
## for each sample:
for(m in seq_along(theta))
{
    k <- 0
    while(TRUE)
    {
        k <- k + 1
        ## sample random variables
        u <- runif(1)
        z <- runif(1)
        ## check for acceptance, then exit the loop
        if(u <= dbeta(z, alpha, beta) / a)
            break
    }
    ## save the z realisation as a theta sample
    theta[m] <- z
    ## and the number of trials
    trials[m] <- k
}
## average number of trials required:
mean(trials)
[1] 2.5756
## estimate posterior mean of theta:
(Etheta <- mean(theta))
[1] 0.7506988
(se.Etheta <- sqrt(var(theta) / M))
[1] 0.001636567
## estimate P(theta < 0.5 | y):
(Ptheta <- mean(theta < 0.5))
[1] 0.0888
(se.Ptheta <- sqrt(var(theta < 0.5) / M))
[1] 0.002844691
```

The estimates of the true values 0.75 and 0.0877, respectively, appear to be also quite good using rejection sampling. This is not surprising, as the resulting samples have the same distributional properties as if we had directly used the R-function `rbeta` as in Example 8.5. Hence, also the standard errors have a similar size. ■

8.4 Markov Chain Monte Carlo

Application of ordinary Monte Carlo methods is difficult if the unknown parameter is of high dimension. However, *Markov chain Monte Carlo* (MCMC) methods will then be a useful alternative. The idea is to simulate a *Markov chain* $\boldsymbol{\theta}^{(1)}, \dots, \boldsymbol{\theta}^{(m)}, \dots,$ (see Sect. 10.1) which is designed in a way such that it converges to the posterior distribution $f(\boldsymbol{\theta} \mid x)$. After convergence we obtain samples from the posterior distribution, which can be used to estimate posterior characteristics as described in Sect. 8.3. However, these samples will typically be dependent, an inherent feature of Markov chains.

Similar to rejection sampling, there is great liberty in the actual design of an MCMC algorithm. In each iteration m, samples are generated from some *proposal distribution*, which can depend on the current state $\boldsymbol{\theta}^{(m)}$ of the Markov chain. Let $f^*(\boldsymbol{\theta} \mid \boldsymbol{\theta}^{(m)})$ denote the probability mass or density function of this proposal distribution. The *Metropolis–Hastings algorithm*, a very general MCMC technique, accepts the proposal $\boldsymbol{\theta}^*$ from $f^*(\boldsymbol{\theta} \mid \boldsymbol{\theta}^{(m)})$ as the new state of the Markov chain with probability

$$\alpha = \min\left\{1, \frac{f(\boldsymbol{\theta}^* \mid x)}{f(\boldsymbol{\theta}^{(m)} \mid x)} \cdot \frac{f^*(\boldsymbol{\theta}^{(m)} \mid \boldsymbol{\theta}^*)}{f^*(\boldsymbol{\theta}^* \mid \boldsymbol{\theta}^{(m)})}\right\}. \tag{8.18}$$

If the proposal is accepted, then $\boldsymbol{\theta}^{(m+1)} = \boldsymbol{\theta}^*$, otherwise $\boldsymbol{\theta}^{(m+1)} = \boldsymbol{\theta}^{(m)}$, i.e. the proposal $\boldsymbol{\theta}^*$ is rejected. The term $f(\boldsymbol{\theta}^* \mid x)/f(\boldsymbol{\theta}^{(m)} \mid x)$ in (8.18) is the posterior ratio while $f^*(\boldsymbol{\theta}^{(m)} \mid \boldsymbol{\theta}^*)/f^*(\boldsymbol{\theta}^* \mid \boldsymbol{\theta}^{(m)})$ is the proposal ratio. The *acceptance probability* (8.18) is the product of the posterior ratio and the proposal ratio, suitably truncated to the unit interval. Under certain regularity conditions, one can show that this algorithm converges to the target distribution $f(\boldsymbol{\theta} \mid x)$, regardless of the specific choice of the proposal distribution $f^*(\boldsymbol{\theta} \mid \boldsymbol{\theta}^{(m)})$. Specification of the regularity conditions requires some insight into limit theory for Markov chains. This is beyond the scope of this book, but at the end of this chapter we give some suitable references for readers interested in MCMC theory.

The speed of convergence and the dependence between successive samples will depend heavily on the choice of the proposal distribution. It is very important in practice to have a good proposal distribution for the specific model at hand. Some special cases of the general Metropolis–Hastings algorithm have special names. If the proposal distribution is symmetric around the current value, i.e. $f^*(\boldsymbol{\theta}^{(m)} \mid \boldsymbol{\theta}^*) = f^*(\boldsymbol{\theta}^* \mid \boldsymbol{\theta}^{(m)})$, one obtains the *Metropolis algorithm* with acceptance probability

$$\alpha = \min\left\{1, \frac{f(\boldsymbol{\theta}^* \mid x)}{f(\boldsymbol{\theta}^{(m)} \mid x)}\right\}.$$

A special case of this is the so-called *random walk proposal*, which is defined as the current value $\boldsymbol{\theta}^{(m)}$ plus a random number variate of a zero-centred symmetric distribution. If the proposal density does not depend on $\boldsymbol{\theta}^{(m)}$, i.e. $f^*(\boldsymbol{\theta} \mid \boldsymbol{\theta}^{(m)}) = f^*(\boldsymbol{\theta})$, then the proposal is called the *independence proposal*.

Another special case leads to the acceptance probability α always equal to one. This is the case if $f^*(\boldsymbol{\theta}^* \mid \boldsymbol{\theta}^{(m)}) = f(\boldsymbol{\theta}^* \mid x)$, i.e. if the proposal density is equal to the posterior density, the target density. At first sight this appears to be of limited value, as we implicitly assumed that direct sampling from the target density is unavailable. However, α will also equal unity if a specific component θ_j of $\boldsymbol{\theta}$ is updated by a sample from its *full conditional distribution* $f(\theta_j \mid x, \boldsymbol{\theta}_{-j})$, where $\boldsymbol{\theta}_{-j}$ denotes the vector $\boldsymbol{\theta}$ without the component θ_j. Because $f(\theta_j \mid x, \boldsymbol{\theta}_{-j}) \propto f(\boldsymbol{\theta} \mid x)$, $j = 1, \dots, p$, the acceptance probability α is still one in this case. Iteratively updating all p components of $\boldsymbol{\theta}$ with samples from their corresponding full conditionals is called *Gibbs sampling*. The approach can be adopted to updating multidimensional blocks of scalar parameters (not just single parameters) of $\boldsymbol{\theta}$ from their respective full conditional distributions.

We note that care must be taken when an improper prior distribution is used, because this may lead to an improper posterior distribution. Impropriety implies that there does not exist a joint density to which the full-conditional distributions correspond. However, the Gibbs sampling output might still look as if the approach was working.

The efficiency of the Metropolis–Hastings algorithm depends crucially on the acceptance rate, i.e. the relative frequency of acceptance (typically assessed after convergence of the Markov chain). However, an acceptance rate close to one is not always good. For example, for random walk proposals, a too large acceptance rate implies that the proposal density is too close around the current value, so the algorithm needs many small steps to explore the target distribution sufficiently. On the other hand, if the acceptance rate of a random walk proposal is too small, large moves are often proposed but rarely accepted. In some cases, the algorithm may even get stuck at a specific value, and subsequent proposals will get rejected for a large number of iterations. For random walk proposals, acceptance rates between 30 and 50 % are typically recommended, which can be easily achieved through appropriate choice of the variance of the proposal distribution. Things are different for independence proposals, where a high acceptance rate is desired, which means that the proposal density is close to the target density.

Example 8.12 (Screening for colon cancer) We now want to apply Gibbs sampling to the problem described in Sect. 1.1.5 and also to compare the likelihood approaches in Example 2.12 and Sect. 2.3.2. For the probability π, we choose a $\text{Be}(0.5, 0.5)$ prior, so the full conditional of π, is

$$\pi \mid \mathbf{Z} \sim \text{Be}\left(0.5 + \sum_{k=0}^{6} k \cdot Z_k, 0.5 + \sum_{k=0}^{6} (6-k) Z_k\right). \tag{8.19}$$

However, we do not know Z_0, but we do know the full conditional distribution of Z_0 for known π (cf. Example 2.12):

$$Z_0 \mid \pi \sim \text{NBin}\big(n, 1-(1-\pi)^N\big) - n, \tag{8.20}$$

where $N = 6$ and $n = 196$. This suggests to implement a Gibbs sampler in R by iteratively simulating from (8.19) and (8.20). Here is the R-code:

```
## data set:
fulldata <- c(NA, 37, 22, 25, 29, 34, 49)
k <- 0:6
n <- sum(fulldata[-1])
## impute start value for Z0 (first element):
fulldata[1] <- 10
## MCMC settings:
nburnin <- 100
niter <- 10000
## where the samples will be saved:
pisamples <- Z0samples <- numeric(niter)
## set the random seed:
set.seed(920)
## do the sampling:
```

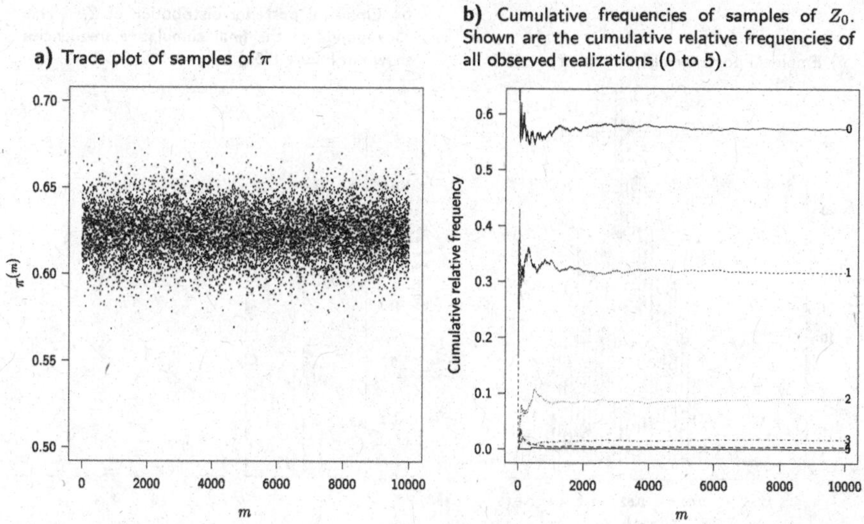

Fig. 8.6 Paths of the Gibbs sampling Markov chain (without burn-in of first 100 iterations). Both plots suggest the convergence of the Markov chain

```
for(i in (- nburnin + 1):niter)
{
    ## draw pi from full conditional
    pi <- rbeta(1, 0.5 + sum(k * fulldata),
                0.5 + sum((6 - k) * fulldata))
    ## draw Z0 from full conditional
    fulldata[1] <- rnbinom(1, size=n, prob=1-(1-pi)^6)
    ## if the burn-in iterations have passed, save the samples
    if(i>0)
    {
        pisamples[i] <- pi
        Z0samples[i] <- fulldata[1]
    }
}
```

Note that the R function `rnbinom` generates a random number from a differently parametrised negative binomial distribution, where the number of non-successes and not the number of trials up to the first success is counted; the latter one is used in the definition of Z_0.

In practice one has to decide when the simulated Markov chain has reached its target distribution. A common approach is to ignore the first few iterations, the so-called *burn-in phase*, so the remaining random numbers can be regarded as samples from the target distribution. In the above code, the `nburnin` first samples are ignored as burn-in. One should still inspect the trace plots of the parameter samples to ensure that the samples at least "visually" converge. Figure 8.6 suggests that this is the case here. We can now compute summary statistics of posterior characteristics of interest, e.g. estimates of quantiles and expectations.

```
summary(pisamples)
   Min. 1st Qu.  Median    Mean 3rd Qu.    Max.
 0.5683  0.6142  0.6240  0.6240  0.6337  0.6776
```

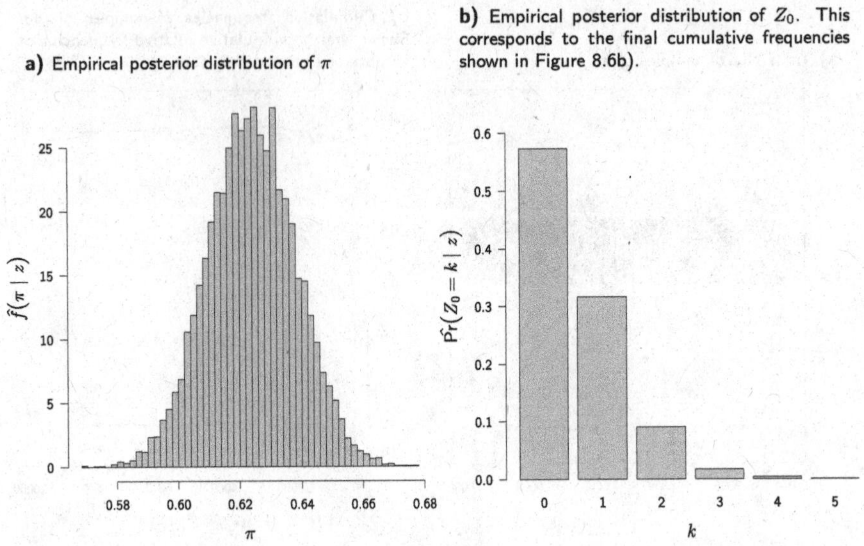

Fig. 8.7 Estimated posterior distributions of π and Z_0

```
summary(Z0samples)
   Min. 1st Qu.  Median    Mean 3rd Qu.    Max.
 0.0000  0.0000  0.0000  0.5638  1.0000  5.0000
```

The estimated marginal posterior distributions of π and Z_0 are shown as histograms in Fig. 8.7. ■

Application of Gibbs sampling requires that random numbers from the full conditional distributions can be generated easily. However, this may not always be the case, so alternative approaches need to be investigated. One idea is to use rejection sampling or other techniques to obtain samples from the full conditional distribution indirectly. However, it is typically easier to use a simple Metropolis–Hastings proposal to update the corresponding component of $\boldsymbol{\theta}$. The latter approach is sometimes called *Metropolis-within-Gibbs*. We illustrate the approach with the following example.

Example 8.13 (Hardy–Weinberg equilibrium) To avoid the assumptions underlying Hardy–Weinberg equilibrium as discussed in Sect. 1.1.4, sometimes a more general model allowing for *Hardy–Weinberg disequilibrium* is used. This introduces a second *disequilibrium parameter* δ in the definition of the probabilities:

$$\pi_1 = \upsilon^2 + \delta, \qquad \pi_2 = 2\upsilon(1-\upsilon) - 2\delta \quad \text{and} \quad \pi_3 = (1-\upsilon)^2 + \delta.$$

If $\delta = 0$, one obtains the usual Hardy–Weinberg equilibrium with parameter υ. Clearly, this extended formulation ensures that the probabilities π_1, π_2 and π_3 add

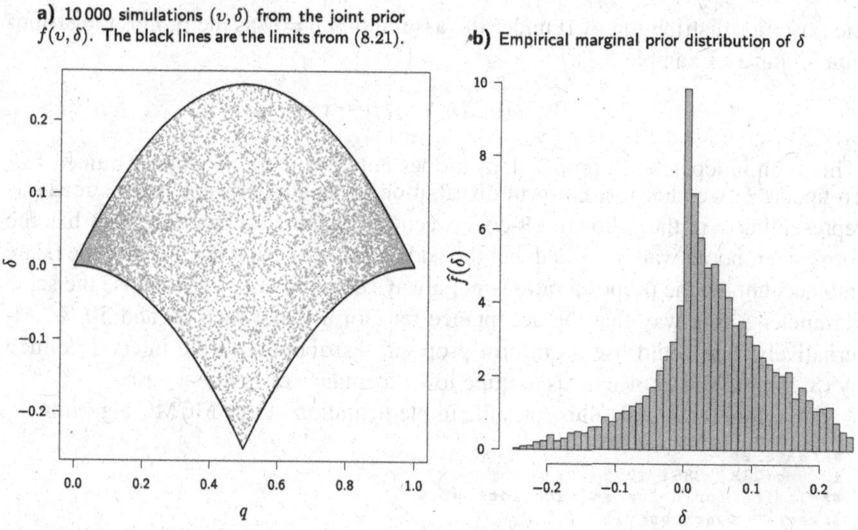
a) 10 000 simulations (υ, δ) from the joint prior $f(\upsilon, \delta)$. The black lines are the limits from (8.21).
b) Empirical marginal prior distribution of δ
0.2
0.1
δ 0.0
−0.1
−0.2
0.0
0.2
0.4
0.6
0.8
1.0
q
10
8
6
4
2
0
$f(\delta)$
−0.2
−0.1
0.0
0.1
0.2
δ

Fig. 8.8 10 000 independent samples from the prior $f(\upsilon, \delta)$ with $\alpha = \beta = 1$, i.e. a marginal uniform prior for υ

up to one, and the additional restriction

$$\max\{-\upsilon^2, -(1-\upsilon)^2\} \leq \delta \leq \upsilon(1-\upsilon) \tag{8.21}$$

ensures that the individual probabilities are all within the unit interval. The corresponding likelihood function of υ and δ is of course only a reparametrisation of a multinomial likelihood. This reparametrisation is useful to investigate the presence of Hardy–Weinberg equilibrium, represented by the null hypothesis $H_0 : \delta = 0$.

A Bayesian analysis starts by choosing appropriate prior distributions for υ and δ. Due to (8.21), these distributions cannot be independent, and we therefore factorise the prior in the form

$$f(\upsilon, \delta) = f(\upsilon) f(\delta \mid \upsilon) \tag{8.22}$$

and select the $\mathrm{Be}(\alpha, \beta)$ distribution for the marginal prior of υ and the uniform distribution on the interval (8.21) for the conditional prior of δ given υ.

Figure 8.8a illustrates for $\alpha = \beta = 1$ the dependence between the two parameters using 10 000 simulations from the prior by first simulating υ^* from $f(\upsilon)$ and subsequently δ^* from $f(\delta \mid \upsilon^*)$. Via (8.22), this gives samples from the joint distribution $f(\upsilon, \delta)$ as well as from the marginal distribution $f(\delta)$. Figure 8.8b shows that the marginal prior of δ is not at all uniform and not even symmetric around zero. The prior probability $\Pr(\delta > 0)$ is approximately 0.75, estimated based on the samples.

Turning to posterior inference based on the MN blood group frequencies in Iceland as described in Sect. 1.1.4, note that the full conditional distributions of υ and δ are not of closed form and not easy to sample from. We therefore use Metropolis–Hastings proposals for both conditional distributions. For υ, it may be useful to use the posterior distribution of υ under the assumption $\delta = 0$ as the proposal distribution (compare Example 6.7):

$$\upsilon \mid \boldsymbol{x} \sim \text{Be}(\alpha + 2x_1 + x_2, \beta + x_2 + 2x_3).$$

This is an independence proposal, as it does not depend on the current value of υ. To update δ, we choose a uniform distribution of length $2s$ (the variable `scale` is representing s in the following R-code), centred at the current value. This has the form of a random walk proposal, but the additional restriction (8.21) has to be taken into account, so the proposal ratio is not always equal to unity. We choose the scale parameter s in a way that the acceptance rate for δ is between 30 and 50 %. Alternatively, one could use a uniform proposal distribution on the interval defined by (8.21), but this choice leads to quite low acceptance rates in our case.

The following R-code illustrates the implementation of this MCMC algorithm.

```
## data set
x <- c(233, 385, 129)
## limits for delta as functions of v
lower <- function(v)
{
    lower <- pmax(-v^2, -(1-v)^2)
    return(lower)
}
upper <- function(v)
{
    upper <- v * (1 - v)
    return(upper)
}
## function to compute probabilities from v and d
myprob <- function(v, d)
{
    p1 <- v^2 + d
    p2 <- 2 * v * (1-v) - 2 * d
    p3 <- (1-v)^2 + d
    p <- c(p1, p2, p3)
    ## if the result is valid, return it,
    ## otherwise NAs:
    if(all(p >= 0) & all(p <= 1))
    {
        return(p)
    } else {
        return(rep(NA, 3))
    }
}
## use a uniform prior on v:
alpha <- 1
beta <- 1
## MCMC sampling setup
scale <- 0.03
niter <- 10000
nburnin <- 100
## Initialise samples and counters
vsamples <- numeric(niter)
dsamples <- numeric(niter)
v <- 0.5
d <- 0
vyes <- 0
dyes <- 0
## start MCMC sampling
for(i in (-nburnin + 1):niter)
{
    ## proposal for v:
```

```
first <- alpha + 2 * x[1] + x[2]
second <- beta + x[2] + 2 * x[3]
vstar <- rbeta(1, first, second)
## is this a valid combination of v and d?
valid <- (d >= lower(vstar)) && (d <= upper(vstar))
## compute the log-posterior ratio
logPostRatio <-
    if(valid)
    {
        ## from the likelihood
        dmultinom(x, prob=myprob(vstar, d), log=TRUE) -
            dmultinom(x, prob=myprob(v, d), log=TRUE) +
                ## from the marginal prior on v
                dbeta(vstar, alpha, beta, log=TRUE) -
                    dbeta(v, alpha, beta, log=TRUE) +
                        ## from the prior on d given v
                        dunif(x=d, lower(vstar), upper(vstar),
                              log=TRUE) -
                            dunif(x=d, min=lower(v), max=upper(v),
                                  log=TRUE)
    } else {
        ## if the combination is not valid, then the likelihood of
        ## the proposal is zero.
        - Inf
    }
## compute the log proposal ratio
logPropRatio <- dbeta(v, first, second, log=TRUE) -
    dbeta(vstar, first, second, log=TRUE)
## hence we obtain the log acceptance probability
logAcc <- logPostRatio + logPropRatio
## decide acceptance
if(log(runif(1)) <= logAcc)
{
    v <- vstar
    ## count acceptances
    if(i > 0)
    {
        vyes <- vyes + 1
    }
}

## proposal for d:
first <- max(d - scale, lower(v))
second <- min(d + scale, upper(v))
dstar <- runif(1, min=first, max=second)
## compute the log posterior ratio
logPostRatio <-
    ## from the likelihood
    dmultinom(x, prob=myprob(v, dstar), log=TRUE) -
        dmultinom(x, prob=myprob(v, d), log=TRUE) +
            ## from the prior on d given v
            dunif(x=dstar, lower(v), upper(v), log=TRUE) -
                dunif(x=d, min=lower(v), max=upper(v), log=TRUE)
## compute the log proposal ratio
logPropRatio <- dunif(d, first,second, log=TRUE) -
    dunif(dstar, first, second, log=TRUE)
## hence we obtain the log acceptance probability
logAcc <- logPostRatio + logPropRatio
## decide acceptance
if(log(runif(1)) <= logAcc)
{
    d <- dstar
    ## count acceptances
    if(i > 0)
    {
        dyes <- dyes + 1
    }
```

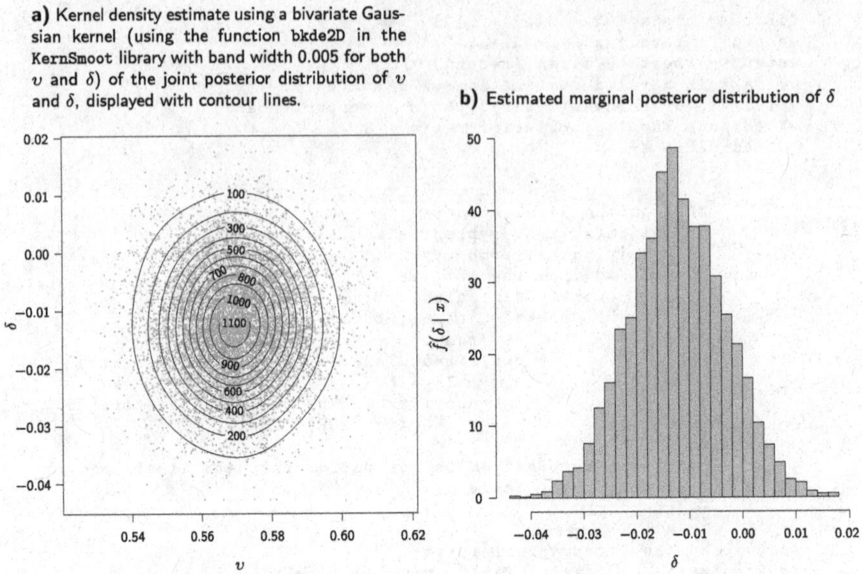

Fig. 8.9 Estimated joint posterior distribution of υ and δ and marginal posterior distribution of δ based on 10 000 MCMC samples

```
        }
        ## if burnin was passed, save the samples
        if(i > 0)
        {
            vsamples[i] <- v
            dsamples[i] <- d
        }
}
```

The empirical acceptance rates turn out to be `vyes/niter` $= 97.6\ \%$ and `dyes/niter` $= 45.6\ \%$ for υ and δ, respectively. Figures 8.9a and 8.9b show the empirical posterior distribution of υ and δ and also the marginal posterior of δ. The estimated posterior means are $\hat{\mathsf{E}}(\upsilon \mid \boldsymbol{x}) = 0.5697$ and $\hat{\mathsf{E}}(\delta \mid \boldsymbol{x}) = -0.0122$. The posterior probability $\Pr(\delta > 0 \mid \boldsymbol{x})$ is estimated as 0.0911, so substantially smaller than a priori, but not decisively close to zero. This suggests that the assumption of Hardy–Weinberg equilibrium for these data may not be completely unreasonable.

Similar results are obtained using a likelihood analysis, as we will show now. The MLEs (with standard errors in brackets) $\hat{\upsilon}_{\text{ML}} = 0.5696$ (0.0125) and $\hat{\delta}_{\text{ML}} = -0.0125$ (0.0089) are easily obtained. If we interpret these estimates from a Bayesian point of view using the second normal approximation described in Sect. 6.6.2, we obtain the posterior probability

$$\Pr(\delta > 0 \mid x) = 1 - \Pr\left\{\frac{\delta - \hat{\delta}_{\text{ML}}}{\text{se}(\hat{\delta}_{\text{ML}})} \leq \frac{-\hat{\delta}_{\text{ML}}}{\text{se}(\hat{\delta}_{\text{ML}})}\right\} \approx 1 - \Phi\left\{\frac{-\hat{\delta}_{\text{ML}}}{\text{se}(\hat{\delta}_{\text{ML}})}\right\} = 0.0803,$$

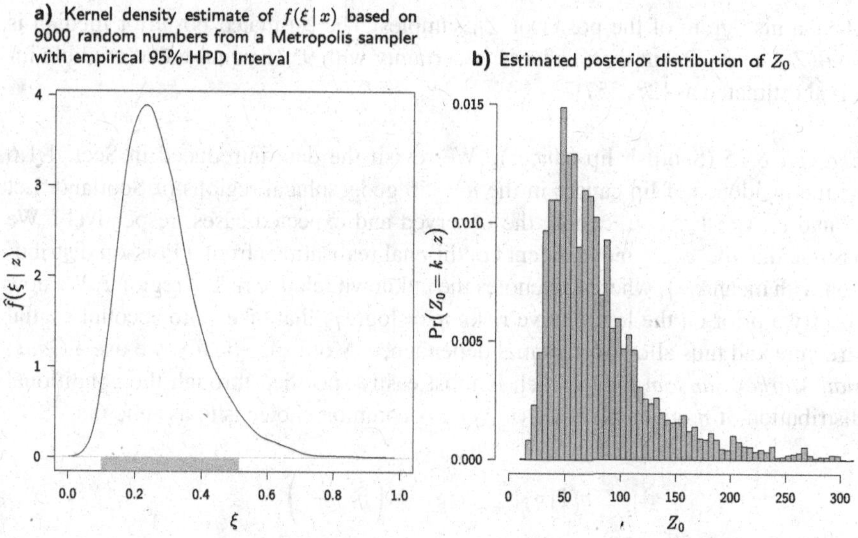
a) Kernel density estimate of $f(\xi \mid x)$ based on 9000 random numbers from a Metropolis sampler with empirical 95%-HPD Interval
b) Estimated posterior distribution of Z_0
$f(\xi \mid z)$
ξ
$\Pr(Z_0 = k \mid z)$
Z_0

Fig. 8.10 Estimated posterior distribution of ξ and Z_0 based on 10 000 MCMC samples

where $\Phi(\cdot)$ denotes the standard normal distribution function. This is again close to the fully Bayes estimate. We will revisit this example in Example 8.16, where we will apply explicit Bayesian model selection in order to decide between the Hardy–Weinberg and the multinomial model. ■

Example 8.14 (Screening for colon cancer) In Example 8.2 we have already numerically derived the posterior distribution of the parameters μ and ρ in the truncated beta-binomial model, cf. Fig. 8.2. To estimate the posterior distribution of the false negative fraction ξ, we now use random numbers from the joint posterior $f(\boldsymbol{\theta} \mid x)$ of $\boldsymbol{\theta} = (\mu, \rho)^\top$, generated using a bivariate Metropolis sampler. Samples $\xi^{(m)} = \xi(\boldsymbol{\theta}^{(m)})$ from the posterior distribution of the transformation ξ can then easily be obtained. Those can in turn be used to compute $Z_0^{(m)} = 196 \cdot \xi^{(m)}/(1 - \xi^{(m)})$ which are samples from the posterior of Z_0.

We choose a normal proposal $h(\boldsymbol{\theta}^* \mid \boldsymbol{\theta}^{(m)})$ with mean equal to the current value and covariance matrix proportional to the negative inverse curvature of the log-posterior at the posterior mode (cf. Sect. 6.6.2). The corresponding proportionality constant is chosen such that the acceptance rates are between 30 and 50 %.

Figure 8.10a displays a kernel density estimate of the posterior of the false negative fraction based on the last 9000 random numbers (1000 burn-in samples have been disregarded). The posterior distribution is quite skewed with estimated mean 0.28 much larger than the empirical median 0.26 or mode 0.20. The mode has been estimated as the mean of the limits of the empirical 1 % HPD interval. The empirical 95 % HPD interval is [0.10, 0.51]. Compared with the MLE $\hat{\xi}_{\mathrm{ML}} = 0.24$ from Example 5.10, both mean and median are larger, and the uncertainty is smaller than the one based on the profile likelihood interval [0.11, 0.55]. Figure 8.10b displays a histogram of the posterior Z_0 samples. The estimated posterior median is $\widehat{\text{Med}}(Z_0 \mid x) = 70$, but there is large uncertainty with 95 % equal-tailed credible interval estimated as $[28, 257]$. ■

Example 8.15 (Scottish lip cancer) We revisit the data introduced in Sect. 1.1.6 on the incidence of lip cancer in the $n = 56$ geographical regions of Scotland. Let x_i and e_i, $i = 1, \dots, n$, denote the observed and expected cases, respectively. We assume that the x_i are independent conditional realisations from a Poisson distribution with mean $e_i \lambda_i$, where λ_i denotes the unknown relative risk in region i. We now specify a prior on the log relative risks $\eta_i = \log(\lambda_i)$ that takes into account spatial structure and thus allows for spatial dependence. More specifically, we use a *Gaussian Markov random field*, which is most easily specified through the conditional distribution of η_i given all other $\{\eta_j\}_{j \neq i}$. A common choice is to assume that

$$\eta_i \mid \{\eta_j\}_{j \neq i}, \sigma^2 \sim \mathrm{N}\left(\bar{\eta}_i, \frac{\sigma^2}{n_i}\right),$$

where $\bar{\eta}_i = n_i^{-1} \sum_{j \sim i} \eta_j$ denotes the mean of the n_i spatially neighbouring regions of region i, and σ^2 is an unknown variance parameter.

To simulate from the posterior distribution, the obvious choice is a Gibbs sampler that iteratively updates the $n + 1$ unknown parameters $\lambda_1, \dots, \lambda_n, \sigma^2$. Due to conditional conjugacy, we use an inverse gamma prior for σ^2, i.e. $\sigma^2 \sim \mathrm{IG}(\alpha, \beta)$ a priori, so the full conditional distribution of σ^2 is again inverse gamma,

$$\sigma^2 \mid \boldsymbol{\eta} \sim \mathrm{IG}\left(\alpha + \frac{n-1}{2}, \beta + \frac{1}{2} \sum_{i \sim j} (\eta_i - \eta_j)^2\right);$$

where the sum in the second (scale) parameter goes over all pairs of neighbouring regions $i \sim j$. Slightly more involved is simulation from the full conditional distribution of λ_i, $i = 1, \dots, n$, which is not of a known form. This problem can be circumvented by using a Metropolis–Hastings step with (conditional) independence proposal

$$\lambda_i \sim \mathrm{G}\left(x_i + \frac{\tilde{\mu}^2}{\tilde{\sigma}^2}, e_i + \frac{\tilde{\mu}}{\tilde{\sigma}^2}\right). \tag{8.23}$$

This choice is motivated by the fact that the conditional prior of $\lambda_i \mid \{\lambda_j\}_{j \neq i}$, by the definition a log-normal distribution, can be well approximated through a gamma distribution $\mathrm{G}(\tilde{\mu}^2/\tilde{\sigma}^2, \tilde{\mu}/\tilde{\sigma}^2)$ with matching moments. The two parameters $\tilde{\mu}$ and $\tilde{\sigma}^2$ are expectation and variance of that log-normal distribution and are given by

$$\tilde{\mu} = \exp\left(\bar{\eta}_i + \frac{\sigma^2}{2n_i}\right) \quad \text{and}$$

$$\tilde{\sigma}^2 = \left\{\exp(\sigma^2/n_i) - 1\right\} \exp(2\bar{\eta}_i + \sigma^2/n_i),$$

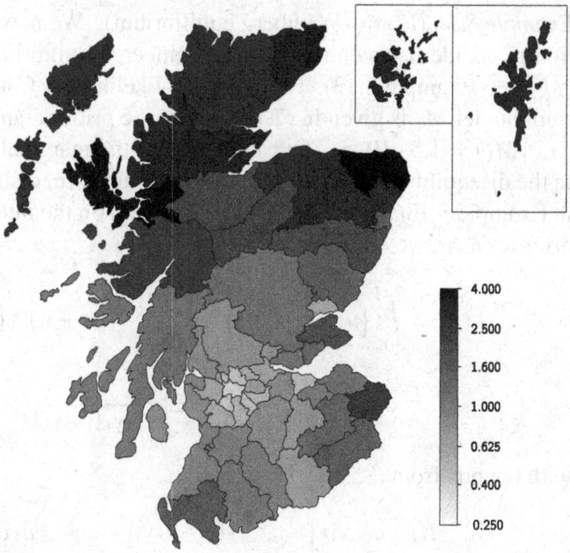

Fig. 8.11 Geographical distribution of the posterior mean estimates $\mathsf{E}(\lambda_i \mid x)$ of the relative risk of lip cancer in Scotland

cf. Appendix A.5.2. The gamma distribution is conjugate to the Poisson likelihood and can be analytically combined to obtain the proposal distribution (8.23) as an approximation to the full conditional of λ_i.

A simulation of length 100 000 with burn-in of 10 000 gave an average acceptance rate of 94 %. For the individual λ_is, the acceptance rate was never below 67 %. We used the prior parameters $\alpha = 1$ and $\beta = 0.01$. Figure 8.11 displays the corresponding posterior mean estimates $\mathsf{E}(\lambda_i \mid x_{1:n})$. Compared with the MLEs shown in Fig. 1.2, obtained from a model without spatial dependence, a much smoother picture can be observed. The empirical Bayes estimates displayed in Fig. 6.13 are also less variable than the MLEs but do not take the spatial structure of the data into account. ■

8.5 Numerical Calculation of the Marginal Likelihood

The calculation of the marginal likelihood $f(x)$ in the non-conjugate case is the greatest challenge in the implementation of Bayesian statistics.

8.5.1 Calculation Through Numerical Integration

In case the necessary integration in (8.1) is not feasible analytically, a natural approach is to try numerical methods of integration.

Example 8.16 (Hardy–Weinberg equilibrium) We now use Bayesian model selection to decide between the Hardy–Weinberg equilibrium (M_1) and disequilibrium (M_2, see Example 8.13). The marginal likelihood of the Hardy–Weinberg equilibrium model M_1 is given in (7.17); under the prior parameters $\alpha = \beta = 1$, we obtain $f(x \mid M_1) = 1.5 \cdot 10^{-5}$. The calculation of the marginal likelihood is more difficult in the disequilibrium model M_2. Using the uniform distribution as a prior for υ like in Example 8.13 and the uniform distribution on the interval bounded by (8.21) as a prior for $\delta \mid \upsilon$, we obtain

$$f(x) = \int_0^1 \bigl\{u(\upsilon) - l(\upsilon)\bigr\}^{-1} \Biggl[\int_{l(\upsilon)}^{u(\upsilon)} \bigl(\upsilon^2 + \delta\bigr)^{x_1} \bigl\{2\upsilon(1-\upsilon) - 2\delta\bigr\}^{x_2} \times \bigl\{(1-\upsilon)^2 + \delta\bigr\}^{x_3}\, d\delta \Biggr] d\upsilon \frac{n!}{x_1!x_2!x_3!}$$

with bounds from (8.21):

$$l(\upsilon) = \max\bigl\{-\upsilon^2, -(1-\upsilon)^2\bigr\} \quad \text{and} \quad u(\upsilon) = \upsilon(1-\upsilon).$$

This integration is feasible only numerically, so we use the R-function `integrate` (see Appendix C.2.1). We obtain $f(x \mid M_2) = 2.1 \cdot 10^{-6}$, which yields a Bayes factor of $\mathrm{BF}_{12} = 7.44$ and thus positive evidence for the Hardy–Weinberg equilibrium model. This is comparable with the results of Example 7.6, where another prior was used in the more general model M_2 through a different (multinomial) parametrisation and its conjugate (Dirichlet) prior. ■

Example 8.17 (Screening for colon cancer) Using likelihood inference, the truncated beta-binomial model has been clearly preferred over the truncated binomial model on the basis of a χ^2 goodness-of-fit test (see Example 5.18). For a Bayesian model selection procedure between the simpler model M_2 of Example 8.1 and the more flexible model M_1 of Example 8.2, we calculate the Bayes factor

$$\mathrm{BF}_{12} = \frac{f(x \mid M_1)}{f(x \mid M_2)},$$

where the numerator and denominator were already used in the calculation of the posterior parameter densities. The necessary integration is again not feasible analytically, and the R-function `integrate` has been applied instead. We obtain $\mathrm{BF}_{12} = 1.62 \cdot 10^{39}$ and therefore again overwhelming evidence for the truncated beta-binomial model compared to the simpler binomial one. ■

8.5.2 Monte Carlo Estimation of the Marginal Likelihood

In more complex models, numerical methods and the Laplace approximation are computationally costly and/or imply non-negligible inaccuracies. An obvious idea then is to adapt (MC)MC methods to not only estimate posterior densities but also marginal likelihoods. Certainly, this involves higher costs and/or lower accuracies as well. We illustrate this fact with three possible procedures.

First, observe that the equation

$$f(x) = \int f(x \mid \boldsymbol{\theta}) f(\boldsymbol{\theta}) \, d\boldsymbol{\theta}$$

allows a direct application of Monte Carlo integration (see Sect. 8.3) using randomly drawn numbers $\boldsymbol{\theta}^{(1)}, \dots, \boldsymbol{\theta}^{(M)}$ from the prior distribution $f(\boldsymbol{\theta})$. The resulting Monte Carlo estimator is given as

$$\hat{f}(x) = \frac{1}{M} \sum_{m=1}^{M} f\left(x \mid \boldsymbol{\theta}^{(m)}\right), \tag{8.24}$$

the arithmetic average of the likelihood values of the random numbers drawn from the prior distribution. We apply this simple approach to Example 8.12.

Example 8.18 (Screening for colon cancer) The parameter $\theta = \pi$ in the considered binomial model is a scalar, and the observed data x are given by $\mathbf{Z} = (Z_1, \dots, Z_k)^\top$. Since Z_0 is not observed, the likelihood function is given in this case by a binomial distribution truncated to positive values (see Example 2.12):

$$f(\mathbf{Z} \mid \pi) = \prod_{k=1}^{N} \left\{ \binom{N}{k} \pi^k (1-\pi)^{N-k} \right\}^{Z_k} / \left\{1 - (1-\pi)^N\right\}^n, \tag{8.25}$$

with $N = 6$ tests per patient and a total number of $n = 196$ positively tested patients. We used Jeffreys' prior Be(0.5, 0.5) as a prior distribution for π and now draw $M = 30\,000$ random numbers from it and then calculate their average likelihood value. In order to assess the variability of the estimator (8.24) we show in Fig. 8.12a estimations for five different simulations as functions of the sample size $m = 1, \dots, M$. For $M = 30\,000$ the estimates range between $\exp(-440.520)$ and $\exp(-440.437)$. ■

The variance of the estimator (8.24) is typically quite high if the likelihood function is distinctly more concentrated than the prior distribution. This means that the likelihood function will take values close to zero at most of the random numbers drawn from the prior distribution but very high values for very few of those random numbers. This is the case in Example 8.18, where the prior has very different mass centres than the likelihood function, see Fig. 8.12b. Therefore, it is advisable to use random numbers drawn from the posterior distribution instead the prior distribution. For a direct application of importance sampling, the posterior distribution $f(\boldsymbol{\theta} \mid x)$ has to be available including the proportionality constant, which is identical to the inverse of the marginal likelihood. But this is precisely the quantity we are intending to calculate, such that importance sampling cannot be applied directly.

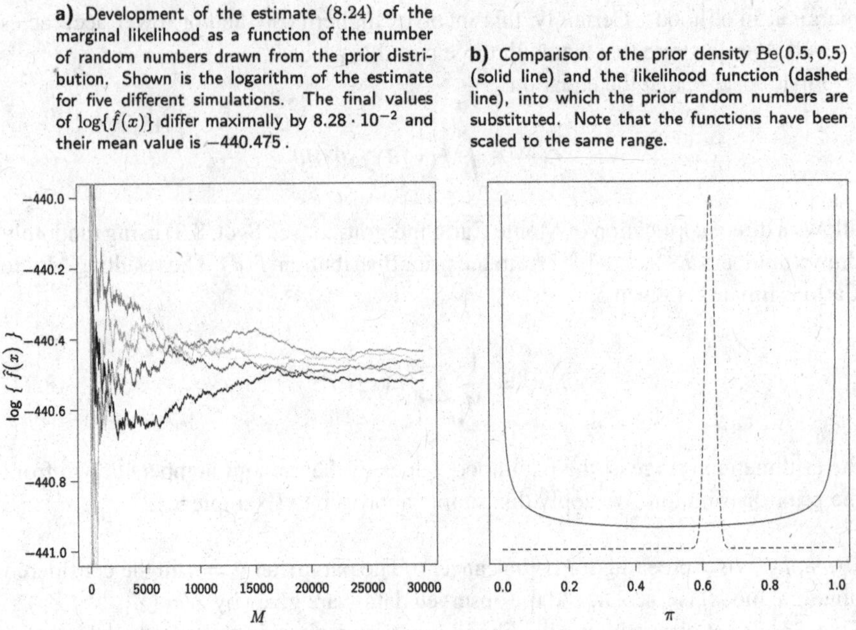
a) Development of the estimate (8.24) of the marginal likelihood as a function of the number of random numbers drawn from the prior distribution. Shown is the logarithm of the estimate for five different simulations. The final values of $\log\{\hat{f}(x)\}$ differ maximally by $8.28 \cdot 10^{-2}$ and their mean value is −440.475.
b) Comparison of the prior density Be(0.5, 0.5) (solid line) and the likelihood function (dashed line), into which the prior random numbers are substituted. Note that the functions have been scaled to the same range.
−440.0
−440.2
−440.4
−440.6
−440.8
−441.0
log { f(x) }
0
5000
10000
15000
20000
25000
30000
M
0.0
0.2
0.4
0.6
0.8
1.0
π

Fig. 8.12 Application of the estimator (8.24) in the binomial model of the colon cancer screening study

The trick here is to instead apply importance sampling to estimate the integral

$$\int f(\boldsymbol{\theta})\, d\boldsymbol{\theta} = \int \frac{f(\boldsymbol{\theta})}{f(\boldsymbol{\theta} \mid x)} f(\boldsymbol{\theta} \mid x)\, d\boldsymbol{\theta},$$

known to be equal to one. Drawing $\boldsymbol{\theta}^{(1)}, \ldots, \boldsymbol{\theta}^{(M)}$ from the posterior distribution $f(\boldsymbol{\theta} \mid x)$, we obtain the following "estimate" of the above integral:

$$1 = \frac{1}{M} \sum_{m=1}^{M} \frac{f(\boldsymbol{\theta}^{(m)})}{f(\boldsymbol{\theta}^{(m)} \mid x)} = \frac{1}{M} \sum_{m=1}^{M} \frac{f(x)}{f(x \mid \boldsymbol{\theta}^{(m)})}.$$

Rearranging provides an estimate of the marginal likelihood:

$$\hat{f}(x) = \left\{ \frac{1}{M} \sum_{m=1}^{M} \frac{1}{f(x \mid \boldsymbol{\theta}^{(m)})} \right\}^{-1}, \tag{8.26}$$

the harmonic average of the likelihood values at the random numbers drawn from the posterior distribution. Unfortunately, the statistical performance of this estimator is again quite poor. Since the variance of the inverse likelihood is in most cases infinite, the estimator is unstable in many applications. However, it is at least simulation-consistent due to the law of large numbers, i.e. $\hat{f}(x)$ converges to $f(x)$ as $M \to \infty$.

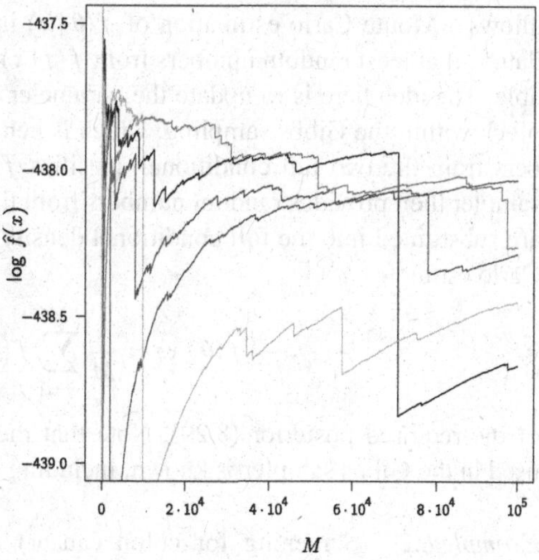

Fig. 8.13 Development of estimate (8.27) of the marginal likelihood in the binomial model of the colon cancer screening study as a function of the number of random numbers drawn from the posterior distribution. Shown is the logarithm of the estimate for five different Markov chains. The sharp changes in the estimates in the course of adding more random numbers are noteworthy. The final values of $\log\{\hat{f}(x)\}$ differ maximally by $6.32 \cdot 10^{-1}$, and their mean value is -438.339

Example 8.19 (Screening for colon cancer) We now take a different route than in Example 8.18 to compute the marginal likelihood for the colon cancer screening model. Using the likelihood (8.25) and applying (8.26), we obtain the estimator

$$\hat{f}(\mathbf{Z}) = M\left[\sum_{m=1}^{M}\{1-(1-\pi^{(m)})^N\}^n \prod_{k=1}^{N}\left\{\binom{N}{k}(\pi^{(m)})^k(1-\pi^{(m)})^{N-k}\right\}^{-Z_k}\right]^{-1}. \tag{8.27}$$

Figure 8.13 shows this estimator of the marginal likelihood as a function of the sample size M, where we drew M random numbers from the (approximate) posterior distribution using a Gibbs sampler as described in Example 8.12. ■

A third approach is based on the formula

$$f(x) = \frac{f(x \mid \boldsymbol{\theta}) f(\boldsymbol{\theta})}{f(\boldsymbol{\theta} \mid x)}, \tag{8.28}$$

which is a simple rearrangement of Bayes' theorem and holds for all $\boldsymbol{\theta} \in \Theta$. It will mostly be evaluated at parameter values $\boldsymbol{\theta}$ in the centre of the posterior distribution, e.g. the posterior mean or the posterior mode since in these regions a higher accuracy of the density estimation can be expected for a fixed number of random draws from the posterior. The denominator $f(\boldsymbol{\theta} \mid x)$ in the above identity is typically only known up to proportionality. Indeed, the corresponding normalising constant $f(x)$ is actually the quantity we intend to calculate. The representation

$$f(\boldsymbol{\theta} \mid x) = \int f(\boldsymbol{\theta} \mid x, z) f(z \mid x)\, dz \tag{8.29}$$

allows a Monte Carlo estimation of $f(\boldsymbol{\theta} \mid x)$ in the case of latent, i.e. unobserved, data z, if at least random numbers from $f(z \mid x)$ and the density $f(\boldsymbol{\theta} \mid x, z)$ are available. The idea here is to update the parameter vector $\boldsymbol{\theta}$, similarly to z, as an entire block within the Gibbs sampling, which is hence implemented using random numbers from the two full conditional densities $f(\boldsymbol{\theta} \mid x, z)$ and $f(z \mid x, \boldsymbol{\theta})$. The Gibbs sampler then provides random numbers from the marginal posterior $f(z \mid x)$, which are substituted into the full conditional density $f(\boldsymbol{\theta} \mid x, z)$ of $\boldsymbol{\theta}$ to obtain the Monte Carlo estimate

$$\hat{f}(\boldsymbol{\theta} \mid x) = \frac{1}{M} \sum_{m=1}^{M} f\left(\boldsymbol{\theta} \mid x, z^{(m)}\right)$$

of the required posterior (8.29). Note that the full conditional density $f(\boldsymbol{\theta} \mid x, z)$ used in the Gibbs sampler is known, including its normalisation constant.

Example 8.20 (Screening for colon cancer) Again, we intend to illustrate this method using the colon cancer screening example. Here $\boldsymbol{\theta}$ corresponds to the unknown probability π, the observed data x is $\mathbf{Z}$, and the unobserved data z is Z_0. The value of the density $f(\pi \mid x)$ at the estimated posterior mean $\hat{\mathsf{E}}(\pi \mid \mathbf{Z}) = 0.624$ can be estimated directly by

$$\frac{1}{M} \sum_{m=1}^{M} f\left(0.624 \mid \mathbf{Z}, \mathrm{Z}_0^{(m)}\right) = 27.889,$$

where we use $M = 100\,000$ simulations of $\mathrm{Z}_0 \mid \mathbf{Z}$ using Gibbs sampling, and (8.19) gives the density $f(\pi \mid x, z)$. Substitution into Eq. (8.28) leads to the Monte Carlo estimate of the log marginal likelihood

$$\log\{\hat{f}(x)\} = \log\{f(x \mid 0.624)\} + \log\{f(0.624)\} - \log(27.889) = -440.469,$$

which is actually quite close to the first estimate $\log\{\hat{f}(x)\} = -440.475$ from Example 8.18. ■

8.6 Exercises

1. Let $X \sim \mathrm{Po}(e\lambda)$ with known e, and assume the prior $\lambda \sim \mathrm{G}(\alpha, \beta)$.
   (a) Compute the posterior expectation of λ.
   (b) Compute the Laplace approximation of this posterior expectation.
   (c) For $\alpha = 0.5$ and $\beta = 0$, compare the Laplace approximation with the exact value, given the observations $x = 11$ and $e = 3.04$, or $x = 110$ and $e = 30.4$. Also compute the relative error of the Laplace approximation.
   (d) Now consider $\theta = \log(\lambda)$. First, derive the posterior density function using the change-of-variables formula (A.11). Second, compute the Laplace approximation of the posterior expectation of $\lambda = \exp(\theta)$ and

compare again with the exact value you have obtained by numerical integration using the R-function `integrate`.

2. In Example 8.3, derive the Laplace approximation (8.9) for the posterior expectation of π using the variance-stabilising transformation.
3. For estimating the odds ratio θ from Example 5.8, we will now use Bayesian inference. We assume independent Be(0.5, 0.5) distributions as priors for the probabilities π_1 and π_2.
   (a) Compute the posterior distributions of π_1 and π_2 for the data given in Table 3.1. Simulate samples from these posterior distributions and transform them into samples from the posterior distributions of θ and $\psi = \log(\theta)$. Use the samples to compute Monte Carlo estimates of the posterior expectations, medians, equal-tailed credible intervals and HPD intervals for θ and ψ. Compare with the results from likelihood inference in Example 5.8.
   (b) Try to compute the posterior densities of θ and ψ analytically. Use the density functions to numerically compute the posterior expectations and HPD intervals. Compare with the Monte Carlo estimates from 3(a).
4. In this exercise we will estimate a Bayesian hierarchical model with MCMC methods. Consider Example 6.31, where we had the following model:

$$\hat{\psi}_i \mid \psi_i \sim \mathrm{N}(\psi_i, \sigma_i^2),$$
$$\psi_i \mid \nu, \tau \sim \mathrm{N}(\nu, \tau^2),$$

   where we assume that the empirical log odds ratios $\hat{\psi}_i$ and corresponding variances $\sigma_i^2 := 1/a_i + 1/b_i + 1/c_i + 1/d_i$ are known for all studies $i = 1, \dots, n$. Instead of empirical Bayes estimation of the hyper-parameters ν and τ^2, we here proceed in a fully Bayesian way by assuming hyper-priors for them. We choose $\nu \sim \mathrm{N}(0, 10)$ and $\tau^2 \sim \mathrm{IG}(1, 1)$.
   (a) Derive the full conditional distributions of the unknown parameters $\psi_1, \dots, \psi_n$, ν and τ^2.
   (b) Implement a Gibbs sampler to simulate from the corresponding posterior distributions.
   (c) For the data given in Table 1.1, compute 95 % credible intervals for $\psi_1, \dots, \psi_n$ and ν. Produce a plot similar to Fig. 6.15 and compare with the results from the empirical Bayes estimation.
5. Let X_i, $i = 1, \dots, n$, denote a random sample from a Po(λ) distribution with gamma prior $\lambda \sim \mathrm{G}(\alpha, \beta)$ for the mean λ.
   (a) Derive closed forms of $\mathsf{E}(\lambda \mid x_{1:n})$ and $\mathrm{Var}(\lambda \mid x_{1:n})$ by computing the posterior distribution of $\lambda \mid x_{1:n}$.
   (b) Approximate $\mathsf{E}(\lambda \mid x_{1:n})$ and $\mathrm{Var}(\lambda \mid x_{1:n})$ by exploiting the asymptotic normality of the posterior (cf. Sect. 6.6.2).
   (c) Consider now the log mean $\theta = \log(\lambda)$. Use the change-of-variables formula (A.11) to compute the posterior density $f(\theta \mid x_{1:n})$.

(d) Let $\alpha = 1, \beta = 1$ and assume that $\bar{x} = 9.9$ has been obtained for $n = 10$ observations from the model. Compute approximate values of $\mathsf{E}(\theta \mid x_{1:n})$ and $\text{Var}(\theta \mid x_{1:n})$ via:
  (i) the asymptotic normality of the posterior,
  (ii) numerical integration (cf. Appendix C.2.1) and
  (iii) Monte Carlo integration.

6. Consider the genetic linkage model from Exercise 5 in Chap. 2. Here we assume a uniform prior on the proportion ϕ, i.e. $\phi \sim \text{U}(0, 1)$. We would like to compute the posterior mean $\mathsf{E}(\phi \mid \boldsymbol{x})$.
   (a) Construct a rejection sampling algorithm to simulate from $f(\phi \mid \boldsymbol{x})$ using the prior density as the proposal density.
   (b) Estimate the posterior mean of ϕ by Monte Carlo integration using $M = 10\,000$ samples from $f(\phi \mid \boldsymbol{x})$. Calculate also the Monte Carlo standard error.
   (c) In 6(b) we obtained samples of the posterior distribution assuming a uniform prior on ϕ. Suppose we now assume a $\text{Be}(0.5, 0.5)$ prior instead of the previous $\text{U}(0, 1) = \text{Be}(1, 1)$. Use the importance sampling weights to estimate the posterior mean and Monte Carlo standard error under the new prior based on the old samples from 6(b).
7. As in Exercise 6, we consider the genetic linkage model from Exercise 5 in Chap. 2. Now, we would like to sample from the posterior distribution of ϕ using MCMC. Using the Metropolis–Hastings algorithm, an arbitrary proposal distribution can be used, and the algorithm will always converge to the target distribution. However, the time until convergence and the degree of dependence between the samples depends on the chosen proposal distribution.
   (a) To sample from the posterior distribution, construct an MCMC sampler based on the following normal independence proposal (cf. approximation 3 in Sect. 6.6.2):

$$\phi^* \sim \text{N}\big(\text{Mod}(\phi \mid \boldsymbol{x}), F^2 \cdot C^{-1}\big),$$

   where $\text{Mod}(\phi \mid \boldsymbol{x})$ denotes the posterior mode, C the negative curvature of the log-posterior at the mode and F a factor to blow up the variance.
   (b) Construct an MCMC sampler based on the following random walk proposal:

$$\phi^* \sim \text{U}\big(\phi^{(m)} - d, \phi^{(m)} + d\big),$$

   where $\phi^{(m)}$ denotes the current state of the Markov chain, and d is a constant.
   (c) Generate $M = 10\,000$ samples from algorithm 7(a), setting $F = 1$ and $F = 10$, and from algorithm 7(b) with $d = 0.1$ and $d = 0.2$. To check the convergence of the Markov chain:
     (i) plot the generated samples to visually check the traces,
     (ii) plot the autocorrelation function using the R-function `acf`,

(iii) generate a histogram of the samples,
(iv) compare the acceptance rates.
What do you observe?

8. Cole et al. (2012) describe a rejection sampling approach to sample from a posterior distribution as a simple and efficient alternative to MCMC. They summarise their approach as:
   I. Define a model with likelihood function $L(\theta; y)$ and prior $f(\theta)$.
   II. Obtain the maximum likelihood estimate $\hat{\theta}_{\text{ML}}$.
   III. To obtain a sample from the posterior:
      (i) Draw θ^* from the prior distribution (note: this must cover the range of the posterior).
      (ii) Compute the ratio $p = L(\theta^*; y)/L(\hat{\theta}_{\text{ML}}; y)$.
      (iii) Draw u from U(0, 1).
      (iv) If $u \leq p$, then accept θ^*. Otherwise, reject θ^* and repeat.
   (a) Using Bayes' rule, write out the posterior density of $f(\theta \mid y)$. In the notation of Sect. 8.3.3, what are the functions $f_X(\theta)$, $f_Z(\theta)$ and $L(\theta; x)$ in the Bayesian formulation?
   (b) Show that the acceptance probability $f_X(\theta^*)/\{af_Z(\theta^*)\}$ is equal to $L(\theta^*; y)/L(\hat{\theta}_{\text{ML}}; y)$. What is a?
   (c) Explain why the inequality $f_X(\theta) \leq af_Z(\theta)$ is guaranteed by the approach of Cole et al. (2012).
   (d) In the model of Exercise 6(c), use the proposed rejection sampling scheme to generate samples from the posterior of ϕ.

8.7 References

An overview of numerical methods in Bayesian inference is given by Evans and Swartz (1995). Tierney and Kadane (1986) have studied application of the Laplace approximation to Bayesian inference in detail, see also Bernardo and Smith (2000, Sect. 5.5.1). Classical texts on stochastic simulation and Monte Carlo inference are Ripley (1987) and Devroye (1986), while Robert and Casella (2004, 2010) are more recent books. An excellent introduction to Markov chain Monte Carlo methods is given in Green (2001), more details can be found in Gilks et al. (1996). Further seminal papers are Tierney (1994) and Besag et al. (1995). Different approaches for the Monte Carlo estimation of the marginal likelihood are described in Newton and Raftery (1994) and Chib (1995).

Prediction

9

Contents

A common problem in statistics concerns the prediction of future data. Consider the following scenario: Let $x_{1:n}$ denote the realisation of a random sample $X_{1:n}$ from a distribution with density (or probability mass) function $f(x; \theta)$ with unknown parameter θ. Our goal is to predict a future independent observation $Y = X_{n+1}$, also from $f(x; \theta)$. Obviously, the observed data $x_{1:n}$ should be taken into account in this task.

In contrast to previous chapters, which were concerned with inference for unknown parameters, that are inherently unobservable, we now want to infer a quantity that we will be able to observe. We are interested in *predictive inference*.

We may want to derive a *point prediction* $\hat{Y}$, but also a *predictive distribution* with *predictive density* $f(y)$. The point prediction is then just a function of the predictive distribution, for example its mean. A 95 % *prediction interval* can be obtained based, for example, on the 2.5 % and 97.5 % quantiles of the predictive distribution.

For simplicity, in the following we will call $f(y)$ a density function even if Y may also be a discrete random variable. The methods are described for scalar Y and θ but can easily be generalised to vectorial $\boldsymbol{Y}$ or $\boldsymbol{\theta}$.

L. Held, D. Sabanés Bové, *Likelihood and Bayesian Inference*,
Statistics for Biology and Health, https://doi.org/10.1007/978-3-662-60792-3_9,

9.1 Plug-in Prediction

A naive approach to compute a predictive distribution is the *plug-in prediction*: We simply use the density function $f(x;\theta)$ underlying our random sample and replace the unknown parameter θ with the MLE $\hat{\theta}_{\text{ML}}$, i.e. $f(y) = f(y;\hat{\theta}_{\text{ML}})$. This leads to the plug-in predictive distribution with density

$$f(y) = f(y;\hat{\theta}_{\text{ML}}), \tag{9.1}$$

where $\hat{\theta}_{\text{ML}} = \hat{\theta}_{\text{ML}}(x_{1:n})$ has been calculated based on the realisation $x_{1:n}$ of the random sample $X_{1:n}$. However, by replacing the true, unknown parameter θ with the MLE $\hat{\theta}_{\text{ML}}$, the uncertainty in estimating θ is ignored. The plug-in prediction therefore produces prediction intervals that are often too narrow. Because of its simplicity, this method is nevertheless commonly used.

Example 9.1 (Scottish lip cancer) Suppose we observe $n = 1$ observation x from a Poisson distribution $X \sim \text{Po}(e_x\lambda)$. Our goal is to predict $Y \sim \text{Po}(e_y\lambda)$, and we assume that both $e_x > 0$ and $e_y > 0$ are known. The MLE of λ is $\hat{\lambda}_{\text{ML}} = x/e_x$, so the plug-in prediction is $Y \sim \text{Po}(x \cdot e_y/e_x)$.

As an illustration, consider Sect. 1.1.6 and assume that in a particular region of Scotland there have been $x = 11$ cases of lip cancer in a five-year period; however, only $e_x = 3.04$ cases have been expected. Our goal is to predict the number of cases Y in the next five years, and for simplicity, we assume that $e_x = e_y$. The plug-in predictive distribution of Y is now a Poisson distribution with mean (and variance) $x \cdot e_y/e_x = x = 11$. The interval [5, 17] is a 95 % prediction interval since $\sum_{y=5}^{17} f(y;\lambda = 11) \approx 0.95$.

However, suppose now that no case ($x = 0$) has been observed in that region. The plug-in prediction is then a degenerate Poisson distribution with mean zero, i.e. predicts with absolute certainty that there will be no case in the future, a rather untenable forecast. ■

This example illustrates that the plug-in prediction is in general unsatisfactory. Instead, one should try to incorporate the uncertainty in the estimation of the unknown parameter θ into the prediction. This goal can be achieved by both likelihood or Bayesian approaches.

9.2 Likelihood Prediction

In this section we introduce two prediction approaches based on the likelihood, which both take into account the uncertainty of the estimated parameters.

9.2.1 Predictive Likelihood

To obtain a predictive distribution of $Y = X_{n+1}$, we introduce the so-called predictive likelihood $L_p(y)$. For this purpose, we first consider the likelihood function $L(\theta; x_{1:n}, y) = f(x, y; \theta)$ with respect to the observations $X_{1:n} = x_{1:n}$ and $Y = y$. Since the realisation $Y = y$ is unobserved, this may be understood as a function of θ *and* y for fixed $x_{1:n}$.

Definition 9.1 (Extended likelihood function) The *extended likelihood function* $L(\theta, y)$ is the likelihood function with respect to the observations $X_{1:n} = x_{1:n}$ and $Y = y$, understood as a function of θ and y for fixed $x_{1:n}$:

$$L(\theta, y) = f(x_{1:n}, y; \theta). \qquad \blacklozenge$$

Now assume that $X_{1:n}$ and Y are independent and distributed according to a density function $f(x; \theta)$. Then we have

$$f(x_{1:n}, y; \theta) = f(y; \theta) \prod_{i=1}^{n} f(x_i; \theta).$$

To determine the predictive likelihood of y, the parameter θ has to be eliminated. Formally, this can be done analogously to obtaining the profile likelihood, see Sect. 5.3.

Definition 9.2 (Predictive likelihood) The *predictive likelihood function*, or shorter the *predictive likelihood*, $L_p(y)$ is obtained by maximising the extended likelihood function $L(\theta, y)$ with respect to θ while y is kept fixed, i.e.

$$L_p(y) = \max_{\theta} L(\theta, y) = L\big(\hat{\theta}(y), y\big).$$

Here the estimate $\hat{\theta}(y)$ is the MLE of θ based on the extended data $x_{1:n}$ and y. $\blacklozenge$

Since Y is a random variable, the predictive likelihood $L_p(y)$ can be viewed as a non-normalised density function. After suitable normalisation, we obtain a proper density function from which we can compute point predictions and prediction intervals. We call the corresponding predictive distribution the *likelihood prediction* with density function $f_p(y) = L_p(y) / \int L_p(u)\, du$.

Example 9.2 (Normal distribution) Let $X_{1:n}$ denote a random sample from an $\mathrm{N}(\mu, \sigma^2)$ distribution, from which a future observation Y has to be predicted. The variance σ^2 is assumed known, the expectation μ unknown. The joint density function of $X_{1:n}$ and Y is

$$\begin{aligned} f(x_{1:n}, y; \mu) &= C \cdot \exp\left[-\frac{1}{2\sigma^2}\left\{\sum_{i=1}^{n}(x_i - \mu)^2 + (y-\mu)^2\right\}\right] \\ &= C \cdot \exp\left[-\frac{1}{2\sigma^2}\left\{\sum_{i=1}^{n}(x_i - \bar{x})^2 + n(\bar{x}-\mu)^2 + (y-\mu)^2\right\}\right]. \end{aligned}$$

From that the extended likelihood function can be derived as

$$L(\mu, y) = \exp\left[-\frac{1}{2\sigma^2}\left\{n(\bar{x}-\mu)^2 + (y-\mu)^2\right\}\right],$$

which is (for fixed y) maximised by

$$\hat{\mu}(y) = \frac{n\bar{x}+y}{n+1}.$$

Substituting $\hat{\mu}(y)$ into $L(\mu, y)$ leads to the predictive likelihood

$$\begin{aligned} L_p(y) = L\big(\hat{\mu}(y), y\big) &= \exp\left[-\frac{1}{2\sigma^2}\left\{n\left(\bar{x} - \frac{n\bar{x}+y}{n+1}\right)^2 + \left(y - \frac{n\bar{x}+y}{n+1}\right)^2\right\}\right] \\ &= \exp\left[-\frac{1}{2\sigma^2}\left\{n\left(\frac{y-\bar{x}}{n+1}\right)^2 + n^2\left(\frac{y-\bar{x}}{n+1}\right)^2\right\}\right] \\ &= \exp\left\{-\frac{1}{2\sigma^2}\frac{n+n^2}{(n+1)^2}(y-\bar{x})^2\right\} \\ &= \exp\left\{-\frac{1}{2\sigma^2}\frac{n}{n+1}(y-\bar{x})^2\right\}, \end{aligned}$$

which can be identified as the kernel of a normal density with expectation $\bar{x}$ and variance $(1+1/n)\sigma^2$. Hence, the likelihood prediction

$$Y \sim \mathrm{N}\big(\bar{x}, \sigma^2(1+1/n)\big) \tag{9.2}$$

differs from the plug-in prediction $Y \sim \mathrm{N}(\bar{x}, \sigma^2)$ only through a larger variance. The likelihood prediction therefore provides somewhat larger prediction intervals since the uncertainty in the MLE $\hat{\mu}_{\mathrm{ML}} = \bar{x}$ has been accounted for. ■

Example 9.3 (Blood alcohol concentration) If we consider the estimated standard deviation $\sigma = 237.8$ of the transformation factors in Sect. 1.1.7 as known, a likelihood prediction interval for the transformation factor of a new individual can be computed from (9.2). In particular, the 95 % prediction interval has limits

$$2449.2 \pm 1.96 \cdot \sqrt{1+1/185} \cdot 237.8 = 1981.9 \quad \text{and } 2916.5,$$

cf. Table 1.3. The point prediction is of course equal to 2449.2. Note that $\sqrt{1+1/185} \approx 1.003$, so very close to one. The 95 % plug-in prediction interval will therefore be essentially the same.

It is interesting that the lower limit of the above prediction interval for TF is close to the currently used transformation factor of $\text{TF}_0 = 2000$ in Switzerland. So the factor $\text{TF}_0 = 2000$ can be justified as the approximate lower limit of a 95 % prediction interval, but not as a useful point prediction. ■

Example 9.4 (Poisson model) We aim to apply the predictive likelihood approach in Example 9.1. The extended likelihood of λ and y is (after elimination of multiplicative constants) given by

$$L(\lambda, y) = \frac{e_y^y}{y!}\lambda^{x+y} \exp\{-(e_x + e_y)\lambda\}.$$

For fixed y, $L(\lambda, y)$ is maximised by

$$\hat{\lambda}(y) = \frac{x+y}{e_x + e_y}.$$

Substituting this into $L(\lambda, y)$ leads to the predictive likelihood

$$L_p(y) = \frac{e_y^y}{y!}\left(\frac{x+y}{e_x+e_y}\right)^{x+y} \exp\{-(x+y)\}.$$

This function can be normalised numerically if the required infinite summation of $L_p(y)$ is truncated at a sufficiently large upper bound $y_{\max}$, for example, at $y_{\max} = 1000$. We then obtain the predictive probability mass function $f_p(y)$, from which the expectation and the variance can be derived. For $x = 11$ and $e_x = e_y = 3.04$, these are given as 11.498 and 22.998, respectively. The mean and variance of the plug-in prediction are 11, cf. Example 9.1. While the mean of the predictive likelihood prediction is fairly close to the mean of the plug-in prediction, the variance of the likelihood prediction is more than twice as large as the variance of the plug-in prediction. The prediction interval $[5, 17]$ has probability 0.95 for the plug-in prediction. In contrast, the interval has only probability 0.84 for the likelihood prediction. ■

9.2.2 Bootstrap Prediction

Another approach to incorporate the uncertainty of the estimate $\hat{\theta}_{\text{ML}}$ into the prediction procedure is as follows: Let $f(\hat{\theta}_{\text{ML}}; \theta)$ denote the density function of the ML estimator $\hat{\theta}_{\text{ML}}$ depending on a random sample $X_{1:n}$ from $f(x;\theta)$ with true but unknown parameter θ. The idea is now to replace the too optimistic predictive distribution (9.1) with the density

$$g(y;\theta) = \int f(y; \hat{\theta}_{\text{ML}}) f(\hat{\theta}_{\text{ML}}; \theta)\, d\hat{\theta}_{\text{ML}} \tag{9.3}$$

such that $\hat{\theta}_{\text{ML}}$ has been eliminated from $f(y; \hat{\theta}_{\text{ML}})$ through integration. If the ML estimator has a discrete distribution, the integral has to be replaced by a sum. The idea to eliminate the unknown parameter by integration from the likelihood function can also be motivated through a Bayesian argumentation, see Sect. 9.3.

A shortcoming of the distribution (9.3) is that $g(y; \theta)$ still depends on the true but unknown parameter θ. A possible remedy to this is to replace θ in $g(y; \theta)$ with the MLE $\hat{\theta}_{\text{ML}} = \hat{\theta}_{\text{ML}}(x_{1:n})$ based on the observed realisation $x_{1:n}$ of the random sample $X_{1:n}$. This leads to the so-called *bootstrap predictive distribution*

$$f(y) = g\big(y; \hat{\theta}_{\text{ML}}(x_{1:n})\big). \tag{9.4}$$

The reasons for calling this prediction approach the *bootstrap prediction* will be discussed after the following two examples.

Example 9.5 (Normal model) As in Example 9.2, let $X_{1:n}$ denote a random sample from an $\text{N}(\mu, \sigma^2)$ distribution with $\theta = \mu$ unknown and σ^2 known. The ML estimator $\hat{\mu}_{\text{ML}} = \bar{X}$ has the distribution $\text{N}(\mu, \sigma^2/n)$. Note that this is true not only asymptotically but in this case also for finite samples. Now we have

$$\begin{aligned}
g(y; \mu) &= \int \frac{1}{\sqrt{2\pi}} \frac{1}{\sigma} \exp\left\{-\frac{1}{2} \frac{(y - \hat{\mu}_{\text{ML}})^2}{\sigma^2}\right\} \\
&\quad \times \frac{1}{\sqrt{2\pi}} \frac{\sqrt{n}}{\sigma} \exp\left\{-\frac{1}{2} \frac{(\hat{\mu}_{\text{ML}} - \mu)^2}{\sigma^2/n}\right\} d\hat{\mu}_{\text{ML}} \\
&= C \int \exp\left[-\frac{1}{2}\left\{\frac{(\hat{\mu}_{\text{ML}} - y)^2}{\sigma^2} + \frac{(\hat{\mu}_{\text{ML}} - \mu)^2}{\sigma^2/n}\right\}\right] d\hat{\mu}_{\text{ML}},
\end{aligned}$$

where $C = \sqrt{n}/(2\pi\sigma^2)$. Setting $\tau^2 = \sigma^2/(1+n)$ and

$$c = \tau^2\left(\frac{y}{\sigma^2} + \frac{n}{\sigma^2}\mu\right) = \frac{y + n\mu}{1+n},$$

we obtain (cf. Appendix B.1.5)

$$\begin{aligned}
g(y; \mu) &= C \int \exp\left[-\frac{1}{2}\left\{\frac{(\hat{\mu}_{\text{ML}} - c)^2}{\tau^2} + \frac{(y - \mu)^2}{\sigma^2(1 + \frac{1}{n})}\right\}\right] d\hat{\mu}_{\text{ML}} \\
&= C \exp\left\{-\frac{1}{2} \frac{(y - \mu)^2}{\sigma^2(1 + \frac{1}{n})}\right\} \underbrace{\int \exp\left\{-\frac{1}{2} \frac{(\hat{\mu}_{\text{ML}} - c)^2}{\tau^2}\right\} d\hat{\mu}_{\text{ML}}}_{=\sqrt{2\pi}\cdot\tau = \sigma\sqrt{2\pi}/\sqrt{1+n}} \\
&= \frac{1}{\sqrt{2\pi}} \frac{1}{\sigma} \frac{1}{\sqrt{1 + \frac{1}{n}}} \exp\left\{-\frac{1}{2} \frac{(y - \mu)^2}{\sigma^2(1 + \frac{1}{n})}\right\},
\end{aligned}$$

i.e. a normal density with expectation μ and variance $(1 + 1/n)\sigma^2$. By replacing μ in $g(y; \mu)$ with $\hat{\mu}_{\text{ML}}(x) = \bar{x}$ we identify the bootstrap prediction as the normal distribution with mean $\bar{x}$ and variance $(1+1/n)\sigma^2$. Therefore, predictive likelihood and bootstrap prediction are in this case identical, cf. Example 9.2. ■

Example 9.6 (Poisson model) Let $X \sim \text{Po}(e_x\lambda)$. Then the MLE is $\hat{\lambda}_{\text{ML}} = x/e_x$. Since X follows the Poisson distribution with parameter $e_x\lambda$, the probability mass function of $\hat{\lambda}_{\text{ML}}$ is

$$f(\hat{\lambda}_{\text{ML}}; \lambda) = \frac{(e_x\lambda)^{e_x\hat{\lambda}_{\text{ML}}}}{(e_x\hat{\lambda}_{\text{ML}})!} \exp(-e_x\lambda) \quad \text{for } \hat{\lambda}_{\text{ML}} = 0, \frac{1}{e_x}, \frac{2}{e_x}, \ldots.$$

For (9.4), we therefore obtain

$$\begin{aligned} f(y) &= \sum_{t=0,\frac{1}{e_x},\ldots} \frac{(e_y t)^y}{y!} \exp(-e_y t) \frac{(e_x\hat{\lambda}_{\text{ML}})^{e_x t}}{(e_x t)!} \exp(-e_x\hat{\lambda}_{\text{ML}}) \\ &= \frac{\exp(-e_x\hat{\lambda}_{\text{ML}})}{y!} \left(\frac{e_y}{e_x}\right)^y \sum_{s=0}^{\infty} \frac{s^y(e_x\hat{\lambda}_{\text{ML}})^s}{s!} \exp\left(-\frac{e_y}{e_x}s\right), \end{aligned}$$

where the last equality is due to the substitution $s = t \cdot e_x$. It is difficult to compute the sum in this equation analytically. An alternative is a Monte Carlo approximation of the integral (9.4):

$$\hat{f}(y) = \frac{1}{m} \sum_{i=1}^{m} f\bigl(y; \hat{\lambda}_{\text{ML}}^{(i)}\bigr). \tag{9.5}$$

Here, $\hat{\lambda}_{\text{ML}}^{(1)}, \ldots, \hat{\lambda}_{\text{ML}}^{(m)}$ denote independent samples from $f(\hat{\lambda}_{\text{ML}}; \lambda)$, where the true parameter λ has been replaced by the MLE $\hat{\lambda}_{\text{ML}}(x)$ based on the actual observation x. This procedure is easily implemented in R as follows:

```
## Data:
x <- 11
ex <- 3.04
ey <- 3.04
## MLE:
(lambdahat <- x/ex)
[1] 3.618421
## bootstrap prediction:
set.seed(1)
m <- 10000
lambdasample <- rpois(m, lambda=lambdahat * ex) / ex
support <- 0:100
ghat <- rep(0, length(support))
for(i in 1:m)
{
    ghat <- ghat + dpois(support, lambda=lambdasample[i] * ey)
}
ghat <- ghat/m
## empirical moments of the predictive distribution:
(e.g <- sum(ghat*support))
[1] 11.0068
```

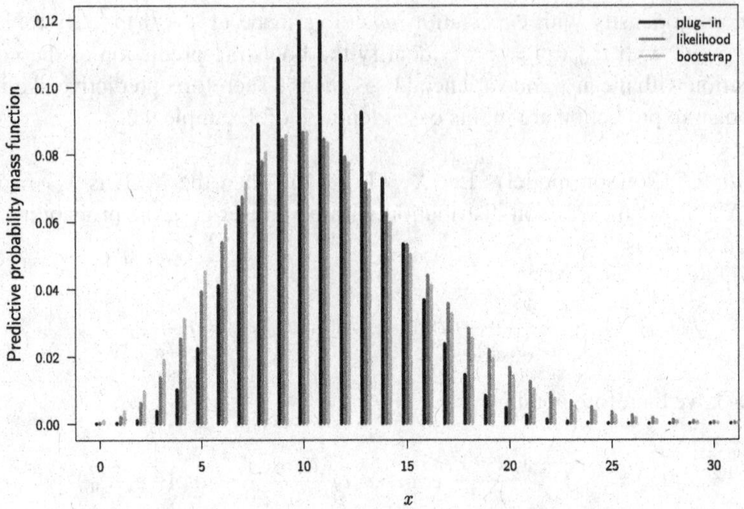

Fig. 9.1 Comparison of the probability mass functions of plug-in, likelihood and bootstrap predictions in the Poisson example (the bars are centred around their x-value)

```
e.g2 <- sum(ghat*support^2)
(var.g <- e.g2 - e.g^2)
[1] 22.07575
```

The empirical mean e.g = 11.0068 of the bootstrap predictive distribution is very close to the mean of the plug-in prediction. The empirical prediction variance var.g = 22.0758 is, however, larger as for the plug-in prediction. The resulting predictive distribution is very similar to the predictive likelihood distribution, cf. Fig. 9.1.

Incidentally, note that we can calculate the expectation and the variance of the prediction Y analytically using the law of total expectation and the law of total variance (cf. Appendix A.3.4). Starting with the expectation and variance belonging to the distribution (9.3), we have

$$\begin{aligned}\mathsf{E}(Y) &= \mathsf{E}\{\mathsf{E}(Y \mid \hat{\lambda}_{\mathrm{ML}})\} = \mathsf{E}(e_y \hat{\lambda}_{\mathrm{ML}}) = e_y \lambda \quad \text{and}\\ \operatorname{Var}(Y) &= \mathsf{E}\{\operatorname{Var}(Y \mid \hat{\lambda}_{\mathrm{ML}})\} + \operatorname{Var}\{\mathsf{E}(Y \mid \hat{\lambda}_{\mathrm{ML}})\}\\ &= \mathsf{E}(e_y \hat{\lambda}_{\mathrm{ML}}) + \operatorname{Var}(e_y \hat{\lambda}_{\mathrm{ML}})\\ &= e_y \lambda + \frac{e_y^2}{e_x} \lambda.\end{aligned}$$

We obtain the corresponding moments of the bootstrap prediction (9.4) by replacing λ with $\hat{\lambda}_{\mathrm{ML}}$, giving

$$\mathsf{E}(Y) = e_y \hat{\lambda}_{\mathrm{ML}} \quad \text{and} \tag{9.6}$$

$$\operatorname{Var}(Y) = e_y \hat{\lambda}_{\mathrm{ML}} + \frac{e_y^2}{e_x} \hat{\lambda}_{\mathrm{ML}}. \tag{9.7}$$

In our case ($e_x = e_y = 3.04$ and $\hat{\lambda}_{\mathrm{ML}} = 11/3.04$) we therefore have $\mathsf{E}(Y) = 11$ and $\mathrm{Var}(Y) = 22$. The empirical estimates above differ slightly from this due to additional Monte Carlo error. ■

Implementation of the Monte Carlo version of the predictive distribution (9.4) requires independent samples from the distribution $f(\hat{\theta}_{\mathrm{ML}}; \theta)$, where the true parameter θ is replaced by the MLE $\hat{\theta}_{\mathrm{ML}}(x_{1:n})$ based on the observed data $x_{1:n}$. If this distribution is unknown, there are two alternatives. First, one can use the asymptotic normal distribution of the ML estimator as an approximation of $f(\hat{\theta}_{\mathrm{ML}}; \theta)$. Another approximate but more accurate approach is based on the bootstrap, see also Sect. 3.2.5. For repeated samples of size n *with* replacement from the original data, the MLE is calculated. The empirical distribution of those MLEs is under certain regularity conditions a good approximation of the true distribution. The term $f(\hat{\theta}_{\mathrm{ML}}; \theta)$ in (9.3) is then replaced with this approximation, and the integral reduces to a sum as in (9.5). Note, however, that a bootstrap approach requires a reasonably large sample size n, it would not be possible in Example 9.6, where only $n = 1$ observation is available.

9.3 Bayesian Prediction

A Bayesian viewpoint offers a very natural approach to the prediction problem.

9.3.1 Posterior Predictive Distribution

Consider a random sample $X_{1:n}$ from a distribution with density $f(x \mid \theta)$. Our goal is to predict a future (independent) observation $Y = X_{n+1}$ from $f(x \mid \theta)$. The *posterior predictive distribution* of $Y \mid x_{1:n}$, which can be calculated easily using elementary rules from probability, is the central quantity of interest:

$$\begin{aligned} f(y \mid x_{1:n}) &= \int f(y, \theta \mid x_{1:n})\, d\theta = \int f(y \mid \theta, x_{1:n}) f(\theta \mid x_{1:n})\, d\theta \\ &= \int f(y \mid \theta) f(\theta \mid x_{1:n})\, d\theta. \end{aligned} \tag{9.8}$$

Note that the last line follows from the assumption of conditional independence of $X_{1:n}$ and Y, given θ. Thus, to compute $f(y \mid x_{1:n})$, we simply need to integrate the product of the likelihood $f(y \mid \theta)$ and the posterior density $f(\theta \mid x_{1:n})$ with respect to θ. The result $f(y \mid x_{1:n})$ is called the posterior predictive distribution in contrast to the *prior predictive distribution*

$$f(y) = \int f(y \mid \theta) f(\theta)\, d\theta. \tag{9.9}$$

Compared to Eq. (9.8), the posterior $f(\theta \mid x_{1:n})$ has been replaced by the prior $f(\theta)$ to obtain the prior predictive distribution, which plays a central role in Bayesian model choice, cf. Sect. 7.2.1.

It is worth emphasising that the posterior predictive distribution $f(y \mid x_{1:n})$ is *not* equal to the naive plug-in prediction $f(y \mid \hat{\theta})$. In fact, $f(y \mid \hat{\theta})$ is obtained by replacing the posterior distribution $f(\theta \mid x_{1:n})$ in (9.8) with a point measure at $\hat{\theta}$, typically the MLE. In contrast, the Bayesian prediction automatically incorporates the uncertainty in the parameter estimation. There is, however, a connection between the two approaches. As we have seen in Sect. 6.6, the posterior distribution is asymptotically normal with mean equal to the MLE and variance approaching zero if the sample size n tends to infinity. Therefore,

$$f(\theta \mid x_{1:n}) \xrightarrow{n\to\infty} \text{point mass distribution in } \hat{\theta}_{\text{ML}}.$$

The Bayesian prediction will therefore converge to the plug-in prediction for increasing sample size n:

$$f(y \mid x_{1:n}) \xrightarrow{n\to\infty} f(y \mid \hat{\theta}_{\text{ML}}).$$

Example 9.7 (Normal model) Let $X_{1:n}$ denote a random sample from an $\text{N}(\mu, \sigma^2)$ distribution with unknown mean μ and known variance σ^2. Our goal is to predict a future-independent observation $Y = X_{n+1}$ from the same distribution. Using Jeffreys' prior $f(\mu) \propto 1$ for μ, we know that the posterior is (cf. Example 6.14)

$$\mu \mid x_{1:n} \sim \text{N}\left(\bar{x}, \frac{\sigma^2}{n}\right).$$

The posterior predictive distribution of $Y \mid x_{1:n}$ has therefore the density

$$\begin{aligned} f(y \mid x_{1:n}) &= \int f(y \mid \mu) f(\mu \mid x_{1:n})\, d\mu \\ &\propto \int \exp\left[-\frac{1}{2}\left\{\frac{(\mu - y)^2}{\sigma^2} + \frac{n(\mu - \bar{x})^2}{\sigma^2}\right\}\right] d\mu. \end{aligned}$$

Using Appendix B.1.5, we have

$$\frac{(\mu - y)^2}{\sigma^2} + \frac{n(\mu - \bar{x})^2}{\sigma^2} = C(\mu - c)^2 + \frac{(y - \bar{x})^2}{(1 + \frac{1}{n})\sigma^2},$$

where $C = (n+1)/\sigma^2$ and $c = (y + n\bar{x})/(n+1)$. Note that the first term in this sum is a quadratic form of μ, whereas the second term does not depend on μ. It then follows that

$$\begin{aligned} f(y \mid x_{1:n}) &\propto \int \exp\left\{-\frac{C}{2}(\mu - c)^2 - \frac{1}{2}\frac{(y - \bar{x})^2}{(1 + \frac{1}{n})\sigma^2}\right\} d\mu \\ &= \exp\left\{-\frac{1}{2}\frac{(y - \bar{x})^2}{(1 + \frac{1}{n})\sigma^2}\right\} \underbrace{\int \exp\left\{-\frac{C}{2}(\mu - c)^2\right\} d\mu}_{=\sqrt{2\pi}\sigma/\sqrt{n+1}} \\ &\propto \exp\left\{-\frac{1}{2}\frac{(y - \bar{x})^2}{(1 + \frac{1}{n})\sigma^2}\right\}. \end{aligned}$$

Table 9.1 Mean and standard deviation of the posterior predictive distribution of the transformation factor

Gender	Number of volunteers	Predictive distribution	
		Mean	Standard deviation
Female	33	2305.5	241.2
Male	152	2473.1	238.5
Total	185	2445.8	238.4

This is the kernel of the normal density with mean $\bar{x}$ and variance $\sigma^2(1+\frac{1}{n})$, so the posterior predictive distribution is

$$Y \mid x_{1:n} \sim \mathrm{N}\left(\bar{x}, \sigma^2\left(1+\frac{1}{n}\right)\right). \tag{9.10}$$

Note that the plug-in prediction is also normal with mean $\bar{x}$, but has smaller variance σ^2. However, the likelihood and the bootstrap prediction give exactly the same result, cf. Examples 9.2 and 9.5.

We note that if the variance is also unknown, application of a reference prior $f(\mu, \sigma^2) \propto \sigma^{-2}$ leads to a t distribution as posterior predictive distribution:

$$Y \mid x_{1:n} \sim \mathrm{t}\left(\bar{x}, \left(1+\frac{1}{n}\right)\frac{\sum_{i=1}^n (x_i - \bar{x})^2}{n-1}, n-1\right),$$

see Exercise 2. ■

Example 9.8 (Blood alcohol concentration) Suppose that instead of Jeffreys' prior as in Example 9.7, we now use a normal prior $\mu \sim \mathrm{N}(\nu, \delta^{-1})$ for the mean μ of a normal random sample with known variance σ^2. The posterior predictive distribution is then

$$Y \mid x_{1:n} \sim \mathrm{N}\left(\mathsf{E}(\mu \mid x_{1:n}), \sigma^2\left(\frac{\delta\sigma^2 + n + 1}{\delta\sigma^2 + n}\right)\right). \tag{9.11}$$

Here $\mathsf{E}(\mu \mid x_{1:n})$ denotes the posterior mean of μ as derived in Example 6.8. Note that the posterior predictive variance in (9.11) is always larger than σ^2 and reduces to $\sigma^2(1+\frac{1}{n})$ for $\delta = 0$ as it should, cf. Eq. (9.10).

We now compute the posterior predictive distribution for the transformation factor both separately for females and males and overall, using a $\mu \sim \mathrm{N}(\nu, \delta^{-1})$ prior and with known standard deviation $\sigma = 237.8$. The mean and standard deviation of the posterior predictive normal distribution (9.11) are shown in Table 9.1. ■

Example 9.9 (Laplace's rule of succession) Laplace's rule of succession describes a famous predictive distribution, derived in the 18th century by Pierre-Simon Laplace (1749–1827). Laplace tried to calculate the probability that the sun will rise tomorrow, given that it has risen every day for the past n days, say. More generally, one may assume that the sun has risen x times in the past n days. Laplace was of course mainly interested in the case $x = n$.

Assume that $X_{1:n}$ is a random sample from the $\mathrm{B}(\pi)$ distribution. Then

$$X = \sum_{i=1}^{n} X_i \sim \mathrm{Bin}(n, \pi).$$

A priori we assume that $\pi \sim \mathrm{Be}(\alpha, \beta)$, so the posterior distribution is $\pi \mid x \sim \mathrm{Be}(\alpha + x, \beta + n - x)$. Our goal is to predict a future observation $Y = X_{n+1} \sim \mathrm{B}(\pi)$. Now $Y \mid x$ is a beta-binomial distribution with parameters 1, $\alpha + x$ and $\beta + n - x$, which is simply a Bernoulli distribution with success probability $\Pr(Y = 1 \mid x) = (\alpha + x)/(\alpha + \beta + n)$, cf. Appendix A.5.1.

Laplace was particularly interested in the case $\alpha = \beta = 1$, i.e. a uniform prior distribution. The posterior predictive probability for success is now

$$\Pr(Y = 1 \mid x) = \frac{x + 1}{n + 2},$$

and therefore

$$\Pr(Y = 0 \mid x) = \frac{n - x + 1}{n + 2}.$$

For example, suppose that the sun has risen in the past $x = n = 1\,000\,000$ days, say. The probability that it will not rise tomorrow (without additional information) is

$$\Pr(Y = 0 \mid x) = \frac{1}{1\,000\,002} \approx 10^{-6}.$$

■

Laplace's rule of succession can be applied also to more interesting situations, as in the following example.

Example 9.10 (Sequential analysis of binary outcomes from a clinical trial) Suppose that a clinical trial is conducted where the primary outcome is the success rate of a novel therapy. The therapy comes with certain side effects, so it is generally agreed upon that a success probability below 0.5 cannot be justified for individual therapy. However, there is some optimism that the success rate π is around 0.75, and this is reflected in a $\mathrm{Be}(6, 2)$ prior for π.

The outcomes Y from patients enrolled enter sequentially in the order shown in Table 9.2, with success coded as 1 if the patient had a successful therapy and 0 otherwise. The predictive probability $\Pr(Y = 1 \mid \text{data so far})$, the probability that the therapy is successful for a future patient given the data so far, may now be used as an early stopping criterion: if it falls below 0.5, then the trial must be stopped. Fortunately, this is not the case for the data shown in Table 9.2. This scenario is of course somewhat simplistic but contains important features of a sequential analysis of clinical trials using predictive probabilities.

■

Table 9.2 Successive change of the posterior predictive probability $\Pr(Y = 1 \mid \text{data so far})$ for success using a Be(6, 8) prior for the success probability at the start of the trial

Patient ID	Patient outcome	$\Pr(Y = 1 \mid \text{data so far})$
	(start of clinical trial)	$6/8 = 0.75$
1	0	$6/9 \approx 0.67$
2	0	$6/10 = 0.60$
3	1	$7/11 \approx 0.64$
4	0	$7/12 \approx 0.58$
5	1	$8/13 \approx 0.62$
6	1	$9/14 \approx 0.64$
7	1	$10/15 \approx 0.67$
8	1	$11/16 \approx 0.69$
9	1	$12/17 \approx 0.71$
10	0	$12/18 \approx 0.67$
11	0	$12/19 \approx 0.63$
⋮	⋮	⋮

9.3.2 Computation of the Posterior Predictive Distribution

In Sect. 7.2.1 we noted that the prior predictive distribution (9.9) is simply the denominator in Bayes' theorem; it can therefore be calculated if likelihood, prior and posterior are known:

$$f(y) = \frac{f(y \mid \theta) f(\theta)}{f(\theta \mid y)},$$

which holds for any θ. A similar formula also holds for the posterior predictive distribution:

$$\begin{aligned} f(y \mid x) &= \frac{f(y \mid x, \theta) f(\theta \mid x)}{f(\theta \mid x, y)} \\ &= \frac{f(y \mid \theta) f(\theta \mid x)}{f(\theta \mid x, y)}, \end{aligned}$$

where the last equation follows from conditional independence of X and Y given θ.

If $f(\theta)$ is conjugate with respect to $f(x \mid \theta)$, then $f(\theta \mid x)$ and $f(\theta \mid x, y)$ belong to the same family of distributions considered. The posterior predictive distribution can then be derived without explicit integration. We will illustrate the procedure in the following example.

Example 9.11 (Poisson model) Let $X \sim \text{Po}(e_x \lambda)$, $Y \sim \text{Po}(e_y \lambda)$ and $\lambda \sim \text{G}(\alpha, \beta)$ a priori, with X and Y being conditionally independent. From Example 6.30 we know that

$$\lambda \mid x \sim \text{G}(\tilde{\alpha}, \tilde{\beta}) \quad \text{and}$$

$$\lambda \mid x, y \sim \mathrm{G}(\tilde{\alpha} + y, \tilde{\beta} + e_y),$$

where $\tilde{\alpha} = \alpha + x$ and $\tilde{\beta} = \beta + e_x$. The posterior predictive distribution is therefore

$$\begin{aligned} f(y \mid x) &= \frac{f(y \mid \lambda) f(\lambda \mid x)}{f(\lambda \mid x, y)} \\ &= \frac{\frac{(e_y\lambda)^y}{y!} \exp(-e_y\lambda) \frac{\tilde{\beta}^{\tilde{\alpha}}}{\Gamma(\tilde{\alpha})} \lambda^{\tilde{\alpha}-1} \exp(-\tilde{\beta}\lambda)}{\frac{(\tilde{\beta}+e_y)^{\tilde{\alpha}+y}}{\Gamma(\tilde{\alpha}+y)} \lambda^{\tilde{\alpha}+y-1} \exp\{-(\tilde{\beta}+e_y)\lambda\}} \\ &= \frac{\tilde{\beta}^{\tilde{\alpha}}}{\Gamma(\tilde{\alpha})} \frac{e_y^y \Gamma(\tilde{\alpha}+y)}{y!} (\tilde{\beta}+e_y)^{-(\tilde{\alpha}+y)}, \end{aligned}$$

i.e. the Poisson-gamma distribution with parameters $\tilde{\alpha}$, $\tilde{\beta}$ and e_y, compare Appendix A.5.1.

An equivalent way to obtain the posterior predictive distribution is to first identify the prior predictive distribution $f(y)$ as the Poisson-gamma distribution with parameters α, β and e_x. The posterior predictive distribution $f(y \mid x)$ can then be obtained by replacing the prior parameters α and β by the posterior parameters $\tilde{\alpha}$ and $\tilde{\beta}$, respectively, and e_x by e_y, compare (9.9) with (9.8).

The mean and variance of this posterior predictive distribution are

$$\mathsf{E}(Y \mid x) = \frac{\alpha + x}{\beta + e_x} e_y \quad \text{and}$$

$$\mathrm{Var}(Y \mid x) = \frac{\alpha + x}{\beta + e_x} e_y \left(1 + \frac{e_y}{\beta + e_x}\right).$$

Under Jeffreys' prior, i.e. $\alpha = 1/2$ and $\beta = 0$ (cf. Table 6.3), these formulas simplify to

$$\mathsf{E}(Y \mid x) = \frac{e_y}{e_x}\left(x + \frac{1}{2}\right) \quad \text{and}$$

$$\mathrm{Var}(Y \mid x) = \left(1 + \frac{e_y}{e_x}\right) \frac{e_y}{e_x}\left(x + \frac{1}{2}\right).$$

In our original example with $x = 11$, $e_x = e_y = 3.04$ we obtain $\mathsf{E}(Y \mid x) = 11.5$ and $\mathrm{Var}(Y \mid x) = 23$. Table 9.3 contains the mean and variance of all predictive distributions considered for the Poisson model example. The results of likelihood and Bayesian approaches to prediction are very close, whereas the bootstrap prediction leads to somewhat smaller values of the predictive mean and variance. Also, the predictive mean of all prediction methods is close to the mean of the plug-in predictive distribution, but the variance is nearly twice as large. For $x = 0$, the Bayesian approach gives $\mathsf{E}(Y \mid x) = 0.5$ and $\mathrm{Var}(Y \mid x) = 1$, again in sharp contrast to the plug-in predictive distribution with mean and variance equal to zero. ■

Table 9.3 Mean and variance of different predictive distributions for $x = 11$ and $e_x = e_y = 3.04$. Exact values are given for the bootstrap prediction based on (9.6) and (9.7)

Predictive distribution	Mean	Variance
Plug-in	11.000	11.000
Likelihood	11.498	22.998
Bootstrap	11.000	22.000
Bayesian	11.500	23.000

9.3.3 Model Averaging

In Sect. 7.2.4 Bayesian model averaging has been introduced, which allows one to combine estimates from different models with respect to their posterior probabilities. Analogously, it is possible to combine different predictions from different models: Let $M_1, \ldots, M_K$ denote the models considered and $f(y \mid x, M_k)$ the posterior predictive density in model M_k, then the model-averaged posterior predictive density is

$$f(y \mid x) = \sum_{k=1}^{K} f(y \mid x, M_k) \cdot \mathsf{Pr}(M_k \mid x), \tag{9.12}$$

where the posterior model probabilities $\mathsf{Pr}(M_k \mid x)$ are appearing as weights similar as in Sect. 7.2.4. Prediction based on model averaging takes therefore model uncertainty into account.

Example 9.12 (Blood alcohol concentration) In Example 7.7 we have considered two models for the blood alcohol transformation data. Model M_1 did not distinguish the two subgroups defined by gender whereas model M_2 did. The posterior model probabilities turned out to be $\mathsf{Pr}(M_1 \mid x) = 0.0463$ and $\mathsf{Pr}(M_2 \mid x) = 0.9537$.

In Example 9.8 we have computed the corresponding posterior predictive distribution under the two models. The posterior predictive mean

$$\mathsf{E}(y \mid x) = \sum_{k=1}^{2} \mathsf{E}(y \mid x, M_k) \cdot \mathsf{Pr}(M_k \mid x)$$

is equal to the model-averaged mean transformation factor shown in Table 7.3. Note that the posterior predictive distribution (9.12) is now a mixture of two normal distributions, with mixture weights equal to $\mathsf{Pr}(M_1 \mid x)$ and $\mathsf{Pr}(M_2 \mid x)$. The mean and standard deviation of the two normal components can be read off from Table 9.1. ■

Predictions based on model averaging are more than a technical gadget using elementary probability. Frequently, one has to assume that a statistical model does not reflect the truth entirely, cf. the discussion at the end of Sect. 7.2.4. Even if there is a *correct* model, it is uncertain that it has been considered as a candidate model. Under uncertainty on which model fits the underlying data, predictions based on model averaging will generally provide better results than predictions based on a single model. For example, the model-averaged expectation

$$\mathsf{E}(y \mid x) = \sum_{k=1}^{K} \mathsf{E}(y \mid x, M_k) \cdot \Pr(M_k \mid x)$$

of the posterior predictive distribution (9.12) minimises the expectation of the posterior predictive squared error-loss $l(a, y) = (a - y)^2$, i.e.

$$\mathsf{E}(y \mid x) = \arg\min_{d} \int l(a, y) f(y \mid x)\, dy.$$

The proof is similar to the proof that the posterior mean minimises the expected squared error loss, see Sect. 6.4.1 for comparison.

9.4 Assessment of Predictions

Prediction of future events is one of the main challenges of statistical methodology. We saw that both likelihood and Bayesian approaches provide a predictive distribution $f(y)$, i.e. a probabilistic prediction in contrast to a deterministic point prediction. But how can we assess the quality of a probabilistic prediction? In this section we will discuss how the performance of a prediction method can be quantified. We will introduce scoring rules that allow us to measure the quality of a probabilistic prediction method through comparison with the actually observed event.

9.4.1 Discrimination and Calibration

In order to judge the performance of a statistical model, the actual observations are compared with the corresponding predictive distributions. At least two aspects are of importance in this comparison: *discrimination* and *calibration*. Discrimination describes how well the model is able to predict different observations with different predictions. Point predictions are central to discrimination, whereas the uncertainty of the prediction may not be considered at all. Calibration, however, takes into account the entire predictive distribution in the sense of a statistical agreement with the actual observations. For example, there should be on average only five out of a hundred observations outside a 95 % prediction interval if the prediction is calibrated.

Any further discussion of these concepts heavily depends on whether the predictive distribution is discrete or continuous. To begin with, we will consider (multiple) binary predictive distributions $f(y_i)$ (indexed with i), which are completely characterised by the corresponding prediction probabilities $\Pr(Y_i = 1) = \pi_i$. Subsequently, we will discuss the univariate continuous case.

For a binary variable $Y_i \in \{0, 1\}$, perfect agreement between the prediction probabilities $\pi_i \in [0, 1]$ and the actually observed realisations $y_i \in \{0, 1\}$ is not achievable. The discrimination of a binary prediction describes in this case the capacity to correctly predict the classification ($y_i = 0$ or $y_i = 1$). Classification will usually be achieved using a threshold π_t assigning the classes $y_i = 0$ or $y_i = 1$ to the probabilities π_i if $\pi_i < \pi_t$ or $\pi_i \geq \pi_t$ is satisfied, respectively. This rule is not specifically connected to the probabilities π_i, one could just as well use $\text{logit}(\pi_i)$ with threshold $\text{logit}(\pi_t)$ for classification.

Calibration on the other hand is directly connected to the prediction probabilities π_i and implies that on average $\pi_i \cdot 100$ % of the events actually occur with prediction probability π_i. For example, events that are predicted with a probability of 80 % should on average occur in four out of five cases. Good discrimination does not necessarily coincide with good calibration, as the following fictitious example shows.

Example 9.13 (Prediction of soccer matches) Suppose we aim to predict the outcome of soccer matches. For simplicity, the outcome of the ith match is dichotomised in home victory ($y_i = 1$) and tie or away win ($y_i = 0$).

From experience it is known that approximately 50 % of all professional soccer matches are won by the home team. A possible strategy therefore is to choose $\pi_i = 0.5$ no matter which teams are playing. Such a prediction is well calibrated but is obviously not discriminating. Since only the frequency of the event of interest comes into play for this strategy, we will be referring to it as the *prevalence prediction*.

A soccer expert may classify all matches into two equally sized groups, where the home victory probability is predicted as $\pi_i = 0.8$ in the first group and as $\pi_i = 0.2$ in the second. Assume that actually all matches of the first group have been home victories, whereas all games in the second group are not. The prediction of this expert has then optimal discrimination but is on the other hand not calibrated since not 80 % but 100 % of the matches in the first group and not 20 % but 0 % in the second group are home victories. We will refer to this prediction as the *expert prediction*.

Both criteria would be satisfied by a perfect prediction of $\pi_i = 1$ for all home victories and $\pi_i = 0$ for all ties and away wins. This prediction will therefore be called the *oracle prediction*. ■

By far the most frequently used measure of discrimination for binary predictions is the so-called *area under the curve* (AUC), also called *c-index*, which is usually defined as the area under the *ROC curve*, where ROC stands for "receiver operating characteristic", a term commonly used in signal detection theory. In general, AUC is a number between zero and one, where only values above 0.5 reflect a certain quality of classification. AUC can also be defined in a different way:

Definition 9.3 (AUC) AUC is the probability that a randomly chosen event i, which actually occurred ($y_i = 1$), has a larger prediction probability than another randomly chosen event j, which actually did not occur ($y_j = 0$):

$$\text{AUC} = \Pr(\pi_i > \pi_j).$$

In the case that different events may have the same probabilities this definition has to be extended to

$$\text{AUC} = \Pr(\pi_i > \pi_j) + \frac{1}{2}\Pr(\pi_i = \pi_j). \qquad \blacklozenge$$

Note that AUC has not necessarily to be defined through the probabilities π_i since one could alternatively use any strictly monotone transformation as, for example, $\text{logit}(\pi_i)$. AUC can be estimated by the *Wilcoxon rank sum statistic*.

To empirically assess calibration of binary predictions, the probabilities have to be grouped. In practice identical or at least very close prediction probabilities are combined in J groups with representative probabilities $\pi_1, \dots, \pi_J$. The groups should be of roughly the same size, and observations with the same prediction probability should not be in different groups. For each π_j, let n_j be the number of predicted events, and $\bar{y}_j$ the relative frequency of the predicted event in the jth group. The total number of predictions is denoted by $N = \sum_{j=1}^{J} n_j$.

Definition 9.4 (Sanders' calibration) *Sanders' calibration* is defined as

$$\text{SC} = \frac{\sum_{j=1}^{J} n_j(\bar{y}_j - \pi_j)^2}{N}. \qquad \blacklozenge$$

Smaller values of SC are better, perfect calibration corresponds to $\text{SC} = 0$. Grouping of the data allows us to define an alternative measure of discrimination.

Definition 9.5 (Murphy's resolution) *Murphy's resolution* is given by

$$\text{MR} = \frac{\sum_{j=1}^{J} n_j(\bar{y}_j - \bar{y})^2}{N},$$

where $\bar{y} = N^{-1}\sum_{i=1}^{N} y_i$ denotes the overall prevalence. ♦

MR is a measure that should be as large as possible. Since it does not take into account the order of the predictions, it is much less commonly used than AUC. In fact a prediction, which always classifies *incorrectly*, will have the largest possible value of MR. Limited to predictions with $\text{AUC} \geq 0.5$, it is nevertheless a sensible measure, and it will be revisited in Sect. 9.4.2.

Example 9.14 (Prediction of soccer matches) The values of AUC, SC and MR are summarised in Table 9.4 for the three different predictions of Example 9.13. Additionally the *inverted oracle prediction*, which classifies all matches incorrectly, is given for illustration. The prevalence prediction is well calibrated ($\text{SC} = 0$) but not discriminating ($\text{AUC} = 0.5$ and $\text{MR} = 0$). The expert prediction is not well calibrated ($\text{SC} = 0.04 > 0$) but always classifies correctly ($\text{AUC} = 1$). The oracle prediction always classifies correctly as well and is perfectly calibrated, whereas the inverted oracle prediction has the worst possible values of AUC and SC. The values

Table 9.4 Comparison of the different predictions in the soccer example using AUC, Sanders' calibration (SC), Murphy's resolution (MR) and the Brier score (BS)

Prediction	AUC	SC	MR	BS
Prevalence	0.5	0	0	0.25
Expert	1	0.04	0.25	0.04
Oracle	1	0	0.25	0
Inverted oracle	0	1	0.25	1

of MR are, however, identical for expert, oracle and inverted oracle prediction since it is assessing only discrimination but not the direction of classification. The values of the Brier score BS in the last column are discussed in Example 9.16. ■

Next, we will discuss continuous predictions with density $f(y)$ and corresponding distribution function $F(y)$. In this case the following quantity is often used as a calibration measure:

Definition 9.6 (PIT) The *probability integral transform* (PIT) is the value of the predictive distribution function $F(y) = \Pr(Y \leq y)$ at the actually observed value y_o:

$$\text{PIT}(y_o) = F(y_o). \qquad \blacklozenge$$

Displaying a *PIT histogram* is a commonly used method to check calibration of a set of predictions because PIT values from perfect predictions follow the standard uniform distribution: $\text{PIT}(Y_o) \sim \text{U}(0, 1)$. Indeed, Y_o has a distribution function F if $Y_o = F^{-1}(U)$ where $U \sim \text{U}(0, 1)$, so $F(Y_o) \sim \text{U}(0, 1)$. However, predictions not equal to the data-generating distribution may also have uniform PIT histograms, as the following example shows.

Example 9.15 (Normal model) Suppose X and Y_o are independent, normally distributed random variables, both with unknown expectation μ and known variance $\sigma^2 = 1$. We want to predict Y_o after observing $X = x$. The plug-in prediction is in this case $Y \mid X = x \sim \text{N}(x, 1)$, whereas the Bayesian prediction with Jeffreys' prior for μ (just as the likelihood or the bootstrap prediction) predicts $Y \mid X = x \sim \text{N}(x, 2)$, cf. Example 9.7.

Based on $M = 10\,000$ independent realisations of X and Y_o (with true $\mu = 0$), PIT values for the plug-in and the Bayesian prediction have been computed as follows and are shown in Fig. 9.2:

```
set.seed(1)
M <- 10000
x <- rnorm(M)
y <- rnorm(M)
pit.plugin <- pnorm(y, mean = x, sd = 1)
pit.bayes <- pnorm(y, mean = x, sd = sqrt(2))
```

The plug-in prediction is obviously not well calibrated since the corresponding PIT histogram in Fig. 9.2 does not resemble a uniform distribution but has a bowl shape, which is typical for predictions with too small variances. The Bayesian prediction $Y \mid X = x \sim \text{N}(x, 2)$ seems to be well calibrated, which can be confirmed

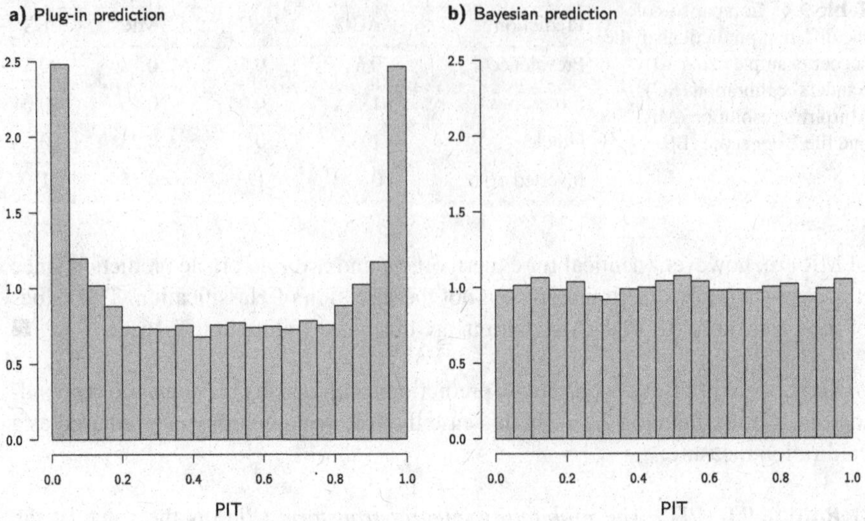

Fig. 9.2 PIT histograms of the plug-in and the Bayesian prediction of $Y_o \sim \mathrm{N}(\mu, 1)$ based on an independent realisation of $X \sim \mathrm{N}(\mu, 1)$

analytically. First, note that

$$\begin{aligned} \mathrm{PIT}(y_o) &= \Pr(Y \leq y_o) \\ &= \Pr\left(\frac{Y-x}{\sqrt{2}} \leq \frac{y_o - x}{\sqrt{2}}\right) \\ &= \Phi\left(\frac{y_o - x}{\sqrt{2}}\right), \end{aligned}$$

where $\Phi(\cdot)$ denotes the distribution function of the standard normal distribution. Now x is a realisation of $X \sim \mathrm{N}(\mu, 1)$, and y_o is a realisation of $Y_o \sim \mathrm{N}(\mu, 1)$. We have $Z = (Y_o - X)/\sqrt{2} \sim \mathrm{N}(0, 1)$ and therefore $\Phi(Z) \sim \mathrm{U}(0, 1)$. Viewed as a function of the random variables X and Y_o, we thus have

$$\mathrm{PIT}(Y_o) = \Phi\left(\frac{Y_o - X}{\sqrt{2}}\right) \sim \mathrm{U}(0, 1).$$

Note that, by definition, the oracle prediction $Y \sim N(0, 1)$, which is equal to the true data-generating distribution of Y_o, also has uniform PIT values. ■

As for binary events, perfect calibration is not sufficient for a good prediction. In the above example both the Bayesian and oracle predictions are well calibrated, but common sense suggests that the oracle prediction is better. Indeed, another aspect of a good continuous prediction is *sharpness*, i.e. how concentrated a predictive distribution is. The oracle has smaller variance (or higher sharpness) than the Bayesian prediction, so should be preferred.

For the assessment of the accuracy of continuous predictions, the focus is often on the point prediction $\hat{Y}$, which is in most cases the expectation $\mathsf{E}(Y)$ of the predictive distribution. A very frequently used criterion is the *squared prediction error*

$$\mathrm{SPE} = (\hat{Y} - y_o)^2.$$

However, the reduction of the predictive distribution to a point prediction ignores the uncertainty of the prediction. The squared prediction error (similarly to the AUC) does not take into account if the underlying predictive distribution is correctly calibrated. In the following section we will introduce scoring rules which allow us to simultaneously quantify calibration and discrimination (for binary predictions) or calibration and sharpness (for continuous predictions).

9.4.2 Scoring Rules

To be able to incorporate the entire predictive distribution in the assessment of the prediction, so-called scoring rules are considered.

Definition 9.7 (Scoring rules) A *scoring rule* $S(f(y), y_o)$ assigns a real number to the probability mass or density function $f(y)$ of a predictive distribution and the actually observed value y_o. ♦

Scoring rules are typically negatively oriented, i.e. smaller values of $S(f(y), y_o)$ reflect a better prediction. It is reasonable to only use proper scoring rules, defined as follows.

Definition 9.8 (Proper scoring rule) A scoring rule $S(f(y), y_o)$ is called *proper* if the expected score $\mathsf{E}\{S(f(y), Y_o)\}$ with respect to the true data-generating distribution $Y_o \sim f_o$ is minimised if the predictive distribution f is equal to the data-generating distribution f_o. If the minimum is unique, then the scoring rule is called *strictly proper*. ♦

For strictly proper scoring rules, it is therefore disadvantageous to deviate from the true data-generating distribution $f_o(y)$ since the expected score is optimal if $f(y) = f_o(y)$.

In practice scoring rules will be applied for as many events as possible in order to rely on a better basis for the assessment of the quality of the prediction. A strictly proper scoring rule can be used for several events since the sum and the average of the scores remain strictly proper.

Definition 9.9 (Scoring rules for binary predictions) Let $Y \sim \mathrm{B}(\pi)$ be the predictive distribution for a binary event, i.e.

$$f(y) = \begin{cases} \pi & \text{for } y = 1, \\ 1 - \pi & \text{for } y = 0. \end{cases}$$

The *Brier score* BS, the *absolute score* AS and the *logarithmic score* LS are defined as

$$\mathrm{BS}\big(f(y), y_o\big) = (y_o - \pi)^2, \tag{9.13}$$

$$\mathrm{AS}\big(f(y), y_o\big) = |y_o - \pi| \quad \text{and} \tag{9.14}$$

$$\mathrm{LS}\big(f(y), y_o\big) = -\log f(y_o), \tag{9.15}$$

respectively. ♦

The Brier score is also known as the *probability score.*

Result 9.1 (Brier score) *The Brier score* (9.13) *is strictly proper.*

Proof To show Result 9.1, let $\mathrm{B}(\pi_o)$ be the true distribution of Y_o. Then the expected Brier score is given by

$$\begin{aligned}\mathsf{E}\big\{\mathrm{BS}\big(f(y), Y_o\big)\big\} &= \mathsf{E}\big\{(Y_o - \pi)^2\big\}\\ &= \mathsf{E}\big(Y_o^2\big) - 2\pi\,\mathsf{E}(Y_o) + \pi^2\\ &= \mathsf{E}(Y_o) - 2\pi\pi_o + \pi^2\\ &= \pi_o - 2\pi\pi_o + \pi^2.\end{aligned}$$

Hence, we have

$$\frac{d\,\mathsf{E}\{\mathrm{BS}(f(y), Y_o)\}}{d\pi} = -2\pi_o + 2\pi,$$

from which we derive the root $\pi = \pi_o$. Inspecting the second derivative

$$\frac{d^2\,\mathsf{E}\{\mathrm{BS}(f(y), Y_o)\}}{d\pi^2} = 2$$

shows that the minimum is unique. Using the true success probability π_o as predictive probability hence gives the minimal expected score. □

In the following we decompose the mean Brier score $\overline{\mathrm{BS}}$—averaged over a series of binary predictions—in two terms measuring calibration and discrimination, respectively. To do so, we consider a series of N binary predictions with predictive probabilities $\pi_1, \ldots, \pi_N$. The corresponding observed events are denoted by $y_1, \ldots, y_N$, and $\bar{y}$ denotes the overall prevalence of the observed binary events.

We first note that the Brier score of the prevalence prediction is $\bar{y}(1 - \bar{y})$ and may be used as an upper bound for useful predictions. In *Murphy's decomposition* the predictions are first grouped as in Sect. 9.4.1 leading to the decomposition

$$\overline{\mathrm{BS}} = \frac{1}{N}\sum_{i=1}^{N}(y_i - \pi_i)^2 = \bar{y}(1 - \bar{y}) - \mathrm{MR} + \mathrm{SC} \tag{9.16}$$

of the mean Brier score. This shows explicitly that the Brier score assesses both discrimination and calibration, through MR and SC, respectively.

Example 9.16 (Prediction of soccer matches) In the last column of Table 9.4 the Brier score of each prediction is given. The Brier score orders the four predictions by combining discrimination and calibration in a sensible way: the oracle prediction is the best, followed by the expert and the prevalence prediction, and the inverted oracle prediction is the worst.

Note that the overall prevalence of matches won by the home team is 50 %, so the upper bound on the Brier score is 0.25, and this is of course the Brier score of the prevalence prediction. It is easily confirmed that Murphy's decomposition is fulfilled for all predictions considered. ■

Result 9.2 (Absolute score) *The absolute score* (9.14) *is not proper.*

Proof To show this result, let $\mathrm{B}(\pi_o)$ be the true distribution of Y_o. Then the expected absolute score is

$$\begin{aligned} \mathsf{E}\{\mathrm{AS}(f(y), Y_o)\} &= \mathsf{E}\{|Y_o - \pi|\} \\ &= (1-\pi)\pi_o + \pi(1-\pi_o) \\ &= \pi(1-2\pi_o) + \pi_o. \end{aligned}$$

The expected score is not minimised by $\pi = \pi_o$ but by

$$\pi = \begin{cases} 0 & \text{for } \pi_o < 1/2, \\ 1 & \text{for } \pi_o > 1/2, \end{cases}$$

so the absolute score is not proper. For $\pi_o = 1/2$, the expected score is 1/2 and hence independent of π. □

Result 9.3 (Logarithmic score) *The logarithmic score* (9.15) *is strictly proper.*

Proof To show the propriety of the logarithmic score, let again $\mathrm{B}(\pi_o)$ be the true distribution of Y_o. The expected logarithmic score is given by

$$\begin{aligned} \mathsf{E}\{\mathrm{LS}(f(y), Y_o)\} &= -\mathsf{E}\{\log f(Y_o)\} \\ &= -\log(\pi)\pi_o - \log(1-\pi)(1-\pi_o). \end{aligned}$$

Hence, we have

$$\frac{d\,\mathsf{E}\{\mathrm{LS}(f(y), Y_o)\}}{d\pi} = -\pi_o/\pi + (1-\pi_o)/(1-\pi),$$

from which we derive the root $\pi = \pi_o$. Through inspection of the second derivative we conclude that this minimum is unique. □

The logarithmic score is strictly proper not only for binary events but for arbitrary probability mass or density functions of Y. For example, if the predictive distribution is normal, i.e. $Y \sim \mathrm{N}(\mu, \sigma^2)$, then the logarithmic score is

$$\mathrm{LS}\big(f(y), y_o\big) = \frac{1}{2}\left\{\log(2\pi) + \log(\sigma^2) + \frac{(y_o - \mu)^2}{\sigma^2}\right\}.$$

As an alternative, one can use the *continuous ranked probability score* (CRPS), which is defined as

$$\mathrm{CRPS}\big(f(y), y_o\big) = \int \big\{F(t) - \mathsf{I}_{[y_o,\infty)}(t)\big\}^2 dt$$

and which is also strictly proper for arbitrary predictions with distribution function $F(y)$. The CRPS is closely related to the Brier score (9.13). For fixed t, we have

$$\mathsf{I}_{[y_o,\infty)}(t) = \begin{cases} 1 & \text{for } y_o \le t, \\ 0 & \text{for } y_o > t, \end{cases}$$

with corresponding success probability $\Pr(Y \le t) = F(t)$. So $\{F(t) - \mathsf{I}_{[y_o,\infty)}(t)\}^2$ is a Brier score, and the CRPS is the integral of the Brier score over all possible thresholds t.

It is possible to show that the CRPS can be written as

$$\mathrm{CRPS}\big(f(y), y_o\big) = \mathsf{E}\big\{|Y_1 - y_o|\big\} - \frac{1}{2}\mathsf{E}\big\{|Y_1 - Y_2|\big\},$$

where Y_1 and Y_2 are independent random variables with density function $f(y)$ or distribution function $F(y)$. This representation allows a simplification of the formula for the CRPS for certain predictive distributions. For example, for a normally distributed prediction $Y \sim \mathrm{N}(\mu, \sigma^2)$, the CRPS is

$$\mathrm{CRPS}\big(f(y), y_o\big) = \sigma\left[\tilde{y}_o\big\{2\Phi(\tilde{y}_o) - 1\big\} + 2\varphi(\tilde{y}_o) - \frac{1}{\sqrt{\pi}}\right], \tag{9.17}$$

where $\tilde{y}_o = (y_o - \mu)/\sigma$, while $\varphi(\cdot)$ and $\Phi(\cdot)$ denote the density and distribution functions, respectively, of the standard normal distribution. A derivation of this result is discussed in Exercise 6.

Example 9.17 (Normal model) For the predictions of Example 9.15, we calculated the logarithmic score and the CRPS for each of the $M = 10\,000$ independent realisations of X and Y_o and averaged these values afterwards. Table 9.5 shows that the plug-in prediction performs worse for both scores and that the Bayesian prediction is always better. This is due to the lack of calibration of the plug-in prediction. The oracle prediction performs best. Table 9.5 also gives the averaged squared prediction error, which is the same for the plug-in and Bayesian predictions since both provide the same point predictions. Differences in the predictive variances are not taken into account. Again, the oracle prediction performs best. ■

Table 9.5 Comparison of the different predictions in the normal model example with respect to the logarithmic score (LS), the continuous ranked probability score (CRPS) and the squared prediction error (SPE)

Prediction	LS	CRPS	SPE
Plug-in	1.9172	0.9196	1.9966
Bayesian	1.7647	0.7950	1.9966
Oracle	1.4097	0.5546	0.9816

9.5 Exercises

1. Five physicians participate in a study to evaluate the effect of a medication for migraine. Physician $i = 1, \ldots, 5$ treats n_i patients with the new medication, and it shows positive effects for y_i of the patients. Let π be the probability that an arbitrary migraine patient reacts positively to the medication. Given that

$$n = (3, 2, 4, 4, 3) \quad \text{and} \quad y = (2, 1, 4, 3, 3),$$

   (a) provide an expression for the likelihood $L(\pi)$ for this study and
   (b) specify a conjugate prior distribution $f(\pi)$ for π and choose appropriate values for its parameters. Using these parameters, derive the posterior distribution $f(\pi \mid n, y)$.
   (c) A sixth physician wants to participate in the study with $n_6 = 5$ patients. Determine the posterior predictive distribution for y_6 (the number of patients out of the five for which the medication will have a positive effect).
   (d) Calculate the likelihood prediction as well.
2. Let $X_{1:n}$ be a random sample from an $\mathrm{N}(\mu, \sigma^2)$ distribution, from which a further observation $Y = X_{n+1}$ is to be predicted. Both the expectation μ and the variance σ^2 are unknown.
   (a) Start by determining the plug-in predictive distribution.
   (b) Calculate the likelihood and bootstrap predictive distributions.
   (c) Derive the Bayesian predictive distribution under the assumption of the reference prior $f(\mu, \sigma^2) \propto \sigma^{-2}$.
3. Derive Eq. (9.11).
4. Prove Murphy's decomposition (9.16) of the Brier score.
5. Investigate if the scoring rule

$$S\big(f(y), y_o\big) = -f(y_o)$$

   is proper for a binary observation Y.
6. For a normally distributed prediction show that it is possible to write the CRPS as in (9.17) using the formula for the expectation of the folded normal distribution in Appendix A.5.2.

9.6 References

A classical monograph on statistical prediction is Geisser (1993). For the theoretical properties and other propositions, we refer to Pawitan (2001, Chap. 16) and for the more detailed description, to Young and Smith (2005, Chap. 10), where Bayesian prediction is discussed as well. Bernardo and Smith (2000) provide an extensive discussion on the Bayesian approach. An introduction into the assessment of predictions is given in O'Hagan et al. (2006, Chap. 8). A theoretical exposition on scoring rules is given in Gneiting and Raftery (2007), and calibration is reviewed in detail in Gneiting et al. (2007).

Markov Models for Time Series Analysis 10

Contents

The statistical analysis of time series is concerned with data which consists of time-ordered sequences of measurements. Such a sequence is usually assumed to be equally spaced, in which case the distance between two successive observations is always constant, for example one day. Otherwise a time series is called unequally spaced.

To introduce some notation, let $x = (x_1, \dots, x_n)$ denote a time series of observations x_t made at times $t = 1, \dots, n$. If the series is not equally spaced, then $x(t_i)$ is the more appropriate notation for the observation made at time t_i, $i = 1, \dots, n$.

The purpose of this chapter is to illustrate how likelihood and Bayesian inference can be employed in the statistical analysis of time series. We do not aim to provide a comprehensive overview of statistical models for time series. Instead we try to

L. Held, D. Sabanés Bové, *Likelihood and Bayesian Inference*,
Statistics for Biology and Health, https://doi.org/10.1007/978-3-662-60792-3_10,

discuss a number of important approaches for time series analysis which are used heavily in various biomedical applications, restricting our attention to models for equally-spaced time series.

We distinguish between observation-driven and parameter-driven models for time series. Both classes aim to take into account dependence between successive observations of a time series. An observation-driven model relates the distribution of the response variable x_t directly to the p last observations $x_{t-1}, \dots, x_{t-p}$. Unknown parameters in an observation-driven model are typically global, i.e. do not depend on time. Here likelihood inference is the inferential method of choice.

In contrast, a parameter-driven model assumes that the observations are conditionally independent given some latent unobserved process, which is typically allowed to change over time. The time-dependent latent process is unknown as well as additional parameters determining the dynamics of this process. Here empirical Bayes and fully Bayes approaches to inference provide useful alternatives to a pure likelihood analysis.

10.1 The Markov Property

A discrete-time *random process* X is a sequence of random variables $X = \{X_1, X_2, \dots\}$ which take values in a so-called *state space* S. For the moment assume that the state space S is finite.

Definition 10.1 (Markov property) The process X satisfies the *Markov property* if

$$\Pr(X_t = i \mid X_1 = x_1, X_2 = x_2, \dots, X_{t-1} = x_{t-1}) = \Pr(X_t = i \mid X_{t-1} = x_{t-1}) \tag{10.1}$$

for all $t \geq 2$ and all states $i, x_1, \dots, x_{t-1} \in S$. We then call X a (first-order) *Markov chain*. ♦

The Markov property implies that, conditional on all observations in the past, the distribution of X_t depends only on the last observation x_{t-1}. If X were a second-order Markov chain, the conditional distribution of X_t would only depend on x_{t-1} and x_{t-2}. This can be easily generalised to k-th order Markov chains. We can also consider a continuous Markov chain with real-valued state space S, in which case the Markov property (10.1) is formulated in terms of conditional density functions:

$$f(x_t \mid X_1 = x_1, X_2 = x_2, \dots, X_{t-1} = x_{t-1}) = f(x_t \mid X_{t-1} = x_{t-1}).$$

10.2 Observation-Driven Models for Categorical Data

Suppose now that the observation x_t made at time $t = 1, \dots, n$ can only take integer values in the finite set $S = \{1, 2, \dots, K\}$. A first-order Markov model is characterised through the transition probabilities

$$\Pr(X_t = j \mid X_{t-1} = i) = p_{ij}$$

of a sequence of discrete random variables $X_1, \dots, X_n$.

The transition probabilities are conveniently summarised in a $K \times K$ *transition matrix* $\mathbf{P} = (p_{ij})_{ij}$. The matrix $\mathbf{P}$ hence has entries between zero and one with all row sums equal to one. The specification of the joint distribution of X is completed with the *initial distribution* $\gamma_i = \Pr(X_1 = i)$ of the first observation x_1.

The formulation implicitly assumes that the distribution of X_t depends only on the previous observation X_{t-1}, but not on past observations beyond X_{t-1}. This is the Markov property, which can be stated explicitly as

$$\Pr(X_t \mid X_{t-1}, \dots, X_1) = \Pr(X_t \mid X_{t-1}).$$

One important feature of Markov chains is that under regularity conditions there exists a *stationary distribution* $\boldsymbol{\pi}$ (with entries $\pi_i = \Pr(X_t = i)$) with the defining feature $\boldsymbol{\pi}^\top = \boldsymbol{\pi}^\top \mathbf{P}$. This formula can be nicely interpreted as follows: If X_{t-1} has some distribution $\boldsymbol{\pi}^\top$ then the distribution of X_t is $\boldsymbol{\pi}^\top \mathbf{P}$. If this distribution is identical to $\boldsymbol{\pi}$ (the distribution of X_{t-1}), then the distribution of X_{t+1}, X_{t+2}, etc. will also be $\boldsymbol{\pi}$, so $\boldsymbol{\pi}$ is the stationary distribution of the Markov chain.

The stationary distribution can be computed via

$$\boldsymbol{\pi}^\top = \mathbf{1}^\top (\mathbf{I} - \mathbf{P} + \mathbf{J})^{-1}, \tag{10.2}$$

here $\mathbf{1}$ is a column vector with ones, $\mathbf{I}$ is the identity matrix and $\mathbf{J}$ is a matrix with only ones, both of dimension $K \times K$. If there are only $K = 2$ states, the stationary distribution can be computed simply via

$$\boldsymbol{\pi} = \begin{pmatrix} p_{21}/(p_{12} + p_{21}) \\ p_{12}/(p_{12} + p_{21}) \end{pmatrix}.$$

10.2.1 Maximum Likelihood Inference

Turning to ML inference, we assume that a realisation $x = (x_1, \dots, x_n)$ of length n with observations $x_t = i_t$ has been observed. We can distinguish a conditional and a full likelihood approach. The conditional likelihood is conditional on the first observation x_1 and takes the form

$$L_C(\mathbf{P}) = p_{i_1,i_2} \cdot p_{i_2,i_3} \cdot \ldots \cdot p_{i_{n-1},i_n} = \prod_{i,j} p_{ij}^{n_{ij}},$$

here n_{ij} denotes the observed number of transitions from state i to state j in x. The corresponding log likelihood

$$l_C(\mathbf{P}) = \sum_{i,j} n_{ij} \log(p_{ij})$$

can now be maximised under the restriction that all rows in **P** must sum to one. Using the Lagrange method (see Appendix B.2.5), this is equivalent to maximizing the function

$$l_c^*(\mathbf{P}) = l_c - \sum_{i=1}^{K} \lambda_i \left(\sum_{j=1}^{K} p_{ij} - 1 \right) = \sum_{i,j} n_{ij} \log(p_{ij}) - \sum_{i=1}^{K} \lambda_i \left(\sum_{j=1}^{K} p_{ij} - 1 \right)$$

with respect to **P**. Now

$$\frac{\partial l_c^*(\mathbf{P})}{\partial p_{ij}} = \frac{n_{ij}}{p_{ij}} - \lambda_i,$$

from which we can derive the ML estimates

$$\hat{p}_{ij} = \frac{n_{ij}}{\lambda_i}.$$

Because of $\sum_{j=1}^{K} \hat{p}_{ij} = 1$ we have

$$1 = \frac{\sum_j n_{ij}}{\lambda_i}$$

so

$$\lambda_i = \sum_j n_{ij} = n_i$$

where n_i denotes the number of observations of $(x_1, \ldots, x_{n-1})$ (ignoring the n-th observation) in state i. We can now re-write the ML estimates as

$$\hat{p}_{ij} = \frac{n_{ij}}{n_i}.$$

This is an intuitive result: The ML estimates of the transition probabilities p_{ij} are simply the empirical proportions of observed transitions from i to j among all transitions from i to any state in $\{1, \ldots, K\}$ (including a "stay" at state i). In analogy to Example 4.22 the corresponding standard errors are equal to $\sqrt{\hat{p}_{ij}(1 - \hat{p}_{ij})/n_i}$.

Incorporation of the likelihood of the first observation x_1 complicates ML estimation. If we assume that X_1 is a realisation from some arbitrary initial distribution $\boldsymbol{\gamma}$, then there is only one observation to estimate $\boldsymbol{\gamma}$. A useful estimate of $\boldsymbol{\gamma}$ can rarely be obtained from one observation, so $\boldsymbol{\gamma}$ is typically assumed to be completely known.

Alternatively one may assume that $\boldsymbol{\gamma} = \boldsymbol{\pi}$, i.e. the initial distribution of X_1 equals the stationary distribution $\boldsymbol{\pi}$ of the Markov chain. The initial distribution is then a function of the transition matrix **P**, so again only **P** needs to be estimated. However, the additional term $\log \pi_{i_1}$ has to be added to the conditional log likelihood $l_c(\mathbf{P})$, here π_i is the i-th component of the stationary distribution $\boldsymbol{\pi} = \boldsymbol{\pi}(\mathbf{P})$. Numerical techniques are now typically needed to compute the full ML estimates. Those will in general (and in particular if n is large) not differ much from the conditional ML estimates, which serve as suitable starting values for optimisation.

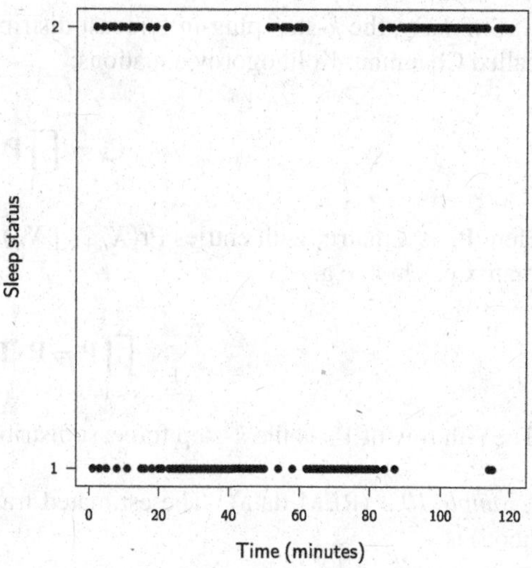

Fig. 10.1 Binary time series of length $n = 120$ minutes with information whether an infant was judged to be in REM sleep ($x_t = 2$) during minute t or not ($x_t = 1$)

Table 10.1 Conditional and full ML estimates and standard errors of the diagonal entries of **P**

Likelihood	$\hat{p}_{11}$	$\text{se}(\hat{p}_{11})$	$\hat{p}_{22}$	$\text{se}(\hat{p}_{11})$
conditional	0.769	0.052	0.741	0.060
full	0.775	0.052	0.734	0.059

Example 10.1 (REM data) Here we present an analysis of a binary time series of length $n = 120$ minutes representing an infant's sleep pattern. The outcome variable x_t reports if an infant was judged to be in rapid eye movement (REM) sleep ($x_t = 2$) during minute t, $x_t = 1$ otherwise. The data are shown Fig. 10.1. Conditional and full ML estimates (based on the assumption that the initial distribution equals the stationary distribution) of the diagonal entries of the 2×2 transition matrix **P** are given in Table 10.1. ■

10.2.2 Prediction

It is possible to compute the k-step forecast distribution $\Pr(X_{n+k}|x_1, \ldots, x_n)$, which, due to the Markov assumption, depends only on x_n:

$$\Pr(X_{n+k} \mid x_1, \ldots, x_n) = \Pr(X_{n+k} \mid x_n).$$

Consider first the one-step forecast distribution of $X_{n+1} \mid x_n$, i.e. $k = 1$. If the transition matrix **P** is known, then the distribution of $X_{n+1} \mid x_n = i$ is given by the i-th row of **P**. This corresponds to the plug-in prediction discussed in Chap. 9 as the estimated transition matrix $\hat{\mathbf{P}}$ is assumed to equal the true transition matrix. Incorporating the uncertainty with respect to the estimation of **P** will involve more efforts, but the difference between the two resulting estimates will typically be small.

Similarly, the k-step plug-in forecast distribution can be calculated with the so-called Chapman–Kolmogorov equations:

$$\mathbf{P}_k = \prod_{i=1}^{k} \mathbf{P}.$$

Here $\mathbf{P}_k$ is a matrix with entries $\Pr(X_{n+k} \mid X_n)$ and the right-hand side refers to the matrix product, e.g.

$$\prod_{i=1}^{2} \mathbf{P} = \mathbf{P} \cdot \mathbf{P}.$$

The i-th row of $\mathbf{P}_k$ is the k-step forecast distribution of $X_{n+k} \mid x_n = i$.

Example 10.2 (REM data) The estimated transition matrix (using the full likelihood) is

$$\hat{\mathbf{P}} = \begin{pmatrix} 0.775 & 0.225 \\ 0.266 & 0.734 \end{pmatrix}. \tag{10.3}$$

The last observation of the observed sequence was $x_{120} = 2$, so the plug-in one-step forecast distribution is the second line of $\mathbf{P}$, i.e. $\Pr(X_{121} = 1 \mid x_{120} = 2) = 0.266$ and $\Pr(X_{121} = 2 \mid x_{120} = 2) = 0.734$. The estimated transition matrix $\hat{\mathbf{P}}_2$ of the 2-step (plug-in) forecast distribution is

$$\hat{\mathbf{P}}_2 = \hat{\mathbf{P}} \cdot \hat{\mathbf{P}} = \begin{pmatrix} 0.660 & 0.340 \\ 0.401 & 0.599 \end{pmatrix}$$

with the second line relevant in our case due to $x_{120} = 2$, e.g. $\Pr(X_{122} = 1 \mid x_{120} = 2) = 0.401$.

For $k \to \infty$, the k-step forecast distribution will converge to the stationary distribution $\boldsymbol{\pi} = (0.541, 0.459)^\top$, regardless of the last observed value:

$$\lim_{k\to\infty} \mathbf{P}_k = \begin{pmatrix} \boldsymbol{\pi}^\top \\ \boldsymbol{\pi}^\top \\ \vdots \\ \boldsymbol{\pi}^\top \end{pmatrix}.$$

Figure 10.2 illustrates that both $\Pr(X_{120+k} = 1 \mid x_{120} = 1)$ and $\Pr(X_{120+k} = 1 \mid x_{120} = 2)$ converge to $\pi_1 = 0.541$. ■

10.2.3 Inclusion of Covariates

It is possible to include covariates $\mathbf{z}_t$ into the model described above. For illustration, consider a first-order Markov chain with $K = 2$ states. A popular approach is to assume a logistic regression model for $\Pr(X_t = 1 \mid X_{t-1} = x_{t-1})$ with explanatory variables $\mathbf{z}_t$ and x_{t-1} (typically coded as 0-1). Alternatively one may model

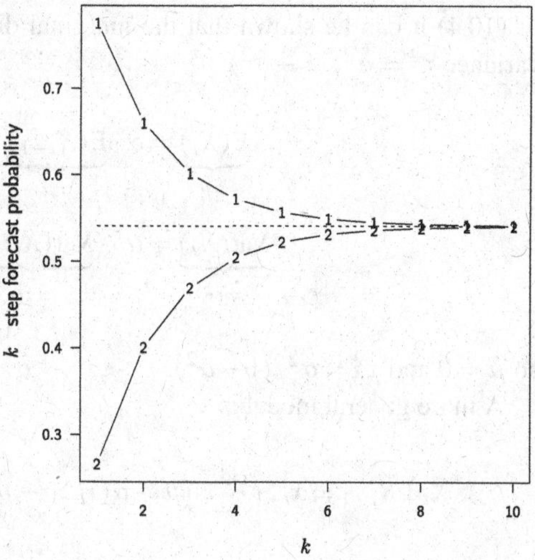

Fig. 10.2 k-step forecast probability $\Pr(X_{120+k} = 1 \mid x_{120} = 1)$ (denoted by "1") and $\Pr(X_{120+k} = 1 \mid x_{120} = 2)$ (denoted by "2") of a first-order Markov chain with estimated transition matrix (10.3). The probability π_1, the first component of the stationary distribution $\boldsymbol{\pi}$, is also shown (*dashed line*)

$\Pr(X_t = 1 \mid X_{t-1} = 0)$ and $\Pr(X_t = 1 \mid X_{t-1} = 1)$ with separate logistic regression models and covariates $\mathbf{z}_t$. We omit details here and point the interested reader to the literature on transition models for longitudinal data.

10.3 Observation-Driven Models for Continuous Data

We now describe selected Markov models for time series with continuous measurements.

10.3.1 The First-Order Autoregressive Model

A simple model for an equally-spaced time series of continuous measurements is the *first-order autoregressive* (AR(1)) *model*

$$X_t \mid X_{t-1} = x_{t-1} \overset{\text{cid}}{\sim} \mathrm{N}(\alpha x_{t-1}, \sigma^2), \quad t = 2, \ldots, n,$$

where $\overset{\text{cid}}{\sim}$ stands for "is conditionally independent distributed as". An equivalent unconditional description of this model is

$$X_t = \alpha X_{t-1} + \epsilon_t \tag{10.4}$$

where the error terms ϵ_t are assumed to be independent mean-zero normal random variables with variance σ^2. It can be easily shown that this model is stationary if $|\alpha| < 1$. Stationarity means that the marginal mean μ and variance τ^2 of X_t are constant, i.e. do not depend on time. Taking expectation and variance on both sides of (10.4) it can be shown that the marginal distribution of X_t has mean zero and variance $\tau^2 = \sigma^2/(1-\alpha^2)$:

$$\underbrace{\mathsf{E}(X_t)}_{=\mu} = \alpha \cdot \underbrace{\mathsf{E}(X_{t-1})}_{=\mu}$$

$$\underbrace{\text{Var}(X_t)}_{=\tau^2} = \alpha^2 \cdot \underbrace{\text{Var}(X_{t-1})}_{=\tau^2} + \underbrace{\text{Var}(\epsilon_t)}_{=\sigma^2},$$

so $\mu = 0$ and $\tau^2 = \sigma^2/(1-\alpha^2)$.

A more general model is

$$X_t \mid X_{t-1} = x_{t-1} \overset{\text{cid}}{\sim} \mathrm{N}\big(\mu + \alpha(x_{t-1} - \mu), \sigma^2\big), \quad t = 2, \dots, n, \tag{10.5}$$

with stationary mean μ.

10.3.2 Maximum Likelihood Inference

Again there are two options for ML estimation of $\boldsymbol{\theta} = (\mu, \alpha, \sigma^2)^\top$. The first approach uses the likelihood conditional on the first observation x_1. Since the distribution of X_t depends only on the observation x_{t-1} from the previous time point, the joint density of $(X_2, \dots, X_n)$, conditional on $X_1 = x_1$ has the form

$$f(x_2, \dots, x_n \mid x_1) = \prod_{t=2}^{n} f(x_t \mid x_{t-1}),$$

here $f(x_t \mid x_{t-1})$ is the density of a $\mathrm{N}(\mu + \alpha(x_{t-1} - \mu), \sigma^2)$ distribution, see (10.5). The corresponding log likelihood can therefore easily be derived as

$$l_c(\boldsymbol{\theta}) = -\frac{n-1}{2}\log\sigma^2 - \sum_{t=2}^{n} \frac{\{x_t - \mu - \alpha(x_{t-1} - \mu)\}^2}{2\sigma^2}. \tag{10.6}$$

The partial derivative with respect to μ is

$$\begin{aligned}
\frac{\partial l_c(\boldsymbol{\theta})}{\partial \mu} &= \frac{\alpha - 1}{\sigma^2}\left[\sum_{t=2}^{n}\{x_t - \mu - \alpha(x_{t-1} - \mu)\}\right] \\
&= \frac{\alpha - 1}{\sigma^2}\left\{\sum_{t=2}^{n}(x_t - \alpha x_{t-1}) + (n-1)\mu(\alpha - 1)\right\}
\end{aligned}$$

from which we obtain the ML estimate of μ for fixed α:

$$\hat{\mu}(\alpha) = \frac{\sum_{t=2}^{n}(x_t - \alpha x_{t-1})}{(1-\alpha)(n-1)}$$
$$= \frac{1}{n-1}\left(\sum_{t=2}^{n-1} x_t + \frac{x_n - \alpha x_1}{1-\alpha}\right). \tag{10.7}$$

Note that this is essentially just the average of the observed time series, only the end-values x_1 and x_n are treated slightly differently.

We can now plug-in (10.7) into (10.6) to obtain the profile log-likelihood (cf. Sect. 5.3) of α and σ^2. Maximisation of this profile log-likelihood with respect to α is independent of σ^2 and can be based on maximizing

$$\sum_{t=2}^{n}\left\{x_t - \hat{\mu}(\alpha) - \alpha\left(x_{t-1} - \hat{\mu}(\alpha)\right)\right\}^2$$

from which we numerically obtain $\hat{\alpha}_{\mathrm{ML}}$ and subsequently $\hat{\mu}_{\mathrm{ML}} = \hat{\mu}(\hat{\alpha}_{\mathrm{ML}})$. This defines residuals $r_t = x_t - \hat{\mu}_{\mathrm{ML}} - \hat{\alpha}_{\mathrm{ML}}(x_{t-1} - \hat{\mu}_{\mathrm{ML}})$, from which the ML estimate of σ^2 can be derived as

$$\hat{\sigma}^2_{\mathrm{ML}} = \sum_{t=2}^{n} \frac{r_t^2}{n-1}.$$

We note that the partial derivative of (10.6) with respect to α is

$$\frac{\partial l_c(\boldsymbol{\theta})}{\partial \alpha} = \frac{1}{\sigma^2}\sum_{t=2}^{n}\left(x_t - \mu - \alpha(x_{t-1} - \mu)\right)(x_{t-1} - \mu)$$
$$= \frac{1}{\sigma^2}\left\{\sum_{t=2}^{n}(x_t - \mu)(x_{t-1} - \mu) - \alpha\sum_{t=2}^{n}(x_{t-1} - \mu)^2\right\}$$

from which we obtain the conditional ML estimate of α for fixed μ:

$$\hat{\alpha}(\mu) = \frac{\sum_{t=2}^{n}(x_t - \mu)(x_{t-1} - \mu)}{\sum_{t=1}^{n-1}(x_t - \mu)^2}. \tag{10.8}$$

This is essentially the classical estimate of the first-order autocorrelation, only the term $(x_n - \mu)^2$ is missing in the sum of the denominator.

A full likelihood approach takes also the observation x_1 into account, assuming that it is a realisation from the stationary distribution $\mathrm{N}(\mu, \sigma^2/(1-\alpha^2))$. Then the term

$$\frac{1}{2}\{\log(1-\alpha^2) - \log(\sigma^2)\} - (1-\alpha^2)(x_1 - \mu)^2/2\sigma^2$$

has to be added to the conditional log likelihood (10.6) and numerical maximisation is necessary to derive the ML estimates.

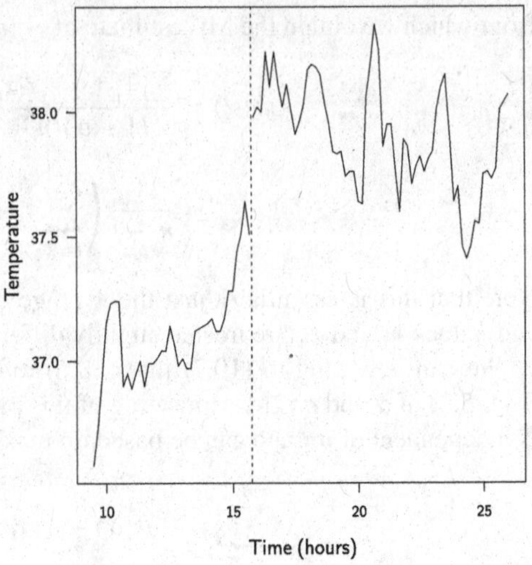

Fig. 10.3 Body temperature (in °C) of a beaver, measured in 10 min intervals. The *dashed line* indicates the start of activity outside the retreat

Table 10.2 Conditional and full ML estimates with standard errors from separate analyses of beaver body temperature inside and outside the retreat

Activity	Likelihood	$\hat{\mu}$	se($\hat{\mu}$)	$\hat{\alpha}$	se($\hat{\alpha}$)	$\hat{\sigma}^2$	se($\hat{\sigma}^2$)
inside	conditional	37.238	0.119	0.834	0.080	0.009	0.002
inside	full	37.073	0.212	0.942	0.062	0.011	0.002
outside	conditional	37.908	0.083	0.797	0.078	0.017	0.003
outside	full	37.916	0.074	0.787	0.075	0.017	0.003

Example 10.3 (Beaver body temperature) Here we consider a study on temperature dynamics of a North American beaver (*Castor canadensis*) in north-central Wisconsin. Body temperature (in degrees Celsius) of a female beaver was measured by telemetry every 10 minutes. The beaver was active outside the retreat from observation $t = 39$ (i.e. at 15:50) onwards. The time series plot in Fig. 10.3 illustrates that this change-point coincides with an upward shift of the mean temperature level.

We first fit separate AR(1) processes to both parts of the time series, using conditional and full likelihood. The results are listed in Table 10.2.

We note that it is useful to transform the autoregressive parameter α to $\tanh^{-1}(\alpha)$ (note that this is Fisher's z-transformation, see Example 4.16) to avoid numerical problems with values of $|\alpha|$ equal to or larger than one. For similar reasons, σ^2 is estimated on a logarithmic scale. In addition, it is also useful to select good starting values for the optimisation procedure. Here we have chosen the overall mean for μ and the empirical first autocorrelation for α. A starting value for the residual variance σ^2 has been derived using an estimate of the marginal variance of the series, which is approximately $\sigma^2/(1-\alpha^2)$.

From Table 10.2 we can see that the estimates of α and σ^2 from the two parts of the time series are fairly similar whereas the level μ appears to be somewhat different. In Example 10.4 we will present an analysis of the whole time series with a level shift (represented by a binary covariate) and a common AR(1) process for the residual time series. ■

We note that there is much more efficient and stable software in R to fit this model. Indeed, the function `arima()` using the argument `order=c(1,0,0)` will produce the same ML estimates as our own implementation above using the full likelihood. *Higher-order autoregressive models* (AR(p)) as well as so-called moving average models (MA(q)) and combinations of both (ARMA(p,q)) can be fitted using the Box–Jenkings modelling framework for time series. The stationarity assumption can be avoided using so-called integrated ARMA(p,q) models, short ARIMA models. Finally, seasonality can be included leading to so-called SARIMA models.

10.3.3 Inclusion of Covariates

A useful extension of model (10.5) allows to include covariates. The AR(1) model will then be

$$X_t \mid X_{t-1} = x_{t-1} \overset{\text{cid}}{\sim} \mathrm{N}\big(\mu + \mathbf{z}_t^\top \boldsymbol{\beta} + \alpha\big(x_{t-1} - \mu - \mathbf{z}_t^\top \boldsymbol{\beta}\big), \sigma^2\big), \quad t = 2, \dots, n,$$

replacing the mean μ in (10.5) with a linear function $\mu + \mathbf{z}_t^\top \boldsymbol{\beta}$ of time-dependent covariates $\mathbf{z}_t$. We note that it is straightforward to incorporate covariates in the R function `arima()` using the argument `xreg`. This is illustrated in a re-analysis of the beaver time series, including information on the times when the beaver was outside the retreat.

Example 10.4 (Beaver body temperature) We now analyse the complete time series on body temperature of the beaver using a binary indicator for activity outside the retreat as covariate z_t with the R function `arima()`. For reference, we also include a model without the covariate (`model2`) and a model without autoregression, thus not allowing for residual correlation (`model3`).

```
library(MASS)
attach(beav2)
model1 <- arima(temp, order = c(1, 0, 0), xreg = activ)
model2 <- arima(temp, order = c(1, 0, 0))
model3 <- arima(temp, order = c(0, 0, 0), xreg = activ)
```

Table 10.3 gives the parameter estimates of the three different models. The full model estimates the mean temperature during activity to be 0.61 °C (SE: 0.14 °C) higher than without activity. The model fits the time series considerably better than an AR(1) model without this covariate. Not allowing for residual correlation gives also a considerably worse model fit, as can be seen from the AIC values in Table 10.3. In addition, the activity estimate is somewhat larger (0.81 °C) while the associated standard error (SE: 0.04 °C) is very small due to ignoring substantial residual correlation. ■

Table 10.3 ML estimates and standard errors of parameters describing beaver body temperature with activity as binary covariate

Model	Covariate	Autoregression	$\hat{\mu}$	se($\hat{\mu}$)	$\hat{\alpha}$	se($\hat{\alpha}$)	$\hat{\beta}$	se($\hat{\beta}$)	AIC
1	yes	yes	37.19	0.12	0.87	0.07	0.61	0.14	−125.55
2	no	yes	37.49	0.35	0.97	0.02			−110.10
3	yes	no	37.10	0.03			0.81	0.04	−21.47

10.3.4 Prediction

Prediction of an AR(1) model is simple if the plug-in approach is employed, i.e. estimated parameters are treated as fixed. Specifically, suppose the data follow a model of the form (10.5) with $x_1, \ldots, x_n$ already observed, then the distribution of X_{n+1} depends only on x_n and has the following form:

$$X_{n+1} \mid X_n = x_n \sim \mathrm{N}\big(\mu + \alpha(x_n - \mu), \sigma^2\big).$$

Similarly, the 2-step predictive distribution of $X_{n+2} \mid X_n = x_n$ is normal with mean

$$\begin{aligned}
\mathsf{E}(X_{n+2} \mid X_n = x_n) &= \mathsf{E}\big\{\mathsf{E}(X_{n+2} \mid X_{n+1}, X_n = x_n)\big\} \\
&= \mathsf{E}\big(\mu + \alpha(X_{n+1} - \mu) \mid X_n = x_n\big) \\
&= \mu + \alpha\big\{\mathsf{E}(X_{n+1} \mid X_n = x_n) - \mu\big\} \\
&= \mu + \alpha\big\{\mu + \alpha(x_n - \mu) - \mu\big\} \\
&= \mu + \alpha^2(x_n - \mu)
\end{aligned}$$

and variance

$$\begin{aligned}
\mathrm{Var}(X_{n+2} \mid X_n = x_n) &= \mathsf{E}\big\{\mathrm{Var}(X_{n+2} \mid X_{n+1}, X_n = x_n)\big\} \\
&\quad + \mathrm{Var}\big\{\mathsf{E}(X_{n+2} \mid X_{n+1}, X_n = x_n)\big\} \\
&= \mathsf{E}(\sigma^2) + \mathrm{Var}\big(\mu + \alpha(X_{n+1} - \mu) \mid X_n = x_n\big) \\
&= \sigma^2 + \alpha^2\sigma^2 \\
&= \sigma^2(1 + \alpha^2).
\end{aligned}$$

Iterating this result gives mean and variance of the k-step predictive distribution as

$$\mathsf{E}(X_{n+k} \mid X_n = x_n) = \mu + \alpha^k(x_n - \mu) \quad \text{and}$$

$$\mathrm{Var}(X_{n+k} \mid X_n = x_n) = \sigma^2 \sum_{j=1}^{k} \alpha^{2(j-1)}.$$

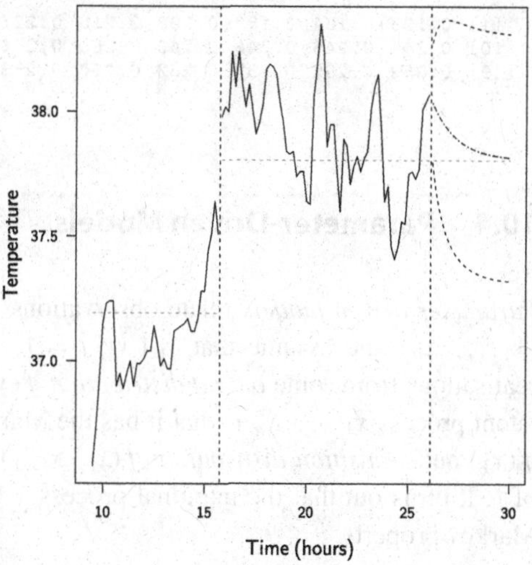

Fig. 10.4 Body temperature (in °C) of a beaver, measured in 10 min intervals. Also shown are predictions with pointwise 95 % prediction intervals for the next 4 hours, assuming that the beaver remains outside the retreat. The *grey lines* are the estimated stationary means $\hat{\mu}$ and $\hat{\mu} + \hat{\beta}$ inside and outside the retreat, respectively

Note that for $k \to \infty$ the predictive distribution of $X_{n+k} \mid X_n = x_n$ will (for $|\alpha| < 1$) converge to the stationary distribution with mean μ and variance $\sigma^2/(1-\alpha^2)$ due to $\alpha^k \to 0$ for $k \to \infty$ and

$$\lim_{k\to\infty} \sum_{j=1}^{k} \alpha^{2(j-1)} = \lim_{k\to\infty} \sum_{j=0}^{k} (\alpha^2)^j = \frac{1}{1-\alpha^2},$$

the limit of the geometric series.

Example 10.5 (Beaver body temperature) Figure 10.4 shows predictions of beaver temperature with pointwise 95 % prediction intervals for the next four hours, assuming that the beaver remains outside the retreat. The predictions are based on the AR(1) model fitted in Example 10.4 with the activity covariate. Note that the point predictions quickly converge to the estimated stationary mean $\hat{\mu} + \hat{\beta}$ outside the retreat.

```
## predict the next four hours in 10 min intervals
n.ahead <- 6*4
p <- predict(model1, newxreg=rep(1, n.ahead), n.ahead=n.ahead)
pred <- p$pred
pred.se <- p$se
round(pred, 3)
Time Series:
Start = 101
End = 124
Frequency = 1
 [1] 38.037 38.007 37.982 37.960 37.940 37.923 37.908 37.895
 [9] 37.884 37.874 37.865 37.858 37.851 37.846 37.841 37.836
[17] 37.832 37.829 37.826 37.823 37.821 37.819 37.818 37.816
round(pred.se, 3)
Time Series:
Start = 101
End = 124
Frequency = 1
```

```
 [1] 0.123 0.164 0.189 0.206 0.218 0.227 0.233 0.238 0.242
[10] 0.244 0.246 0.248 0.249 0.250 0.251 0.251 0.252 0.252
[19] 0.252 0.252 0.252 0.253 0.253 0.253
```

■

10.4 Parameter-Driven Models

Parameter-driven models relate observations $y = (y_1, \dots, y_n)$ to latent states $x = (x_1, \dots, x_n)$ and assume that $y_t \mid x_t$, $t = 1, \dots, n$, are conditionally independent realisations from some *output distribution* $f(y_t \mid x_t)$. A classical assumption on the latent process $x_1, \dots, x_n$ is that it has the Markov property with *initial distribution* $f(x_1)$ and *transition distribution* $f(x_t \mid x_{t-1})$, usually assumed to be independent of t. It turns out that the marginal process y (integrating out x) does not have the Markov property.

10.4.1 The Likelihood Function

The output distribution $f(y_t \mid x_t)$, the initial distribution $f(x_1)$ and the transition distribution $f(x_t \mid x_{t-1})$ may depend on unknown parameters $\boldsymbol{\theta}$ to be estimated from the data. ML estimation will be based on the likelihood function

$$L(\boldsymbol{\theta}) = f(y; \boldsymbol{\theta}) = \int f(y, x; \boldsymbol{\theta}) dx,$$

which is difficult to compute if the dimension of x is large. However, the likelihood can also be expressed in the form

$$L(\boldsymbol{\theta}) = f(y_1; \boldsymbol{\theta}) \cdot \prod_{t=2}^{n} f(y_t \mid y_{\leq(t-1)}; \boldsymbol{\theta}) \tag{10.9}$$

where $y_{\leq t} = (y_1, \dots, y_t)$. In this form the likelihood is easier to calculate. In the following we suppress the dependence on $\boldsymbol{\theta}$ to simplify notation.

The first term on the right-hand side of (10.9) can be computed as

$$f(y_1) = \int f(y_1 \mid x_1) f(x_1) dx_1.$$

The remaining terms $f(y_t \mid y_{\leq(t-1)})$, $t = 2, \dots, n$, can be written as

$$f(y_t \mid y_{\leq(t-1)}) = \int f(y_t \mid x_t) f(x_t \mid y_{\leq(t-1)}) dx_t. \tag{10.10}$$

The first term $f(y_t \mid x_t)$ in (10.10) is known and the second term $f(x_t \mid y_{\le(t-1)})$ can be computed recursively via the *forward pass algorithm*: Suppose $f(x_{t-1} \mid y_{\le(t-2)})$ is already available. First compute

$$\begin{aligned} f(x_{t-1} \mid y_{\le(t-1)}) &= f(x_{t-1} \mid y_{t-1}, y_{\le(t-2)}) \\ &\propto f(x_{t-1}, y_{t-1} \mid y_{\le(t-2)}) \\ &= f(y_{t-1} \mid x_{t-1}) f(x_{t-1} \mid y_{\le(t-2)}) \end{aligned}$$

with subsequent normalisation and then compute

$$f(x_t \mid y_{\le(t-1)}) = \int f(x_t \mid x_{t-1}) \cdot f(x_{t-1} \mid y_{\le(t-1)}) dx_{t-1}. \tag{10.11}$$

Iteration gives $f(x_t \mid y_{\le(t-1)})$ for all $t = 2, \ldots, n$.

10.4.2 The Posterior Distribution

Consider now $\boldsymbol{\theta}$ as fixed. Of central interest is often the posterior distribution of the latent states, i.e.

$$\begin{aligned} f(x \mid y) &\propto f(y \mid x) f(x) \\ &= \prod_{t=1}^{n} f(y_t \mid x_t) \cdot f(x_1) \cdot \prod_{t=2}^{n} f(x_t \mid x_{t-1}) \\ &= f(y_1 \mid x_1) \cdot f(x_1) \cdot \prod_{t=2}^{n} \{ f(y_t \mid x_t) \cdot f(x_t \mid x_{t-1}) \}. \end{aligned} \tag{10.12}$$

A crucial property for most of the following algorithms is that $x \mid y$ inherits the Markov property from x, though its transition probabilities are a function of y and therefore time-dependent. In fact, it is possible to show that

$$f(x \mid y) = f(x_1 \mid y) \cdot \prod_{t=2}^{n} f(x_t \mid x_{t-1}, y_{\ge t}) \tag{10.13}$$

$$= f(x_n \mid y) \cdot \prod_{t=n-1}^{1} f(x_t \mid x_{t+1}, y_{\le t}) \tag{10.14}$$

where $y_{\ge t} = (y_t, \ldots, y_n)$ and $y_{\le t} = (y_1, \ldots, y_t)$ is as defined above.

Proof To show (10.13), note that

$$
\begin{aligned}
f(x \mid y) &= f(x_1 \mid y) \cdot \prod_{t=2}^{n} f(x_t \mid x_{\leq(t-1)}, y) \\
&= f(x_1 \mid y) \cdot \prod_{t=2}^{n} f(x_t \mid x_{t-1}, y) \\
&= f(x_1 \mid y) \cdot \prod_{t=2}^{n} f(x_t \mid x_{t-1}, y_{\geq t})
\end{aligned}
$$

where both lines follow from the Markov property of (x, y). Equation (10.14) can be shown with similar arguments, using the fact that every Markov chain retains the Markov property with time reversed. □

Equation (10.14) can be used to simulate from $f(x \mid y)$ using the *conditional distribution method* by first sampling x_n from $f(x_n \mid y)$ and subsequently sampling x_t from $f(x_t \mid x_{t+1}, y_{\leq t}), t = n-1, \ldots, 1$. The required distributions can be calculated as follows. First note that

$$
\begin{aligned}
f(x_t \mid x_{t+1}, y_{\leq t}) &\propto f(x_{t+1} \mid x_t, y_{\leq t}) \cdot f(x_t \mid y_{\leq t}) \\
&\propto f(x_{t+1} \mid x_t) \cdot f(x_t \mid y_{\leq t}).
\end{aligned}
$$

Now $f(x_{t+1} \mid x_t)$ is the transition distribution of the underlying Markov chain x and $f(x_t \mid y_{\leq t})$ can be computed with the forward pass algorithm. Of course, application of this equation requires appropriate normalisation, either numerically or analytically.

The recursive algorithm to simulate from $f(x \mid y)$ is now as follows:

1. Compute $f(x_1 \mid y_1)$ as the product $f(y_1 \mid x_1) \cdot f(x_1)$ with subsequent normalisation. Record $f(x_1 \mid y_1)$.
2. For $t = 2, \ldots, n$ compute
   a. $f(x_t \mid y_{\leq(t-1)})$ from Eq. (10.11) and then
   b. $f(x_t \mid y_{\leq t})$ as the product $f(x_t \mid y_{\leq(t-1)}) \cdot f(y_t \mid x_t)$ with subsequent normalisation. Record $f(x_t \mid y_{\leq t})$.
3. Simulate x_n from $f(x_n \mid y_{\leq n}) = f(x_n \mid y)$. Record x_n.
4. For $t = n-1, \ldots, 1$ compute $f(x_t \mid x_{t+1}, y) = f(x_t \mid x_{t+1}, y_{\leq t})$ as the product $f(x_{t+1} \mid x_t) \cdot f(x_t \mid y_{\leq t})$ with subsequent normalisation. Simulate x_t from $f(x_t \mid x_{t+1}, y_{\leq t})$ and record it.

The obtained sequence $(x_1, \ldots, x_n)$ is then a sample from $f(x \mid y)$. Since the method combines filtering forward in time (step 1 and 2) with samples backward in time (step 3 and 4), it is often called *forward filtering backward sampling algorithm*.

If the state space S of x is finite, then the following modification of steps 3 and 4 can be used to compute the marginal posterior distribution $f(x_t \mid y), t = n, \ldots, 1$:

3. Record $f(x_n \mid y_{\leq n}) = f(x_n \mid y)$.
4. For $t = n-1, \ldots, 1$ compute $f(x_t \mid x_{t+1}, y) = f(x_t \mid x_{t+1}, y_{\leq t})$ as the product $f(x_{t+1} \mid x_t) \cdot f(x_t \mid y_{\leq t})$ with subsequent normalisation. Then compute and record

$$\begin{aligned} f(x_t \mid y) &= \sum_{x_{t+1} \in S} f(x_t \mid x_{t+1}, y) \cdot f(x_{t+1} \mid y) \\ &= \sum_{x_{t+1} \in S} f(x_t \mid x_{t+1}, y_{\leq t}) \cdot f(x_{t+1} \mid y). \end{aligned}$$

This algorithm will be applied to the REM data in Example 10.10.

10.5 Hidden Markov Models

We will now discuss a generalisation of Markov chain models, so-called *hidden Markov models*, which are a special case of parameter-driven models. The idea is to separate the states of the Markov chain and the actual observations, which were previously assumed to be identical. Applications of hidden Markov models are manifold, for example in DNA-sequencing and speech recognition.

We assume that the latent process $X = (X_1, \ldots, X_n)$ follows a (homogeneous) Markov chain on a discrete state space $S = \{1, \ldots, K\}$ with transition matrix $\mathbf{P} = (p_{ij})_{ij}$ where $p_{ij} = \Pr(X_t = j \mid X_{t-1} = i)$ and initial distribution $\boldsymbol{\pi}$ with entries $\pi_i = \Pr(X_1 = i)$ (often defined through $\boldsymbol{\pi}^\top = \boldsymbol{\pi}^\top \mathbf{P}$). The observations $y_t \mid x_t$ are now assumed to be conditionally independent from a distribution $f(y_t \mid x_t)$ with unknown parameters $\boldsymbol{\theta}$, for example a discrete distribution with support S and misclassification probabilities $\theta_r = \Pr(y_t \neq r \mid x_t = r)$, or a Poisson distribution with unknown rate parameters.

If all misclassification probabilities are zero then $y_t = x_t$ so the hidden Markov model reduces to an ordinary Markov model. This makes it clear that hidden Markov models are more general than ordinary Markov models and provide a more flexible framework for statistical modelling.

Example 10.6 (Noisy binary channel) This example is taken from unpublished lecture notes by Julian Besag. The following time series of binary observations represents the observed data:

$$y = (2, 2, 2, 1, 2, 2, 1, 1, 1, 1, 1, 2, 1, 1, 1, 2, 1, 2, 2, 2).$$

The support of y_t and the state space of x_t is $S = \{1, 2\}$, the length of the time series is $n = 20$. Also assume that the transition matrix $\mathbf{P}$ of the underlying hidden Markov chain X is known:

$$\mathbf{P} = \begin{pmatrix} p_{11} = 0.75 & 1 - p_{11} = 0.25 \\ 1 - p_{22} = 0.25 & p_{22} = 0.75 \end{pmatrix}. \tag{10.15}$$

The corresponding stationary distribution is $\pi = (0.5, 0.5)^\top$. The distribution of y_t given x_t is also assumed to be known with misclassification probabilities

$$\theta_1 = \Pr(y_t = 2 \mid x_t = 1) = 0.2 \quad \text{and} \tag{10.16}$$

$$\theta_2 = \Pr(y_t = 1 \mid x_t = 2) = 0.2. \tag{10.17}$$

The hyperparameters $\mathbf{P}$ and $\boldsymbol{\theta}$ are thus assumed to be known, so the goal of statistical inference reduces to the restoration of the sequence x given the observations y. ■

10.5.1 The Viterbi Algorithm

The common approach is to find the sequence $\hat{x}_{\text{MAP}}$ which maximises the posterior distribution $f(x \mid y)$, as given in (10.12). However, there are 2^n different possible sequences x, which makes direct evaluation of $f(x \mid y)$ for all x computationally challenging, if n is too large.

A more efficient algorithm to find the posterior mode is the *Viterbi* (1967) *algorithm*, a recursive algorithm which finds the MAP estimate in $O(K^2 \cdot n)$ steps. The algorithm proceeds as follows. First note that, due to (10.12) the unnormalised log posterior can be written in the form

$$G(x) = g_1(x_1) + \sum_{t=2}^{n} g_t(x_t, x_{t-1})$$

with

$$g_1(x_1) = \log f(y_1 \mid x_1) + \log f(x_1) \quad \text{and}$$

$$g_t(x_t, x_{t-1}) = \log f(y_t \mid x_t) + \log f(x_t \mid x_{t-1}).$$

Let x^* denote the MAP estimate and suppose the t-th position of x^* is $x_t^* = k$. Now define

$$G_{t,k}(x_1, \dots, x_{t-1}) = g_1(x_1) + g_2(x_2, x_1) + \dots + g_t(k, x_{t-1}) \quad \text{and}$$

$$H_{t,k}(x_{t+1}, \dots, x_n) = g_{t+1}(x_{t+1}, k) + g_{t+2}(x_{t+2}, x_{t+1}) + \dots + g_n(x_n, x_{n-1}),$$

for $k \in \{1, \dots, K\}$. Note that $G_{t,k}(x_1, \dots, x_{t-1})$ is a function of the states before time t, while $H_{t,k}(x_{t+1}, \dots, x_n)$ is a function of the states after time t, so maximisation of $G(x)$ (assuming $x_t^* = k$) can be done by separately maximizing $G_{t,k}$ and $H_{t,k}$. Note also that

$$G_{t,l}(x_1, \dots, x_{t-2}, k) = G_{t-1,k}(x_1, \dots, x_{t-2}) + g_t(l, k)$$

holds.

The Viterbi algorithm is then based on the following recursion:

1. Compute $G_{1,i}^* = g_1(x_1 = i)$ for $i = 1, \dots, K$.

Table 10.4 Restoration of a noisy binary channel

Estimate of x	Posterior probability $\Pr(x \mid y)$
$\hat{x}_{\text{MAP}_1} = (2\,2\,2\,2\,2\,2\,1\,1\,1\,1\,1\,1\,1\,1\,1\,1\,1\,2\,2\,2)$	0.0304
$\hat{x}_{\text{MAP}_2} = (2\,2\,2\,2\,2\,2\,1\,1\,1\,1\,1\,1\,1\,1\,2\,2\,2\,2\,2)$	0.0304
$\hat{x}_{\text{MPM}} = (2\,2\,2\,2\,2\,2\,1\,1\,1\,1\,1\,1\,1\,1\,1\,2\,1\,2\,2\,2)$	0.0135
$y = (2\,2\,2\,1\,2\,2\,1\,1\,1\,1\,1\,2\,1\,1\,1\,2\,1\,2\,2\,2)$	0.0027

2. Compute for $i = 1, \dots, K$

$$G^*_{t,i} = \max_l \left\{ G^*_{t-1,l} + g_t(i,l) \right\} = G^*_{t-1,l^*} + g_2(k, l^*).$$

It can be shown that the sequence $x^* = \{x^*_1, x^*_2, \dots, x^*_n\}$ where $x^*_1 = \arg\max_i G^*_{1,i}$ and $x^*_t = \arg\max_i G^*_{t,i}$, $t = 2, \dots, n$, is the MAP estimate $\hat{x}_{\text{MAP}}$.

Example 10.7 (Noisy binary channel) Table 10.4 gives several estimates of the underlying sequence x. It is interesting to note that the posterior mode is not unique, there are two estimates of x with identical posterior probability 0.0304. The Viterbi algorithm will give one of them, depending on the selection of the maximum in the case of ties. Note that the observed sequence y has a considerably smaller posterior probability of 0.0027.

```
auxiliary <- function(y, m, theta) {
    n <- length(y)
    probs <- matrix(NA, ncol = m, nrow = n)
    for (i in 1:m) probs[, i] <- theta[i, y]
    return(probs)
}
viterbi <- function(y, K, theta, P, delta = NULL) {
    if (is.null(delta))
        delta <- solve(t(diag(K) - P + 1), rep(1, K))
    n <- length(y)
    probs <- auxiliary(y, K, theta)
    xi <- matrix(0, n, K)
    foo <- delta * probs[1, ]
    xi[1, ] <- foo/sum(foo)
    for (i in 2:n) {
        foo <- apply(xi[i - 1, ] * P, 2, max) * probs[i, ]
        xi[i, ] <- foo/sum(foo)
    }
    map <- numeric(n)
    map[n] <- which.max(xi[n, ])
    for (i in (n - 1):1) {
        map[i] <- which.max(P[, map[i + 1]] * xi[i, ])
    }
    return(map)
}
noisy.channel <- c(2, 2, 2, 1, 2, 2, 1, 1, 1, 1,
                   1, 2, 1, 1, 1, 2, 1, 2, 2, 2)
P <- matrix(c(0.75, 0.25, 0.25, 0.75), ncol = 2, byrow = T)
eps <- 0.2
theta <- matrix(c(1 - eps, eps, eps, 1 - eps), ncol = 2, byrow = T)
(map <- viterbi(noisy.channel, K = 2, theta, P))
 [1] 2 2 2 2 2 2 1 1 1 1 1 1 1 1 2 2 2 2 2
```

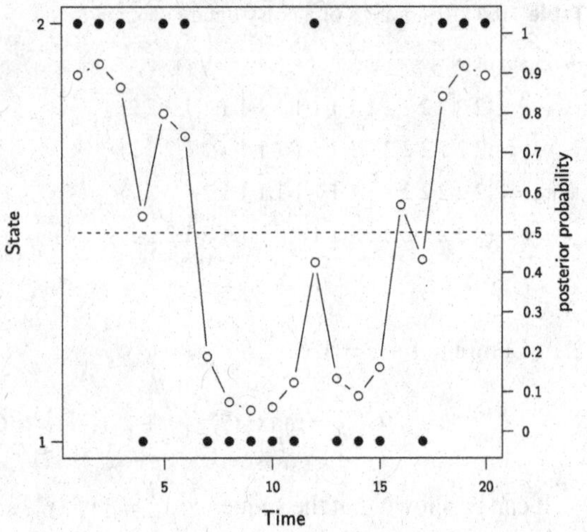

Fig. 10.5 Restoration of a binary signal. Shown are the observed data (*filled circles*) and the corresponding posterior probabilities $\Pr(x_t = 2 \mid y)$ (*solid circle*) based on fixed (known) parameters for the transition and observation model

Table 10.4 also gives the *marginal posterior mode* $\hat{x}_{\text{MPM}}$, where each element x_t, $t = 1, \ldots, 20$, has marginal posterior probability $\Pr(x_t \mid y) \geq 0.5$. This has been computed using the algorithm outlined at the end of Sect. 10.4.2. Note that that $\hat{x}_{\text{MPM}}$ and $\hat{x}_{\text{MAP}}$ do not necessarily coincide, a phenomenon which was discussed in more generality in Sect. 6.5.4 in Chap. 6.

Figure 10.5 displays the marginal posterior probabilities $\Pr(x_t = 2 \mid y)$ together with the observed sequence y. One can see that the MPM estimate does not coincide with the data at positions $t = 4$ and 12. ■

In reality often certain hyperparameters such as the transition matrix $\mathbf{P}$ or parameters determining the distribution of $f(y_i \mid x_i)$ may also be unknown. To estimate such unknown hyperparameters, a likelihood approach, maximizing the likelihood (10.9), can be used. Direct maximisation is often feasible if the length n of the time series is not too large. The Baum–Welch algorithm is an alternative EM algorithm to solve this task. We omit details here.

Example 10.8 (Noisy binary channel) Estimation of both transition matrix and misclassification probabilities $\boldsymbol{\theta} = (\theta_1, \theta_2)^\top$ gives the parameter estimates

$$\hat{\mathbf{P}}_{\text{ML}} = \begin{pmatrix} 0.825 & 0.175 \\ 0.143 & 0.857 \end{pmatrix}$$

and $\hat{\boldsymbol{\theta}}_{\text{ML}} = (0.230, 0.197)^\top$. A simpler model with the restriction $\theta = \theta_1 = \theta_2$ gives

$$\hat{\mathbf{P}}_{\text{ML}} = \begin{pmatrix} 0.815 & 0.185 \\ 0.140 & 0.860 \end{pmatrix}$$

and $\hat{\theta}_{\text{ML}} = 0.213$ with virtually no difference in the value of the log likelihood function.

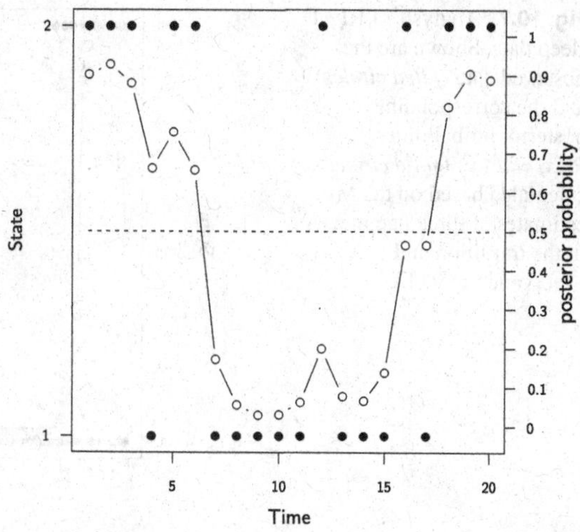

Fig. 10.6 Restoration of a binary signal. Shown are the observed data (*filled circles*) and the corresponding posterior probabilities $\Pr(x_t = 2 \mid y)$ (*solid circle*), calculated based on the ML estimates of the three parameters in the transition and observation model

We have recomputed the MPM estimate based on the estimates of the simpler three-parameter model. The result is shown in Fig. 10.6. The MPM estimate differs only at position $t = 16$ from the one shown in Fig. 10.5, which has been computed with fixed transition matrix (10.15) and misclassification probabilities (10.16) and (10.17). ■

Example 10.9 (REM data) As a second example, we fit a hidden Markov model to the REM data. Estimation of both transition matrix and the unknown parameter vector $\boldsymbol{\theta} = (\theta_1, \theta_2)^\top$ gives the parameter estimates

$$\hat{\mathbf{P}}_{\mathrm{ML}} = \begin{pmatrix} 0.947 & 0.053 \\ 0.028 & 0.972 \end{pmatrix}$$

and $\hat{\boldsymbol{\theta}}_{\mathrm{ML}} = (0, 0.235)^\top$. Note that the ML estimate of θ_1 is on the boundary of the parameter space. The alternative model with the restriction $\theta = \theta_1 = \theta_2$ gives

$$\hat{\mathbf{P}}_{\mathrm{ML}} = \begin{pmatrix} 0.952 & 0.048 \\ 0.042 & 0.958 \end{pmatrix}$$

and $\hat{\theta}_{\mathrm{ML}} = 0.122$ with a log-likelihood difference of 3.87 at 1 degree of freedom ($p = 0.049$). This indicates moderate evidence against the simpler model. Figure 10.7 gives the corresponding marginal posterior probabilities. ■

10.5.2 Bayesian Inference for Hidden Markov Models

Alternatively, a Bayesian approach may be employed using additional hyper-priors on the unknown parameters and simulation from the joint posterior distribution

$$f(x, \boldsymbol{\theta} \mid y) \propto f(y \mid x, \boldsymbol{\theta}) \cdot f(x \mid \boldsymbol{\theta}) \cdot f(\boldsymbol{\theta})$$

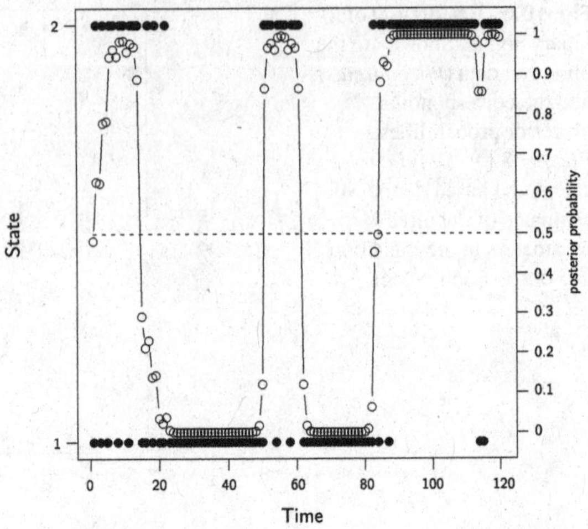

Fig. 10.7 Analysis of REM sleep data. Shown are the observed data (*filled circles*) and the corresponding posterior probabilities $\Pr(x_t = 2 \mid y)$ (*solid circles*), calculated based on the ML estimates of three parameters in the transition and observation model

with Markov chain Monte Carlo (MCMC) techniques (cf. Sect. 8.4). Specifically, we select suitable starting values $x^{(1)}$ and $\boldsymbol{\theta}^{(1)}$ for x and $\boldsymbol{\theta}$ and sample iteratively $x^{(s)}$ from $f(x \mid \boldsymbol{\theta}^{(s-1)}, y)$ and $\boldsymbol{\theta}^{(s)}$ from $f(\boldsymbol{\theta} \mid x^{(s)}, y)$ in turn for $s = 2, \ldots, S$.

Sampling from $f(x \mid \boldsymbol{\theta}, y)$ can be efficiently done using the forward filtering backward sampling algorithm described at the end of Sect. 10.4.2. The parameter vector $\boldsymbol{\theta}$ consists of the transition matrix $\mathbf{P}$ and additional unknown parameters in the observation model. Regarding $\mathbf{P}$ it is convenient to work with conjugate Dirichlet priors for each row of $\mathbf{P}$, which reduce to two independent beta priors for the case of $K = 2$ states. If the observation model is Bernoulli, then independent beta priors are also convenient for the misclassification probabilities θ_r.

Example 10.10 (REM data) We reconsider the REM data using a fully Bayesian approach. We use independent Be(9, 1) priors for p_{11} and p_{22} as well as independent Be(1, 9) priors for θ_1 and θ_2 such that the expected prior probability of a jump from one state to the other and the expected misclassification probability are both equal to 10 %. We collected 1000 samples after a burn-in of 100 iterations.

Figure 10.8 gives histograms of the posterior distribution of the parameters p_{11}, p_{22}, θ_1 and θ_2. Note that the mode of θ_2 is clearly larger than zero while the posterior distribution of θ_1 seems to peak at zero. This corresponds to the ML estimates reported earlier.

Finally Fig. 10.9 shows the posterior probabilities $\Pr(x_t = 2 \mid y)$ based on the empirical posterior samples of x. Note that these posterior probabilities fully incorporate posterior uncertainty with respect to the hyperparameters, in contrast to the analysis conditional on ML estimates. ■

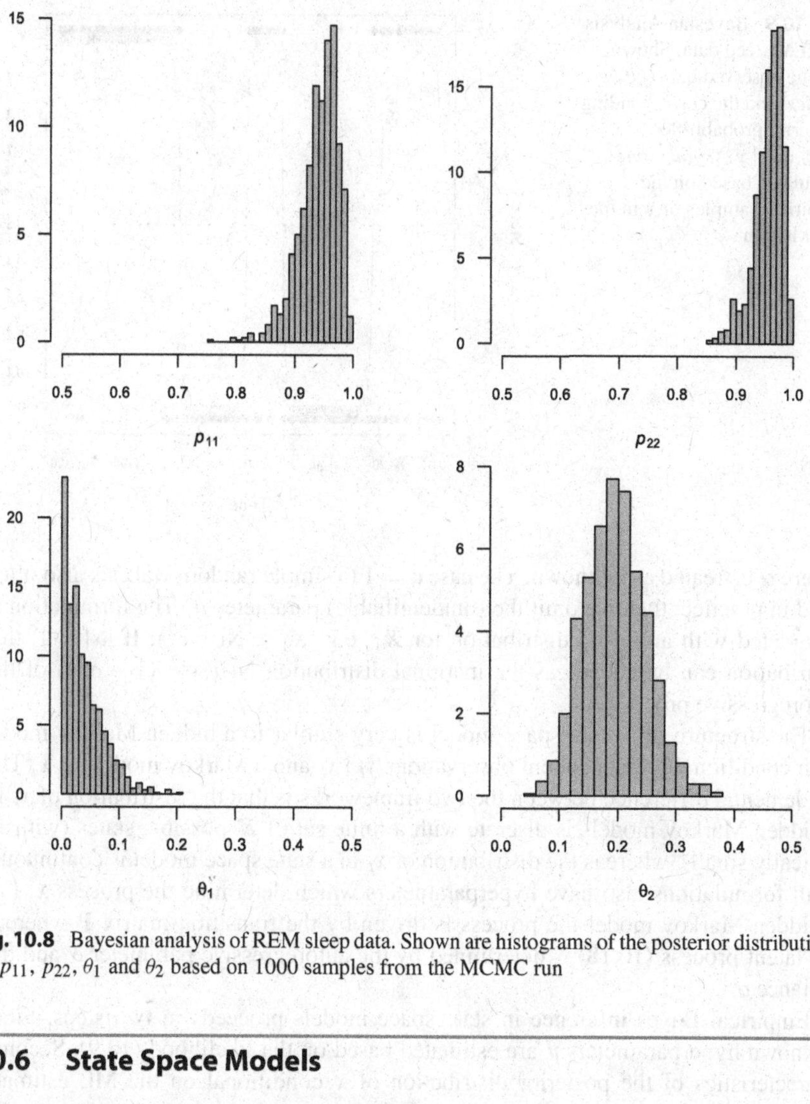

Fig. 10.8 Bayesian analysis of REM sleep data. Shown are histograms of the posterior distribution of p_{11}, p_{22}, θ_1 and θ_2 based on 1000 samples from the MCMC run

10.6 State Space Models

State space models are parameter-driven models, where the latent vector x follows an autoregressive Gaussian process and the mean of the distribution of $y_t \mid x_t$ depends in some way on x_t. For example, if y_t is a continuous normal measurement then x_t is often simply the mean of y_t. If y_t is binary, then a logit model with $\pi_t = \mathsf{E}(y_t \mid x_t) = 1/(1+\exp(-x_t))$ could be used.

The latent Gaussian process can also be multivariate, in which case the distribution of $y_t \mid x_t$ will have a form as in a generalised linear model. However, for simplicity we will assume in the following that x_t is scalar, i.e.

$$X_t \mid X_{t-1} = x_{t-1} \sim \mathrm{N}\big(\mu + \alpha(x_{t-1} - \mu), \sigma^2\big), \tag{10.18}$$

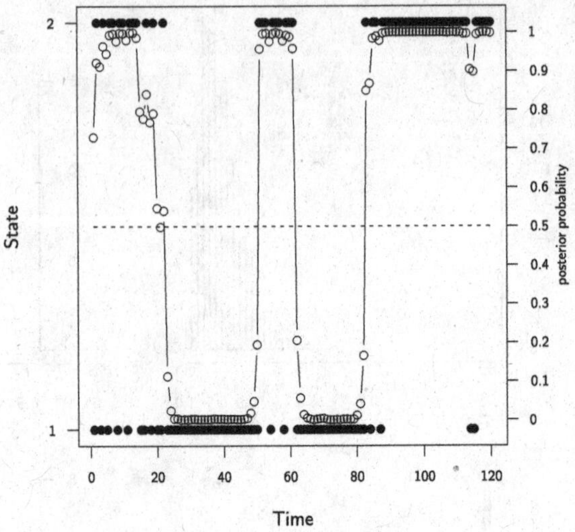

Fig. 10.9 Bayesian Analysis of REM sleep data. Shown are the observed data (*filled circles*) and the corresponding posterior probabilities $\Pr(x_t = 2 \mid y)$ (*solid circle*), calculated based on the empirical samples of x in the MCMC run

where α is treated as unknown. The case $\alpha = 1$ (a simple random walk) is also often used in practice, then we omit the (unidentifiable) parameter μ. The formulation is completed with an initial distribution for X_1, e.g. $X_1 \sim \mathrm{N}(\nu, \tau^2)$. If $|\alpha| < 1$, this distribution can be chosen as the marginal distribution $\mathrm{N}(\mu, \sigma^2/(1-\alpha^2))$ of the autoregressive process.

The structure of a state space model is very similar to a hidden Markov model with conditionally independent observations $y_t \mid x_t$ and a Markov model for x. The fundamental difference between the two frameworks is that the distribution of x_t in a hidden Markov models is discrete with a finite set of K possible states (with K typically small), whereas the distribution of x_t in a state space model is continuous. Both formulations also have hyperparameters which determine the process x. For a hidden Markov model the process is driven by the transition matrix $\mathbf{P}$ whereas the latent process (10.18) is determined by the autoregressive parameter α and the variance σ^2.

Empirical Bayes inference in state space models proceeds in two steps. First, unknown hyperparameters $\boldsymbol{\theta}$ are estimated based on the likelihood (10.9). Second, characteristics of the posterior distribution of x conditional on the ML estimate $\hat{\boldsymbol{\theta}}_{\mathrm{ML}}$ are computed using the methods described in Sect. 10.4.2. Conceptually the inferential process is identical to inference in hidden Markov models.

Fully Bayesian inference can be performed with MCMC, but recently the integrated nested Laplace approximations (INLA) method has been proposed as a suitable alternative. INLA is an extension of the Laplace approximation described in Sect. 8.2. We will show in the following how INLA can be used to fit state space models to the REM data.

Example 10.11 (REM data) As in Example 10.2 we use a binary observation model for the response y_t but now assume that the latent logit-transformed response probability $x_t = \mathrm{logit}(\pi_t)$ follows an AR(1) process with unknown autoregressive pa-

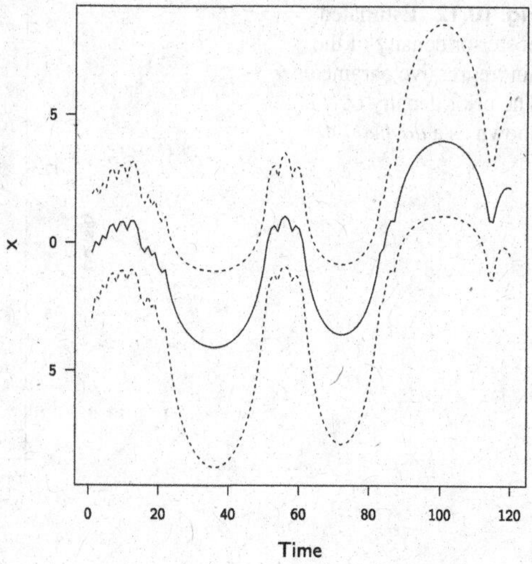

Fig. 10.10 Estimated latent AR(1) process x with 95 % credible intervals

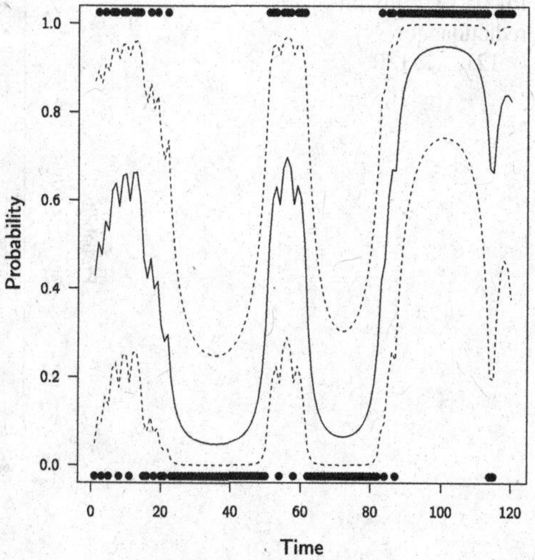

Fig. 10.11 Estimated response probability π with 95 % credible intervals. The observed data are shown with *dots*

rameter α. Alternatively, a random walk ($\alpha = 1$) could be considered, which gives very similar results. The estimated latent process x is shown in Fig. 10.10 (posterior median within 95 % credible intervals), while Fig. 10.11 displays the corresponding response probability π_t and shows how the fitted curve smoothes the observed data. Finally, Fig. 10.12 displays the estimated posterior density of the autoregressive parameter α, which concentrates around 0.95. For comparison, the prior density is also shown, which is bounded away from $\alpha = 1$. The prior on α has been chosen such that $\text{logit}((\alpha + 1)/2) \sim \text{N}(0, 1/0.15)$, while the prior used for the variance σ^2 was $\sigma^2 \sim \text{IG}(1, 0.005)$. Finally, Fig. 10.13 shows the fit with predictions for the next 10 time points $t = 121, \ldots, 130$.

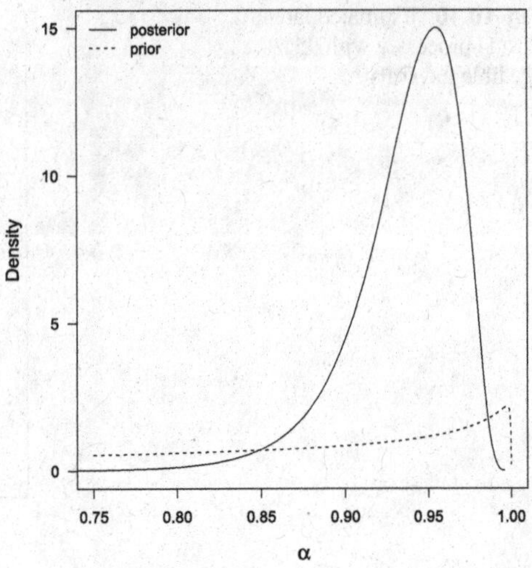

Fig. 10.12 Estimated posterior density of the autoregressive parameter α. The prior density of α is shown as a *dashed line*

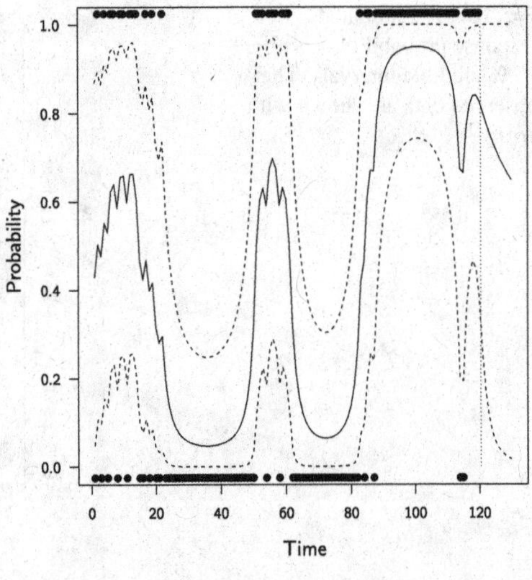

Fig. 10.13 Fit with predictions for $t = 121, \ldots, 130$

```
library(INLA)
rem.data <- c(1, 2, 1, 2, 1, 2, 2, 1, 2, 2, 1, 2, 2, 2, 1,
              1, 2, 1, 2, 1, 1, 2, 1, 1, 1, 1, 1, 1, 1, 1,
              1, 1, 1, 1, 1, 1, 1, 1, 1, 1, 1, 1, 1, 1, 1,
              1, 1, 1, 1, 1, 2, 2, 2, 1, 2, 2, 2, 1, 2, 2,
              2, 1, 1, 1, 1, 1, 1, 1, 1, 1, 1, 1, 1, 1, 1,
              1, 1, 1, 1, 1, 1, 1, 2, 1, 2, 2, 1, 2, 2, 2,
              2, 2, 2, 2, 2, 2, 2, 2, 2, 2, 2, 2, 2, 2, 2,
              2, 2, 2, 2, 2, 2, 2, 2, 1, 1, 2, 2, 2, 2, 2)
len <- length(rem.data)
time <- c(1:len)
mydata <- list(y = rem.data - 1, t = time)
formula1 <- y ~ f(t, model = "ar1",
                  hyper = list(rho = list(initial = 5, prior = "normal
                     ",
```

```
                                    c(0, 0.15)),
                              prec = list(prior = "loggamma",
                                          param = c(1, 0.005))))
model1 <- inla(formula1, family = "binomial", data = mydata,
               control.predictor = list(compute = TRUE))
npred <- 10
mydata.pred <- list(y = c(rem.data - 1, rep(NA, npred)),
                    t = c(time, c((len + 1):(len + npred))))
model1.pred <- inla(formula1, family = "binomial", data = mydata.pred,
                    control.predictor = list(compute = TRUE),
                    Ntrials = rep(1, len + npred))
```

■

10.7 Exercises

1. Derive formula (10.2) for the stationary distribution of a Markov chain.
2. A Markov chain model can be used to model the disease history of stroke patients in the first few months after the stroke. We consider three states (Chen et al. 1999): a favourable state (moderate to light disability), an unfavourable state (no cortical function or strong disability) and death. The probability that a patient in the favourable state dies within the next month was estimated to be 0.017, whereas for a patient in the unfavourable state, the probability to die in the next month was 0.085. The probability to transition from the unfavourable state to the favourable state within one month was 0.136, whereas the probability for the opposite transition was 0.017.
   (a) Derive the transition matrix.
   (b) Given that a patient is initially in the unfavourable state, what is the probability the he will be in the favourable state two months later?
   (c) Since transitions from the state "death" to other states are impossible, this state is called an *absorbing state*. What does this imply for the k-step forecast distribution if $k \to \infty$?
3. Fit two second-order autoregressive (AR(2)) models to the beaver body temperatures time series using the function `arima()`, one model with the covariate indicating activity outside the retreat and one without covariates. Compare the estimated coefficients and AIC values of these models and the models considered in Example 10.4.
4. Use INLA to fit a simple random walk to the REM data. Plot the estimated latent process x and the estimated response probability π as a function of time, both with 95 % credible intervals (see Figs. 10.10 and 10.11 for the AR(1) model). Also add the corresponding lines for the AR(1) model for comparison. Hint: Fig. 10.10 was produced with the following code:

```
quantities <- c("mean", "0.025quant", "0.975quant")
matplot(time, model1$summary.linear.predictor[,quantities],
        type = "l", lty = rep(c(1, 2, 2), 2),
        xlab = "Time", ylab = expression(x))
```

10.8 References

Cox (1981) has proposed to distinguish between observation-driven and parameter-driven models for time series. The theory of Markov chains (and more general Markov processes) is nicely summarised in various textbooks, for example in Grimmett and Stirzaker (2001, Chap. 6) and includes a proof that every Markov chain retains the Markov property with time reversed. The EEG time series is taken from Carlin and Polson (1992), the beaver time series from Reynolds (1994). Transition models for longitudinal data are described in Diggle et al. (2002). Classical time series analysis is discussed in much more detail in Diggle (1990) and Shumway and Stoffer (2017). The conditional distribution method is described in Devroye (1986, Chap. 11). The Baum–Welch algorithm is described in detail in Baum et al. (1970).

Probabilities, Random Variables and Distributions

A

Contents

This appendix gives important definitions and results from probability theory in a compact and sometimes slightly simplifying way. The reader is referred to Grimmett and Stirzaker (2001) for a comprehensive and more rigorous introduction to probability theory. We start with the notion of probability and then move on to random variables and expectations. Important limit theorems are described in Appendix A.4.

L. Held, D. Sabanés Bové, *Likelihood and Bayesian Inference*,
Statistics for Biology and Health, https://doi.org/10.1007/978-3-662-60792-3,

A.1 Events and Probabilities

Any experiment involving randomness can be modelled with *probabilities*. Probabilities $\Pr(A)$ between zero and one are assigned to *events* A such as "It will be raining tomorrow" or "I will suffer a heart attack in the next year". The *certain event* has probability one, while the *impossible event* has probability zero.

Any event A has a disjoint, *complementary event* A^c such that $\Pr(A) + \Pr(A^c) = 1$. For example, if A is the event that "It will be raining tomorrow", then A^c is the event that "It will not be raining tomorrow". More generally, a series of events $A_1, A_2, \ldots, A_n$ is called a *partition* if the events are pairwise disjoint and if $\Pr(A_1) + \cdots + \Pr(A_n) = 1$.

A.1.1 Conditional Probabilities and Independence

Conditional probabilities $\Pr(A \mid B)$ are calculated to update the probability $\Pr(A)$ of a particular event under the additional information that a second event B has occurred. They can be calculated via

$$\Pr(A \mid B) = \frac{\Pr(A, B)}{\Pr(B)}, \tag{A.1}$$

where $\Pr(A, B)$ is the probability that both A and B occur. This definition is only sensible if the occurrence of B is possible, i.e. if $\Pr(B) > 0$. Rearranging this equation gives $\Pr(A, B) = \Pr(A \mid B) \Pr(B)$, but $\Pr(A, B) = \Pr(B \mid A) \Pr(A)$ must obviously also hold. Equating and rearranging these two formulas gives Bayes' theorem:

$$\Pr(A \mid B) = \frac{\Pr(B \mid A) \Pr(A)}{\Pr(B)}, \tag{A.2}$$

which will be discussed in more detail in Appendix A.1.2.

Two events A and B are called *independent* if the occurrence of B does not change the probability of A occurring:

$$\Pr(A \mid B) = \Pr(A).$$

From (A.1) it follows that under independence

$$\Pr(A, B) = \Pr(A) \cdot \Pr(B)$$

and

$$\Pr(B \mid A) = \Pr(B).$$

Conditional probabilities behave like ordinary probabilities if the conditional event is fixed, so $\Pr(A \mid B) + \Pr(A^c \mid B) = 1$, for example. It then follows that

$$\Pr(B) = \Pr(B \mid A) \Pr(A) + \Pr(B \mid A^c) \Pr(A^c) \tag{A.3}$$

and more generally

$$\Pr(B) = \Pr(B \mid A_1)\Pr(A_1) + \Pr(B \mid A_2)\Pr(A_2) + \cdots + \Pr(B \mid A_n)\Pr(A_n) \quad \text{(A.4)}$$

if $A_1, A_2, \ldots, A_n$ is a partition. This is called the *law of total probability*.

A.1.2 Bayes' Theorem

Let A and B denote two events A, B with $0 < \Pr(A) < 1$ and $\Pr(B) > 0$. Then

$$\begin{aligned}\Pr(A \mid B) &= \frac{\Pr(B \mid A) \cdot \Pr(A)}{\Pr(B)} \\ &= \frac{\Pr(B \mid A) \cdot \Pr(A)}{\Pr(B \mid A) \cdot \Pr(A) + \Pr(B \mid A^c) \cdot \Pr(A^c)}. \end{aligned} \quad \text{(A.5)}$$

For a general partition $A_1, A_2, \ldots, A_n$ with $\Pr(A_i) > 0$ for all $i = 1, \ldots, n$, we have that

$$\Pr(A_j \mid B) = \frac{\Pr(B \mid A_j) \cdot \Pr(A_j)}{\sum_{i=1}^{n} \Pr(B \mid A_i) \cdot \Pr(A_i)} \quad \text{(A.6)}$$

for each $j = 1, \ldots, n$.

For only $n = 2$ events $\Pr(A)$ and $\Pr(A^c)$ in the denominator of (A.6), i.e. Eq. (A.5), a formulation using *odds* $\omega = \pi/(1 - \pi)$ rather than probabilities π provides a simpler version without the uncomfortable sum in the denominator of (A.5):

$$\underbrace{\frac{\Pr(A \mid B)}{\Pr(A^c \mid B)}}_{\text{Posterior Odds}} = \underbrace{\frac{\Pr(A)}{\Pr(A^c)}}_{\text{Prior Odds}} \cdot \underbrace{\frac{\Pr(B \mid A)}{\Pr(B \mid A^c)}}_{\text{Likelihood Ratio}}.$$

Odds ω can be easily transformed back to probabilities $\pi = \omega/(1 + \omega)$.

A.2 Random Variables

A.2.1 Discrete Random Variables

We now consider possible real-valued realisations x of a *discrete random variable* X. The *probability mass function* of a discrete random variable X,

$$f(x) = \Pr(X = x),$$

describes the *distribution* of X by assigning probabilities for the events $\{X = x\}$. The set of all realisations x with truly positive probabilities $f(x) > 0$ is called the *support* of X.

A *multivariate discrete random variable* $\boldsymbol{X} = (X_1, \dots, X_p)$ has realisations $\boldsymbol{x} = (x_1, \dots, x_p)$ in $\mathbb{R}^p$ and the probability mass function

$$f(\boldsymbol{x}) = \Pr(\boldsymbol{X} = \boldsymbol{x}) = \Pr(X_1 = x_1, \dots, X_p = x_p).$$

Of particular interest is often the joint bivariate distribution of two discrete random variables X and Y with probability mass function $f(x, y)$. The *conditional probability mass function* $f(x \mid y)$ is then defined via (A.1):

$$f(x \mid y) = \frac{f(x, y)}{f(y)}. \tag{A.7}$$

Bayes' theorem (A.2) now translates to

$$f(x \mid y) = \frac{f(y \mid x) f(x)}{f(y)}. \tag{A.8}$$

Similarly, the law of total probability (A.4) now reads

$$f(y) = \sum_x f(y \mid x) f(x), \tag{A.9}$$

where the sum is over the support of X. Combining (A.7) and (A.9) shows that the argument x can be removed from the joint density $f(x, y)$ via summation to obtain the *marginal probability mass function* $f(y)$ of Y:

$$f(y) = \sum_x f(x, y),$$

again the sum over the support of X.

Two random variables X and Y are called *independent* if the events $\{X = x\}$ and $\{Y = y\}$ are independent for all x and y. Under independence, we can therefore factorise the joint distribution into the product of the marginal distributions:

$$f(x, y) = f(x) \cdot f(y).$$

If X and Y are independent and g and h are arbitrary real-valued functions, then $g(X)$ and $h(Y)$ are also independent.

A.2.2 Continuous Random Variables

A *continuous random variable* X is usually defined with its *density function* $f(x)$, a non-negative real-valued function. The density function is the derivative of the *distribution function* $F(x) = \Pr(X \leq x)$, and therefore $\int_{-\infty}^{x} f(u)\, du = F(x)$ and $\int_{-\infty}^{+\infty} f(x)\, dx = 1$. In the following we assume that this derivative exists. We then have

$$\Pr(a \le X \le b) = \int_a^b f(x)\,dx$$

for any real numbers a and b, which determines the *distribution* of X.

Under suitable regularity conditions, the results derived in Appendix A.2.1 for probability mass functions also hold for density functions with replacement of summation by integrals. Suppose that $f(x, y)$ is the joint density function of two random variables X and Y, i.e.

$$\Pr(X \le x, Y \le y) = \int_{-\infty}^{y} \int_{-\infty}^{x} f(u, v)\,du\,dv.$$

The *marginal density function* $f(y)$ of Y is

$$f(y) = \int f(x, y)\,dx,$$

and the law of total probability now reads

$$f(y) = \int f(y \mid x) f(x)\,dx.$$

Bayes' theorem (A.8) still reads

$$f(x \mid y) = \frac{f(y \mid x) f(x)}{f(y)}. \tag{A.10}$$

If X and Y are independent continuous random variables with density functions $f(x)$ and $f(y)$, then

$$f(x, y) = f(x) \cdot f(y)$$

for all x and y.

A.2.3 The Change-of-Variables Formula

When the transformation $g(\cdot)$ is one-to-one and differentiable, there is a formula for the probability density function $f_Y(y)$ of $Y = g(X)$ directly in terms of the probability density function $f_X(x)$ of a continuous random variable X. This is known as the *change-of-variables* formula:

$$\begin{aligned} f_Y(y) &= f_X\{g^{-1}(y)\} \cdot \left| \frac{dg^{-1}(y)}{dy} \right| \\ &= f_X(x) \cdot \left| \frac{dg(x)}{dx} \right|^{-1}, \end{aligned} \tag{A.11}$$

where $x = g^{-1}(y)$. As an example, we will derive the density function of a log-normal random variable $Y = \exp(X)$, where $X \sim \mathrm{N}(\mu, \sigma^2)$. From $y = g(x) = \exp(x)$ we find that the inverse transformation is $x = g^{-1}(y) = \log(y)$, which has the derivative $dg^{-1}(y)/dy = 1/y$. Using the formula (A.11), we get the density function

$$\begin{aligned} f_Y(y) &= f_X\{g^{-1}(y)\} \cdot \left|\frac{dg^{-1}(y)}{dy}\right| \\ &= \frac{1}{\sqrt{2\pi\sigma^2}} \exp\left\{-\frac{1}{2}\frac{(\log(y)-\mu)^2}{\sigma^2}\right\}\left|\frac{1}{y}\right| \\ &= \frac{1}{\sigma}\frac{1}{y}\varphi\left\{\frac{\log(y)-\mu}{\sigma}\right\}, \end{aligned}$$

where $\varphi(x)$ is the standard normal density function.

Consider a multivariate continuous random variable $\boldsymbol{X}$ that takes values in $\mathbb{R}^p$ and a one-to-one and differentiable transformation $\boldsymbol{g}: \mathbb{R}^p \to \mathbb{R}^p$, which yields $\boldsymbol{Y} = \boldsymbol{g}(\boldsymbol{X})$. As defined in Appendix B.2.2, let $\boldsymbol{g}'(\boldsymbol{x})$ and $(\boldsymbol{g}^{-1})'(\boldsymbol{y})$ be the Jacobian matrices of the transformation $\boldsymbol{g}$ and its inverse $\boldsymbol{g}^{-1}$, respectively. Then the multivariate change-of-variables formula is:

$$\begin{aligned} f_{\boldsymbol{Y}}(\boldsymbol{y}) &= f_{\boldsymbol{X}}\{\boldsymbol{g}^{-1}(\boldsymbol{y})\} \cdot \left|(\boldsymbol{g}^{-1})'(\boldsymbol{y})\right| \\ &= f_{\boldsymbol{X}}(\boldsymbol{x}) \cdot \left|\boldsymbol{g}'(\boldsymbol{x})\right|^{-1}, \end{aligned} \tag{A.12}$$

where $\boldsymbol{x} = \boldsymbol{g}^{-1}(\boldsymbol{y})$ and the absolute values of the determinants of the Jacobian matrices are meant. As an example, we will derive the density function of the multivariate normal distribution from a location and scale transformation of univariate standard normal random variables. Consider $\boldsymbol{X} = (X_1, \dots, X_p)^\top$ with $X_i \overset{\text{iid}}{\sim} \mathrm{N}(0, 1)$, $i = 1, \dots, p$. So the joint density function is

$$f_{\boldsymbol{X}}(\boldsymbol{x}) = \prod_{i=1}^{p} (2\pi)^{-1/2} \exp\left(-\frac{1}{2}x_i^2\right) = (2\pi)^{-p/2} \exp\left(-\frac{1}{2}\boldsymbol{x}^\top\boldsymbol{x}\right),$$

which is of course the density function the $\mathrm{N}_p(\boldsymbol{0}, \boldsymbol{I})$ distribution. Let the location and scale transformation be defined by $\boldsymbol{y} = \boldsymbol{g}(\boldsymbol{x}) = \boldsymbol{A}\boldsymbol{x} + \boldsymbol{\mu}$, where $\boldsymbol{A}$ is an invertible $p \times p$ matrix, and $\boldsymbol{\mu}$ is a p-dimensional vector of real numbers. Then the inverse transformation is given by $\boldsymbol{x} = \boldsymbol{g}^{-1}(\boldsymbol{y}) = \boldsymbol{A}^{-1}(\boldsymbol{y} - \boldsymbol{\mu})$ with Jacobian $(\boldsymbol{g}^{-1})'(\boldsymbol{y}) = \boldsymbol{A}^{-1}$. Therefore, the density function of the multivariate continuous random variable $\boldsymbol{Y} = \boldsymbol{g}(\boldsymbol{X})$ is

$$\begin{aligned} f_{\boldsymbol{Y}}(\boldsymbol{y}) &= f_{\boldsymbol{X}}\{\boldsymbol{g}^{-1}(\boldsymbol{y})\} \cdot \left|(\boldsymbol{g}^{-1})'(\boldsymbol{y})\right| \\ &= (2\pi)^{-p/2} \exp\left\{-\frac{1}{2}\boldsymbol{g}^{-1}(\boldsymbol{y})^\top \boldsymbol{g}^{-1}(\boldsymbol{y})\right\}\left|\boldsymbol{A}^{-1}\right| \end{aligned}$$

$$= (2\pi)^{-p/2}|A|^{-1} \exp\left\{-\frac{1}{2}(y-\mu)^\top \left(AA^\top\right)^{-1}(y-\mu)\right\}, \qquad (A.13)$$

where we have used $|A^{-1}| = |A|^{-1}$ and $(A^{-1})^\top A^{-1} = (AA^\top)^{-1}$ in the last step. If we define $\Sigma = AA^\top$ and compare (A.13) with the density function for the multivariate normal distribution $\mathrm{N}_p(\mu, \Sigma)$ from Appendix A.5.3, we find that the corresponding kernels match. The normalising constant can also be matched by noting that

$$|\Sigma|^{-1/2} = \left|AA^\top\right|^{-1/2} = \left(|A||A|\right)^{-1/2} = |A|^{-1}.$$

A.2.4 Multivariate Normal Distributions

Marginal and conditional distributions of a multivariate normal distribution (compare Appendix A.5.3) are also normal. Let the multivariate random variables X and Y be jointly multivariate normal with

$$\text{mean} \quad \mu = \begin{pmatrix} \mu_X \\ \mu_Y \end{pmatrix} \quad \text{and covariance matrix} \quad \Sigma = \begin{pmatrix} \Sigma_{XX} & \Sigma_{XY} \\ \Sigma_{YX} & \Sigma_{YY} \end{pmatrix},$$

where μ and Σ are partitioned according to the dimensions of X and Y. Then the marginal distributions of X and Y are

$$X \sim \mathrm{N}(\mu_X, \Sigma_{XX}),$$
$$Y \sim \mathrm{N}(\mu_Y, \Sigma_{YY}).$$

The corresponding conditional distributions are

$$Y \mid X = x \sim \mathrm{N}(\mu_{Y|X=x}, \Sigma_{Y|X=x}), \quad \text{where}$$
$$\mu_{Y|X=x} = \mu_Y + \Sigma_{YX}\Sigma_{XX}^{-1}(x - \mu_X) \quad \text{and}$$
$$\Sigma_{Y|X=x} = \Sigma_{YY} - \Sigma_{YX}\Sigma_{XX}^{-1}\Sigma_{XY}.$$

If X and Y are jointly multivariate normal and uncorrelated, i.e. $\Sigma_{XY} = \mathbf{0}$, then X and Y are also independent.

A linear transformation AX of a multivariate normal random variable $X \sim \mathrm{N}_p(\mu, \Sigma)$ is again normal: $AX \sim \mathrm{N}_q(A\mu, A\Sigma A^\top)$. Here A denotes a $q \times p$ matrix of full rank $q \leq p$.

Sometimes quadratic forms of a multivariate normal random variable $X \sim \mathrm{N}_p(\mu, \Sigma)$ are also of interest. One useful formula is for the expectation of $Q = X^\top AX$, where A is a $p \times p$ square matrix:

$$\mathsf{E}(Q) = \mathrm{tr}(A\Sigma) + \mu^\top A\mu.$$

A.3 Expectation, Variance and Covariance

This section describes fundamental summaries of the distribution of random variables, namely their expectation and variance. For multivariate random variables, the association between the components is summarised by their covariance or correlation. Moreover, we list some useful inequalities.

A.3.1 Expectation

For a continuous random variable X with density function $f(x)$, the *expectation* or *mean value* of X is the real number

$$\mathsf{E}(X) = \int x f(x)\,dx. \tag{A.14}$$

The expectation of a function $h(X)$ of X is

$$\mathsf{E}\{h(X)\} = \int h(x) f(x)\,dx. \tag{A.15}$$

Note that the expectation of a random variable does not necessarily exist; we then say that X has an *infinite expectation*. If the integral exists, then X has a *finite expectation*. For a discrete random variable X with probability mass function $f(x)$, the integral in (A.14) and (A.15) has to be replaced with a sum over the support of X.

For any real numbers a and b, we have

$$\mathsf{E}(a \cdot X + b) = a \cdot \mathsf{E}(X) + b.$$

For any two random variables X and Y, we have

$$\mathsf{E}(X + Y) = \mathsf{E}(X) + \mathsf{E}(Y),$$

i.e. the expectation of a sum of random variables equals the sum of the expectations of the random variables. If X and Y are independent, then

$$\mathsf{E}(X \cdot Y) = \mathsf{E}(X) \cdot \mathsf{E}(Y).$$

The expectation of a p-dimensional random variable $\boldsymbol{X} = (X_1, \dots, X_p)^\top$ is the vector $\mathsf{E}(\boldsymbol{X})$ with components equal to the individual expectations:

$$\mathsf{E}(\boldsymbol{X}) = \bigl(\mathsf{E}(X_1), \dots, \mathsf{E}(X_p)\bigr)^\top.$$

The expectation of a real-valued function $h(\boldsymbol{X})$ of a p-dimensional random variable $\boldsymbol{X} = (X_1, \dots, X_p)^\top$ is

$$\mathsf{E}\{h(\boldsymbol{X})\} = \int h(\boldsymbol{X}) f(\boldsymbol{X})\,d\boldsymbol{x}. \tag{A.16}$$

A.3.2 Variance

The *variance* of a random variable X is defined as

$$\mathrm{Var}(X) = \mathsf{E}\{X - \mathsf{E}(X)\}^2.$$

The variance of X may also be written as

$$\mathrm{Var}(X) = \mathsf{E}(X^2) - \mathsf{E}(X)^2$$

and

$$\mathrm{Var}(X) = \frac{1}{2}\mathsf{E}\{(X_1 - X_2)^2\},$$

where X_1 and X_2 are independent copies of X, i.e. X, X_1 and X_2 are independent and identically distributed. The square root $\sqrt{\mathrm{Var}(X)}$ of the variance of X is called the *standard deviation*.

For any real numbers a and b, we have

$$\mathrm{Var}(a \cdot X + b) = a^2 \cdot \mathrm{Var}(X).$$

A.3.3 Moments

Let k denote a positive integer. The *kth moment* m_k of a random variable X is defined as

$$m_k = \mathsf{E}(X^k),$$

whereas the *kth central moment* is

$$c_k = \mathsf{E}\{(X - m_1)^k\}.$$

The expectation $\mathsf{E}(X)$ of a random variable X is therefore its first moment, whereas the variance $\mathrm{Var}(X)$ is the second central moment. Those two quantities are therefore often referred to as "the first two moments". The third and fourth moments of a random variable quantify its skewness and kurtosis, respectively. If the kth moment of a random variable exists, then all lower moments also exist.

A.3.4 Conditional Expectation and Variance

For continuous random variables, the *conditional expectation* of Y given $X = x$,

$$\mathsf{E}(Y \mid X = x) = \int y f(y \mid x)\, dy, \tag{A.17}$$

is the expectation of Y with respect to the conditional density of Y given $X = x$, $f(y \mid x)$. It is a real number (if it exists). For discrete random variables, the integral in (A.17) has to be replaced with a sum, and the conditional density has to be replaced with the conditional probability mass function $f(y \mid x) = \Pr(Y = y \mid X = x)$. The conditional expectation can be interpreted as the mean of Y, if the value x of the random variable X is already known. The *conditional variance* of Y given $X = x$,

$$\operatorname{Var}(Y \mid X = x) = \mathsf{E}\big[\{Y - \mathsf{E}(Y \mid X = x)\}^2 \mid X = x\big], \tag{A.18}$$

is the variance of Y, conditional on $X = x$. Note that the outer expectation is again with respect to the conditional density of Y given $X = x$.

However, if we now consider the realisation x of the random variable X in $g(x) = \mathsf{E}(Y \mid X = x)$ as unknown, then $g(X)$ is a function of the random variable X and therefore also a random variable. This is the *conditional expectation* of Y given X. Similarly, the random variable $h(X)$ with $h(x) = \operatorname{Var}(Y \mid X = x)$ is called the *conditional variance* of Y given $X = x$. Note that the nomenclature is somewhat confusing since only the addendum "of Y given X" indicates that we do not refer to the numbers (A.17) nor (A.18), respectively, but to the corresponding random variables.

These definitions give rise to two useful results. The *law of total expectation* states that the expectation of the conditional expectation of Y given X equals the (ordinary) expectation of Y for any two random variables X and Y:

$$\mathsf{E}(Y) = \mathsf{E}\{\mathsf{E}(Y \mid X)\}. \tag{A.19}$$

Equation (A.19) is also known as the *law of iterated expectations*. The *law of total variance* provides a useful decomposition of the variance of Y:

$$\operatorname{Var}(Y) = \mathsf{E}\{\operatorname{Var}(Y \mid X)\} + \operatorname{Var}\{\mathsf{E}(Y \mid X)\}. \tag{A.20}$$

These two results are particularly useful if the first two moments of $Y \mid X = x$ and X are known. Calculation of expectation and variance of Y via (A.19) and (A.20) is then often simpler than directly based on the marginal distribution of Y.

A.3.5 Covariance

Let $(X, Y)^\top$ denote a bivariate random variable with joint probability mass or density function $f_{X,Y}(x, y)$. The *covariance* of X and Y is defined as

$$\begin{aligned}\operatorname{Cov}(X, Y) &= \mathsf{E}\big[\{X - \mathsf{E}(X)\}\{Y - \mathsf{E}(Y)\}\big] \\ &= \mathsf{E}(XY) - \mathsf{E}(X)\,\mathsf{E}(Y),\end{aligned}$$

where $\mathsf{E}(XY) = \int xy f_{X,Y}(x, y)\,dx\,dy$, see (A.16). Note that $\operatorname{Cov}(X, X) = \operatorname{Var}(X)$ and $\operatorname{Cov}(X, Y) = 0$ if X and Y are independent.

For any real numbers a, b, c and d, we have

$$\mathrm{Cov}(a \cdot X + b, c \cdot Y + d) = a \cdot c \cdot \mathrm{Cov}(X, Y).$$

The *covariance matrix* of a p-dimensional random variable $\boldsymbol{X} = (X_1, \dots, X_p)^\top$ is

$$\mathrm{Cov}(\boldsymbol{X}) = \mathsf{E}\big[\{\boldsymbol{X} - \mathsf{E}(\boldsymbol{X})\} \cdot \{\boldsymbol{X} - \mathsf{E}(\boldsymbol{X})\}^\top\big].$$

The covariance matrix can also be written as

$$\mathrm{Cov}(\boldsymbol{X}) = \mathsf{E}(\boldsymbol{X} \cdot \boldsymbol{X}^\top) - \mathsf{E}(\boldsymbol{X}) \cdot \mathsf{E}(\boldsymbol{X})^\top$$

and has entry $\mathrm{Cov}(X_i, X_j)$ in the ith row and jth column. In particular, on the diagonal of $\mathrm{Cov}(\boldsymbol{X})$ there are the variances of the components of $\boldsymbol{X}$.

If $\boldsymbol{X}$ is a p-dimensional random variable and $\boldsymbol{A}$ is a $q \times p$ matrix, we have

$$\mathrm{Cov}(\boldsymbol{A} \cdot \boldsymbol{X}) = \boldsymbol{A} \cdot \mathrm{Cov}(\boldsymbol{X}) \cdot \boldsymbol{A}^\top.$$

In particular, for the bivariate random variable $(X, Y)^\top$ and matrix $\boldsymbol{A} = (1, 1)$, we have

$$\mathrm{Var}(X + Y) = \mathrm{Var}(X) + \mathrm{Var}(Y) + 2 \cdot \mathrm{Cov}(X, Y).$$

If X and Y are independent, then

$$\mathrm{Var}(X + Y) = \mathrm{Var}(X) + \mathrm{Var}(Y) \quad \text{and}$$

$$\mathrm{Var}(X \cdot Y) = \mathsf{E}(X)^2 \,\mathrm{Var}(Y) + \mathsf{E}(Y)^2 \,\mathrm{Var}(X) + \mathrm{Var}(X)\,\mathrm{Var}(Y).$$

A.3.6 Correlation

The *correlation* of X and Y is defined as

$$\mathrm{Corr}(X, Y) = \frac{\mathrm{Cov}(X, Y)}{\sqrt{\mathrm{Var}(X)\,\mathrm{Var}(Y)}},$$

as long as the variances $\mathrm{Var}(X)$ and $\mathrm{Var}(Y)$ are positive.

An important property of the correlation is

$$\big|\mathrm{Corr}(X, Y)\big| \le 1, \tag{A.21}$$

which can be shown with the *Cauchy–Schwarz inequality* (after Augustin Louis Cauchy, 1789–1857, and Hermann Amandus Schwarz, 1843–1921). This inequality states that for two random variables X and Y with finite second moments $\mathsf{E}(X^2)$ and $\mathsf{E}(Y^2)$,

$$\mathsf{E}(X \cdot Y)^2 \le \mathsf{E}(X^2)\,\mathsf{E}(Y^2). \tag{A.22}$$

Applying (A.22) to the random variables $X - \mathsf{E}(X)$ and $Y - \mathsf{E}(Y)$, one obtains

$$\mathsf{E}[\{X - \mathsf{E}(X)\}\{Y - \mathsf{E}(Y)\}]^2 \le \mathsf{E}[\{X - \mathsf{E}(X)\}^2]\,\mathsf{E}[\{Y - \mathsf{E}(Y)\}^2],$$

from which

$$\operatorname{Corr}(X, Y)^2 = \frac{\operatorname{Cov}(X, Y)^2}{\operatorname{Var}(X)\operatorname{Var}(Y)} \le 1,$$

i.e. (A.21), easily follows. If $Y = a \cdot X + b$ for some $a > 0$ and b, then $\operatorname{Corr}(X, Y) = +1$. If $a < 0$, then $\operatorname{Corr}(X, Y) = -1$.

Let $\boldsymbol{\Sigma}$ denote the covariance matrix of a p-dimensional random variable $\boldsymbol{X} = (X_1, \dots, X_p)^\top$. The *correlation matrix* $\boldsymbol{R}$ of $\boldsymbol{X}$ can be obtained via

$$\boldsymbol{R} = \boldsymbol{S}\boldsymbol{\Sigma}\boldsymbol{S},$$

where $\boldsymbol{S}$ denotes the diagonal matrix with entries equal to the standard deviations $\sqrt{\operatorname{Var}(X_i)}$, $i = 1, \dots, n$, of the components of $\boldsymbol{X}$. A correlation matrix $\boldsymbol{R}$ has the entry $\operatorname{Corr}(X_i, X_j)$ in the ith row and jth column. In particular, the diagonal elements are all one.

A.3.7 Jensen's Inequality

Let X denote a random variable with finite expectation $\mathsf{E}(X)$, and $g(x)$ a *convex* function (if the second derivative $g''(x)$ exists, this is equivalent to $g''(x) \ge 0$ for all $x \in \mathbb{R}$). Then

$$\mathsf{E}\{g(X)\} \ge g\{\mathsf{E}(X)\}.$$

If $g(x)$ is even *strictly convex* ($g''(x) > 0$ for all real x) and X is not a constant, i.e. not *degenerate*, then

$$\mathsf{E}\{g(X)\} > g\{\mathsf{E}(X)\}.$$

For *(strictly) concave* functions $g(x)$ (if the second derivative $g''(x)$ exists, this is equivalent to the fact that for all $x \in \mathbb{R}$, $g''(x) \le 0$ and $g''(x) < 0$, respectively), the analogous results

$$\mathsf{E}\{g(X)\} \le g\{\mathsf{E}(X)\}$$

and

$$\mathsf{E}\{g(X)\} < g\{\mathsf{E}(X)\},$$

respectively, can be obtained.

A.3.8 Kullback–Leibler Discrepancy and Information Inequality

Let $f_X(x)$ and $f_Y(y)$ denote two density or probability functions, respectively, of random variables X and Y. The quantity

$$\mathsf{D}(f_X \parallel f_Y) = \mathsf{E}\left[\log\left\{\frac{f_X(X)}{f_Y(X)}\right\}\right] = \mathsf{E}\left[\log\{f_X(X)\}\right] - \mathsf{E}\left[\log\{f_Y(X)\}\right]$$

is called the *Kullback–Leibler discrepancy* from f_X to f_Y (after Solomon Kullback, 1907–1994, and Richard Leibler, 1914–2003) and quantifies effectively the "distance" between f_X and f_Y. However, note that in general

$$\mathsf{D}(f_X \parallel f_Y) \neq \mathsf{D}(f_Y \parallel f_X)$$

since $\mathsf{D}(f_X \parallel f_Y)$ is not symmetric in f_X and f_Y, so $\mathsf{D}(\cdot \parallel \cdot)$ is not a distance in the usual sense.

If X and Y have equal support, then the *information inequality* holds:

$$\mathsf{D}(f_X \parallel f_Y) = \mathsf{E}\left[\log\left\{\frac{f_X(X)}{f_Y(X)}\right\}\right] \geq 0,$$

where equality holds if and only if $f_X(x) = f_Y(x)$ for all $x \in \mathbb{R}$.

A.4 Convergence of Random Variables

After a definition of the different modes of convergence of random variables, several limit theorems are described, which have important applications in statistics.

A.4.1 Modes of Convergence

Let $X_1, X_2, \ldots$ be a sequence of random variables. We say:

1. $X_n \to X$ converges *in rth mean*, $r \geq 1$, written as $X_n \xrightarrow{r} X$, if $\mathsf{E}(|X_n^r|) < \infty$ for all n and

$$\mathsf{E}\left(|X_n - X|^r\right) \to 0 \quad \text{as } n \to \infty.$$

The case $r = 2$, called *convergence in mean square*, is often of particular interest.

2. $X_n \to X$ converges *in probability*, written as $X_n \xrightarrow{P} X$, if

$$\mathsf{Pr}\left(|X_n - X| > \varepsilon\right) \to 0 \quad \text{as } n \to \infty \text{ for all } \varepsilon > 0.$$

3. $X_n \to X$ converges *in distribution*, written as $X_n \xrightarrow{D} X$, if

$$\mathsf{Pr}(X_n \leq x) \to \mathsf{Pr}(X \leq x) \quad \text{as } n \to \infty$$

for all points $x \in \mathbb{R}$ at which the distribution function $F_X(x) = \mathsf{Pr}(X \leq x)$ is continuous.

The following relationships between the different modes of convergence can be established:

$$X_n \xrightarrow{r} X \implies X_n \xrightarrow{P} X \text{ for any } r \geq 1,$$

$$X_n \xrightarrow{P} X \implies X_n \xrightarrow{D} X,$$

$$X_n \xrightarrow{D} c \implies X_n \xrightarrow{P} c,$$

where $c \in \mathbb{R}$ is a constant.

A.4.2 Continuous Mapping and Slutsky's Theorem

The *continuous mapping theorem* states that any continuous function $g : \mathbb{R} \to \mathbb{R}$ is limit-preserving for convergence in probability and convergence in distribution:

$$X_n \xrightarrow{P} X \implies g(X_n) \xrightarrow{P} g(X),$$

$$X_n \xrightarrow{D} X \implies g(X_n) \xrightarrow{D} g(X).$$

Slutsky's theorem states that the limits of $X_n \xrightarrow{D} X$ and $Y_n \xrightarrow{P} a \in \mathbb{R}$ are preserved under addition and multiplication:

$$X_n + Y_n \xrightarrow{D} X + a$$

$$X_n \cdot Y_n \xrightarrow{D} a \cdot X.$$

A.4.3 Law of Large Numbers

Let $X_1, X_2, \dots$ be a sequence of independent and identically distributed random variables with finite expectation μ. Then

$$\frac{1}{n} \sum_{i=1}^{n} X_i \xrightarrow{P} \mu \quad \text{as } n \to \infty.$$

A.4.4 Central Limit Theorem

Let $X_1, X_2, \dots$ denote a sequence of independent and identically distributed random variables with mean $\mu = \mathsf{E}(X_i) < \infty$ and finite, non-zero variance ($0 < \text{Var}(X_i) = \sigma^2 < \infty$). Then, as $n \to \infty$,

$$\frac{1}{\sqrt{n\sigma^2}} \left(\sum_{i=1}^{n} X_i - n\mu \right) \xrightarrow{D} Z,$$

where $Z \sim \mathrm{N}(0, 1)$. A more compact notation is

$$\frac{1}{\sqrt{n\sigma^2}}\left(\sum_{i=1}^{n} X_i - n\mu\right) \overset{a}{\sim} \mathrm{N}(0, 1),$$

where $\overset{a}{\sim}$ stands for "is asymptotically distributed as".

If $\boldsymbol{X}_1, \boldsymbol{X}_2, \ldots$ denotes a sequence of independent and identically distributed p-dimensional random variables with mean $\boldsymbol{\mu} = \mathsf{E}(\boldsymbol{X}_i)$ and finite, positive definite covariance matrix $\boldsymbol{\Sigma} = \mathrm{Cov}(\boldsymbol{X}_i)$, then, as $n \to \infty$,

$$\frac{1}{\sqrt{n}}\left(\sum_{i=1}^{n} \boldsymbol{X}_i - n\boldsymbol{\mu}\right) \xrightarrow{D} \boldsymbol{Z},$$

where $\boldsymbol{Z} \sim \mathrm{N}_p(\boldsymbol{0}, \boldsymbol{\Sigma})$ denotes a p-dimensional normal distribution with expectation $\boldsymbol{0}$ and covariance matrix $\boldsymbol{\Sigma}$, compare Appendix A.5.3. In more compact notation we have

$$\frac{1}{\sqrt{n}}\left(\sum_{i=1}^{n} \boldsymbol{X}_i - n\boldsymbol{\mu}\right) \overset{a}{\sim} \mathrm{N}_p(\boldsymbol{0}, \boldsymbol{\Sigma}).$$

A.4.5 Delta Method

Consider $T_n = \frac{1}{n}\sum_{i=1}^{n} X_i$, where the X_is are independent and identically distributed random variables with finite expectation μ and variance σ^2. Suppose $g(\cdot)$ is (at least in a neighbourhood of μ) continuously differentiable with derivative g' and $g'(\mu) \neq 0$. Then

$$\sqrt{n}\{g(T_n) - g(\mu)\} \overset{a}{\sim} \mathrm{N}\left(0, g'(\mu)^2 \cdot \sigma^2\right)$$

as $n \to \infty$.

Somewhat simplifying, the delta method states that

$$g(Z) \overset{a}{\sim} \mathrm{N}\left(g(\nu), g'(\nu)^2 \cdot \tau^2\right)$$

if $Z \overset{a}{\sim} \mathrm{N}(\nu, \tau^2)$.

Now consider $\boldsymbol{T}_n = \frac{1}{n}(\boldsymbol{X}_1 + \cdots + \boldsymbol{X}_n)$, where the p-dimensional random variables $\boldsymbol{X}_i$ are independent and identically distributed with finite expectation $\boldsymbol{\mu}$ and covariance matrix $\boldsymbol{\Sigma}$. Suppose that $\boldsymbol{g} : \mathbb{R}^p \to \mathbb{R}^q$ $(q \leq p)$ is a mapping continuously differentiable in a neighbourhood of $\boldsymbol{\mu}$ with $q \times p$ Jacobian matrix $\boldsymbol{D}$ (cf. Appendix B.2.2) of full rank q. Then

$$\sqrt{n}\{\boldsymbol{g}(\boldsymbol{T}_n) - \boldsymbol{g}(\boldsymbol{\mu})\} \overset{a}{\sim} \mathrm{N}_q\left(\boldsymbol{0}, \boldsymbol{D}\boldsymbol{\Sigma}\boldsymbol{D}^\top\right)$$

as $n \to \infty$.

Somewhat simplifying, the multivariate delta method states that if $\boldsymbol{Z} \overset{a}{\sim} \mathrm{N}_p(\boldsymbol{\nu}, \mathcal{T})$, then

$$g(\boldsymbol{Z}) \overset{a}{\sim} \mathrm{N}_q\big(g(\boldsymbol{\nu}), \boldsymbol{D}\mathcal{T}\boldsymbol{D}^\top\big).$$

A.5 Probability Distributions

In this section we summarise the most important properties of the probability distributions used in this book. A random variable is denoted by X, and its probability or density function is denoted by $f(x)$. The probability or density function is defined for values in the support $\mathcal{T}$ of each distribution and is always zero outside of $\mathcal{T}$. For each distribution, the mean $\mathsf{E}(X)$, variance $\mathrm{Var}(X)$ and mode $\mathrm{Mod}(X)$ are listed, if appropriate.

In the first row we list the name of the distribution, an abbreviation and the core of the corresponding R-function (e.g. `norm`), indicating the parametrisation implemented in R. Depending on the first letter, these functions can be conveniently used as follows:

`r` stands for *r*andom and generates independent random numbers or vectors from the distribution considered. For example, `rnorm(n, mean = 0, sd = 1)` generates n random numbers from the standard normal distribution.

`d` stands for *d*ensity and returns the probability and density function, respectively. For example, `dnorm(x)` gives the density of the standard normal distribution.

`p` stands for *p*robability and gives the distribution function $F(x) = \Pr(X \leq x)$ of X. For example, if X is standard normal, then `pnorm(0)` returns 0.5, while `pnorm(1.96)` is $0.975002 \approx 0.975$.

`q` stands for *q*uantile and gives the quantile function. For example, `qnorm(0.975)` is $1.959964 \approx 1.96$.

For some distributions, not all four options may be available. The first argument of each function is not listed since it depends on the particular function used. It is either the number n of random variables generated, a value x in the domain $\mathcal{T}$ of the random variable or a probability $p \in [0, 1]$. The arguments x and p can be vectors, as well as some parameter values. The option `log = TRUE` is useful to compute the log of the density, distribution or quantile function. For example, multiplication of very small numbers, which may cause numerical problems, can be replaced by addition of the log numbers and subsequent application of the exponential function `exp()` to the obtained sum.

With the option `lower.tail = FALSE`, available in `p`- and `q`-type functions, the *upper* tail of the distribution function $\Pr(X > x)$ and the upper quantile z with $\Pr(X > z) = p$, respectively, are returned. Further details can be found in the documentation to each function, e.g. by typing `?rnorm`.

A.5.1 Univariate Discrete Distributions

Table A.1 gives some elementary facts about the most important univariate discrete distributions used in this book. The function `sample` can be applied in various settings, for example to simulate discrete random variables with finite support or for resampling. Functions for the beta-binomial distribution (except for the quantile function) are available in the package `VGAM`. The density and random number generator functions of the noncentral hypergeometric distribution are available in the package `MCMCpack`.

Table A.1 Univariate discrete distributions

Urn model: `sample(x, size, replace = FALSE, prob = NULL)`

A sample of size `size` is drawn from an urn with elements `x`. The corresponding sample probabilities are listed in the vector `prob`, which does not need to be normalised. If `prob = NULL`, all elements are equally likely. If `replace = TRUE`, these probabilities do not change after the first draw. The default, however, is `replace = FALSE`, in which case the probabilities are updated draw by draw. The call `sample(x)` takes a random sample of size `length(x)` without replacement, hence returns a random permutation of the elements of `x`. The call `sample(x, replace = TRUE)` returns a random sample from the empirical distribution function of `x`, which is useful for (nonparametric) bootstrap approaches.

Bernoulli: $\mathrm{B}(\pi)$ `_binom(..., size = 1, prob = π)`

$0 < \pi < 1$ $\mathcal{T} = \{0, 1\}$

$f(x) = \pi^x (1-\pi)^{1-x}$ $\mathrm{Mod}(X) = \begin{cases} 0, & \pi \le 0.5, \\ 1, & \pi \ge 0.5. \end{cases}$

$\mathsf{E}(X) = \pi$ $\mathrm{Var}(X) = \pi(1-\pi)$

If $X_i \sim \mathrm{B}(\pi)$, $i = 1, \dots, n$, are independent, then $\sum_{i=1}^n X_i \sim \mathrm{Bin}(n, \pi)$.

Binomial: $\mathrm{Bin}(n, \pi)$ `_binom(..., size = n, prob = π)`

$0 < \pi < 1, n \in \mathbb{N}$ $\mathcal{T} = \{0, \dots, n\}$

$f(x) = \binom{n}{x} \pi^x (1-\pi)^{n-x}$ $\mathrm{Mod}(X) = \begin{cases} \lfloor z_m = (n+1)\pi \rfloor, & z_m \notin \mathbb{N}, \\ z_m - 1 \text{ and } z_m, & \text{else}. \end{cases}$

$\mathsf{E}(X) = n\pi$ $\mathrm{Var}(X) = n\pi(1-\pi)$

The case $n = 1$ corresponds to a Bernoulli distribution with success probability π. If $X_i \sim \mathrm{Bin}(n_i, \pi)$, $i = 1, \dots, n$, are independent, then $\sum_{i=1}^n X_i \sim \mathrm{Bin}(\sum_{i=1}^n n_i, \pi)$.

Geometric: $\mathrm{Geom}(\pi)$ `_geom(..., prob = π)`

$0 < \pi < 1$ $\mathcal{T} = \mathbb{N}$

$f(x) = \pi(1-\pi)^{x-1}$

$\mathsf{E}(X) = 1/\pi$ $\mathrm{Var}(X) = (1-\pi)/\pi^2$

Caution: The functions in `R` relate to the random variable $X - 1$, i.e. the number of *failures* until a success has been observed. If $X_i \sim \mathrm{Geom}(\pi)$, $i = 1, \dots, n$, are independent, then $\sum_{i=1}^n X_i \sim \mathrm{NBin}(n, \pi)$.

Table A.1 (Continued)

Hypergeometric: HypGeom(n, N, M)	`_hyper(..., m = M, n = N − M, k = n)`
$N \in \mathbb{N}, M \in \{0, \ldots, N\}, n \in \{1, \ldots, N\}$	
$\mathcal{T} = \{\max(0, n + M - N), \ldots, \min(n, M)\}$	
$f(x) = C \cdot \binom{M}{x}\binom{N-M}{n-x}$	$C = \binom{N}{n}^{-1}$
$\text{Mod}(X) = \begin{cases} \lfloor x_m \rfloor, & x_m \notin \mathbb{N}, \\ x_m - 1 \text{ and } x_m, & \text{else.} \end{cases}$	$x_m = \frac{(n+1)(M+1)}{(N+2)}$
$\mathsf{E}(X) = n\frac{M}{N}$	$\text{Var}(X) = n\frac{M}{N}\frac{N-M}{N}\frac{(N-n)}{(N-1)}$
Noncentral hypergeometric: NCHypGeom(n, N, M, θ)	`MCMCpack::_noncenhypergeom(..., n1 = M, n2 = N − M, m1 = n, psi = θ)`
$N \in \mathbb{N}, M \in \{0, \ldots, N\}, n \in \{0, \ldots, N\}, \theta \in \mathbb{R}_+$	
$\mathcal{T} = \{\max(0, n + M - N), \ldots, \min(n, M)\}$	
$f(x) = C \cdot \binom{M}{x}\binom{N-M}{n-x}\theta^x$	$C = \{\sum_{x \in \mathcal{T}} \binom{M}{x}\binom{N-M}{n-x}\theta^x\}^{-1}$
$\text{Mod}(X) = \left\lfloor \frac{-2c}{b + \text{sign}(b)\sqrt{b^2 - 4ac}} \right\rfloor$	$a = \theta - 1$, $b = (M + n + 2)\theta + N - M - n$, $c = \theta(M + 1)(n + 1)$
This distribution arises if $X \sim \text{Bin}(M, \pi_1)$ independent of $Y \sim \text{Bin}(N - M, \pi_2)$ and $Z = X + Y$, then $X \mid Z = n \sim \text{NCHypGeom}(n, N, M, \theta)$ with the odds ratio $\theta = \frac{\pi_1(1-\pi_2)}{(1-\pi_1)\pi_2}$. For $\theta = 1$, this reduces to HypGeom(n, N, M).	
Negative binomial: NBin(r, π)	`_nbinom(..., size = r, prob = π)`
$0 < \pi < 1, r \in \mathbb{N}$	$\mathcal{T} = \{r, r + 1, \ldots\}$
$f(x) = \binom{x-1}{r-1}\pi^r(1 - \pi)^{x-r}$	$\text{Mod}(X) = \begin{cases} \lfloor z_m = 1 + \frac{r-1}{\pi} \rfloor, & z_m \notin \mathbb{N}, \\ z_m - 1 \text{ and } z_m, & \text{else.} \end{cases}$
$\mathsf{E}(X) = \frac{r}{\pi}$	$\text{Var}(X) = \frac{r(1-\pi)}{\pi^2}$
Caution: The functions in R relate to the random variable $X - r$, i.e. the number of *failures* until r successes have been observed. The NBin$(1, \pi)$ distribution is a geometric distribution with parameter π. If $X_i \sim \text{NBin}(r_i, \pi)$, $i = 1, \ldots, n$, are independent, then $\sum_{i=1}^n X_i \sim \text{NBin}(\sum_{i=1}^n r_i, \pi)$.	
Poisson: Po(λ)	`_pois(..., lambda = λ)`
$\lambda > 0$	$\mathcal{T} = \mathbb{N}_0$
$f(x) = \frac{\lambda^x}{x!}\exp(-\lambda)$	$\text{Mod}(X) = \begin{cases} \lfloor \lambda \rfloor, & \lambda \notin \mathbb{N}, \\ \lambda - 1, \lambda, & \text{else.} \end{cases}$
$\mathsf{E}(X) = \lambda$	$\text{Var}(X) = \lambda$
If $X_i \sim \text{Po}(\lambda_i)$, $i = 1, \ldots, n$, are independent, then $\sum_{i=1}^n X_i \sim \text{Po}(\sum_{i=1}^n \lambda_i)$. Moreover, $X_1 \mid \{X_1 + X_2 = n\} \sim \text{Bin}(n, \lambda_1/(\lambda_1 + \lambda_2))$.	

Table A.1 (Continued)

Poisson-gamma: $\text{PoG}(\alpha, \beta, \nu)$	`_nbinom(..., size` $= \alpha$, `prob` $= \beta/(\beta+\nu)$)
$\alpha, \beta > 0$	$\mathcal{T} = \mathbb{N}_0$
$f(x) = C \cdot \frac{\Gamma(\alpha+x)}{x!}\left(\frac{\nu}{\beta+\nu}\right)^x$	$C = \left(\frac{\beta}{\beta+\nu}\right)^\alpha \frac{1}{\Gamma(\alpha)}$
$\text{Mod}(X) = \begin{cases} \lceil \frac{\nu(\alpha-1)}{\beta} - 1 \rceil, & \alpha\nu > \beta + \nu, \\ 0, 1 & \alpha\nu = \beta + \nu, \\ 0, & \alpha\nu < \beta + \nu. \end{cases}$	
$\mathsf{E}(X) = \nu\frac{\alpha}{\beta}$	$\text{Var}(X) = \alpha\frac{\nu}{\beta}(1 + \frac{\nu}{\beta})$
The gamma function $\Gamma(x)$ is described in Appendix B.2.1. The Poisson-gamma distribution generalises the negative binomial distribution, since $X + \alpha \sim \text{NBin}(\alpha, \frac{\beta}{\beta+\nu})$, if $\alpha \in \mathbb{N}$. In R there is only one function for both distributions.	
Beta-binomial: $\text{BeB}(n, \alpha, \beta)$	`VGAM::_betabinom.ab(..., size` $= n$, `shape1` $= \alpha$, `shape2` $= \beta$)
$\alpha, \beta > 0, n \in \mathbb{N}$	$\mathcal{T} = \{0, \dots, n\}$
$f(x) = \binom{n}{x}\frac{\text{B}(\alpha+x, \beta+n-x)}{\text{B}(\alpha,\beta)}$	
$\text{Mod}(X) = \begin{cases} \lfloor x_m \rfloor, & x_m \notin \mathbb{N}, \\ x_m - 1 \text{ and } x_m, & \text{else.} \end{cases}$	$x_m = \frac{(n+1)(\alpha-1)}{\alpha+\beta-2}$
$\mathsf{E}(X) = n\frac{\alpha}{\alpha+\beta}$	$\text{Var}(X) = n\frac{\alpha\beta}{(\alpha+\beta)^2}\frac{(\alpha+\beta+n)}{(\alpha+\beta+1)}$
The beta function $\text{B}(x, y)$ is described in Appendix B.2.1. The $\text{BeB}(n, 1, 1)$ distribution is a discrete uniform distribution with support $\mathcal{T}$ and $f(x) = (n+1)^{-1}$. For $n = 1$, the beta-binomial distribution $\text{BeB}(1, \alpha, \beta)$ reduces to the Bernoulli distribution $\text{B}(\pi)$ with success probability $\pi = \alpha/(\alpha + \beta)$.	

A.5.2 Univariate Continuous Distributions

Table A.2 gives some elementary facts about the most important univariate continuous distributions used in this book. The density and random number generator functions of the inverse gamma distribution are available in the package `MCMCpack`. The distribution and quantile function (as well as random numbers) can be calculated with the corresponding functions of the gamma distribution. Functions relating to the general t distribution are available in the package `sn`. The functions `_t(..., df` $= \alpha$) available by default in `R` cover the standard t distribution. The log-normal, folded normal, Gumbel and the Pareto distributions are available in the package `VGAM`. The gamma–gamma distribution is currently not available.

A.5.3 Multivariate Distributions

Table A.3 gives details about the most important multivariate probability distributions used in this book. Multivariate random variables $\boldsymbol{X}$ are always given in bold face. Note that there is no distribution or quantile function available in `R` for the

Table A.2 Univariate continuous distributions

Uniform: $\mathrm{U}(a, b)$	`_unif(..., min = a, max = b)`
$b > a$	$\mathcal{T} = (a, b)$
$f(x) = 1/(b-a)$	
$\mathsf{E}(X) = (a+b)/2$	$\mathrm{Var}(X) = (b-a)^2/12$
X is standard uniform if $a = 0$ and $b = 1$.	
Beta: $\mathrm{Be}(\alpha, \beta)$	`_beta(..., shape1 = α, shape2 = β)`
$\alpha, \beta > 0$	$\mathcal{T} = (0, 1)$
$f(x) = \mathrm{B}(\alpha, \beta)^{-1} x^{\alpha-1}(1-x)^{\beta-1}$	$\mathrm{Mod}(X) = \frac{\alpha-1}{\alpha+\beta-2}$ if $\alpha, \beta > 1$
$\mathsf{E}(X) = \frac{\alpha}{\alpha+\beta}$	$\mathrm{Var}(X) = \frac{\alpha\beta}{(\alpha+\beta)^2(\alpha+\beta+1)}$
The beta function $\mathrm{B}(x, y)$ is described in Appendix B.2.1. The $\mathrm{Be}(1, 1)$ distribution is a standard uniform distribution. If $X \sim \mathrm{Be}(\alpha, \beta)$ then $1 - X \sim \mathrm{Be}(\beta, \alpha)$ and $\beta/\alpha \cdot X/(1-X) \sim \mathrm{F}(2\alpha, 2\beta)$.	
Exponential: $\mathrm{Exp}(\lambda)$	`_exp(..., rate = λ)`
$\lambda > 0$	$\mathcal{T} = \mathbb{R}^+$
$f(x) = \lambda \exp(-\lambda x)$	$\mathrm{Mod}(X) = 0$
$\mathsf{E}(X) = 1/\lambda$	$\mathrm{Var}(X) = 1/\lambda^2$
If $X_i \sim \mathrm{Exp}(\lambda)$, $i = 1, \ldots, n$, are independent, then $\sum_{i=1}^n X_i \sim \mathrm{G}(n, \lambda)$.	
Weibull: $\mathrm{Wb}(\mu, \alpha)$	`_weibull(..., shape = α, scale = μ)`
$\mu, \alpha > 0$	$\mathcal{T} = \mathbb{R}_+$
$f(x) = \frac{\alpha}{\mu}(\frac{x}{\mu})^{\alpha-1} \exp\{-(\frac{x}{\mu})^\alpha\}$	$\mathrm{Mod}(X) = \mu \cdot (1 - \frac{1}{\alpha})^{\frac{1}{\alpha}}$
$\mathsf{E}(X) = \mu \cdot \Gamma(\frac{1}{\alpha} + 1)$	$\mathrm{Var}(X) = \mu \cdot \{\Gamma(\frac{2}{\alpha} + 1) - \Gamma(\frac{1}{\alpha} + 1)^2\}$
The gamma function $\Gamma(x)$ is described in Appendix B.2.1. The $\mathrm{Wb}(\mu, 1)$ distribution is the exponential distribution with parameter $1/\mu$. Although it was discovered earlier in the 20th century, the Weibull distribution is named after the Swedish engineer Waloddi Weibull, 1887–1979, who described it in 1951.	
Gamma: $\mathrm{G}(\alpha, \beta)$	`_gamma(..., shape = α, rate = β)`
$\alpha, \beta > 0$	$\mathcal{T} = \mathbb{R}^+$
$f(x) = \frac{\beta^\alpha}{\Gamma(\alpha)} x^{\alpha-1} \exp(-\beta x)$	$\mathrm{Mod}(X) = \frac{\alpha-1}{\beta}$ if $\alpha > 1$
$\mathsf{E}(X) = \alpha/\beta$	$\mathrm{Var}(X) = \alpha/\beta^2$
The gamma function $\Gamma(x)$ is described in Appendix B.2.1. β is an inverse scale parameter, which means that $Y = c \cdot X$ has the gamma distribution $\mathrm{G}(\alpha, \beta/c)$. The $\mathrm{G}(1, \beta)$ is the exponential distribution with parameter β. For $\alpha = d/2$ and $\beta = 1/2$, one obtains the chi-squared distribution $\chi^2(d)$. If $X_i \sim \mathrm{G}(\alpha_i, \beta)$, $i = 1, \ldots, n$, are independent, then $\sum_{i=1}^n X_i \sim \mathrm{G}(\sum_{i=1}^n \alpha_i, \beta)$.	
Inverse gamma: $\mathrm{IG}(\alpha, \beta)$	`MCMCpack::_invgamma(..., shape = α, scale = β)`
$\alpha, \beta > 0$	$\mathcal{T} = \mathbb{R}^+$
$f(x) = \frac{\beta^\alpha}{\Gamma(\alpha)} x^{-(\alpha+1)} \exp(-\beta/x)$	$\mathrm{Mod}(X) = \frac{\beta}{\alpha+1}$
$\mathsf{E}(X) = \frac{\beta}{\alpha-1}$ if $\alpha > 1$	$\mathrm{Var}(X) = \frac{\beta^2}{(\alpha-1)^2(\alpha-2)}$ if $\alpha > 2$
The gamma function $\Gamma(x)$ is described in Appendix B.2.1. If $X \sim \mathrm{G}(\alpha, \beta)$ then $1/X \sim \mathrm{IG}(\alpha, \beta)$.	

Table A.2 (Continued)

Gamma–gamma: $\mathrm{Gg}(\alpha, \beta, \delta)$	
$\alpha, \beta, \delta > 0$	$\mathcal{T} = \mathbb{R}^+$
$f(x) = \frac{\beta^\alpha}{\mathrm{B}(\alpha,\delta)} \frac{x^{\delta-1}}{(\beta+x)^{\alpha+\delta}}$	$\mathrm{Mod}(X) = \frac{(\delta-1)\beta}{\alpha+1}$ if $\delta > 1$
$\mathsf{E}(X) = \frac{\delta\beta}{\alpha-1}$ if $\alpha > 1$	$\mathrm{Var}(X) = \frac{\beta^2(\delta^2+\delta(\alpha-1))}{(\alpha-1)^2(\alpha-2)}$ if $\alpha > 2$
The beta function $\mathrm{B}(x, y)$ is described in Appendix B.2.1. A gamma–gamma random variable $Y \sim \mathrm{Gg}(\alpha, \beta, \delta)$ is generated by the mixture $X \sim \mathrm{G}(\alpha, \beta)$, $Y \mid \{X = x\} \sim \mathrm{G}(\delta, x)$.	
Chi-squared: $\chi^2(d)$	`_chisq(..., df = d)`
$d > 0$	$\mathcal{T} = \mathbb{R}^+$
$f(x) = \frac{(\frac{1}{2})^{\frac{d}{2}}}{\Gamma(\frac{d}{2})} x^{\frac{d}{2}-1} \exp(-x/2)$	$\mathrm{Mod}(X) = d - 2$ if $d > 2$
$\mathsf{E}(X) = d$	$\mathrm{Var}(X) = 2d$
The gamma function $\Gamma(x)$ is described in Appendix B.2.1. If $X_i \sim \mathrm{N}(0, 1)$, $i = 1, \ldots, n$, are independent, then $\sum_{i=1}^n X_i^2 \sim \chi^2(n)$.	
Normal: $\mathrm{N}(\mu, \sigma^2)$	`_norm(..., mean = μ, sd = σ)`
$\mu \in \mathbb{R}, \sigma^2 > 0$	$\mathcal{T} = \mathbb{R}$
$f(x) = \frac{1}{\sqrt{2\pi\sigma^2}} \exp\{-\frac{1}{2}\frac{(x-\mu)^2}{\sigma^2}\}$	$\mathrm{Mod}(X) = \mu$
$\mathsf{E}(X) = \mu$	$\mathrm{Var}(X) = \sigma^2$
X is standard normal if $\mu = 0$ and $\sigma^2 = 1$, i.e. $f(x) = \varphi(x) = \frac{1}{\sqrt{2\pi}} \exp(-\frac{1}{2}x^2)$. If X is standard normal, then $\sigma X + \mu \sim \mathrm{N}(\mu, \sigma^2)$.	
Log-normal: $\mathrm{LN}(\mu, \sigma^2)$	`VGAM::_lnorm(..., meanlog = μ, sdlog = σ)`
$\mu \in \mathbb{R}, \sigma^2 > 0$	$\mathcal{T} = \mathbb{R}^+$
$f(x) = \frac{1}{\sigma}\frac{1}{x}\varphi\{\frac{\log(x)-\mu}{\sigma}\}$	$\mathrm{Mod}(X) = \exp(\mu - \sigma^2)$
$\mathsf{E}(X) = \exp(\mu + \sigma^2/2)$	$\mathrm{Var}(X) = \{\exp(\sigma^2) - 1\}\exp(2\mu + \sigma^2)$
If X is normal, i.e. $X \sim \mathrm{N}(\mu, \sigma^2)$, then $\exp(X) \sim \mathrm{LN}(\mu, \sigma^2)$.	
Folded normal: $\mathrm{FN}(\mu, \sigma^2)$	`VGAM::_foldnorm(..., mean = μ, sd = σ)`
$\mu \in \mathbb{R}, \sigma^2 > 0$	$\mathcal{T} = \mathbb{R}_0^+$
$f(x) = \frac{1}{\sigma}\{\varphi(\frac{x-\mu}{\sigma}) + \varphi(\frac{x+\mu}{\sigma})\}$	$\mathrm{Mod}(X) \begin{cases} = 0, & \lvert\mu\rvert \leq \sigma, \\ \in (0, \lvert\mu\rvert), & \lvert\mu\rvert > \sigma. \end{cases}$
$\mathsf{E}(X) = 2\sigma\varphi(\mu/\sigma) + \mu(2\Phi(\mu/\sigma) - 1)$	$\mathrm{Var}(X) = \sigma^2 + \mu^2 - \mathsf{E}(X)^2$
If X is normal, i.e. $X \sim \mathrm{N}(\mu, \sigma^2)$, then $\lvert X\rvert \sim \mathrm{FN}(\mu, \sigma^2)$. The mode can be calculated numerically if $\lvert\mu\rvert > \sigma$. If $\mu = 0$, one obtains the *half normal* distribution with $\mathsf{E}(X) = \sigma\sqrt{2/\pi}$ and $\mathrm{Var}(X) = \sigma^2(1 - 2/\pi)$.	

Table A.2 (Continued)

Student (t): $\mathrm{t}(\mu, \sigma^2, \alpha)$	`sn::_st(..., xi` $=\mu$, `omega` $=\sigma$, `nu` $=\alpha$`)`
$\mu \in \mathbb{R}, \sigma^2, \alpha > 0$	$\mathcal{T} = \mathbb{R}$
$f(x) = C \cdot \{1 + \frac{1}{\alpha\sigma^2}(x-\mu)^2\}^{-\frac{\alpha+1}{2}}$	$C = \{\sqrt{\alpha\sigma^2} B(\frac{\alpha}{2}, \frac{1}{2})\}^{-1}$
$\mathrm{Mod}(X) = \mu$	
$\mathsf{E}(X) = \mu$ if $\alpha > 1$	$\mathrm{Var}(X) = \sigma^2 \cdot \frac{\alpha}{\alpha-2}$ if $\alpha > 2$
The *Cauchy distribution* $\mathrm{C}(\mu, \sigma^2)$ arises for $\alpha = 1$, in which case both $\mathsf{E}(X)$ and $\mathrm{Var}(X)$ do not exist. If X is standard t distributed with degrees of freedom α, i.e. $X \sim \mathrm{t}(0, 1, \alpha) =: \mathrm{t}(\alpha)$, then $\sigma X + \mu \sim \mathrm{t}(\mu, \sigma^2, \alpha)$. A t distributed random variable $X \sim \mathrm{t}(\mu, \sigma^2, \alpha)$ converges in distribution to a normal random variable $Y \sim \mathrm{N}(\mu, \sigma^2)$ as $\alpha \to \infty$.	
F: $\mathrm{F}(\alpha, \beta)$	`_f(..., df1` $=\alpha$, `df2` $=\beta$`)`
$\alpha, \beta > 0$	$\mathcal{T} = \mathbb{R}^+$
$f(x) = C \cdot \frac{1}{x}(1 + \frac{\beta}{\alpha x})^{-\alpha/2}(1 + \frac{\alpha x}{\beta})^{-\beta/2}$	$C = \mathrm{B}(\alpha/2, \beta/2)^{-1}$
$\mathrm{Mod}(X) = \frac{(\alpha-2)\beta}{\alpha(\beta+2)}$ if $\alpha > 2$	
$\mathsf{E}(X) = \frac{\beta}{\beta-2}$ if $\beta > 2$	$\mathrm{Var}(X) = \frac{2\beta^2(\alpha+\beta-2)}{\alpha(\beta-2)^2(\beta-4)}$ if $\beta > 4$
The beta function $\mathrm{B}(x, y)$ is described in Appendix B.2.1. If $X \sim \chi^2(\alpha)$ and $Y \sim \chi^2(\beta)$ are independent, then $Z = \frac{X/\alpha}{Y/\beta}$ follows an F distribution, i.e. $Z \sim \mathrm{F}(\alpha, \beta)$.	
Logistic: $\mathrm{Log}(\mu, \sigma^2)$	`_logis(..., location` $=\mu$, `scale` $=\sigma$`)`
$\mu \in \mathbb{R}, \sigma > 0$	$\mathcal{T} = \mathbb{R}$
$f(x) = \frac{1}{\sigma}\exp(-\frac{x-\mu}{\sigma})\{1 + \exp(-\frac{x-\mu}{\sigma})\}^{-2}$	$\mathrm{Mod}(X) = \mu$
$\mathsf{E}(X) = \mu$	$\mathrm{Var}(X) = \sigma^2 \cdot \pi^2/3$
X has the standard logistic distribution if $\mu = 0$ and $\sigma = 1$. In this case the distribution function has the simple form $F(x) = \{1 + \exp(-x)\}^{-1}$, and the quantile function is the logit function $F^{-1}(p) = \log\{p/(1-p)\}$.	
Gumbel: $\mathrm{Gu}(\mu, \sigma^2)$	`VGAM::_gumbel(..., location` $=\mu$, `scale` $=\sigma$`)`
$\mu \in \mathbb{R}, \sigma > 0$	$\mathcal{T} = \mathbb{R}$
$f(x) = \frac{1}{\sigma}\exp\{-\frac{x-\mu}{\sigma} - \exp(-\frac{x-\mu}{\sigma})\}$	$\mathrm{Mod}(X) = \mu$
$\mathsf{E}(X) = \mu + \sigma\gamma$	$\mathrm{Var}(X) = \sigma^2 \frac{\pi^2}{2}$
The Euler–Mascheroni constant γ is defined as $\gamma = \lim_{n\to\infty} \sum_{k=1}^{n} \frac{1}{k} - \log(n) \approx 0.577215$. If $X \sim \mathrm{Wb}(\mu, \alpha)$, then $-\log(X) \sim \mathrm{Gu}(-\log(\mu), 1/\alpha^2)$. Thus the Gumbel distribution (named after the German mathematician Emil Julius Gumbel, 1891–1966) is also known as the log-Weibull distribution.	
Pareto: $\mathrm{Par}(\alpha, \beta)$	`VGAM::_pareto(..., shape` $=\alpha$, `scale` $=\beta$`)`
$\alpha > 0, \beta > 0$	
$f(x) = \begin{cases} \alpha\beta^\alpha x^{-(\alpha+1)} & \text{if } x \geq \beta, \\ 0 & \text{else} \end{cases}$	$\mathrm{Mod}(X) = \beta$
$\mathsf{E}(X) = \frac{\alpha\beta}{\alpha-1}$ if $\alpha > 1$	$\mathrm{Var}(X) = \frac{\alpha\beta^2}{(\alpha-1)^2(\alpha-2)}$ if $\alpha > 2$
A Pareto random variable $Y \sim \mathrm{Par}(\alpha, \beta)$ is generated by the mixture $X \sim \mathrm{G}(\alpha, \beta)$, $Y \mid \{X = x\} \sim \mathrm{Exp}(x) + \beta$. The Pareto distribution is named after the Italian economist Vilfredo Pareto, 1848–1923.	

multinomial, Dirichlet, Wishart and inverse Wishart distributions. The density function and random variable generator functions for Dirichlet, Wishart and inverse Wishart are available in the package `MCMCpack`. The package `mvtnorm` offers the density, distribution, and quantile functions of the multivariate normal distribution, as well as a random number generator function. The multinomial-Dirichlet and the normal-gamma distributions are currently not available in R.

Table A.3 Multivariate distributions

Multinomial: $\mathrm{M}_k(n, \boldsymbol{\pi})$	`_multinom(..., size` $= n$, `prob` $= \boldsymbol{\pi}$`)`
$\boldsymbol{\pi} = (\pi_1, \ldots, \pi_k)^\top, n \in \mathbb{N}$	$\boldsymbol{x} = (x_1, \ldots, x_k)^\top$
$\pi_i \in (0, 1), \sum_{j=1}^k \pi_j = 1$	$x_i \in \mathbb{N}_0, \sum_{j=1}^k x_j = n$
$f(\boldsymbol{x}) = \frac{n!}{\prod_{j=1}^k x_j!} \prod_{j=1}^k \pi_j^{x_j}$	
$\mathsf{E}(X_i) = n\pi_i$	$\mathrm{Var}(X_i) = n\pi_i(1-\pi_i)$
$\mathsf{E}(\boldsymbol{X}) = n\boldsymbol{\pi}$	$\mathrm{Cov}(\boldsymbol{X}) = n\{\mathrm{diag}(\boldsymbol{\pi}) - \boldsymbol{\pi}\boldsymbol{\pi}^\top\}$

The binomial distribution is a special case of the multinomial distribution: If $X \sim \mathrm{Bin}(n, \pi)$, then $(X, n-X)^\top \sim \mathrm{M}_2(n, (\pi, 1-\pi)^\top)$.

Dirichlet: $\mathrm{D}_k(\boldsymbol{\alpha})$	`MCMCpack::_dirichlet(..., alpha` $= \boldsymbol{\alpha}$`)`
$\boldsymbol{\alpha} = (\alpha_1, \ldots, \alpha_k)^\top$	$\boldsymbol{x} = (x_1, \ldots, x_k)^\top$
$\alpha_i > 0$	$x_i \in (0, 1), \sum_{j=1}^k x_j = 1$
$f(\boldsymbol{x}) = C \cdot \prod_{j=1}^k x_j^{\alpha_j - 1}$	$C = \Gamma(\sum_{j=1}^k \alpha_j) / \prod_{j=1}^k \Gamma(\alpha_j)$
$\mathsf{E}(X_i) = \frac{\alpha_i}{\sum_{j=1}^k \alpha_j}$	$\mathrm{Var}(X_i) = \frac{\mathsf{E}(X_i)\{1-\mathsf{E}(X_i)\}}{1+\sum_{j=1}^k \alpha_j}$
$\mathsf{E}(\boldsymbol{X}) = \boldsymbol{\alpha}(\boldsymbol{e}_k^\top \boldsymbol{\alpha})^{-1}$ where $\boldsymbol{e}_k^\top = (1, \ldots, 1)$	$\mathrm{Cov}(\boldsymbol{X}) = (1 + \boldsymbol{e}_k^\top \boldsymbol{\alpha})^{-1} \cdot [\mathrm{diag}\{\mathsf{E}(\boldsymbol{X})\} - \mathsf{E}(\boldsymbol{X})\,\mathsf{E}(\boldsymbol{X})^\top]$
$\mathrm{Mod}(\boldsymbol{X}) = \left(\frac{\alpha_1 - 1}{d}, \ldots, \frac{\alpha_k - 1}{d}\right)^\top$ where $d = \sum_{i=1}^k \alpha_i - k$ if $\alpha_i > 1$ for all i	

The gamma function $\Gamma(x)$ is described in Appendix B.2.1. The beta distribution is a special case of the Dirichlet distribution: If $X \sim \mathrm{Be}(\alpha, \beta)$, then $(X, 1-X)^\top \sim \mathrm{D}_2((\alpha, \beta)^\top)$.

Multinomial-Dirichlet: $\mathrm{MD}_k(n, \boldsymbol{\alpha})$	
$\boldsymbol{\alpha} = (\alpha_1, \ldots, \alpha_k)^\top, n \in \mathbb{N}$	$\boldsymbol{x} = (x_1, \ldots, x_k)^\top$
$\alpha_i > 0$	$x_i \in \mathbb{N}_0, \sum_{j=1}^k x_j = n$
$f(\boldsymbol{x}) = C \cdot \frac{\prod_{j=1}^k \Gamma(\alpha_j^*)}{\Gamma(\sum_{j=1}^k \alpha_j^*) \prod_{j=1}^k x_j!}$	$C = n!\Gamma(\sum_{j=1}^k \alpha_j) / \prod_{j=1}^k \Gamma(\alpha_j)$
$\alpha_j^* = \alpha_j + x_j$	$\boldsymbol{\pi} = \boldsymbol{\alpha}(\boldsymbol{e}_k^\top \boldsymbol{\alpha})^{-1}$
$\mathsf{E}(X_i) = n\pi_i$	$\mathrm{Var}(X_i) = \frac{\sum_{j=1}^k \alpha_j^*}{1+\sum_{j=1}^k \alpha_j} n\pi_i(1-\pi_i)$
$\mathsf{E}(\boldsymbol{X}) = n\boldsymbol{\pi}$	$\mathrm{Cov}(\boldsymbol{X}) = \frac{\sum_{j=1}^k \alpha_j^*}{1+\sum_{j=1}^k \alpha_j} n\{\mathrm{diag}(\boldsymbol{\pi}) - \boldsymbol{\pi}\boldsymbol{\pi}^\top\}$

The gamma function $\Gamma(x)$ is described in Appendix B.2.1. The beta-binomial distribution is a special case of the multinomial-Dirichlet distribution: If $X \sim \mathrm{BeB}(n, \alpha, \beta)$, then $(X, n-X)^\top \sim \mathrm{MD}_2(n, (\alpha, \beta)^\top)$.

Table A.3 (Continued)

Multivariate normal: $\mathrm{N}_k(\boldsymbol{\mu}, \boldsymbol{\Sigma})$	`mtvnorm::_mvnorm(..., mean =` $\boldsymbol{\mu}$`, sigma =` $\boldsymbol{\Sigma}$`)`
$\boldsymbol{\mu} = (\mu_1, \ldots, \mu_k)^\top \in \mathbb{R}^k$	$\boldsymbol{x} = (x_1, \ldots, x_k)^\top \in \mathbb{R}^k$
$\boldsymbol{\Sigma} \in \mathbb{R}^{k \times k}$ symmetric and positive definite	
$f(\boldsymbol{x}) = C \cdot \exp\{-\frac{1}{2}(\boldsymbol{x}-\boldsymbol{\mu})^\top \boldsymbol{\Sigma}^{-1}(\boldsymbol{x}-\boldsymbol{\mu})\}$	$C = \{(2\pi)^k \lvert\boldsymbol{\Sigma}\rvert\}^{-\frac{1}{2}}$
$\mathsf{E}(\boldsymbol{X}) = \boldsymbol{\mu}$	$\mathrm{Cov}(\boldsymbol{X}) = \boldsymbol{\Sigma}$
$\mathrm{Mod}(\boldsymbol{X}) = \boldsymbol{\mu}$	
Normal-gamma: $\mathrm{NG}(\mu, \lambda, \alpha, \beta)$	
$\mu \in \mathbb{R}, \lambda, \alpha, \beta > 0$	$\boldsymbol{x} = (x_1, x_2)^\top, x_1 \in \mathbb{R}, x_2 \in \mathbb{R}^+$
$f(x_1, x_2) = f(x_1 \mid x_2) f(x_2)$	where $X_2 \sim \mathrm{G}(\alpha, \beta)$, and $X_1 \mid X_2 \sim \mathrm{N}(\mu, (\lambda \cdot x_2)^{-1})$
$\mathsf{E}(X_1) = \mathrm{Mod}(X_1) = \mu$	$\mathrm{Var}(X_1) = \frac{\beta}{\lambda(\alpha-1)}$
$\mathrm{Cov}(X_1, X_2) = 0$	$\mathrm{Mod}(\boldsymbol{X}) = (\mu, (\alpha - 1/2)/\beta)^\top$ if $\alpha > 1/2$
The marginal distribution of X_1 is a t distribution: $X_1 \sim \mathrm{t}(\mu, \beta/(\alpha\lambda), 2\alpha)$.	
Wishart: $\mathrm{Wi}_k(\alpha, \boldsymbol{\Sigma})$	`MCMCpack::_wish(..., v =` α`, S =` $\boldsymbol{\Sigma}$`)`
$\alpha \geq k$, $\boldsymbol{\Sigma} \in \mathbb{R}^{k \times k}$ positive definite	$\boldsymbol{X} \in \mathbb{R}^{k \times k}$ positive definite
$f(\boldsymbol{X}) = C \cdot \lvert\boldsymbol{X}\rvert^{\frac{\alpha-k-1}{2}} \exp\{-\frac{1}{2}\operatorname{tr}(\boldsymbol{\Sigma}^{-1}\boldsymbol{X})\}$	$\operatorname{tr}(\boldsymbol{A}) = \sum_{i=1}^k a_{ii}$
$C = \{2^{\alpha k/2} \lvert\boldsymbol{\Sigma}\rvert^{\alpha/2} \Gamma_k(\alpha/2)\}^{-1}$	$\Gamma_k(\alpha/2) = \pi^{k(k-1)/4} \prod_{j=1}^k \Gamma(\frac{\alpha+1-j}{2})$
$\mathsf{E}(\boldsymbol{X}) = \alpha\boldsymbol{\Sigma}$	$\mathrm{Mod}(\boldsymbol{X}) = (\alpha - k - 1)\boldsymbol{\Sigma}$ if $\alpha \geq k+1$
If $\boldsymbol{X}_i \sim \mathrm{N}_k(\boldsymbol{0}, \boldsymbol{\Sigma})$, $i = 1, \ldots, n$, are independent, then $\sum_{i=1}^n \boldsymbol{X}_i \boldsymbol{X}_i^\top \sim \mathrm{Wi}_k(n, \boldsymbol{\Sigma})$. For a diagonal element X_{ii} of the matrix $\boldsymbol{X} = (X_{ij})$, we have $X_{ii} \sim \mathrm{G}(\alpha/2, 1/(2\sigma_{ii}))$, where σ_{ii} is the corresponding diagonal element of $\boldsymbol{\Sigma} = (\sigma_{ij})$.	
Inverse Wishart: $\mathrm{IWi}_k(\alpha, \boldsymbol{\Psi})$	`MCMCpack::_iwish(..., v =` α`, S =` $\boldsymbol{\Psi}$`)`
$\alpha \geq k$, $\boldsymbol{\Psi} \in \mathbb{R}^{k \times k}$ positive definite	$\boldsymbol{X} \in \mathbb{R}^{k \times k}$ positive definite
$f(\boldsymbol{X}) = C \cdot \lvert\boldsymbol{X}\rvert^{-\frac{\alpha+k+1}{2}} \exp\{-\frac{1}{2}\operatorname{tr}(\boldsymbol{X}^{-1}\boldsymbol{\Psi})\}$	
$C = \lvert\boldsymbol{\Psi}\rvert^{\alpha/2}\{2^{\alpha k/2}\Gamma_k(\alpha/2)\}^{-1}$	$\Gamma_k(\alpha/2) = \pi^{k(k-1)/4} \prod_{j=1}^k \Gamma(\frac{\alpha+1-j}{2})$
$\mathsf{E}(\boldsymbol{X}) = \{(\alpha - k - 1)\boldsymbol{\Psi}\}^{-1}$ if $\alpha > k+1$	$\mathrm{Mod}(\boldsymbol{X}) = (\alpha + k + 1)^{-1}\boldsymbol{\Psi}$
If $\boldsymbol{X} \sim \mathrm{Wi}_k(\alpha, \boldsymbol{\Sigma})$, then $\boldsymbol{X}^{-1} \sim \mathrm{IWi}_k(\alpha, \boldsymbol{\Sigma}^{-1})$. For a diagonal element X_{ii} of the matrix $\boldsymbol{X} = (X_{ij})$, we have $X_{ii} \sim \mathrm{IG}((\alpha - k + 1)/2, \psi_{ii}/2)$, where ψ_{ii} is the corresponding diagonal element of $\boldsymbol{\Psi} = (\psi_{ij})$.	

Some Results from Matrix Algebra and Calculus

B

Contents

This appendix contains several results and definitions from linear algebra and mathematical calculus that are important in statistics.

B.1 Some Matrix Algebra

Matrix algebra is important for handling multivariate random variables and parameter vectors. After discussing important operations on square matrices, we describe several useful tools for inversion of matrices and for combining two quadratic forms. We assume that the reader is familiar with basic concepts of linear algebra.

B.1.1 Trace, Determinant and Inverse

Trace $\operatorname{tr}(\boldsymbol{A})$, determinant $|\boldsymbol{A}|$ and inverse $\boldsymbol{A}^{-1}$ are all operations on square matrices

$$\boldsymbol{A} = \begin{pmatrix} a_{11} & a_{12} & \cdots & a_{1n} \\ a_{21} & a_{22} & \cdots & a_{2n} \\ \vdots & \vdots & & \vdots \\ a_{n1} & a_{n2} & \cdots & a_{nn} \end{pmatrix} = (a_{ij})_{1 \le i,j \le n} \in \mathbb{R}^{n \times n}.$$

L. Held, D. Sabanés Bové, *Likelihood and Bayesian Inference*,
Statistics for Biology and Health, https://doi.org/10.1007/978-3-662-60792-3,

While the first two produce a scalar number, the last one produces again a square matrix. In this section we will review the most important properties of these operations.

The *trace* operation is the simplest one of the three mentioned in this section. It computes the sum of the diagonal elements of the matrix:

$$\operatorname{tr}(\boldsymbol{A}) = \sum_{i=1}^{n} a_{ii}.$$

From this one can easily derive some important properties of the trace operation, as for example invariance to transpose of the matrix $\operatorname{tr}(\boldsymbol{A}) = \operatorname{tr}(\boldsymbol{A}^\top)$, invariance to the order in a matrix product $\operatorname{tr}(\boldsymbol{A}\boldsymbol{B}) = \operatorname{tr}(\boldsymbol{B}\boldsymbol{A})$ or additivity $\operatorname{tr}(\boldsymbol{A}+\boldsymbol{B}) = \operatorname{tr}(\boldsymbol{A}) + \operatorname{tr}(\boldsymbol{B})$.

A non-zero *determinant* indicates that the corresponding system of linear equations has a unique solution. For example, the 2×2 matrix

$$\boldsymbol{A} = \begin{pmatrix} a & b \\ c & d \end{pmatrix} \tag{B.1}$$

has the determinant

$$|\boldsymbol{A}| = \begin{vmatrix} a & b \\ c & d \end{vmatrix} = ad - bc.$$

Sarrus' scheme is a method to compute the determinant of 3×3 matrices:

$$\begin{vmatrix} a_{11} & a_{12} & a_{13} \\ a_{21} & a_{22} & a_{23} \\ a_{31} & a_{32} & a_{33} \end{vmatrix}$$
$$= a_{11}a_{22}a_{33} + a_{12}a_{23}a_{31} + a_{13}a_{21}a_{32} - a_{31}a_{22}a_{13} - a_{32}a_{23}a_{11} - a_{33}a_{21}a_{12}.$$

For the calculation of the determinant of higher-dimensional matrices, similar schemes exist. Here, it is crucial to observe that the determinant of a triangular matrix is the product of its diagonal elements and that the determinant of the product of two matrices is the product of the individual determinants. Hence, using some form of Gaussian elimination/triangularisation, the determinant of any square matrix can be calculated. The command `det(A)` in R provides the determinant; however, for very high dimensions, a log transformation may be necessary to avoid numerical problems as illustrated in the following example. To calculate the likelihood of a vector $\boldsymbol{y}$ drawn from a multivariate normal distribution $\mathrm{N}_n(\boldsymbol{\mu}, \boldsymbol{\Sigma})$, one would naively use

```
set.seed(15)
n <- 500
mu <- rep(1, n)
Sigma <- 0.5^abs(outer(1:n, 1:n, "-"))
y <- rnorm(n, mu, sd=0.1)
1/sqrt(det(2*pi*Sigma)) *
    exp(-1/2 * t(y-mu) %*% solve(Sigma) %*% (y-mu))
     [,1]
[1,]    0
```

The result is approximately correct because the involved determinant is returned as infinity, i.e. the likelihood is zero through division by infinity. Using a log transformation, which is by default provided by the function `determinant`, and back-transforming after the calculation provides the correct result:

```
exp(-1/2 * (n * log(2*pi) +
            determinant(Sigma)$modulus +
            t(y-mu) %*% solve(Sigma) %*% (y-mu)))
              [,1]
[1,] 7.390949e-171
attr(,"logarithm")
[1] TRUE
```

The *inverse* $\boldsymbol{A}^{-1}$ of a square matrix $\boldsymbol{A}$, if it exists, satisfies

$$\boldsymbol{A}\boldsymbol{A}^{-1} = \boldsymbol{A}^{-1}\boldsymbol{A} = \boldsymbol{I},$$

where $\boldsymbol{I}$ denotes the identity matrix. The inverse exists if and only if the determinant of $\boldsymbol{A}$ is non-zero. For the 2×2 matrix from Appendix B.1, the inverse is

$$\boldsymbol{A}^{-1} = \frac{1}{|\boldsymbol{A}|}\begin{pmatrix} d & -b \\ -c & a \end{pmatrix}.$$

Using the inverse, it is easy to find solutions of systems of linear equations of the form $\boldsymbol{A}\boldsymbol{x} = \boldsymbol{b}$ since simply $\boldsymbol{x} = \boldsymbol{A}^{-1}\boldsymbol{b}$. In R, this close relationship is reflected by the fact that the command `solve(A)` returns the inverse $\boldsymbol{A}^{-1}$ and the command `solve(A,b)` returns the solution of $\boldsymbol{A}\boldsymbol{x} = \boldsymbol{b}$. See also the example in the next section.

B.1.2 Cholesky Decomposition

A matrix $\boldsymbol{B}$ is *positive definite* if $\boldsymbol{x}^\top \boldsymbol{B}\boldsymbol{x} > 0$ for all $\boldsymbol{x} \neq \boldsymbol{0}$. A symmetric matrix $\boldsymbol{A}$ is positive definite if and only if there exists an upper triangular matrix

$$\boldsymbol{G} = \begin{pmatrix} g_{11} & g_{12} & \cdots & \cdots & g_{1n} \\ 0 & g_{22} & \cdots & \cdots & g_{2n} \\ \vdots & 0 & \ddots & & \vdots \\ \vdots & & \ddots & \ddots & \vdots \\ 0 & \cdots & \cdots & 0 & g_{nn} \end{pmatrix}$$

such that

$$\boldsymbol{G}^\top \boldsymbol{G} = \boldsymbol{A}.$$

$\boldsymbol{G}$ is called the *Cholesky square root* of the matrix $\boldsymbol{A}$.

The *Cholesky decomposition* is a numerical method to determine the Cholesky root $\boldsymbol{G}$. The command `chol(A)` in R provides the matrix $\boldsymbol{G}$.

Using the Cholesky decomposition in the calculation of the multivariate normal likelihood from above not only leads to the correct result without running into numerical problems

```
U <- chol(Sigma)
x <- forwardsolve(t(U), y - mu)
(2*pi)^(-n/2) / prod(diag(U)) * exp(- sum(x^2)/2)
[1] 7.390949e-171
```

it also speeds up the calculation considerably

```
system.time(1/sqrt(det(2 * pi * Sigma)) * exp(-1/2 *
    t(y - mu) %*% solve(Sigma) %*% (y - mu)))
   user  system elapsed
  0.312   0.008   0.321
system.time({
    U <- chol(Sigma)
    x <- forwardsolve(t(U), y - mu)
    (2 * pi)^(-n/2)/prod(diag(U)) * exp(-sum(x^2)/2)
})
   user  system elapsed
  0.056   0.000   0.058
```

Here, the function `system.time` returns (several different) CPU times of the current R process. Usually, the last of them is of interest because it displays the elapsed wall-clock time.

B.1.3 Inversion of Block Matrices

Let a matrix A be partitioned into four blocks:

$$A = \begin{pmatrix} A_{11} & A_{12} \\ A_{21} & A_{22} \end{pmatrix},$$

where A_{11}, A_{12} have the same numbers of rows, and A_{11}, A_{21} the same numbers of columns. If A, A_{11} and A_{22} are quadratic and invertible, then the inverse A^{-1} satisfies

$$A^{-1} = \begin{pmatrix} B^{-1} & -B^{-1}A_{12}A_{22}^{-1} \\ -A_{22}^{-1}A_{21}B^{-1} & A_{22}^{-1} + A_{22}^{-1}A_{21}B^{-1}A_{12}A_{22}^{-1} \end{pmatrix}$$
$$\text{with } B = A_{11} - A_{12}A_{22}^{-1}A_{21} \tag{B.2}$$

or alternatively

$$A^{-1} = \begin{pmatrix} A_{11}^{-1} + A_{11}A_{12}C^{-1}A_{21}A_{11}^{-1} & -A_{11}^{-1}A_{12}C^{-1} \\ -C^{-1}A_{21}A_{11}^{-1} & C^{-1} \end{pmatrix}$$
$$\text{with } C = A_{22} - A_{21}A_{11}^{-1}A_{12}. \tag{B.3}$$

B.1.4 Sherman–Morrison Formula

Let A be an invertible $n \times n$ matrix and u, v column vectors of length n, such that $1 + v^\top A^{-1} u \neq 0$. The Sherman–Morrison formula provides a closed form for the inverse of the sum of the matrix A and the dyadic product $uv^\top$. It depends on the (original) inverse A^{-1} as follows:

$$\left(A + uv^\top\right)^{-1} = A^{-1} - \frac{A^{-1}uv^\top A^{-1}}{1 + v^\top A^{-1}u}.$$

B.1.5 Combining Quadratic Forms

Let x, a and b be $n \times 1$ vectors, and A, B symmetric $n \times n$ matrices such that the inverse of the sum $C = A + B$ exists. Then we have that

$$\begin{aligned}&(x - a)^\top A(x - a) + (x - b)^\top B(x - b)\\ &= (x - c)^\top C(x - c) + (a - b)^\top AC^{-1}B(a - b),\end{aligned} \tag{B.4}$$

where $c = C^{-1}(Aa + Bb)$. In particular, if $n = 1$, (B.4) simplifies to

$$A(x - a)^2 + B(x - b)^2 = C(x - c)^2 + \frac{AB}{C}(a - b)^2 \tag{B.5}$$

with $C = A + B$ and $c = (Aa + Bb)/C$.

B.2 Some Results from Mathematical Calculus

In this section we describe the gamma and beta functions, which appear in many density formulas. Multivariate derivatives are defined, which are used mostly for Taylor expansions in this book, and these are also explained in this section. Moreover, two results for integration and optimisation and the explanation of the Landau notation used for asymptotic statements are included.

B.2.1 The Gamma and Beta Functions

The gamma function is defined as

$$\Gamma(x) = \int_0^\infty t^{x-1} \exp(-t)\, dt$$

for all $x \in \mathbb{R}$ except zero and negative integers. It is implemented in R in the function `gamma()`. The function `lgamma()` returns $\log\{\Gamma(x)\}$. The gamma function is said to interpolate the factorial $x!$ because $\Gamma(x+1) = x!$ for non-negative integers x.

The beta function is related to the gamma function as follows:

$$\mathrm{B}(x, y) = \frac{\Gamma(x)\Gamma(y)}{\Gamma(x+y)}.$$

It is implemented in R in the function `beta()`. The function `lbeta()` returns $\log\{\mathrm{B}(x, y)\}$.

B.2.2 Multivariate Derivatives

Let f be a real-valued function defined on $\mathbb{R}^m$, i.e.

$$f\colon \quad \mathbb{R}^m \longrightarrow \mathbb{R},$$
$$\boldsymbol{x} \longrightarrow f(\boldsymbol{x}),$$

and $\boldsymbol{g} : \mathbb{R}^p \to \mathbb{R}^q$ a generally vector-valued mapping with function values $\boldsymbol{g}(\boldsymbol{x}) = (g_1(\boldsymbol{x}), \dots, g_q(\boldsymbol{x}))^\top$. Derivatives can easily be extended to such a mutivariate setting. Here, the symbol d for a univariate derivative is replaced by the symbol ∂ for partial derivatives. The notation f' and f'' for the first and second univariate derivatives can be generalised to the multivariate setting as well:

1. The vector of the partial derivatives of f with respect to the m components,

$$\boldsymbol{f}'(\boldsymbol{x}) = \left(\frac{\partial}{\partial x_1} f(\boldsymbol{x}), \dots, \frac{\partial}{\partial x_m} f(\boldsymbol{x})\right)^\top,$$

   is called the *gradient* of f at $\boldsymbol{x}$. Obviously, we have $\boldsymbol{f}'(\boldsymbol{x}) \in \mathbb{R}^m$. We also denote this vector as $\frac{\partial}{\partial \boldsymbol{x}} f(\boldsymbol{x})$. The ith component of the gradient can be interpreted as the slope of f in the ith direction (parallel to the coordinate axis).

   For example, if $f(\boldsymbol{x}) = \boldsymbol{a}^\top \boldsymbol{x}$ for some m-dimensional vector $\boldsymbol{a}$, we have $\boldsymbol{f}'(\boldsymbol{x}) = \boldsymbol{a}$. If $f(\boldsymbol{x})$ is a quadratic form, i.e. $f(\boldsymbol{x}) = \boldsymbol{x}^\top \boldsymbol{A}\boldsymbol{x}$ for some $m \times m$ matrix $\boldsymbol{A}$, we obtain $\boldsymbol{f}'(\boldsymbol{x}) = (\boldsymbol{A} + \boldsymbol{A}^\top)\boldsymbol{x}$. If $\boldsymbol{A}$ is symmetric, this reduces to $\boldsymbol{f}'(\boldsymbol{x}) = 2\boldsymbol{A}\boldsymbol{x}$.

2. Taking the derivatives of every component of the gradient with respect to each coordinate x_j of $\boldsymbol{x}$ gives the *Hessian matrix* or *Hessian*

$$\boldsymbol{f}''(\boldsymbol{x}) = \left(\frac{\partial^2}{\partial x_i \, \partial x_j} f(\boldsymbol{x})\right)_{1 \le i, j \le m}$$

   of f at $\boldsymbol{x}$. This matrix $\boldsymbol{f}''$ has dimension $m \times m$. The Hessian hence contains all second-order derivatives of f. Each second-order derivative exists under the regularity condition that the corresponding first-order partial derivative is continuously differentiable. The matrix $\boldsymbol{f}''(\boldsymbol{x})$ quantifies the curvature of f at $\boldsymbol{x}$. We also denote this matrix as $\frac{\partial}{\partial \boldsymbol{x}} \frac{\partial}{\partial \boldsymbol{x}^\top} f(\boldsymbol{x})$ or $\frac{\partial}{\partial \boldsymbol{x}} \boldsymbol{f}'(\boldsymbol{x})$.

3. The $q \times p$ *Jacobian matrix* of $\boldsymbol{g}$,

$$\boldsymbol{g}'(\boldsymbol{x}) = \left(\frac{\partial g_i(\boldsymbol{x})}{\partial x_j} \right)_{\substack{1 \le i \le q \\ 1 \le j \le p}},$$

is the matrix representation of the derivative of $\boldsymbol{g}$ at $\boldsymbol{x}$. We also denote this matrix as $\frac{\partial}{\partial \boldsymbol{x}} \boldsymbol{g}(\boldsymbol{x})$. The Jacobian matrix provides a linear approximation of $\boldsymbol{g}$ at $\boldsymbol{x}$. Note that the Hessian corresponds to the Jacobian matrix of $\boldsymbol{f}'(\boldsymbol{x})$, where $p = q = m$.

The chain rule for differentiation can also be generalised to vector-valued functions. Let $\boldsymbol{h} : \mathbb{R}^q \to \mathbb{R}^r$ be another differentiable function such that its Jacobian $\boldsymbol{h}'(\boldsymbol{y})$ is an $r \times q$ matrix. Define the composition of $\boldsymbol{g}$ and $\boldsymbol{h}$ as $\boldsymbol{k}(\boldsymbol{x}) = \boldsymbol{h}\{\boldsymbol{g}(\boldsymbol{x})\}$, so $\boldsymbol{k} : \mathbb{R}^p \to \mathbb{R}^r$. The multivariate chain rule now states that

$$\boldsymbol{k}'(\boldsymbol{x}) = \boldsymbol{h}'\{\boldsymbol{g}(\boldsymbol{x})\} \cdot \boldsymbol{g}'(\boldsymbol{x}).$$

The entry in the ith row and jth column of this $r \times p$ Jacobian is the partial derivative of $\boldsymbol{k}(\boldsymbol{x})$ of the ith component $k_i(\boldsymbol{x})$ with respect to the jth coordinate x_j:

$$\frac{\partial k_i(\boldsymbol{x})}{\partial x_j} = \sum_{l=1}^{q} \frac{\partial h_i\{\boldsymbol{g}(\boldsymbol{x})\}}{\partial y_l} \frac{\partial g_l(\boldsymbol{x})}{\partial x_j}.$$

B.2.3 Taylor Approximation

A differentiable function can be approximated with a polynomial using Taylor's theorem: For a real interval I, any $(n+1)$ times continuously differentiable real-valued function f satisfies for $a, x \in I$ that

$$f(x) = \sum_{k=0}^{n} \frac{f^{(k)}(a)}{k!}(x-a)^k + \frac{f^{(n+1)}(\xi)}{(n+1)!}(x-a)^{n+1},$$

where ξ is between a and x, and $f^{(k)}$ denotes the kth derivative of f. Moreover, we have that

$$f(x) = \sum_{k=0}^{n} \frac{f^{(k)}(a)}{k!}(x-a)^k + o\big(|x-a|^n\big) \quad \text{as } x \to a.$$

Therefore, any function f that is n times continuously differentiable in the neighbourhood of a can be approximated up to an error of order $o(|x-a|^n)$ (see Appendix B.2.6) by the nth-order *Taylor polynomial* $\sum_{k=0}^{n} \frac{f^{(k)}(a)}{k!}(x-a)^k$.

The Taylor formula can be extended to a multi-dimensional setting for real-valued functions f that are $(n+1)$-times continuously differentiable on an open subset M of $\mathbb{R}^m$: let $\boldsymbol{a} \in M$; then for all $\boldsymbol{x} \in M$ with $S(\boldsymbol{x}, \boldsymbol{a}) = \{\boldsymbol{a} + t(\boldsymbol{x} - \boldsymbol{a}) \mid t \in [0, 1]\} \subset M$, there exists a $\boldsymbol{\xi} \in S(\boldsymbol{x}, \boldsymbol{a})$ with

$$f(\boldsymbol{x}) = \sum_{|\boldsymbol{k}| \le n} \frac{D^{\boldsymbol{k}} f(\boldsymbol{a})}{\boldsymbol{k}!} (\boldsymbol{x} - \boldsymbol{a})^{\boldsymbol{k}} + \sum_{|\boldsymbol{k}| = n+1} \frac{D^{\boldsymbol{k}} f(\boldsymbol{\xi})}{\boldsymbol{k}!} (\boldsymbol{x} - \boldsymbol{a})^{\boldsymbol{k}},$$

where $\boldsymbol{k} \in \mathbb{N}_0^m$, $|\boldsymbol{k}| = k_1 + \cdots + k_m$, $\boldsymbol{k}! = \prod_{i=1}^m k_i!$, $(\boldsymbol{x} - \boldsymbol{a})^{\boldsymbol{k}} = \prod_{i=1}^m (x_i - a_i)^{k_i}$, and

$$D^{\boldsymbol{k}} f(\boldsymbol{x}) = \frac{d^{|\boldsymbol{k}|}}{dx_1^{k_1} \cdots dx_m^{k_m}} f(\boldsymbol{x}).$$

In particular, the second-order Taylor polynomial of f around $\boldsymbol{a}$ is given by

$$\sum_{|\boldsymbol{k}| \le 2} \frac{D^{\boldsymbol{k}} f(\boldsymbol{a})}{\boldsymbol{k}!} (\boldsymbol{x} - \boldsymbol{a})^{\boldsymbol{k}} = f(\boldsymbol{a}) + (\boldsymbol{x} - \boldsymbol{a})^\top \boldsymbol{f}'(\boldsymbol{a}) + \frac{1}{2} (\boldsymbol{x} - \boldsymbol{a})^\top \boldsymbol{f}''(\boldsymbol{a}) (\boldsymbol{x} - \boldsymbol{a}) \quad \text{(B.6)}$$

where $\boldsymbol{f}'(\boldsymbol{a}) = (\frac{\partial}{\partial x_1} f(\boldsymbol{a}), \ldots, \frac{\partial}{\partial x_m} f(\boldsymbol{a}))^\top \in \mathbb{R}^m$ is the gradient, and $\boldsymbol{f}''(\boldsymbol{a}) = (\frac{\partial^2}{\partial x_i \partial x_j} f(\boldsymbol{a}))_{1 \le i,j \le m} \in \mathbb{R}^{m \times m}$ is the Hessian (see Appendix B.2.2).

B.2.4 Leibniz Integral Rule

Let a, b and f be real-valued functions that are continuously differentiable in t. Then the *Leibniz integral rule* is

$$\begin{aligned} &\frac{\partial}{\partial t} \int_{a(t)}^{b(t)} f(x, t)\, dx \\ &= \int_{a(t)}^{b(t)} \frac{\partial}{\partial t} f(x, t)\, dx - f\{a(t), t\} \cdot \frac{d}{dt} a(t) + f\{b(t), t\} \cdot \frac{d}{dt} b(t). \end{aligned}$$

This rule is also known as differentiation under the integral sign.

B.2.5 Lagrange Multipliers

The method of *Lagrange multipliers* (named after Joseph-Louis Lagrange, 1736–1813) is a technique in mathematical optimisation that allows one to reformulate an optimisation problem with constraints such that potential extrema can be determined using the gradient descent method. Let M be an open subset of $\mathbb{R}^m$, $\boldsymbol{x}_0 \in M$, and

f, g be continuously differentiable functions from M to $\mathbb{R}$ with $g(\boldsymbol{x}_0) = 0$ and non-zero gradient of g at $\boldsymbol{x}_0$, i.e.

$$\frac{\partial}{\partial \boldsymbol{x}} g(\boldsymbol{x}_0) = \left(\frac{\partial}{\partial x_1} g(\boldsymbol{x}_0), \ldots, \frac{\partial}{\partial x_m} g(\boldsymbol{x}_0) \right)^\top \neq \boldsymbol{0}.$$

Suppose that $\boldsymbol{x}_0$ is a local maximum or minimum of $f|_{g^{-1}(0)}$ (the function f constrained to the zeros of g in M). Then there exists a Lagrange multiplier $\lambda \in \mathbb{R}$ with

$$\frac{\partial}{\partial \boldsymbol{x}} f(\boldsymbol{x}_0) = \lambda \cdot \frac{\partial}{\partial \boldsymbol{x}} g(\boldsymbol{x}_0).$$

A similar result holds for multiple constraints: Let g_i, $i = 1, \ldots, k$ $(k < m)$, be continuously differentiable functions from M to $\mathbb{R}$ with $g_i(\boldsymbol{x}_0) = 0$ for all i, and linearly independent gradients $\frac{\partial}{\partial \boldsymbol{x}} g_i(\boldsymbol{x}_0)$, $i = 1, \ldots, k$. Assume that $\boldsymbol{x}_0$ is a local maximum or minimum of $f|_{g_1^{-1}(0) \cap \cdots \cap g_k^{-1}(0)}$. Then there exist Lagrange multipliers $\lambda_1, \ldots, \lambda_k \in \mathbb{R}$, such that

$$\frac{\partial}{\partial \boldsymbol{x}} f(\boldsymbol{x}_0) = \sum_{i=1}^{k} \lambda_i \cdot \frac{\partial}{\partial \boldsymbol{x}} g_i(\boldsymbol{x}_0).$$

B.2.6 Landau Notation

Paul Bachmann (1837–1920) and Edmund Landau (1877–1938) introduced two nowadays popular notations allowing to compare functions with respect to growth rates. The Landau notation is also often called big-O and little-o notation. For example, let f, g be real-valued functions defined on the interval (a, ∞) with $a \in \mathbb{R}$.

1. We call $f(x)$ little-o of $g(x)$ as $x \to \infty$,

$$f(x) = o\{g(x)\} \quad \text{as } x \to \infty,$$

if for every $\varepsilon > 0$, there exists a real number $\delta(\varepsilon) > a$ such that the inequality $|f(x)| \leq \varepsilon |g(x)|$ is true for $x > \delta(\varepsilon)$. If $g(x)$ is non-zero for all x larger than a certain value, this is equivalent to

$$\lim_{x \to \infty} \frac{f(x)}{g(x)} = 0.$$

Hence, the asymptotic growth of the function f is slower than that of g, and $f(x)$ vanishes for large values of x compared to $g(x)$.

The same notation can be defined for limits $x \to x_0$ with $x_0 \geq a$:

$$f(x) = o\{g(x)\} \quad \text{as } x \to x_0$$

means that for any $\varepsilon > 0$, there exists a real number $\delta(\varepsilon) > 0$ such that the inequality $|f(x)| \le \varepsilon |g(x)|$ is true for all $x > a$ with $|x - x_0| < \delta(\varepsilon)$. If g does not vanish on its domain, this is again equivalent to

$$\lim_{\substack{x \to x_0 \\ x > a}} \frac{f(x)}{g(x)} = 0.$$

2. We call $f(x)$ big-O of $g(x)$ as $x \to \infty$,

$$f(x) = O\{g(x)\} \quad \text{as } x \to \infty,$$

if there exist constants $Q > 0$ and $R > a$ such that the inequality $|f(x)| \le Q|g(x)|$ is true for all $x > R$. Again, an equivalent condition is

$$\limsup_{x \to \infty} \left| \frac{f(x)}{g(x)} \right| < \infty$$

if $g(x) \neq 0$ for all x larger than a certain value. The function f is, as $x \to \infty$, asymptotically as at most of the same order as the function g, i.e. it grows at most as fast as g.

Analogously to little-o, the notation big-O for limits $x \to x_0$ can also be defined:

$$f(x) = O\{g(x)\} \quad \text{as } x \to x_0$$

denotes that there exist constants $Q, \delta > 0$, such that the inequality $|f(x)| \le Q|g(x)|$ is true for all $x > a$ with $|x - x_0| < \delta$. If g does not vanish on its domain, this is again equivalent to

$$\limsup_{\substack{x \to x_0 \\ x > a}} \left| \frac{f(x)}{g(x)} \right| < \infty.$$

Some Numerical Techniques

C

Contents

Numerical methods have been steadily gaining importance in statistics. In this appendix we summarise the most important techniques.

C.1 Optimisation and Root Finding Algorithms

Numerical methods for optimisation and for finding roots of functions are very frequently used in likelihood inference. This section provides an overview of the commonly used approaches.

C.1.1 Motivation

Optimisation aims to determine the maximum of a function over a certain range of values. Consider a univariate function $g(\theta)$ defined for values $\theta \in \Theta$. Then we want to find the value θ^* such that $g(\theta^*) \geq g(\theta)$ for all values $\theta \in \Theta$. Note that the maximum θ^* might not be unique. Even if the global maximum θ^* is unique, there might be different local maxima in subsets of Θ, in which case the function $g(\theta)$ is called *multimodal*. Otherwise, it is called *unimodal*.

Of course, maximising $g(\theta)$ is equivalent to minimising $h(\theta) = -g(\theta)$. Therefore, we can restrict our presentation to the first case.

For suitably differentiable functions $g(\theta)$, the first derivative $g'(\theta)$ is very useful to find the extrema of $g(\theta)$. This is because every maximum θ^* is also a root of the first derivative, i.e. $g'(\theta^*) = 0$. The second derivative $g''(\theta)$ evaluated at such a

L. Held, D. Sabanés Bové, *Likelihood and Bayesian Inference*,
Statistics for Biology and Health, https://doi.org/10.1007/978-3-662-60792-3,

root may then be used to decide whether it is a maximum of $g(\theta)$, in which case we have $g''(\theta^*) < 0$. For a minimum, we have $g''(\theta^*) > 0$, and for a saddle point, $g''(\theta^*) = 0$. Note that a univariate search for a root of $g'(\theta)$ corresponds to a univariate optimisation problem: it is equivalent to the minimisation of $|g'(\theta)|$.

One example where numerical optimisation methods are often necessary is solving the score equation $S(\theta) = 0$ to find the maximum likelihood estimate $\theta^* = \hat{\theta}_{\mathrm{ML}}$ as the root. Frequently, it is impossible to do that analytically as in the following application.

Example C.1 Let $X \sim \mathrm{Bin}(N, \pi)$. Suppose that observations are however only available of $Y = X \mid \{X > 0\}$ (Y follows a *truncated binomial distribution*). Consequently, we have the following probability mass for the observations $k = 1, 2, \dots, N$:

$$\begin{aligned}\Pr(X = k \mid X > 0) &= \frac{\Pr(X = k)}{\Pr(X > 0)} = \frac{\Pr(X = k)}{1 - \Pr(X = 0)} \\ &= \frac{\binom{N}{k}\pi^k (1-\pi)^{N-k}}{1 - (1-\pi)^N}.\end{aligned}$$

The log-likelihood kernel and score functions for π are then given by

$$\begin{aligned}l(\pi) &= k \cdot \log(\pi) + (N-k) \cdot \log(1-\pi) - \log\{1 - (1-\pi)^N\}, \\ S(\pi) &= \frac{k}{\pi} - \frac{N-k}{1-\pi} - \frac{1}{1-(1-\pi)^N} \cdot \{-N(1-\pi)^{N-1}\}(-1) \\ &= \frac{k}{\pi} - \frac{N-k}{1-\pi} - \frac{N(1-\pi)^{N-1}}{1-(1-\pi)^N}.\end{aligned}$$

The solution of the score equation $S(\pi) = 0$ corresponds to the solution of

$$-k + k(1-\pi)^N + \pi \cdot N = 0.$$

Since finding closed forms for the roots of such a polynomial equation of degree N is only possible up to $N = 3$, for larger N, numerical methods are necessary to solve the score equation. ■

In the following we describe the most important algorithms. We focus on continuous univariate functions.

C.1.2 Bisection Method

Suppose that we have $g'(a_0) \cdot g'(b_0) < 0$ for two points $a_0, b_0 \in \Theta$. The intermediate value theorem (e.g. Clarke 1971, p. 284) then guarantees that there exists at least one root $\theta^* \in [a_0, b_0]$ of $g'(\theta)$. The *bisection method* searches for a root θ^* with the following iterative algorithm:

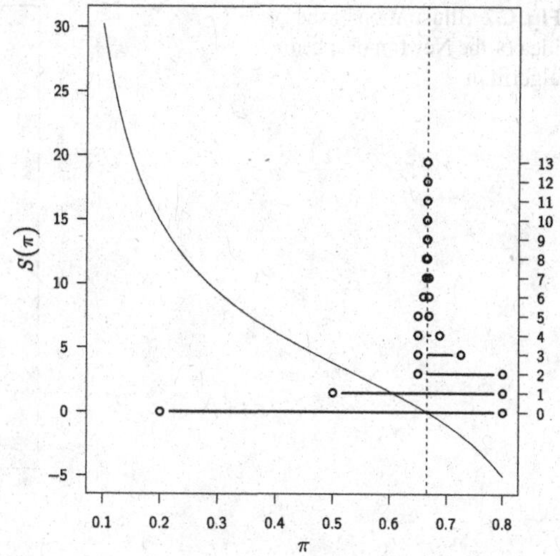

Fig. C.1 Bisection method for the score function $S(\pi)$ of Example C.1 for $y = 4$ and $n = 6$. Shown are the iterated intervals with corresponding index. After 14 iterations convergence was reached based on the relative approximate error ($\varepsilon = 10^{-4}$) with approximated root $\pi = 0.6657$. This approximation is slightly smaller than the maximum likelihood estimate $\hat{\pi} = 4/6 \approx 0.6667$ if Y would be considered exactly binomially distributed

1. Let $\theta^{(0)} = (a_0 + b_0)/2$ and $t = 1$.
2. Reduce the interval containing the root to

$$[a_t, b_t] = \begin{cases} [a_{t-1}, \theta^{(t-1)}] & \text{if } g'(a_{t-1}) \cdot g'(\theta^{(t-1)}) < 0, \\ [\theta^{(t-1)}, b_{t-1}] & \text{otherwise.} \end{cases}$$

3. Calculate the midpoint of the interval, $\theta^{(t)} = \frac{1}{2}(a_t + b_t)$.
4. If convergence has not yet been reached, set $t \leftarrow t + 1$ and go back to step 2; otherwise, $\theta^{(t)}$ is the final approximation of the root θ^*.

Note that the bisection method is a so-called *bracketing method*: if the initial conditions are satisfied, the root can always be determined. Moreover, no second-order derivatives $g''(\theta)$ are necessary. A stopping criterion to assess convergence is, for example, based on the relative approximate error: The convergence is stated when

$$\frac{|\theta^{(t)} - \theta^{(t-1)}|}{|\theta^{(t-1)}|} < \varepsilon.$$

The bisection method is illustrated in Fig. C.1 using Example C.1 with $y = 4$ and $n = 6$.

There exist several optimised bracketing methods. One of them is Brent's method, which has been proposed in 1973 by Richard Peirce Brent (1946–) and combines the bisection method with linear and quadratic interpolation of the inverse function. It results in a faster convergence rate for sufficiently smooth functions. If only linear interpolation is used, then Brent's method is equivalent to the secant method described in Appendix C.1.4.

Brent's method is implemented in the R-function `uniroot(f, interval, tol, ...)`, which searches in a given interval $[a_0, b_0]$ (`interval = c(`a_0, b_0`)`) for a root of the function `f`. The convergence criterion can be controlled with the option

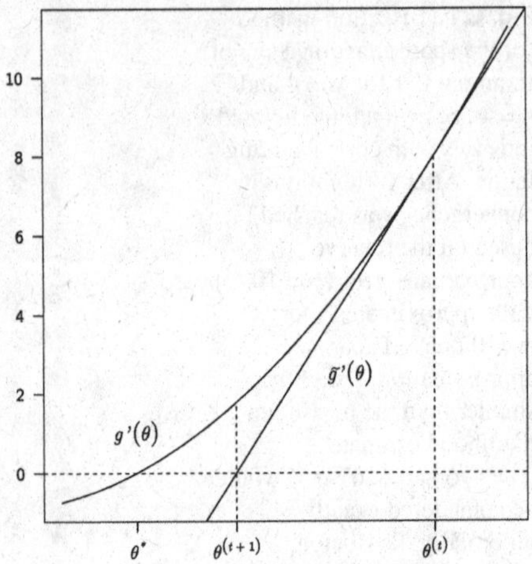

Fig. C.2 Illustration of the idea of the Newton–Raphson algorithm

`tol`. The result of a call is a list with four elements containing the approximated root (`root`), the value of the function at the root (`froot`), the number of performed iterations (`iter`) and the estimated deviation from the true root (`estim.prec`). The function `f` may take more than one argument, but the value θ needs to be the first. Further arguments, which apply for all values of θ, can be passed to `uniroot` as additional named parameters. This is formally hinted at through the "…" argument at the end of the list of arguments of `uniroot`.

C.1.3 Newton–Raphson Method

A faster root finding method for sufficiently smooth functions $g(\theta)$ is the *Newton–Raphson algorithm*, which is named after the Englishmen Isaac Newton (1643–1727) and Joseph Raphson (1648–1715). Suppose that $g'(\theta)$ is differentiable with root θ^* and $g''(\theta^*) \neq 0$, i.e. θ^* is either a local minimum or maximum of $g(\theta)$.

In every iteration t of the Newton–Raphson algorithm, the derivative $g'(\theta)$ is approximated using a linear Taylor expansion around the current approximation $\theta^{(t)}$ of the root θ^*:

$$g'(\theta) \approx \tilde{g}'(\theta) = g'\big(\theta^{(t)}\big) + g''\big(\theta^{(t)}\big)\big(\theta - \theta^{(t)}\big).$$

The function $g'(\theta)$ is hence approximated using the tangent line to $g'(\theta)$ at $\theta^{(t)}$. The idea is then to approximate the root of $g'(\theta)$ by the root of $\tilde{g}'(\theta)$:

$$\tilde{g}'(\theta) = 0 \quad \Longleftrightarrow \quad \theta = \theta^{(t)} - \frac{g'(\theta^{(t)})}{g''(\theta^{(t)})}.$$

The iterative procedure is thus defined as follows (see Fig. C.2 for an illustration):

1. Start with a value $\theta^{(0)}$ for which the second derivative is non-zero, $g''(\theta^0) \neq 0$, i.e. the function $g(\theta)$ is curved at $\theta^{(0)}$.

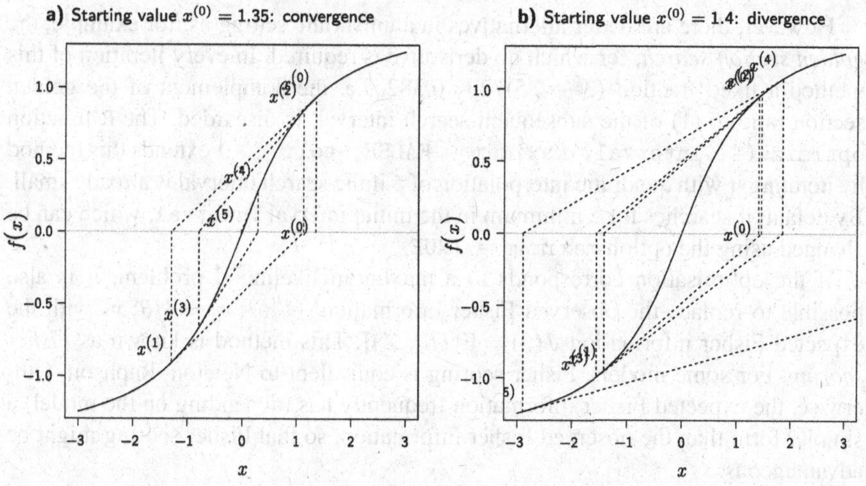

Fig. C.3 Convergence is dependent on the starting value as well: two searches for a root of $f(x) = \arctan(x)$. The *dashed lines* correspond to the *tangent lines* and the *vertical lines* between their roots and the function $f(x)$. While for $x^{(0)} = 1.35$, in (**a**) convergence to the true root $\theta^* = 0$ is obtained, (**b**) illustrates that for $x^{(0)} = 1.4$, the algorithm diverges

2. Set the next approximation to

$$\theta^{(t+1)} = \theta^{(t)} - \frac{g'(\theta^{(t)})}{g''(\theta^{(t)})}. \tag{C.1}$$

3. If the convergence has not yet been reached go back to step 2; otherwise, $\theta^{(t+1)}$ is the final approximation of the root θ^* of $g'(\theta)$.

The convergence of the Newton–Raphson algorithm depends on the form of the function $g(\theta)$ and the starting value $\theta^{(0)}$ (see Fig. C.3 for the importance of the latter). If, however, $g'(\theta)$ is two times differentiable, convex and has a root, the algorithm converges irrespective of the starting point.

Another perspective on the Newton–Raphson method is the following. An approximation of g through a second-order Taylor expansion around $\theta^{(t)}$ is given by

$$g(\theta) \approx \tilde{g}(\theta) = g(\theta^{(t)}) + g'(\theta^{(t)})(\theta - \theta^{(t)}) + \frac{1}{2}g''(\theta^{(t)})(\theta - \theta^{(t)})^2.$$

Minimising this quadratic approximation through

$$\begin{aligned} \tilde{g}'(\theta) = 0 \quad &\Leftrightarrow \quad g'(\theta^{(t)}) + g''(\theta^{(t)})(\theta - \theta^{(t)}) = 0 \\ &\Leftrightarrow \quad \theta = \theta^{(t)} - \frac{g'(\theta^{(t)})}{g''(\theta^{(t)})} \end{aligned}$$

leads to the same algorithm.

However, there are better alternatives in a univariate setting as, for example, the *golden section search*, for which no derivative is required. In every iteration of this method a fixed fraction $(3-\sqrt{5})/2 \approx 0.382$, i.e. the complement of the golden section ratio to (1) of the subsequent search interval, is discarded. The R-function `optimize(f, interval, maximum = FALSE, tol, ...)` extends this method by iterating it with quadratic interpolation of `f` if the search interval is already small. By default it searches for a minimum in the initial interval `interval`, which can be changed using the option `maximum = TRUE`.

If the optimisation corresponds to a maximum likelihood problem, it is also possible to replace the observed Fisher information $-l''(\theta; x) = I(\theta; x)$ with the expected Fisher information $J(\theta) = \mathsf{E}\{I(\theta; X)\}$. This method is known as *Fisher scoring*. For some models, Fisher scoring is equivalent to Newton–Raphson. Otherwise, the expected Fisher information frequently has (depending on the model) a simpler form than the observed Fisher information, so that Fisher scoring might be advantageous.

The Newton–Raphson algorithm is easily generalised to multivariate real-valued functions $g(\boldsymbol{x})$ taking arguments $\boldsymbol{x} \in \mathbb{R}^n$ (see Appendix B.2.2 for the definitions of the multivariate derivatives). Using the $n \times n$ Hessian $\boldsymbol{g}''(\boldsymbol{\theta}^{(t)})$ and the $n \times 1$ gradient $\boldsymbol{g}'(\boldsymbol{\theta}^{(t)})$, the update is defined as follows:

$$\boldsymbol{\theta}^{(t+1)} = \boldsymbol{\theta}^{(t)} - \left\{\boldsymbol{g}''(\boldsymbol{\theta}^{(t)})\right\}^{-1} \cdot \boldsymbol{g}'(\boldsymbol{\theta}^{(t)}).$$

Extensions of the multivariate Newton–Raphson algorithm are *quasi-Newton methods*, which are not using the exact Hessian but calculate a positive definite approximation based on the successively calculated gradients. Consequently, Fisher scoring is a quasi-Newton method. The gradients in turn do not need to be specified as functions but can be approximated numerically, for example, through

$$\frac{\partial g(\boldsymbol{\theta})}{\partial \theta_i} \approx \frac{g(\boldsymbol{\theta} + \varepsilon \cdot \boldsymbol{e}_i) - g(\boldsymbol{\theta} - \varepsilon \cdot \boldsymbol{e}_i)}{2\varepsilon}, \quad i = 1, \dots, n, \tag{C.2}$$

where the ith basis vector $\boldsymbol{e}_i$ has zero components except for the ith component, which is 1, and ε is small. The derivative in the ith direction, which would be obtained from (C.2) by letting $\varepsilon \to 0$, is hence replaced by a finite approximation.

In general, multivariate optimisation of a function `fn` in R is done using the function `optim(par, fn, gr, method, lower, upper, control, hessian = FALSE, ...)`, where the gradient function can be specified using the optional argument `gr`. By default, however, the gradient-free *Nelder–Mead method* is used, which is robust against discontinuous functions but is in exchange rather slow. One can choose the quasi-Newton methods sketched above by assigning the value `BFGS` or `L-BFGS-B` to the argument `method`. For these methods, gradients are numerically approximated by default. For the method `L-BFGS-B`, it is possible to specify a search rectangle through the vectors `lower` and `upper`. The values assigned in the list `control` can have important consequences. For example, the maximum number of iterations is set with `maxit`. An overall scaling is applied to the values of the function using `fnscale`, for example, `control = list(fnscale = -1)` implies that `-fn` is minimised, i.e. `fn` is maximised. This and the option `hessian = TRUE` are necessary to maximise a log-likelihood and to obtain the numerically obtained curvature at the estimate. The function `optim` returns a list containing the optimal function value (`value`), its corresponding x-value (`par`) and, if indicated, the curvature (`hessian`) as well as the important convergence message: the algorithm has obtained convergence if and only if the `convergence` code is 0. A more recent implementation and extension of `optim(...)` is the function `optimr(...)` in the package `optimr`. In the latter function, additional and multiple optimisation methods can be specified.

C.1.4 Secant Method

A disadvantage of the Newton–Raphson and Fisher scoring methods is that they require the second derivative $g''(\theta)$, which in some cases may be very difficult to determine. The idea of the *secant method* is to replace $g''(\theta^{(t)})$ in (C.1) with the approximation

$$\tilde{g}''(\theta^{(t)}) = \frac{g'(\theta^{(t)}) - g'(\theta^{(t-1)})}{\theta^{(t)} - \theta^{(t-1)}},$$

whereby the update is

$$\theta^{(t+1)} = \theta^{(t)} - \frac{\theta^{(t)} - \theta^{(t-1)}}{g'(\theta^{(t)}) - g'(\theta^{(t-1)})} \cdot g'(\theta^{(t)}).$$

Note that this method requires the specification of two starting points $\theta^{(0)}$ and $\theta^{(1)}$. The convergence is slower than for the Newton–Raphson method but faster than for the bisection method. The secant method is illustrated in Fig. C.4 for Example C.1 with $y = 4$ and $n = 6$.

C.2 Integration

In statistics we frequently need to determine the definite integral

$$I = \int_a^b f(x)\,dx.$$

of a univariate function $f(x)$. Unfortunately, there are only few functions $f(x)$ for which the primitive $F(x) = \int^x f(u)\,du$ is available in closed form such that we could calculate

$$I = F(b) - F(a)$$

directly. Otherwise, we need to apply numerical methods to obtain an approximation for I.

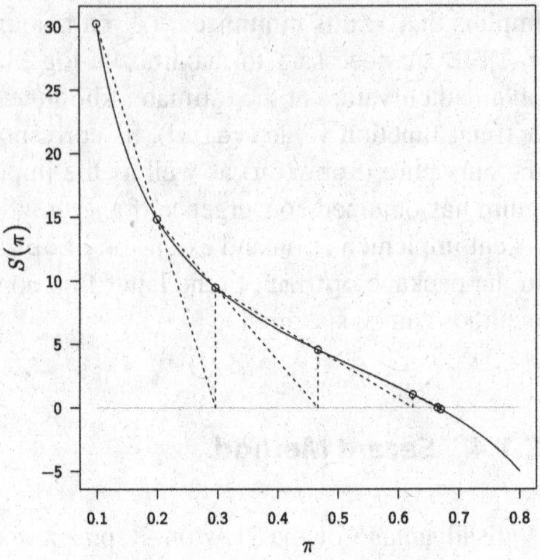

Fig. C.4 Secant method for Example C.1 with observation $y = 4$ and $n = 6$. The starting points have been chosen as $\theta^{(0)} = 0.1$ and $\theta^{(1)} = 0.2$

For example, already the familiar density function of the standard normal distribution

$$\varphi(x) = \frac{1}{\sqrt{2\pi}} \exp\left(-\frac{1}{2}x^2\right),$$

does not have a primitive. Therefore, the distribution function $\Phi(x)$ of the standard normal distribution can only be specified in the general integral form as

$$\Phi(x) = \int_{-\infty}^{x} \varphi(u)\, du,$$

and its calculation needs to rely on numerical methods.

C.2.1 Newton–Cotes Formulas

The *Newton–Cotes formulas*, which are named after Newton and Roger Cotes (1682–1716), are based on the piecewise integration of $f(x)$:

$$I = \int_a^b f(x)\, dx = \sum_{i=0}^{n-1} \int_{x_i}^{x_{i+1}} f(x)\, dx \tag{C.3}$$

over the decomposition of the interval $[a, b]$ into $n - 1$ pieces using the knots $x_0 = a < x_1 < \cdots < x_{n-1} < x_n = b$. Each summand $T_i = \int_{x_i}^{x_{i+1}} f(x)\, dx$ in (C.3) is then approximated as follows: the function f is evaluated at the $m + 1$ equally-spaced interpolation points $x_{i0} = x_i < x_{i1} < \cdots < x_{i,m-1} < x_{i,m} = x_{i+1}$. The resulting $m + 1$ points $(x_{ij}, f(x_{ij}))$ can be interpolated using a polynomial p_i of degree m satisfying $p_i(x_{ij}) = f(x_{ij})$ for $j = 0, \ldots, m$. Therefore, we obtain an approximation of

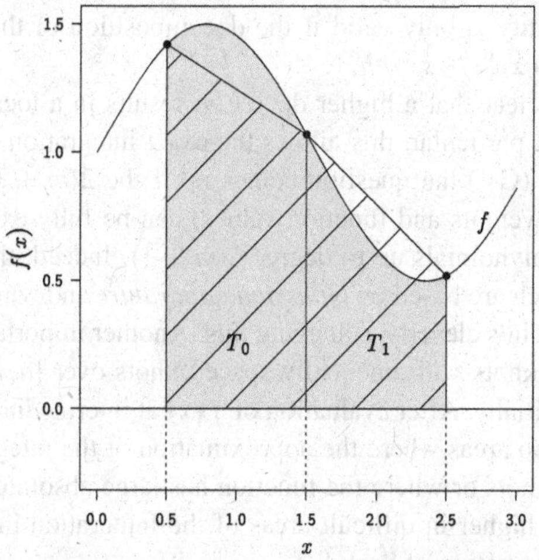

Fig. C.5 Illustration of the trapezoidal rule for $f(x) = \cos(x)\sin(x) + 1$, $a = 0.5$, $b = 2.5$ and $n = 3$ and decomposition into two parts, $[0.5, 1.5]$ and $[1.5, 2.5]$. The *solid grey areas* T_0 and T_1 are approximated by the corresponding areas of the hatched trapezoids. The function f in this case can be integrated analytically resulting in $I = \frac{1}{4}\{\cos(2a) - \cos(2b)\} + (b - a)$. Substituting the bounds leads to $I = 2.0642$. The approximation $I \approx 1 \cdot \{\frac{1}{2} f(0.5) + f(1.5) + \frac{1}{2} f(2.5)\} = 2.0412$ is hence inaccurate (relative error of $(2.0642 - 2.0412)/2.0642 = 0.0111$)

the function $f(x)$ in the interval $[x_i, x_{i+1}]$, which we can integrate analytically:

$$T_i \approx \int_{x_i}^{x_{i+1}} p_i(x)\,dx = \sum_{j=0}^{m} w_{ij} f(x_{ij}), \tag{C.4}$$

where the w_{ij} are weighting the function values at the interpolation points x_{ij} and are available in closed form.

Choosing, for example, as interpolation points the bounds $x_{i0} = x_i$ and $x_{i1} = x_{i+1}$ of each interval implies interpolation polynomials of degree 1, i.e. a straight line through the end points. Each summand T_i is then approximated by the area of a trapezoid (see Fig. C.5),

$$T_i \approx \frac{1}{2}(x_{i+1} - x_i) \cdot f(x_i) + \frac{1}{2}(x_{i+1} - x_i) \cdot f(x_{i+1}),$$

and the weights are in this case given by $w_{i0} = w_{i1} = \frac{1}{2}(x_{i+1} - x_i)$. The Newton–Cotes formula for $m = 1$ is in view of this also called the *trapezoidal rule*. Substitution into (C.3) leads to

$$I \approx \sum_{i=0}^{n-1} \frac{1}{2}(x_{i+1} - x_i)\{f(x_i) + f(x_{i+1})\} = h\left\{\frac{1}{2} f(x_0) + \sum_{i=1}^{n-1} f(x_i) + \frac{1}{2} f(x_n)\right\}, \tag{C.5}$$

where the last identity is only valid if the decomposition of the interval $[a, b]$ is equally-spaced with $x_{i+1} - x_i = h$.

Intuitively, it is clear that a higher degree m results in a locally better approximation of $f(x)$. In particular, this allows the exact integration of polynomials up to degree m. From (C.4) the question comes up if the $2(m + 1)$ degrees of freedom (from $m + 1$ weights and function values) can be fully exploited in order to exactly integrate polynomials up to degree $2(m + 1)$. Indeed, there exist sophisticated methods, which are based on *Gaussian quadrature* and which choose weights and interpolation points cleverly, achieving this. Another important extension is the adaptive choice of knots with unequally spaced knots over $[a, b]$. Here, only few knots are chosen initially. After evaluation of $f(x)$ at intermediate points more new knots are assigned to areas where the approximation of the integral varies strongly when introducing knots or where the function has large absolute value. Hence, the density of knots is higher in difficult areas of the integration interval, parallelling the Monte Carlo integration of Sect. 8.3.

The R function `integrate(f, lower, upper, rel.tol, abs.tol, ...)` implements such an adaptive method for the integration of `f` on the interval between `lower` and `upper`. Improper integrals with boundaries `-Inf` or `Inf` can also be calculated (by mapping the interval to $[0, 1]$ through substitution and then applying the algorithm for bounded intervals). The function `f` must be vectorised, i.e. accepting a vector as a first argument and return a vector of same length. Hence, the following will fail:

```
f <- function(x) 2
try(integrate(f, 0, 1))
```

The function `Vectorize` is helpful here, because it converts a given function to a vectorised version.

```
fv <- Vectorize(f)
integrate(fv, 0, 1)
2 with absolute error < 2.2e-14
```

The desired accuracy of `integrate` is specified through the options `rel.tol` and `abs.tol` (by default typically set to $1.22 \cdot 10^{-4}$). The returned object is a list containing among other things the approximated value of the integral (`value`), the estimation of the absolute error (`abs.error`) and the convergence message (`message`), which reads `OK` in case of convergence. However, one has to keep in mind that, for example, singularities in the interior of the integration interval could be missed. Then, it is advisable to calculate the integral piece by piece so that the singularities lie on the boundaries of the integral pieces and can hence be handled appropriately by the `integrate` function.

Multidimensional integrals of multivariate real-valued functions $f(\boldsymbol{x})$ over a multidimensional rectangle $\mathcal{A} \subset \mathbb{R}^n$,

$$I = \int_{\mathcal{A}} f(\boldsymbol{x})\, d\boldsymbol{x} = \int_{a_1}^{b_1} \int_{a_2}^{b_2} \cdots \int_{a_n}^{b_n} f(\boldsymbol{x})\, dx_n \cdots dx_2\, dx_1,$$

are handled by the R function `adaptIntegrate(f, lowerLimit, upperLimit, tol = 1e-05, fDim = 1, maxEval = 0, absError = 0)` contained in the package `cubature`. Here the function $f(x)$ is given by `f` and `lowerLimit = c(`$a_1, \dots, a_n$`)`, `upperLimit = c(`$b_1, \dots, b_n$`)`. One can specify a maximum number of function evaluations using `maxEval` (default is 0, meaning the unlimited number of function evaluations). Otherwise, the integration routine stops when the estimated error is less than the absolute error requested (`absError`) or when the estimated error is less in absolute value than `tol` times the integral. The returned object is again a list containing the approximated value of the integral and some further information, see the help page `?adaptIntegrate` for details. Note that such integration routines are only useful for a moderate dimension (say, up to $n = 20$). Higher dimensions require, for example, MCMC approaches.

C.2.2 Laplace Approximation

The Laplace approximation is a method to approximate integrals of the form

$$I_n = \int_{-\infty}^{+\infty} \exp\{-nk(u)\}\, du, \tag{C.6}$$

where $k(u)$ is a convex and twice differentiable function with minimum at $u = \tilde{u}$. Such integrals appear, for example, when calculating characteristics of posterior distributions. For $u = \tilde{u}$, we thus have $\frac{dk(\tilde{u})}{du} = 0$ and $\kappa = \frac{d^2k(\tilde{u})}{du^2} > 0$. A second-order Taylor expansion of $k(u)$ around $\tilde{u}$ gives $k(u) \approx k(\tilde{u}) + \frac{1}{2}\kappa(u - \tilde{u})^2$, so (C.6) can be approximately written as

$$\begin{aligned} I_n &\approx \exp\{-nk(\tilde{u})\} \int_{-\infty}^{+\infty} \underbrace{\exp\left\{-\frac{1}{2}n\kappa(u - \tilde{u})^2\right\}}_{\text{kernel of } \mathrm{N}(u \,|\, \tilde{u}, (n\kappa)^{-1})} du \\ &= \exp\{-nk(\tilde{u})\} \cdot \sqrt{\frac{2\pi}{n\kappa}}. \end{aligned}$$

In the multivariate case we consider the integral

$$I_n = \int_{\mathbb{R}^p} \exp\{-nk(\boldsymbol{u})\}\, d\boldsymbol{u}$$

and obtain the approximation

$$I_n \approx \left(\frac{2\pi}{n}\right)^{\frac{p}{2}} |\boldsymbol{K}|^{-\frac{1}{2}} \exp\{-nk(\tilde{\boldsymbol{u}})\},$$

where $\boldsymbol{K}$ denotes the $p \times p$ Hessian of $k(\boldsymbol{u})$ at $\tilde{\boldsymbol{u}}$, and $|\boldsymbol{K}|$ is the determinant of $\boldsymbol{K}$.

Notation

A	event or set
$\lvert A \rvert$	cardinality of a set A
$x \in A$	x is an element of A
$x \notin A$	x is not an element of A
A^c	complement of A
$A \cap B$	joint event: A and B
$A \cup B$	union event: A and/or B
$A \dot{\cup} B$	disjoint event: either A or B
$A \subset B$	A is a subset of B
$\mathsf{Pr}(A)$	probability of A
$\mathsf{Pr}(A \mid B)$	conditional probability of A given B
X	random variable
$\boldsymbol{X}$	multivariate random variable
$X_{1:n}, \boldsymbol{X}_{1:n}$	random sample
$X = x$	event that X equals realisation x
$f_X(x)$	density (or probability mass) function of X
$f_{X,Y}(x, y)$	joint density function of X and Y
$f_{Y \mid X}(y \mid x)$	conditional density of Y given $X = x$
$F_X(x)$	distribution function of X
$F_{X,Y}(x, y)$	joint distribution function of X and Y
$F_{Y \mid X}(y \mid x)$	conditional distribution function of Y given $X = x$
$\mathcal{T}$	support of a random variable or vector
$\mathsf{E}(X)$	expectation of X
$\mathrm{Var}(X)$	variance of X
m_k	kth moment of X
c_k	kth central moment of X
$\mathrm{Cov}(\boldsymbol{X})$	covariance matrix of $\boldsymbol{X}$
$\mathrm{Mod}(X)$	mode of X
$\mathrm{Med}(X)$	median of X
$\mathrm{Cov}(X, Y)$	covariance of X and Y
$\mathrm{Corr}(X, Y)$	correlation of X and Y
$\mathsf{D}(f_X \parallel f_Y)$	Kullback–Leibler discrepancy from f_X to f_Y
$\mathsf{E}(Y \mid X = x)$	conditional expectation of Y given $X = x$
$\mathrm{Var}(Y \mid X = x)$	conditional variance of Y given $X = x$

L. Held, D. Sabanés Bové, *Likelihood and Bayesian Inference*,
Statistics for Biology and Health, https://doi.org/10.1007/978-3-662-60792-3,

$X_n \xrightarrow{r} X$	convergence in rth mean
$X_n \xrightarrow{D} X$	convergence in distribution
$X_n \xrightarrow{P} X$	convergence in probability
$X \sim F$	X distributed as F
$X_n \overset{\text{a}}{\sim} F$	X_n asymptotically distributed as F
$X_i \overset{\text{iid}}{\sim} F$	X_i independent and identically distributed as F
$X_i \overset{\text{ind}}{\sim} F_i$	X_i independent with distribution F_i
$\boldsymbol{A} \in \mathbb{R}^{a \times b}$	$a \times b$ matrix with entries $a_{ij} \in \mathbb{R}$
$\boldsymbol{a} \in \mathbb{R}^k$	vector with k entries $a_i \in \mathbb{R}$
$\dim(\boldsymbol{a})$	dimension of a vector $\boldsymbol{a}$
$\lvert\boldsymbol{A}\rvert$	determinant of $\boldsymbol{A}$
$\boldsymbol{A}^\top$	transpose of $\boldsymbol{A}$
$\operatorname{tr}(\boldsymbol{A})$	trace of $\boldsymbol{A}$
$\boldsymbol{A}^{-1}$	inverse of $\boldsymbol{A}$
$\operatorname{diag}(\boldsymbol{a})$	diagonal matrix with $\boldsymbol{a}$ on diagonal
$\operatorname{diag}\{a_i\}_{i=1}^k$	diagonal matrix with $a_1, \ldots, a_k$ on diagonal
$\mathbf{I}$	identity matrix
$\mathbf{1}, \mathbf{0}$	ones and zeroes vectors
$\mathsf{I}_A(x)$	indicator function of a set A
$\lceil x \rceil$	least integer not less than x
$\lfloor x \rfloor$	integer part of x
$\lvert x \rvert$	absolute value of x
$\log(x)$	natural logarithm function
$\exp(x)$	exponential function
$\operatorname{logit}(x)$	logit function $\log\{x/(1-x)\}$
$\operatorname{sign}(x)$	sign function with value 1 for $x > 0$, 0 for $x = 0$ and -1 for $x < 0$
$\varphi(x)$	standard normal density function
$\Phi(x)$	standard normal distribution function
$x!$	factorial of non-negative integer x
$\binom{n}{x}$	binomial coefficient $\frac{n!}{x!(n-x)!}$ $(n \geq x)$
$\Gamma(x)$	Gamma function
$\mathrm{B}(x, y)$	Beta function
$f'(x), \frac{d}{dx} f(x)$	first derivative of $f(x)$
$f''(x), \frac{d^2}{dx^2} f(x)$	second derivative of $f(x)$
$\boldsymbol{f}'(\boldsymbol{x}), \frac{\partial}{\partial x_i} f(\boldsymbol{x})$	gradient, (which contains) partial first derivatives of $f(\boldsymbol{x})$
$\boldsymbol{f}''(\boldsymbol{x}), \frac{\partial^2}{\partial x_i \, \partial x_j} f(\boldsymbol{x})$	Hessian, (which contains) partial second derivatives of $f(\boldsymbol{x})$
$\arg\max_{x \in A} f(x)$	argument of the maximum of $f(x)$ from A
$\mathbb{R}$	set of all real numbers
$\mathbb{R}^+$	set of all positive real numbers
$\mathbb{R}_0^+$	set of all positive real numbers and zero

$\mathbb{R}^p$	set of all p-dimensional real vectors
$\mathbb{N}$	set of natural numbers
$\mathbb{N}_0$	set of natural numbers and zero
(a, b)	set of real numbers $a < x < b$
$(a, b]$	set of real numbers $a < x \leq b$
$[a, b]$	set of real numbers $a \leq x \leq b$
$a \pm x$	$a - x$ and $a + x$, where $x > 0$
θ	scalar parameter
$\boldsymbol{\theta}$	vectorial parameter
$\hat{\theta}$	estimator of θ
$\text{se}(\hat{\theta})$	standard error of $\hat{\theta}$
z_γ	γ quantile of the standard normal distribution
$t_\gamma(\alpha)$	γ quantile of the standard t distribution with α degrees of freedom
$\chi^2_\gamma(\alpha)$	γ quantile of the χ^2 distribution with α degrees of freedom
$\hat{\theta}_{\text{ML}}$	maximum likelihood estimate of θ
$X_{1:n} = (X_1, \ldots, X_n)$	random sample of univariate random variables
$X = (X_1, \ldots, X_n)$	sample of univariate random variables
$\bar{X}$	arithmetic mean of the sample
S^2	sample variance
$X = \{X_1, X_2, \ldots\}$	random process
$\boldsymbol{X}_{1:n} = (\boldsymbol{X}_1, \ldots, \boldsymbol{X}_n)$	random sample of multivariate random variables
$f(x; \theta)$	density function parametrised by θ
$L(\theta), l(\theta)$	likelihood and log-likelihood function
$\tilde{L}(\theta), \tilde{l}(\theta)$	relative likelihood and log-likelihood
$L_p(\theta), l_p(\theta)$	profile likelihood and log-likelihood
$L_p(y), l_p(y)$	predictive likelihood and log-likelihood
$S(\theta)$	score function
$\boldsymbol{S}(\boldsymbol{\theta})$	score vector
$I(\theta)$	Fisher information
$\boldsymbol{I}(\boldsymbol{\theta})$	Fisher information matrix
$I(\hat{\theta}_{\text{ML}})$	observed Fisher information
$J(\theta)$	per unit expected Fisher information
$J_{1:n}(\theta)$	expected Fisher information from a random sample $X_{1:n}$
$\chi^2(d)$	chi-squared distribution
$\text{B}(\pi)$	Bernoulli distribution
$\text{Be}(\alpha, \beta)$	beta distribution
$\text{BeB}(n, \alpha, \beta)$	beta-binomial distribution
$\text{Bin}(n, \pi)$	binomial distribution
$\text{C}(\mu, \sigma^2)$	Cauchy distribution
$\text{D}_k(\boldsymbol{\alpha})$	Dirichlet distribution
$\text{Exp}(\lambda)$	exponential distribution
$\text{F}(\alpha, \beta)$	F distribution

$\mathrm{FN}(\mu, \sigma^2)$	folded normal distribution
$\mathrm{G}(\alpha, \beta)$	gamma distribution
$\mathrm{Geom}(\pi)$	geometric distribution
$\mathrm{Gg}(\alpha, \beta, \delta)$	gamma–gamma distribution
$\mathrm{Gu}(\mu, \sigma^2)$	Gumbel distribution
$\mathrm{HypGeom}(n, N, M)$	hypergeometric distribution
$\mathrm{IG}(\alpha, \beta)$	inverse gamma distribution
$\mathrm{IWi}_k(\alpha, \boldsymbol{\Psi})$	inverse Wishart distribution
$\mathrm{LN}(\mu, \sigma^2)$	log-normal distribution
$\mathrm{Log}(\mu, \sigma^2)$	logistic distribution
$\mathrm{M}_k(n, \boldsymbol{\pi})$	multinomial distribution
$\mathrm{MD}_k(n, \boldsymbol{\alpha})$	multinomial-Dirichlet distribution
$\mathrm{NCHypGeom}(n, N, M, \theta)$	noncentral hypergeometric distribution
$\mathrm{N}(\mu, \sigma^2)$	normal distribution
$\mathrm{N}_k(\boldsymbol{\mu}, \boldsymbol{\Sigma})$	multivariate normal distribution
$\mathrm{NBin}(r, \pi)$	negative binomial distribution
$\mathrm{NG}(\mu, \lambda, \alpha, \beta)$	normal-gamma distribution
$\mathrm{Par}(\alpha, \beta)$	Pareto distribution
$\mathrm{Po}(\lambda)$	Poisson distribution
$\mathrm{PoG}(\alpha, \beta, \nu)$	Poisson-gamma distribution
$\mathrm{t}(\mu, \sigma^2, \alpha)$	Student (t) distribution
$\mathrm{U}(a, b)$	uniform distribution
$\mathrm{Wb}(\mu, \alpha)$	Weibull distribution
$\mathrm{Wi}_k(\alpha, \boldsymbol{\Sigma})$	Wishart distribution

References

Akaike, H. (1974). A new look at the statistical model identification. *IEEE Transactions on Automatic Control, 19*(6), 716–723.

Bartlett, M. S. (1937). Properties of sufficiency and statistical tests. *Proceedings of the Royal Society of London. Series A, Mathematical and Physical Sciences, 160*(901), 268–282.

Baum, L. E., Petrie, T., Soules, G. & Weiss, N. (1970). A maximization technique occurring in the statistical analysis of probabilistic functions of Markov chains. *The Annals of Mathematical Statistics, 41*, 164–171.

Bayarri, M. J., & Berger, J. O. (2004). The interplay of Bayesian and frequentist analysis. *Statistical Science, 19*(1), 58–80.

Bayes, T. (1763). An essay towards solving a problem in the doctrine of chances. *Philosophical Transactions of the Royal Society, 53*, 370–418.

Berger, J. O., & Sellke, T. (1987). Testing a point null hypothesis: irreconcilability of P values and evidence (with discussion). *Journal of the American Statistical Association, 82*, 112–139.

Bernardo, J. M., & Smith, A. F. M. (2000). *Bayesian theory*. Chichester: Wiley.

Besag, J., Green, P. J., Higdon, D., & Mengersen, K. (1995). Bayesian computation and stochastic systems. *Statistical Science, 10*, 3–66.

Box, G. E. P. (1980). Sampling and Bayes' inference in scientific modelling and robustness (with discussion). *Journal of the Royal Statistical Society, Series A, 143*, 383–430.

Box, G. E. P., & Tiao, G. C. (1973). *Bayesian inference in statistical analysis*. Reading: Addison-Wesley.

Brown, L. D., Cai, T. T., & DasGupta, A. (2001). Interval estimation for a binomial proportion. *Statistical Science, 16*(2), 101–133.

Buckland, S. T., Burnham, K. P., & Augustin, N. H. (1997). Model selection: an integral part of inference. *Biometrics, 53*(2), 603–618.

Burnham, K. P., & Anderson, D. R. (2002). *Model selection and multimodel inference: a practical information-theoretic approach* (2nd ed.). New York: Springer.

Carlin, B. P., & Louis, T. A. (2008). *Bayesian methods for data analysis* (3rd ed.). Boca Raton: Chapman & Hall/CRC.

Carlin, B. P. & Polson, N. G. (1992). Monte Carlo Bayesian methods for discrete regression models and categorical time series. In J. M. Bernardo, J. O. Berger, A. Dawid & A. Smith (Eds.), *Bayesian Statistics 4* (pp. 577–586). Oxford: Oxford University Press.

Casella, G., & Berger, R. L. (2001). *Statistical inference* (2nd ed.). Pacific Grove: Duxbury Press.

Chen, P.-L., Bernard, E. J. & Sen, P. K. (1999). A Markov chain model used in analyzing disease history applied to a stroke study. *Journal of Applied Statistics, 26*(4), 413–422.

Chib, S. (1995). Marginal likelihood from the Gibbs output. *Journal of the American Statistical Association, 90*, 1313–1321.

Chihara, L., & Hesterberg, T. (2019). *Mathematical statistics with resampling and* R (2nd ed.). Hoboken: Wiley.

Claeskens, G., & Hjort, N. L. (2008). *Model selection and model averaging*. Cambridge: Cambridge University Press.

L. Held, D. Sabanés Bové, *Likelihood and Bayesian Inference*,
Statistics for Biology and Health, https://doi.org/10.1007/978-3-662-60792-3,

Clarke, D. A. (1971). *Foundations of analysis: with an introduction to logic and set theory*. New York: Appleton-Century-Crofts.

Clayton, D. G., & Bernardinelli, L. (1992). Bayesian methods for mapping disease risk. In P. Elliott, J. Cuzick, D. English & R. Stern (Eds.), *Geographical and environmental epidemiology: methods for small-area studies* (pp. 205–220). Oxford: Oxford University Press. Chap. 18.

Clayton, D. G., & Kaldor, J. M. (1987). Empirical Bayes estimates of age-standardized relative risks for use in disease mapping. *Biometrics*, *43*(3), 671–681.

Clopper, C. J., & Pearson, E. S. (1934). The use of confidence or fiducial limits illustrated in the case of the binomial. *Biometrika*, *26*(4), 404–413.

Cole, S. R., Chu, H., Greenland, S., Hamra, G., & Richardson, D. B. (2012). Bayesian posterior distributions without Markov chains. *American Journal of Epidemiology*, *175*(5), 368–375.

Collins, R., Yusuf, S., & Peto, R. (1985). Overview of randomised trials of diuretics in pregnancy. *British Medical Journal*, *290*(6461), 17–23.

Connor, J. T., & Imrey, P. B. (2005). Proportions, inferences, and comparisons. In P. Armitage & T. Colton (Eds.), *Encyclopedia of biostatistics* (2nd ed., pp. 4281–4294). Chichester: Wiley.

Cox, D. R. (1981). Statistical analysis of time series. Some recent developments. *Scandinavian Journal of Statistics*, *8*(2), 93–115.

Davison, A. C. (2003). *Statistical models*. Cambridge: Cambridge University Press.

Davison, A. C., & Hinkley, D. V. (1997). *Bootstrap methods and their applications*. Cambridge: Cambridge University Press.

Devroye, L. (1986). *Non-uniform random variate generation*. New York: Springer. Available at http://luc.devroye.org/rnbookindex.html.

Diggle, P. J. (1990). *Time series: a biostatistical introduction*. Oxford: Oxford University Press.

Diggle, P. J., Heagerty, P. J., Liang, K.-Y. & Zeger, S. L. (2002). *Analysis of longitudinal data* (2nd ed.). Oxford: Oxford University Press.

Edwards, A. W. F. (1992). *Likelihood* (2nd ed.). Baltimore: Johns Hopkins University Press.

Edwards, W., Lindman, H., & Savage, L. J. (1963). Bayesian statistical inference in psychological research. *Psychological Review*, *70*, 193–242.

Evans, M., & Swartz, T. (1995). Methods for approximating integrals in statistics with special emphasis on Bayesian integration problems. *Statistical Science*, *10*(3), 254–272.

Falconer, D. S., & Mackay, T. F. C. (1996). *Introduction to quantitative genetics* (4th ed.). Harlow: Longmans Green.

Geisser, S. (1993). *Predictive inference: an introduction*. London: Chapman & Hall/CRC.

Gilks, W. R., Richardson, S., & Spiegelhalter, D. J. (Eds.) (1996). *Markov chain Monte Carlo in practice*. Boca Raton: Chapman & Hall/CRC.

Gneiting, T., Balabdaoui, F., & Raftery, A. E. (2007). Probabilistic forecasts, calibration and sharpness. *Journal of the Royal Statistical Society. Series B (Methodological)*, *69*, 243–268.

Gneiting, T., & Raftery, A. E. (2007). Strictly proper scoring rules, prediction, and estimation. *Journal of the American Statistical Association*, *102*, 359–378.

Good, I. J. (1995). When batterer turns murderer. *Nature*, *375*(6532), 541.

Good, I. J. (1996). When batterer becomes murderer. *Nature*, *381*(6532), 481.

Goodman, S. N. (1999). Towards evidence-based medical statistics. 2.: The Bayes factor. *Annals of Internal Medicine*, *130*, 1005–1013.

Green, P. J. (2001). A primer on Markov chain Monte Carlo. In O. E. Barndorff-Nielson, D. R. Cox & C. Klüppelberg (Eds.), *Complex stochastic systems* (pp. 1–62). Boca Raton: Chapman & Hall/CRC.

Grimmett, G., & Stirzaker, D. (2001). *Probability and random processes* (3rd ed.). Oxford: Oxford University Press.

Held, L. (2008). *Methoden der statistischen Inferenz: Likelihood und Bayes*. Heidelberg: Spektrum Akademischer Verlag.

Iten, P. X. (2009). Ändert das ändern des Strassenverkehrgesetzes das Verhalten von alkohol- und drogenkonsumierenden Fahrzeuglenkern? – Erfahrungen zur 0.5-Promillegrenze und zur Nulltoleranz für 7 Drogen in der Schweiz. *Blutalkohol, 46*, 309–323.

Iten, P. X., & Wüst, S. (2009). Trinkversuche mit dem Lion Alcolmeter 500 – Atemalkohol versus Blutalkohol I. *Blutalkohol, 46*, 380–393.

Jeffreys, H. (1961). *Theory of probability* (3rd ed.). Oxford: Oxford University Press.

Jones, O., Maillardet, R., & Robinson, A. (2014). *Introduction to scientific programming and simulation using R* (2nd ed.). Boca Raton: Chapman & Hall/CRC.

Kass, R. E., & Raftery, A. E. (1995). Bayes factors. *Journal of the American Statistical Association, 90*(430), 773–795.

Kass, R. E., & Wasserman, L. (1995). A reference Bayesian test for nested hypotheses and its relationship to the Schwarz criterion. *Journal of the American Statistical Association, 90*(431), 928–934.

Kirkwood, B. R., & Sterne, J. A. C. (2003). *Essential medical statistics* (2nd ed.). Malden: Blackwell Publishing Limited.

Lange, K. (2002). *Mathematical and statistical methods for genetic analysis* (2nd ed.). New York: Springer.

Lee, P. M. (2012). *Bayesian statistics: an introduction* (4th ed.). Chichester: Wiley.

Lehmann, E. L., & Casella, G. (1998). *Theory of point estimation* (2nd ed.). New York: Springer.

Lloyd, C. J., & Frommer, D. (2004). Estimating the false negative fraction for a multiple screening test for bowel cancer when negatives are not verified. *Australian & New Zealand Journal of Statistics, 46*(4), 531–542.

Merz, J. F., & Caulkins, J. P. (1995). Propensity to abuse—propensity to murder? *Chance, 8*(2), 14.

Millar, R. B. (2011). *Maximum likelihood estimation and inference: with examples in R, SAS and ADMB*. New York: Wiley.

Mossman, D., & Berger, J. O. (2001). Intervals for posttest probabilities: a comparison of 5 methods. *Medical Decision Making, 21*, 498–507.

Newcombe, R. G. (2013). *Confidence intervals for proportions and related measures of effect size*. Boca Raton: CRC Press.

Newton, M. A., & Raftery, A. E. (1994). Approximate Bayesian inference with the weighted likelihood bootstrap. *Journal of the Royal Statistical Society. Series B (Methodological), 56*, 3–48.

O'Hagan, A., Buck, C. E., Daneshkhah, A., Eiser, J. R., Garthwaite, P. H., Jenkinson, D. J., Oakley, J. E., & Rakow, T. (2006). *Uncertain judgements: eliciting experts' probabilities*. Chichester: Wiley.

O'Hagan, A., & Forster, J. (2004). *Bayesian inference* (2nd ed.). London: Arnold.

Patefield, W. M. (1977). On the maximized likelihood function. *Sankhyā: The Indian Journal of Statistics, Series B, 39*(1), 92–96.

Pawitan, Y. (2001). *In all likelihood: statistical modelling and inference using likelihood*. New York: Oxford University Press.

Rao, C. R. (1973). *Linear Statistical Inference and Its Applications*. New York: Wiley.

Reynolds, P. S. (1994). Time-series analyses of beaver body temperatures. In N. Lange, L. Ryan, D. Billard, L. Brillinger, L. Conquest & J. Greenhouse (Eds.), *Case studies in biometry* (pp. 211–228). New York: Wiley. Chap. 11.

Ripley, B. D. (1987). *Stochastic simulation*. Chichester: Wiley.

Robert, C. P. (2001). *The Bayesian choice* (2nd ed.). New York: Springer.

Robert, C. P., & Casella, G. (2004). *Monte Carlo statistical methods* (2nd ed.). New York: Springer.

Robert, C. P., & Casella, G. (2010). *Introducing Monte Carlo methods with R*. New York: Springer.

Royall, R. M. (1997). *Statistical evidence: a likelihood paradigm*. London: Chapman & Hall/CRC.

Schwarz, G. (1978). Estimating the dimension of a model. *The Annals of Statistics, 6*(2), 461–464.

Seber, G. A. F. (1982). Capture–recapture methods. In S. Kotz & N. L. Johnson (Eds.), *Encyclopedia of statistical sciences* (pp. 367–374). Chichester: Wiley.

Sellke, T., Bayarri, M. J., & Berger, J. O. (2001). Calibration of p values for testing precise null hypotheses. *American Statistician*, *55*, 62–71.

Shumway, R. H., & Stoffer, D. S. (2017). *Time series analysis and its applications: with R examples* (4th ed.). Cham: Springer International Publishing.

Spiegelhalter, D. J., Best, N. G., Carlin, B. R., & van der Linde, A. (2002). Bayesian measures of model complexity and fit. *Journal of the Royal Statistical Society. Series B (Methodological)*, *64*, 583–616.

Stone, M. (1977). An asymptotic equivalence of choice of model by cross-validation and Akaike's criterion. *Journal of the Royal Statistical Society. Series B (Methodological)*, *39*(1), 44–47.

Tierney, L. (1994). Markov chain for exploring posterior distributions. *The Annals of Statistics*, *22*, 1701–1762.

Tierney, L., & Kadane, J. B. (1986). Accurate approximations for posterior moments and marginal densities. *Journal of the American Statistical Association*, *81*(393), 82–86.

Venzon, D. J., & Moolgavkar, S. H. (1988). A method for computing profile-likelihood-based confidence intervals. *Journal of the Royal Statistical Society. Series C. Applied Statistics*, *37*, 87–94.

Young, G. A., & Smith, R. L. (2005). *Essentials of statistical inference*. Cambridge: Cambridge University Press.

Index

L. Held, D. Sabanés Bové, *Likelihood and Bayesian Inference*,
Statistics for Biology and Health, https://doi.org/10.1007/978-3-662-60792-3,